Electric Machinery and
Power System Fundamentals

McGraw-Hill Series in Electrical and Computer Engineering

Stephen W. Director, University of Michigan, Ann Arbor, *Senior Consulting Editor*

Circuits and Systems
Communications and Signal Processing
Computer Engineering
Control Theory and Robotics
Electromagnetics
Electronics and VLSI Circuits
Introductory
Power
Antennas, Microwaves, and Radar

Previous Consulting Editors

Electric Machinery and Power System Fundamentals

Stephen J. Chapman
BAE SYSTEMS Australia

Boston   Burr Ridge, IL   Dubuque, IA   Madison, WI   New York   San Francisco   St. Louis
Bangkok   Bogotá   Caracas   Kuala Lumpur   Lisbon   London   Madrid   Mexico City
Milan   Montreal   New Delhi   Santiago   Seoul   Singapore   Sydney   Taipei   Toronto

McGraw-Hill Higher Education

*A Division of The **McGraw-Hill** Companies*

ELECTRIC MACHINERY AND POWER SYSTEM FUNDAMENTALS

Published by McGraw-Hill, a business unit of The McGraw-Hill Companies, Inc., 1221 Avenue of the Americas, New York, NY 10020. Copyright © 2002 by The McGraw-Hill Companies, Inc. All rights reserved. No part of this publication may be reproduced or distributed in any form or by any means, or stored in a database or retrieval system, without the prior written consent of The McGraw-Hill Companies, Inc., including, but not limited to, in any network or other electronic storage or transmission, or broadcast for distance learning.

Some ancillaries, including electronic and print components, may not be available to customers outside the United States.

This book is printed on acid-free paper.

1 2 3 4 5 6 7 8 9 0 DOC/DOC 0 9 8 7 6 5 4 3 2 1

ISBN 0–07–229135–4
ISBN 0–07–112179–X (ISE)

General manager: *Thomas E. Casson*
Publisher: *Elizabeth A. Jones*
Sponsoring editor: *Catherine Fields Shultz*
Developmental editors: *Heather E. Sabol/Emily J. Gray*
Executive marketing manager: *John Wannemacher*
Project manager: *Mary E. Powers*
Production supervisor: *Enboge Chong*
Designer: *David W. Hash*
Supplement producer: *Tammy Juran*
Media technology senior producer: *Phillip Meek*
Compositor: *GAC/Indianapolis*
Typeface: *10/12 Times Roman*
Printer: *R. R. Donnelley & Sons Company/Crawfordsville, IN*

Cover images: © *Corbis Corporation/Dale O'Dell, 2001/ © Photo Disc/Andrew Wakeford, 2001*

Library of Congress Cataloging-in-Publication Data

Chapman, Stephen J.
    Electric machinery and power system fundamentals / Stephen J. Chapman. — 1st ed.
        p.   cm. — (McGraw-Hill series in electrical and computer engineering.)
    Includes index.
    ISBN 0–07–229135–4
    1. Electric machinery. 2. Electric power systems. I. Title. II. Series.

TK2000 .C455   2002
621.31′042—dc21   00-066824
                                                           CIP

INTERNATIONAL EDITION ISBN 0–07–112179–X
Copyright © 2002. Exclusive rights by The McGraw-Hill Companies, Inc., for manufacture and export. This book cannot be re-exported from the country to which it is sold by McGraw-Hill. The International Edition is not available in North America.

www.mhhe.com

This work is dedicated with love to my wife, Rosa, on the occasion of the birth of our eighth child, Devorah Rina Chapman.

S.J.C.

ABOUT THE AUTHOR

Stephen J. Chapman received a B.S. in electrical engineering from Louisiana State University (1975), an M.S.E. in electrical engineering from the University of Central Florida (1979), and pursued further graduate studies at Rice University.

From 1975 to 1980, he served as an officer in the U.S. Navy, assigned to teach electrical engineering at the U.S. Naval Nuclear Power School in Orlando, Florida. From 1980 to 1982, he was affiliated with the University of Houston, where he ran the power systems program in the College of Technology.

From 1982 to 1988 and from 1991 to 1995, he served as a member of the technical staff of the Massachusetts Institute of Technology's Lincoln Laboratory, both at the main facility in Lexington, Massachusetts, and at the field site on Kwajalein Atoll in the Republic of the Marshall Islands. While there, he did research in radar signal processing systems. He ultimately became the leader of four large operational range instrumentation radars at the Kwajalein field site (TRADEX, ALTAIR, ALCOR, and MMW).

From 1988 to 1991, Mr. Chapman was a research engineer in Shell Development Company in Houston, Texas, where he did seismic signal processing research. He was also affiliated with the University of Houston, where he continued to teach on a part-time basis.

Mr. Chapman is currently manager of technical systems for BAE SYSTEMS Australia, in Melbourne, Australia. In this position, he provides technical direction and design authority for the work of younger engineers within the company.

Mr. Chapman is a senior member of the Institute of Electrical and Electronics Engineers (and several of its component societies). He is also a member of the Institution of Engineers (Australia).

BRIEF CONTENTS

CONTENTS

CHAPTER **8**
DC Motors   355

CHAPTER **11**

Introduction to Power-Flow Studies   512

CHAPTER **12**

Symmetrical Faults   555

CHAPTER **13**

Unsymmetrical Faults   591

Preface

The basic purpose of this book is to teach machinery and power system principles at a level suitable to the one-semester or two-quarter junior-level survey course found in many electrical engineering programs. It retains the core topics from my *Electric Machinery Fundamentals,*[1] and adds material on transmission lines, power system representation, power-flow studies, and fault analysis.

Many electrical engineering programs traditionally offered a two-semester sequence covering electrical machinery and power systems, with one semester devoted to machines and one to power systems. Unfortunately, the pressure in recent years to cram more and more into the EE curriculum has caused this sequence to be scaled back to a shorter survey course at many schools. This shorter course attempts to select specific topics from the machinery/power spectrum, and to give students the "flavor" of the field. Its content varies widely from school to school, which means that a text intended for the course must include far more material than can be covered, structured so that the topics can be chosen on an *a la carte* basis. This text is intended to fulfil that requirement.

Another problem with traditional power books is that they have tended to be heavily dominated by "how to" recipes, at the expense of explaining basic principles. For example, most power system textbooks devote one or two entire chapters to calculating the series inductance and shunt capacitance of transmission lines for different geometries. In the real world, though, very few engineers ever do this. Instead, they look up the per-mile characteristics of transmission lines in prepared lookup tables. In contrast, this book derives transmission line inductance and capacitance for one simple case, and uses those equations to describe the physical principles that operate on all transmission lines. It then concentrates on how to *use* the transmission line values, without spending the two chapters on how to *calculate* them.

Another example occurs with the bus admittance matrix $\mathbf{Y}_{bus}$ and the bus impedance matrix $\mathbf{Z}_{bus}$ of a power system. Traditional power texts teach a number of ways to build $\mathbf{Y}_{bus}$ and $\mathbf{Z}_{bus}$, often taking up most of two chapters with this material. This book teaches a *single* way to derive $\mathbf{Y}_{bus}$, and calculates $\mathbf{Z}_{bus}$ as the inverse of $\mathbf{Y}_{bus}$ using MATLAB. This is not the most efficient way to perform the calculation for systems of 1000 busses, but it works well for any power system that could be reasonably included in a classroom exercise. The time saved by not doing all the different versions of the calculations is spent on actually using $\mathbf{Y}_{bus}$ and $\mathbf{Z}_{bus}$ to solve power-flow and fault current problems. This is consistent with the way most engineers actually work, since the mechanics of building $\mathbf{Y}_{bus}$ and $\mathbf{Z}_{bus}$ are usually buried

---

[1] *Electric Machinery Fundamentals*, third edition, by Stephen J. Chapman, McGraw-Hill, 1999

inside commercial power system software packages, and the engineers concentrate on the power-flow and fault current calculations that result.

Another example occurs with power-flow calculations. Most books spend a great deal of time introducing the details of network representations with special components such as tap-changing or phase-shifting transformers. They introduce multiple techniques to solve the resulting systems of equations: Gauss-Seidel iteration and Newton-Raphson iteration. They also tend to include only partial examples because the complete calculations are just too difficult and time-consuming to do by hand. This book introduces only the simplest solution technique: Gauss-Seidel iteration. By using MATLAB, the examples can be complete solutions, because the mathematics is no longer tedious. In addition, the book introduces a functional MATLAB power-flow analysis program, which the students can use to solve significant power-flow problems and learn about system behavior. The end of chapter problems then support "What if?" exercises. For example, one problem examines the effect of capacitors on voltage and current levels in lines. Other problems examine the steady-state effects of adding or removing transmission lines from a power system.

In general, if there are multiple ways to solve a problem, this book introduces only the most straightforward method. It provides software to ease calculations, exposing the underlying principles. In this way, the book attempts to compress the course material while preserving its essence.

THE CONTENT OF THIS BOOK

This book is divided into three major sections: basic principles, power system components, and power systems. The basic principles are covered in Chapters 1 and 2. Chapter 1 provides an introduction to basic electrical and mechanical machinery concepts. Chapter 2 is a review of basic three-phase circuit theory, which may be skipped if the topic has been adequately covered in earlier courses.

Power system components are covered in Chapters 3 through 9. Chapter 3 covers transformers. Chapter 4 covers basic AC machinery principles, while Chapter 5 covers the operation of synchronous generators and motors. Chapter 6 is an optional chapter that discusses the operation of synchronous generators in parallel with large power systems. Chapter 7 covers induction motors, Chapter 8 covers DC machines, and Chapter 9 covers transmission lines. Chapter 4 is a prerequisite for Chapters 5 and 7, and Chapter 5 is a prerequisite for Chapter 6. Otherwise, these component chapters are largely independent, so an instructor can pick and choose the topics of interest among them.

The operation of power systems as a whole is covered in Chapters 10 through 13. Chapter 10 deals with power system representations (one-line diagrams, symbols, etc.) and the basic equations of power systems. Chapter 11 covers the power-flow problem. Chapter 12 covers symmetrical three-phase faults, and Chapter 13 covers unsymmetrical faults. Chapter 10 is a prerequisite for the remaining power system material. After that chapter, an instructor can choose to cover power flows, faults, or both, in any order, since the two topics are structured to be independent.

SUPPLEMENTAL MATERIALS

Supplemental materials supporting the book are available from the book's World Wide Web site, at URL www.mhhe.com/chapman. The materials available at that address include MATLAB source code, pointers to sites of interest to machinery and power students, a list of errata in the text, and the MATLAB-based tools used in the book.

ACKNOWLEDGMENTS

I would like to thank my editors at McGraw-Hill for their patience. This book was a long time coming! I would also like to thank the reviewers, who contributed greatly to the quality of this edition. The reviewers of this edition were John G. Ciezki, Naval Postgraduate School; Fred I. Denny, Louisiana State University; John Lukowski, Michigan Technological University; and Stephen A. Sebo, The Ohio State University.

Finally, I would like to thank my wife Rosa and our children Avi, David, Rachel, Aaron, Sarah, Naomi, Shira, and Devorah for their forbearance during the writing process. I am looking forward to spending more time with them.

Stephen J. Chapman
Melbourne, Victoria, Australia

Mechanical and Electromagnetic Fundamentals

1.1 | ELECTRIC MACHINES AND POWER SYSTEMS

An **electric machine** is a device that can convert either mechanical energy to electrical energy or electrical energy to mechanical energy. When such a device is used to convert mechanical energy to electrical energy, it is called a **generator.** When it converts electrical energy to mechanical energy, it is called a **motor.** Since any given electric machine can convert power in either direction, any machine can be used as either a generator or a motor. Almost all practical motors and generators convert energy from one form to another through the action of a magnetic field, and only machines using magnetic fields to perform such conversions are considered in this book.

Generators are connected to motors and other loads such as lighting by power systems. A **power system** is a network of components designed to efficiently transmit and distribute the energy produced by generators to the locations where it is used. Power systems are the glue that holds the modern world together.

The major components of modern power systems are described below.

Generators

Generators produce the electrical energy distributed by a power system. Almost all of the generators in use today produce electrical energy by converting mechanical energy to electrical energy through the action of a magnetic field. The mechanical energy that the generator converts to electrical form comes from a **prime mover,** which is the device that spins the generator. Prime movers are usually some form of steam or water turbine, but diesel engines are sometimes used in remote locations.

There are many possible energy sources for the prime mover. The most common sources are water, coal, natural gas, oil, and nuclear energy. Water power is perhaps the best source, since it is nonpolluting and there is no cost for the "fuel." Hydroelectric power is a renewable resource—as long as it rains, there will be "fuel" available. However, almost all desirable sites for dams have already been developed, so there is not much scope for future growth in water power. In January 2000, hydroelectric power represented less than 8 percent of the United States' total electricity generation,[1] and the fraction is likely to shrink in the future.

Nuclear power is also an excellent source of nonpolluting energy. Nuclear power plants are extremely expensive to build, and they require elaborate safety systems and expensive training. Once built, though, the cost of the nuclear "fuel" is low, and nuclear power produces no hydrocarbon emissions to pollute the atmosphere. In January 2000, nuclear power plants provided about 20 percent of U.S. energy needs. If a utility has hydroelectric and nuclear generators, it will run them at full power all of the time because the fuel costs of these two sources are so low.

Unfortunately, the future of nuclear power is clouded by concerns over nuclear reactor safety. The incidents at Three-Mile Island and Chernobyl have left an indelible impression on public consciousness, and there is a strong and growing worldwide opposition to nuclear power. Few new nuclear plants are being built, and that situation will not change for the foreseeable future.

Coal is the most common source of energy for electrical power generation. In January 2000, more than 50 percent of U.S. generation was coal fired. Coal is a relatively cheap fuel. Unfortunately, coal is also one of the most polluting fuel sources. Stack scrubbers and other antipollution features have to be included in coal-fired plants to control pollution. The emission-reduction targets set by the 1997 Kyoto Protocol on Climate Change are a strong incentive to not use coal-fired generation, or at least to use it more efficiently.

Natural gas is a much better and cleaner energy source than coal. It is relatively cheap, and it burns cleanly with little pollution. In places like the southern United States, where large supplies of natural gas are available, it is the fuel of choice. In January 2000, about 13 percent of U.S. generation was gas fired. The principal disadvantage of natural gas is that it is relatively hard to transport over long distances.

Oil is a bit more polluting than natural gas, but it is easier to transport. Unfortunately, it is also much more expensive. Oil supplied less than 3 percent of U.S. generation in January 2000.

Coal, oil, and gas all have the additional disadvantage that they are nonrenewable energy sources.

Other sources of electric power include wind turbines and solar energy. These are both renewable resources, but they are not yet economical compared to other sources of electrical energy. Fuel cells and nuclear fusion are also possibilities on the

---

[1]The generation statistics quoted here come for the website of the U.S. Energy Information Administration of the Department of Energy. The URL is http://www.eia.doe.gov. It is a fascinating site—check it out for yourself.

distant horizon. All of these sources together amount to much less than 1 percent of U.S. generation.

Unfortunately, there is no perfect source of energy for electricity generation. All possible sources have their advantages and disadvantages, and all will be used in varying degrees for the foreseeable future.

Transformers

Transformers convert AC electrical energy at one voltage level into AC electrical energy at another voltage level. They are essential for the operation of a modern power system, since transformers allow power to be transmitted with minimal losses over long distances.

Modern generators generate electrical power at voltages of 13.8 to 24 kV, while transmission lines operate at *much* higher voltages to reduce transmission losses. Loads consume electrical power at many different voltage levels varying from about 110 V in a typical home up to about 4160 V in large industrial plants. Transformers are the glue that holds the entire system together by increasing line voltages for transmission over long distances, and then reducing the voltages to the levels required by the end users. They make modern power systems possible, and furthermore, they do it with very high efficiency.

Power Lines

Power lines connect generators to loads, transmitting electrical power from one to the other with minimal losses. Power lines are usually divided into two categories: transmission lines and distribution lines. Transmission lines are designed to transmit electrical power efficiently over long distances. They run at very high voltages to reduce the resistive (I^2R) losses in the lines. Standard transmission line voltages in the United States are 115, 138, 230, 345, 500, and 765 kV.

Once the power reaches the vicinity of the user, its voltage is stepped down, and the power is supplied through distribution lines to the final customers. Distribution lines carry much less power than transmission lines, and for shorter distances, so they can operate at lower voltages without prohibitive losses. In the United States, distribution line voltages vary from 4.16 to 34.5 kV, with voltages around 13.8 kV quite common. Distribution lines supply power directly to a customer's home or plant, where it is stepped down again to the final voltage required by the user.

Loads

There are many types of loads on a modern power system. The most important loads are motors, electric lighting, and electronic products (computers, televisions, phones, fax machines, and so forth).

Protective Devices

In addition to these major components, power systems include a wide variety of devices designed to protect the system. These devices include current, voltage, and power sensors, relays, fuses, and circuit breakers.

There are two common types of failures in a power system: **overloads** and **faults.** Overloads are conditions in which some or all components in the power system are supplying more power than they can safely handle. Overloads can happen because the total demand on the power system simply exceeds the ability of the system to supply power. However, it is more common for overloads to occur in localized parts of the power system because of changes elsewhere within the system. For example, two parallel transmission lines may be sharing the task of providing power to a city. If one of them is disconnected for some reason, the remaining line will supply the total power needed by the city. This may cause the line to be overloaded.

If an overload occurs on a power system, it should be corrected, but power systems are robust enough that operators usually have several minutes to correct the problem before damage occurs.

Faults are conditions in which one or more of the phases in a power system are shorted to ground or to each other. (Faults also occur if a phase is open circuited.) When a short circuit occurs, very large currents flow, and these currents can damage the power system unless they are stopped quickly. Unlike overloads, faults must be cleared *immediately,* so relays are designed to automatically open circuit breakers and isolate faults as soon as they are detected.

The Structure of This Book

This book provides a survey of electric machinery and power systems. It is divided into three parts. Part 1 (Chapters 1 and 2) covers fundamental principles common to both machinery and power systems, including the basics of mechanical rotation, magnetism, and three-phase circuits. Part 2 introduces the major components of power systems, such as transformers, generators, motors, and transmission lines. Part 3 deals with the analysis and operation of power systems themselves.

1.2 | A NOTE ON UNITS AND NOTATION

The design and study of electric machines and power systems are among the oldest areas of electrical engineering. Study began in the latter part of the nineteenth century. At that time, electrical units were being standardized internationally, and these units came to be universally used by engineers. Volts, amperes, ohms, watts, and similar units, which are part of the metric system of units, have long been used to describe electrical quantities in machines.

In English-speaking countries, though, mechanical quantities had long been measured with the English system of units (inches, feet, pounds, etc.). This practice was followed in the study of machines. Therefore, for many years the electrical and mechanical quantities of machines have been measured with different systems of units.

In 1954, a comprehensive system of units based on the metric system was adopted as an international standard. This system of units became known as the *Système International* (SI) and has been adopted throughout most of the world. The

United States is practically the sole holdout—even Britain and Canada have switched over to SI.

The SI units will inevitably become standard in the United States as time goes by, and professional societies such as the Institute of Electrical and Electronics Engineers (IEEE) have standardized on metric units for all work. However, many people have grown up using English units, and this system will remain in daily use for a long time. Engineering students and working engineers in the United States today must be familiar with both sets of units, since they will encounter both throughout their professional lives. Therefore, this book includes problems and examples using both SI and English units. The emphasis in the examples is on SI units, but the older system is not entirely neglected.

Notation

In this book, vectors, electrical phasors, and other complex values are shown in bold face (e.g., **F**), while scalars are shown in italic face (e.g., *R*). In addition, a special font is used to represent magnetic quantities such as magnetomotive force (e.g., $\mathcal{F}$).

1.3 | ROTATIONAL MOTION, NEWTON'S LAW, AND POWER RELATIONSHIPS

Almost all electric machines rotate about an axis, called the *shaft* of the machine. Because of the rotational nature of machinery, it is important to have a basic understanding of rotational motion. This section contains a brief review of the concepts of distance, velocity, acceleration, Newton's law, and power as they apply to rotating machinery. For a more detailed discussion of the concepts of rotational dynamics, see References 2, 4, and 5.

In general, a three-dimensional vector is required to completely describe the rotation of an object in space. However, machines normally turn on a fixed shaft, so their rotation is restricted to one angular dimension. Relative to a given end of the machine's shaft, the direction of rotation can be described as either *clockwise* (CW) or *counterclockwise* (CCW). For the purpose of this volume, a counterclockwise angle of rotation is assumed to be positive, and a clockwise one is assumed to be negative. For rotation about a fixed shaft, all the concepts in this section reduce to scalars.

Each major concept of rotational motion is defined below and is related to the corresponding idea from linear motion.

Angular Position θ

The angular position θ of an object is the angle at which it is oriented, measured from some arbitrary reference point. Angular position is usually measured in radians or degrees. It corresponds to the linear concept of distance along a line.

Angular Velocity ω

Angular velocity (or speed) is the rate of change in angular position with respect to time. It is assumed positive if the rotation is in a counterclockwise direction. Angular velocity is the rotational analog of the concept of velocity on a line. One-dimensional linear velocity along a line is defined as the rate of change of the displacement along the line (r) with respect to time

$$v = \frac{dr}{dt} \tag{1-1}$$

Similarly, angular velocity ω is defined as the rate of change of the angular displacement θ with respect to time.

$$\omega = \frac{d\theta}{dt} \tag{1-2}$$

If the units of angular position are radians, then angular velocity is measured in radians per second.

In dealing with ordinary electric machines, engineers often use units other than radians per second to describe shaft speed. Frequently, the speed is given in revolutions per second or revolutions per minute. Because speed is such an important quantity in the study of machines, it is customary to use different symbols for speed when it is expressed in different units. By using these different symbols, any possible confusion as to the units intended is minimized. The following symbols are used in this book to describe angular velocity:

ω_m angular velocity expressed in radians per second

f_m angular velocity expressed in revolutions per second

n_m angular velocity expressed in revolutions per minute

The subscript m on these symbols indicates a mechanical quantity, as opposed to an electrical quantity. If there is no possibility of confusion between mechanical and electrical quantities, the subscript is often left out.

These measures of shaft speed are related to each other by the following equations:

$$n_m = 60f_m \tag{1-3a}$$

$$f_m = \frac{\omega_m}{2\pi} \tag{1-3b}$$

Angular Acceleration α

Angular acceleration is the rate of change in angular velocity with respect to time. It is assumed positive if the angular velocity is increasing in an algebraic sense. Angular acceleration is the rotational analog of the concept of acceleration on a line. Just as one-dimensional linear acceleration is defined by the equation

$$a = \frac{dv}{dt} \tag{1-4}$$

angular acceleration is defined by

$$\alpha = \frac{d\omega}{dt} \tag{1-5}$$

If the units of angular velocity are radians per second, then angular acceleration is measured in radians per second squared.

Torque τ

In linear motion, a *force* applied to an object causes its velocity to change. In the absence of a net force on the object, its velocity is constant. The greater the force applied to the object, the more rapidly its velocity changes.

There exists a similar concept for rotation. When an object is rotating, its angular velocity is constant unless a *torque* is present on it. The greater the torque on the object, the more rapidly the angular velocity of the object changes.

What is torque? It can loosely be called the "twisting force" on an object. Intuitively, torque is fairly easy to understand. Imagine a cylinder that is free to rotate about its axis. If a force is applied to the cylinder in such a way that its line of action passes through the axis (Figure 1–1a), then the cylinder will not rotate. However, if the same force is placed so that its line of action passes to the right of the axis (Figure 1–1b), then the cylinder will tend to rotate in a counterclockwise direction. The torque or twisting action on the cylinder depends on (1) the magnitude of the applied force and (2) the distance between the axis of rotation and the line of action of the force.

Figure 1–1 | (a) A force applied to a cylinder so that it passes through the axis of rotation. $\tau = 0$. (b) A force applied to a cylinder so that its line of action misses the axis of rotation. Here τ is counterclockwise.

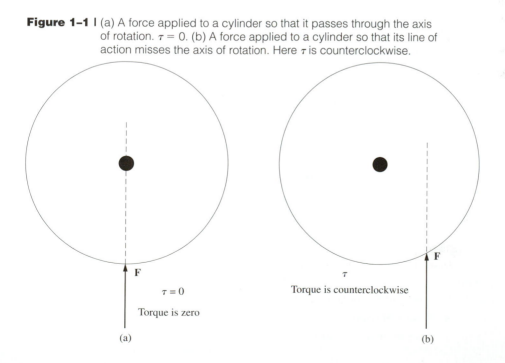

The torque on an object is defined as the product of the force applied to the object and the smallest distance between the line of action of the force and the object's axis of rotation. If **r** is a vector pointing from the axis of rotation to the point of application of the force, and if **F** is the applied force, then the torque can be described as

$$\tau = (\text{force applied})(\text{perpendicular distance})$$
$$= (F)(r \sin \theta)$$
$$= rF \sin \theta \tag{1–6}$$

where θ is the angle between the vector **r** and the vector **F**. The direction of the torque is clockwise if it would tend to cause a clockwise rotation and counterclockwise if it would tend to cause a counterclockwise rotation (Figure 1–2).

The units of torque are newton-meters in SI units and pound-feet in the English system.

Figure 1–2 | Derivation of the equation for the torque on an object.

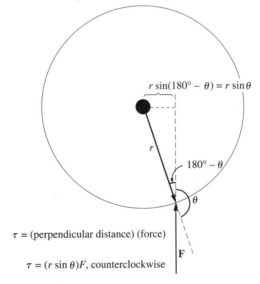

$$r \sin(180° - \theta) = r \sin \theta$$

r

$180° - \theta$

θ

$\tau = (\text{perpendicular distance}) (\text{force})$

F

$\tau = (r \sin \theta)F, \text{ counterclockwise}$

Newton's Law of Rotation

Newton's law for objects moving along a straight line describes the relationship between the force applied to an object and its resulting acceleration. This relationship is given by the equation

$$F = ma \tag{1–7}$$

where

$F =$ net force applied to an object

$m =$ mass of the object

$a =$ resulting acceleration

In SI units, force is measured in newtons, mass in kilograms, and acceleration in meters per second squared. In the English system, force is measured in pounds, mass in slugs, and acceleration in feet per second squared.

A similar equation describes the relationship between the torque applied to an object and its resulting angular acceleration. This relationship, called *Newton's law of rotation,* is given by the equation

$$\tau = J\alpha \tag{1–8}$$

where τ is the net applied torque in newton-meters or pound-feet and α is the resulting angular acceleration in radians per second squared. The term J serves the same purpose as an object's mass in linear motion. It is called the *moment of inertia* of the object and is measured in kilogram-meters squared or slug-feet squared. Calculation of the moment of inertia of an object is beyond the scope of this book. For information about it see Ref. 2.

Work *W*

For linear motion, work is defined as the application of a *force* through a *distance*. In equation form,

$$W = \int F \, dr \tag{1–9}$$

where it is assumed that the force is collinear with the direction of motion. For the special case of a constant force applied collinearly with the direction of motion, this equation becomes just

$$W = Fr \tag{1–10}$$

The units of work are joules in SI and foot-pounds in the English system.

For rotational motion, work is the application of a *torque* through an *angle*. Here the equation for work is

$$W = \int \tau \, d\theta \tag{1–11}$$

and if the torque is constant,

$$W = \tau\theta \tag{1–12}$$

Power *P*

Power is the rate of doing work, or the increase in work per unit time. The equation for power is

$$P = \frac{dW}{dt} \tag{1–13}$$

It is usually measured in joules per second (watts), but also can be measured in foot-pounds per second or in horsepower.

By this definition, and assuming that force is constant and collinear with the direction of motion, power is given by

$$P = \frac{dW}{dt} = \frac{d}{dt}(Fr) = F\left(\frac{dr}{dt}\right) = Fv \tag{1-14}$$

Similarly, assuming constant torque, power in rotational motion is given by

$$P = \frac{dW}{dt} = \frac{d}{dt}(\tau\theta) = \tau\left(\frac{d\theta}{dt}\right) = \tau\omega$$
$$P = \tau\omega \tag{1-15}$$

Equation (1–15) is very important in the study of electric machinery, because it can describe the mechanical power on the shaft of a motor or generator.

Equation (1–15) is the correct relationship among power, torque, and speed if power is measured in watts, torque in newton-meters, and speed in radians per second. If other units are used to measure any of the above quantities, then a constant must be introduced into the equation for unit conversion factors. It is still common in U.S. engineering practice to measure torque in pound-feet, speed in revolutions per minute, and power in either watts or horsepower. If the appropriate conversion factors are included in each term, then Equation (1–15) becomes

$$P \text{ (watts)} = \frac{\tau \text{ (lb-ft) } n \text{ (r/min)}}{7.04} \tag{1-16}$$

$$P \text{ (horsepower)} = \frac{\tau \text{ (lb-ft) } n \text{ (r/min)}}{5252} \tag{1-17}$$

where torque is measured in pound-feet and speed is measured in revolutions per minute.

1.4 | THE MAGNETIC FIELD

As previously stated, magnetic fields are the fundamental mechanism by which energy is converted from one form to another in motors, generators, and transformers. Four basic principles describe how magnetic fields are used in these devices:

1. A current-carrying wire produces a magnetic field in the area around it.
2. A time-changing magnetic field induces a voltage in a coil of wire if it passes through that coil. (This is the basis of *transformer action*.)
3. A current-carrying wire in the presence of a magnetic field has a force induced on it. (This is the basis of *motor action*.)
4. A moving wire in the presence of a magnetic field has a voltage induced in it. (This is the basis of *generator action*.)

This section describes and elaborates on the production of a magnetic field by a current-carrying wire, while later sections of this chapter explain the remaining three principles.

Production of a Magnetic Field

The basic law governing the production of a magnetic field by a current is Ampere's law:

$$\oint \mathbf{H} \cdot d\mathbf{l} = I_{net} \tag{1–18}$$

where $\mathbf{H}$ is the magnetic field intensity produced by the current I_{net}, and dl is a differential element of length along the path of integration. In SI units, I is measured in amperes and H is measured in ampere-turns per meter. To better understand the meaning of this equation, it is helpful to apply it to the simple example in Figure 1–3. Figure 1–3 shows a rectangular core with a winding of N turns of wire wrapped about one leg of the core. If the core is composed of iron or certain other similar metals (collectively called *ferromagnetic materials*), essentially all the magnetic field produced by the current will remain inside the core, so the path of integration in Ampere's law is the mean path length of the core l_c. The current passing within the path of integration I_{net} is then Ni, since the coil of wire cuts the path of integration N times while carrying current i. Ampere's law thus becomes

$$Hl_c = Ni \tag{1–19}$$

Here H is the magnitude of the magnetic field intensity vector $\mathbf{H}$. Therefore, the magnitude of the magnetic field intensity in the core due to the applied current is

$$H = \frac{Ni}{l_c} \tag{1–20}$$

Figure 1–3 | A simple magnetic core.

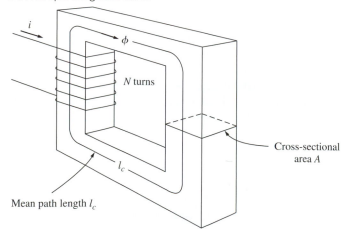

The magnetic field intensity $\mathbf{H}$ is in a sense a measure of the "effort" that a current is putting into the establishment of a magnetic field. The strength of the magnetic field flux produced in the core also depends on the material of the core. The relationship between the magnetic field intensity **H** and the resulting magnetic flux density **B** produced within a material is given by

$$\mathbf{B} = \mu\mathbf{H} \tag{1–21}$$

where

H = magnetic field intensity

μ = magnetic *permeability* of material

B = resulting magnetic flux density produced

The actual magnetic flux density produced in a piece of material is thus given by a product of two terms:

H, representing the effort exerted by the current to establish a magnetic field

μ, representing the relative ease of establishing a magnetic field in a given material

The units of magnetic field intensity are ampere-turns per meter, the units of permeability are henrys per meter, and the units of the resulting flux density are webers per square meter, known as teslas (T).

The permeability of free space is called μ_0, and its value is

$$\mu_0 = 4\pi \times 10^{-7} \text{ H/m} \tag{1–22}$$

The permeability of any other material compared to the permeability of free space is called its *relative permeability:*

$$\mu_r = \frac{\mu}{\mu_0} \tag{1–23}$$

Relative permeability is a convenient way to compare the magnetizability of materials. For example, the steels used in modern machines have relative permeabilities of 2000 to 6000 or even more. This means that, for a given amount of current, 2000 to 6000 times more flux is established in a piece of steel than in a corresponding area of air. (The permeability of air is essentially the same as the permeability of free space.) Obviously, the metals in a transformer or motor core play an extremely important part in increasing and concentrating the magnetic flux in the device.

Also, because the permeability of iron is so much higher than that of air, the great majority of the flux in an iron core like that in Figure 1–3 remains inside the core instead of traveling through the surrounding air, which has much lower permeability. The small leakage flux that does leave the iron core is very important in determining the flux linkages between coils and the self-inductances of coils in transformers and motors.

In a core such as the one shown in Figure 1–3, the magnitude of the flux density is given by

$$B = \mu H = \frac{\mu N i}{l_c} \tag{1–24}$$

Now the total flux in a given area is given by

$$\phi = \int_A \mathbf{B} \cdot d\mathbf{A} \tag{1–25a}$$

where $d\mathbf{A}$ is the differential unit of area. If the flux density vector is perpendicular to a plane of area A, and if the flux density is constant throughout the area, then this equation reduces to

$$\phi = BA \tag{1–25b}$$

Thus, the total flux in the core in Figure 1–3 due to the current i in the winding is

$$\phi = BA = \frac{\mu NiA}{l_c} \tag{1–26}$$

where A is the cross-sectional area of the core.

Magnetic Circuits

In Equation (1–26) we see that the *current* in a coil of wire wrapped around a core produces a magnetic flux in the core. This is in some sense analogous to a voltage in an electric circuit producing a current flow. It is possible to define a "magnetic circuit" whose behavior is governed by equations analogous to those for an electric circuit. The magnetic circuit model of magnetic behavior is often used in the design of electric machines and transformers to simplify the otherwise quite complex design process.

In a simple electric circuit such as the one shown in Figure 1–4a, the voltage source V drives a current I around the circuit through a resistance R. The relationship between these quantities is given by Ohm's law:

$$V = IR$$

Figure 1–4 | (a) A simple electric circuit. (b) The magnetic circuit analog to a transformer core.

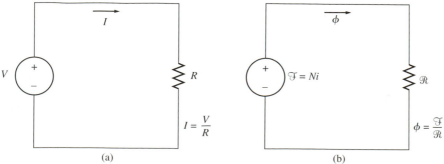

(a)   (b)

In the electric circuit, it is the voltage or electromotive force that drives the current flow. By analogy, the corresponding quantity in the magnetic circuit is called the *magnetomotive force* (mmf). The magnetomotive force of the magnetic circuit is equal to the effective current flow applied to the core, or

$$\mathcal{F} = Ni \qquad (1\text{--}27)$$

where $\mathcal{F}$ is the symbol for magnetomotive force, measured in ampere-turns.

Like the voltage source in the electric circuit, the magnetomotive force in the magnetic circuit has a polarity associated with it. The *positive* end of the mmf source is the end from which the flux exits, and the *negative* end of the mmf source is the end at which the flux reenters. The polarity of the mmf from a coil of wire can be determined from a modification of the right-hand rule: If the fingers of the right hand curl in the direction of the current flow in a coil of wire, then the thumb will point in the direction of the positive mmf (see Figure 1–5).

Figure 1–5 | Determining the polarity of a magnetomotive force source in a magnetic circuit.

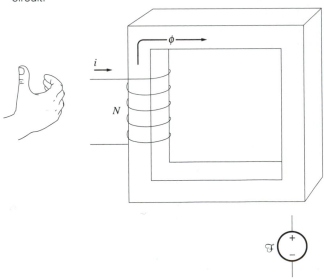

In an electric circuit, the applied voltage causes a current I to flow. Similarly, in a magnetic circuit, the applied magnetomotive force causes flux ϕ to be produced. The relationship between voltage and current in an electric circuit is Ohm's law ($V = IR$); similarly, the relationship between magnetomotive force and flux is

$$\boxed{\mathcal{F} = \phi \mathcal{R}} \qquad (1\text{--}28)$$

where

$\mathcal{F}$ = magnetomotive force of circuit

ϕ = flux of circuit

$\mathcal{R}$ = *reluctance* of circuit

The reluctance of a magnetic circuit is the counterpart of electrical resistance, and its units are ampere-turns per weber.

There is also a magnetic analog of conductance. Just as the conductance of an electric circuit is the reciprocal of its resistance, the *permeance* $\mathcal{P}$ of a magnetic circuit is the reciprocal of its reluctance:

$$\mathcal{P} = \frac{1}{\mathcal{R}} \tag{1-29}$$

The relationship between magnetomotive force and flux can thus be expressed as

$$\phi = \mathcal{F}\mathcal{P} \tag{1-30}$$

Under some circumstances, it is easier to work with the permeance of a magnetic circuit than with its reluctance.

What is the reluctance of the core in Figure 1–3? The resulting flux in this core is given by Equation (1–26):

$$\phi = BA = \frac{\mu NiA}{l_c} \tag{1-26}$$

$$= Ni\left(\frac{\mu A}{l_c}\right)$$

$$\phi = \mathcal{F}\left(\frac{\mu A}{l_c}\right) \tag{1-31}$$

By comparing Equation (1–31) with Equation (1–28), we see that the reluctance of the core is

$$\mathcal{R} = \frac{l_c}{\mu A} \tag{1-32}$$

Reluctances in a magnetic circuit obey the same rules as resistances in an electric circuit. The equivalent reluctance of a number of reluctances in series is just the sum of the individual reluctances:

$$\mathcal{R}_{eq} = \mathcal{R}_1 + \mathcal{R}_2 + \mathcal{R}_3 + \cdots \tag{1-33}$$

Similarly, reluctances in parallel combine according to the equation

$$\frac{1}{\mathcal{R}_{eq}} = \frac{1}{\mathcal{R}_1} + \frac{1}{\mathcal{R}_2} + \frac{1}{\mathcal{R}_3} + \cdots \tag{1-34}$$

Permeances in series and parallel obey the same rules as electrical conductances.

Calculations of the flux in a core performed by using the magnetic circuit concepts are *always* approximations—at best, they are accurate to within about 5 percent of the real answer. There are a number of reasons for this inherent inaccuracy:

1. The magnetic circuit concept assumes that all flux is confined within a magnetic core. Unfortunately, this is not quite true. The permeability of a ferromagnetic core is 2000 to 6000 times that of air, but a small fraction of the flux escapes from the core into the surrounding low-permeability air. This flux outside the core is called *leakage flux,* and it plays a very important role in electric machine design.

2. The calculation of reluctance assumes a certain mean path length and cross-sectional area for the core. These assumptions are not really very good, especially at corners.

3. In ferromagnetic materials, the permeability varies with the amount of flux already in the material. This nonlinear effect is described in detail below. It adds yet another source of error to magnetic circuit analysis, since the reluctances used in magnetic circuit calculations depend on the permeability of the material.

4. If there are air gaps in the flux path in a core, the effective cross-sectional area of the air gap will be larger than the cross-sectional area of the iron core on either side. The extra effective area is caused by the "fringing effect" of the magnetic field at the air gap (Figure 1–6).

Figure 1–6 I The fringing effect of a magnetic field at an air gap. Note the increased cross-sectional area of the air gap compared with the cross-sectional area of the metal.

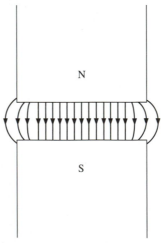

It is possible to partially offset these inherent sources of error by using a "corrected" or "effective" mean path length and the cross-sectional area instead of the actual physical length and area in the calculations.

There are many inherent limitations to the concept of a magnetic circuit, but it is still the easiest design tool available for calculating fluxes in practical machinery design. Exact calculations using Maxwell's equations are just too difficult, and they are not needed anyway, since satisfactory results may be achieved with this approximate method.

The following examples illustrate basic magnetic circuit calculations. Note that in these examples the answers are given to three significant digits.

EXAMPLE 1–1

A ferromagnetic core is shown in Figure 1–7a. Three sides of this core are of uniform width, while the fourth side is somewhat thinner. The depth of the core (into the page) is 10 cm, and the other dimensions are shown in the figure. There is a 200-turn coil wrapped around the left side of the core. Assuming relative permeability μ_r of 2500, how much flux will be produced by a 1-A input current?

■ **Solution**

We will solve this problem twice, once by hand and once by a MATLAB program, and show that both approaches yield the same answer.

Three sides of the core have the same cross-sectional areas, while the fourth side has a different area. Thus, the core can be divided into two regions: (1) the single thinner side and (2) the other three sides taken together. The magnetic circuit corresponding to this core is shown in Figure 1–7b.

The mean path length of region 1 is 45 cm, and the cross-sectional area is 10×10 cm $= 100$ cm^2. Therefore, the reluctance in the first region is

$$\mathcal{R}_1 = \frac{l_1}{\mu A_1} = \frac{l_1}{\mu_r \mu_0 A_1} \tag{1–32}$$

$$= \frac{0.45 \text{ m}}{(2500)(4\pi \times 10^{-7})(0.01 \text{ m}^2)}$$

$$= 14{,}300 \text{ A-turns/Wb}$$

The mean path length of region 2 is 130 cm, and the cross-sectional area is 15×10 cm $= 150$ cm^2. Therefore, the reluctance in the second region is

$$\mathcal{R}_2 = \frac{l_2}{\mu A_2} = \frac{l_2}{\mu_r \mu_0 A_2} \tag{1–32}$$

$$= \frac{1.3 \text{ m}}{(2500)(4\pi \times 10^{-7})(0.015 \text{ m}^2)}$$

$$= 27{,}600 \text{ A-turns/Wb}$$

Therefore, the total reluctance in the core is

$$\mathcal{R}_{eq} = \mathcal{R}_1 + \mathcal{R}_2$$

$$= 14{,}300 \text{ A-turns/Wb} + 27{,}600 \text{ A-turns/Wb}$$

$$= 41{,}900 \text{ A-turns/Wb}$$

The total magnetomotive force is

$$\mathcal{F} = Ni = (200 \text{ turns})(1.0 \text{ A}) = 200 \text{ A-turns}$$

Figure 1–7 | (a) The ferromagnetic core of Example 1–1. (b) The magnetic circuit
corresponding to (a).

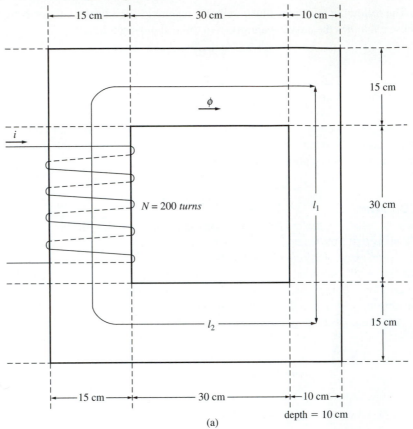

(a)

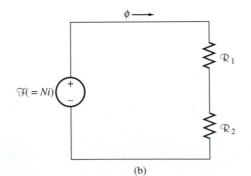

(b)

The total flux in the core is given by

$$\phi = \frac{\mathcal{F}}{\mathcal{R}} = \frac{200 \text{ A-turns}}{41,900 \text{ A-turns/Wb}}$$
$$= 0.0048 \text{ Wb}$$

This calculation can be performed by using a MATLAB script file, if desired. A simple script to calculate the flux in the core is shown below.

```
% M-file: ex1_1.m
% M-file to calculate the flux in Example 1-1.
l1 = 0.45;                    % Length of region 1
l2 = 1.3;                     % Length of region 2
a1 = 0.01;                    % Area of region 1
a2 = 0.015;                   % Area of region 2
ur = 2500;                    % Relative permeability
u0 = 4*pi*1E-7;               % Permeability of free space
n = 200;                      % Number of turns on core
i = 1;                        % Current in amps

% Calculate the first reluctance
r1 = l1 / (ur * u0 * a1);
disp (['r1 = ' num2str(r1)]);

% Calculate the second reluctance
r2 = l2 / (ur * u0 * a2);
disp (['r2 = ' num2str(r2)]);

% Calculate the total reluctance
rtot = r1 + r2;

% Calculate the mmf
mmf = n * i;

% Finally, get the flux in the core
flux = mmf / rtot;

% Display result
disp (['Flux = ' num2str(flux)]);
```

When this program is executed, the results are:

```
» ex1_1
r1 = 14323.9449
r2 = 27586.8568
Flux = 0.004772
```

This program produces the same answer as our hand calculations to the number of significant digits in the problem. ■

EXAMPLE 1–2

Figure 1–8a shows a ferromagnetic core whose mean path length is 40 cm. There is a small gap of 0.05 cm in the structure of the otherwise whole core. The cross-sectional area of the core is 12 cm², the relative permeability of the core is 4000, and the coil of wire on the core has 400 turns. Assume that fringing in the air gap increases the effective cross-sectional area of the air gap by 5 percent. Given this information, find (a) the total reluctance of the flux path (iron plus air gap) and (b) the current required to produce a flux density of 0.5 T in the air gap.

Figure 1–8 | (a) The ferromagnetic core of Example 1–2. (b) The magnetic circuit corresponding to (a).

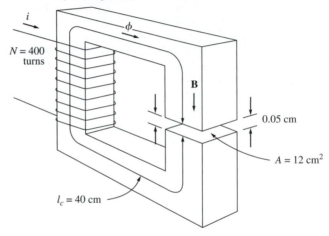

(a)

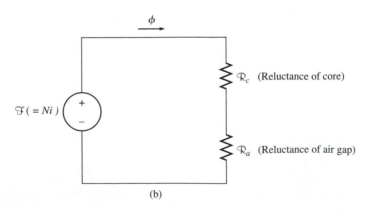

(b)

■ Solution

The magnetic circuit corresponding to this core is shown in Figure 1–8b.

(a) The reluctance of the core is

$$\mathcal{R}_c = \frac{l_c}{\mu A_c} = \frac{l_c}{\mu_r \mu_0 A_c} \tag{1–32}$$

$$= \frac{0.4 \text{ m}}{(4000)(4\pi \times 10^{-7})(0.0012 \text{ m}^2)}$$

$$= 66{,}300 \text{ A-turns/Wb}$$

The effective area of the air gap is $1.05 \times 12 \text{ cm}^2 = 12.6 \text{ cm}^2$, so the reluctance of the air gap is

$$\mathcal{R}_a = \frac{l_a}{\mu_0 A_a} \tag{1–32}$$

$$= \frac{0.0005 \text{ m}}{(4\pi \times 10^{-7})(0.00126 \text{ m}^2)}$$

$$= 316{,}000 \text{ A-turns/Wb}$$

Therefore, the total reluctance of the flux path is

$$\mathcal{R}_{eq} = \mathcal{R}_c + \mathcal{R}_a$$

$$= 66{,}300 \text{ A-turns/Wb} + 316{,}000 \text{ A-turns/Wb}$$

$$= 382{,}300 \text{ A-turns/Wb}$$

Note that the air gap contributes most of the reluctance even though it is 800 times shorter than the core.

(b) Equation (1–28) states that

$$\mathcal{F} = \phi \mathcal{R} \tag{1–28}$$

Since the flux $\phi = BA$ and $\mathcal{F} = Ni$, this equation becomes

$$Ni = BA\mathcal{R}$$

so

$$i = \frac{BA\mathcal{R}}{N}$$

$$= \frac{(0.5 \text{ T})(0.00126 \text{ m}^2)(383{,}200 \text{ A-turns/Wb})}{400 \text{ turns}}$$

$$= 0.602 \text{ A}$$

Notice that, since the *air-gap* flux was required, the effective air-gap area was used in the above equation. ■

EXAMPLE 1-3

Figure 1–9a shows a simplified rotor and stator for a dc motor. The mean path length of the stator is 50 cm, and its cross-sectional area is 12 cm². The mean path length of the rotor is 5 cm, and its cross-sectional area also may be assumed to be 12 cm². Each air gap between the rotor and the stator is 0.05 cm wide, and the cross-sectional area of each air gap (including fringing) is 14 cm². The iron of the core has a relative permeability of 2000, and there are 200 turns of wire on the core. If the current in the wire is adjusted to be 1 A, what will the resulting flux density in the air gaps be?

Figure 1–9 | (a) A simplified diagram of a rotor and stator for a dc motor. (b) The magnetic circuit corresponding to (a).

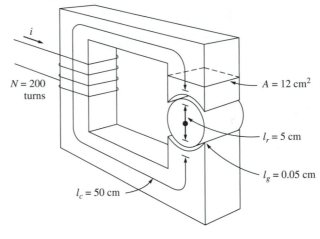

$N = 200$ turns

$A = 12\ \text{cm}^2$

$l_r = 5\ \text{cm}$

$l_g = 0.05\ \text{cm}$

$l_c = 50\ \text{cm}$

(a)

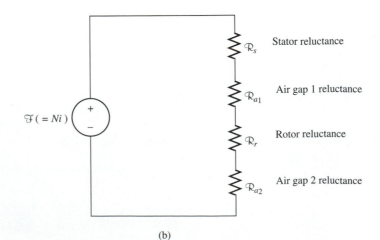

$\mathcal{R}_s$ Stator reluctance

$\mathcal{R}_{a1}$ Air gap 1 reluctance

$\mathcal{F}\ (= Ni\ )$

$\mathcal{R}_r$ Rotor reluctance

$\mathcal{R}_{a2}$ Air gap 2 reluctance

(b)

■ Solution

To determine the flux density in the air gap, it is necessary to first calculate the magnetomotive force applied to the core and the total reluctance of the flux path. With this information, the total flux in the core can be found. Finally, knowing the cross-sectional area of the air gaps enables the flux density to be calculated.

The reluctance of the stator is

$$\mathcal{R}_s = \frac{l_s}{\mu_r \mu_0 A_s}$$

$$= \frac{0.5 \text{ m}}{(2000)(4\pi \times 10^{-7})(0.0012 \text{ m}^2)}$$

$$= 166,000 \text{ A-turns/Wb}$$

The reluctance of the rotor is

$$\mathcal{R}_r = \frac{l_r}{\mu_r \mu_0 A_r}$$

$$= \frac{0.05 \text{ m}}{(2000)(4\pi \times 10^{-7})(0.0012 \text{ m}^2)}$$

$$= 16,600 \text{ A-turns/Wb}$$

The reluctance of the air gaps is

$$\mathcal{R}_a = \frac{l_a}{\mu_r \mu_0 A_a}$$

$$= \frac{0.0005 \text{ m}}{(1)(4\pi \times 10^{-7})(0.0014 \text{ m}^2)}$$

$$= 284,000 \text{ A-turns/Wb}$$

The magnetic circuit corresponding to this machine is shown in Figure 1–9b. The total reluctance of the flux path is thus

$$\mathcal{R}_{eq} = \mathcal{R}_s + \mathcal{R}_{a1} + \mathcal{R}_r + \mathcal{R}_{a2}$$

$$= 166,000 + 284,000 + 16,600 + 284,000 \text{ A-turns/Wb}$$

$$= 751,000 \text{ A-turns/Wb}$$

The net magnetomotive force applied to the core is

$$\mathcal{F} = Ni = (200 \text{ turns})(1.0 \text{ A}) = 200 \text{ A-turns}$$

Therefore, the total flux in the core is

$$\phi = \frac{\mathcal{F}}{\mathcal{R}} = \frac{200 \text{ A-turns}}{751,000 \text{ A-turns/Wb}}$$

$$= 0.00266 \text{ Wb}$$

Finally, the magnetic flux density in the motor's air gap is

$$B = \frac{\phi}{A} = \frac{0.000266 \text{ Wb}}{0.0014 \text{ m}^2} = 0.19 \text{ T} \qquad ■$$

Magnetic Behavior of Ferromagnetic Materials

Earlier in this section, magnetic permeability was defined by the equation

$$\mathbf{B} = \mu\mathbf{H} \tag{1–21}$$

It was explained that the permeability of ferromagnetic materials is very high, up to 6000 times the permeability of free space. In that discussion and in the examples that followed, the permeability was assumed to be constant regardless of the magneto-motive force applied to the material. Although permeability is constant in free space, this most certainly is *not* true for iron and other ferromagnetic materials.

To illustrate the behavior of magnetic permeability in a ferromagnetic material, apply a direct current to the core shown in Figure 1–3, starting with 0 A and slowly working up to the maximum permissible current. When the flux produced in the core is plotted versus the magnetomotive force producing it, the resulting plot looks like Figure 1–10a. This type of plot is called a *saturation curve* or a *magnetization curve.* At first, a small increase in the magnetomotive force produces a huge increase in the resulting flux. After a certain point, though, further increases in the magnetomotive force produce relatively smaller increases in the flux. Finally, an increase in the magnetomotive force produces almost no change at all. The region of this figure in which the curve flattens out is called the *saturation region,* and the core is said to be *saturated.* In contrast, the region where the flux changes very rapidly is called the *unsaturated region* of the curve, and the core is said to be *unsaturated.* The transition region between the unsaturated region and the saturated region is sometimes called the *knee* of the curve. Note that the flux produced in the core is linearly related to the applied magnetomotive force in the unsaturated region, and approaches a constant value regardless of magnetomotive force in the saturated region.

Another closely related plot is shown in Figure 1–10b. Figure 1–10b is a plot of magnetic flux density **B** versus magnetizing intensity **H**. From Equations (1–20) and (1–25b),

$$H = \frac{Ni}{l_c} = \frac{\mathcal{F}}{l_c} \tag{1–20}$$

$$\phi = BA \tag{1–25b}$$

it is easy to see that *magnetizing intensity is directly proportional to magnetomotive force* and *magnetic flux density is directly proportional to flux* for any given core. Therefore, the relationship between B and H has the same shape as the relationship between flux and magnetomotive force. The slope of the curve of flux density versus magnetizing intensity at any value of H in Figure 1–10b is by definition the permeability of the core at that magnetizing intensity. The curve shows that the permeability is large and relatively constant in the unsaturated region and then gradually drops to a very low value as the core becomes heavily saturated.

Figure 1–10c is a magnetization curve for a typical piece of steel shown in more detail and with the magnetizing intensity on a logarithmic scale. Only with the magnetizing intensity shown logarithmically can the huge saturation region of the curve fit onto the graph.

Figure 1–10 | (a) Sketch of a DC magnetization curve for a ferromagnetic core. (b) The magnetization curve expressed in terms of flux density and magnetizing intensity. (c) A detailed magnetization curve for a typical piece of steel. (d) A plot of relative permeability μ_r as a function of magnetizing intensity H for a typical piece of steel.

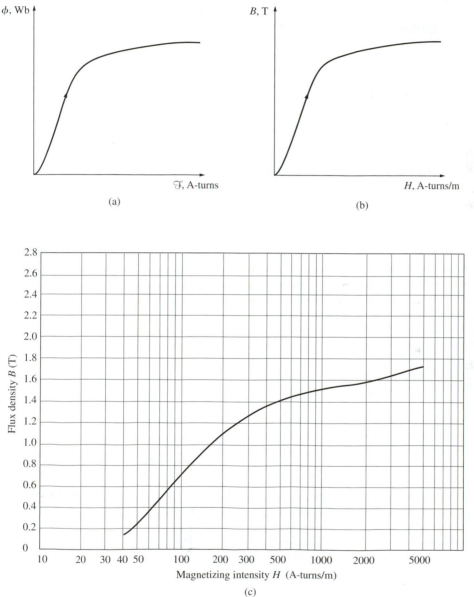

(a)

(b)

(c)

(continued)

Figure 1–10 | *(continued)*

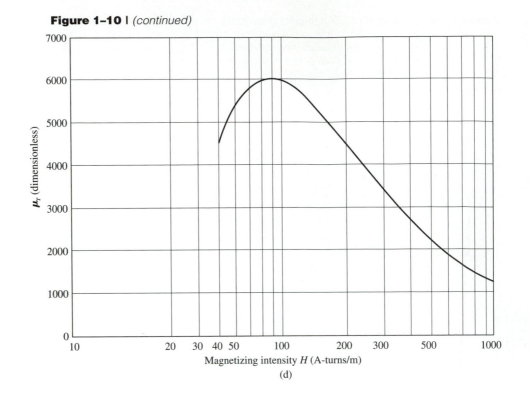

Magnetizing intensity H (A-turns/m)

(d)

The advantage of using a ferromagnetic material for cores in electric machines and transformers is that one gets many times more flux for a given magnetomotive force with iron than with air. However, if the resulting flux has to be proportional, or nearly so, to the applied magnetomotive force, then the core *must* be operated in the unsaturated region of the magnetization curve.

Since real generators and motors depend on magnetic flux to produce voltage and torque, they are designed to produce as much flux as possible. As a result, most real machines operate near the knee of the magnetization curve, and the flux in their cores is not linearly related to the magnetomotive force producing it. This nonlinearity accounts for many of the peculiar behaviors of machines that will be explained in future chapters. We will use MATLAB to calculate solutions to problems involving the nonlinear behavior of real machines.

EXAMPLE 1–4

Find the relative permeability of the typical ferromagnetic material whose magnetization curve is shown in Figure 1–10c at (a) $H = 50$, (b) $H = 100$, (c) $H = 500$, and (d) $H = 1000$ A-turns/m.

■ Solution

The permeability of a material is given by

$$\mu = \frac{B}{H}$$

and the relative permeability is given by

$$\mu_r = \frac{\mu}{\mu_0} \qquad (1\text{--}23)$$

Thus, it is easy to determine the permeability at any given magnetizing intensity.

(a) At $H = 50$ A-turns/m, $B = 0.25$ T, so

$$\mu = \frac{B}{H} = \frac{0.25 \text{ T}}{50 \text{ A-turns/m}} = 0.0050 \text{ H/m}$$

and

$$\mu_r = \frac{\mu}{\mu_0} = \frac{0.0050 \text{ H/m}}{4\pi \times 10^{-7} \text{ H/m}} = 3980$$

(b) At $H = 100$ A-turns/m, $B = 0.72$ T, so

$$\mu = \frac{B}{H} = \frac{0.72 \text{ T}}{100 \text{ A-turns/m}} = 0.0072 \text{ H/m}$$

and

$$\mu_r = \frac{\mu}{\mu_0} = \frac{0.0072 \text{ H/m}}{4\pi \times 10^{-7} \text{ H/m}} = 5730$$

(c) At $H = 500$ A-turns/m, $B = 1.40$ T, so

$$\mu = \frac{B}{H} = \frac{1.40 \text{ T}}{500 \text{ A-turns/m}} = 0.0028 \text{ H/m}$$

and

$$\mu_r = \frac{\mu}{\mu_0} = \frac{0.0028 \text{ H/m}}{4\pi \times 10^{-7} \text{ H/m}} = 2230$$

(d) At $H = 1000$ A-turns/m, $B = 1.51$ T, so

$$\mu = \frac{B}{H} = \frac{1.51 \text{ T}}{1000 \text{ A-turns/m}} = 0.00151 \text{ H/m}$$

and

$$\mu_r = \frac{\mu}{\mu_0} = \frac{0.00151 \text{ H/m}}{4\pi \times 10^{-7} \text{ H/m}} = 1200 \qquad ■$$

Notice that as the magnetizing intensity is increased, the relative permeability first increases and then starts to drop off. The relative permeability of a typical ferromagnetic material as a function of the magnetizing intensity is shown in

Figure 1–10d. This shape is fairly typical of all ferromagnetic materials. It can easily be seen from the curve for μ_r versus H that the assumption of constant relative permeability made in Examples 1–1 to 1–3 is valid only over a relatively narrow range of magnetizing intensities (or magnetomotive forces).

In the following example, the relative permeability is not assumed constant. Instead, the relationship between B and H is given by a graph.

EXAMPLE 1–5

A square magnetic core has a mean path length of 55 cm and a cross-sectional area of 150 cm². A 200-turn coil of wire is wrapped around one leg of the core. The core is made of a material having the magnetization curve shown in Figure 1–10c.

(a) How much current is required to produce 0.012 Wb of flux in the core?

(b) What is the core's relative permeability at that current level?

(c) What is its reluctance?

■ Solution

(a) The required flux density in the core is

$$B = \frac{\phi}{A} = \frac{1.012 \text{ Wb}}{0.015 \text{ m}^2} = 0.8 \text{ T}$$

From Figure 1–10c, the required magnetizing intensity is

$$H = 115 \text{ A-turns/m}$$

From Equation (1–20), the magnetomotive force needed to produce this magnetizing intensity is

$$\mathscr{F} = Ni = Hl_c$$
$$= (115 \text{ A-turns/m})(0.55 \text{ m}) = 63.25 \text{ A-turns}$$

so the required current is

$$i = \frac{\mathscr{F}}{N} = \frac{63.25 \text{ A-turns}}{200 \text{ turns}} = 0.316 \text{ A}$$

(b) The core's permeability at this current is

$$\mu = \frac{B}{H} = \frac{0.8 \text{ T}}{115 \text{ A-turns/m}} = 0.00696 \text{ H/m}$$

Therefore, the relative permeability is

$$\mu_r = \frac{\mu}{\mu_0} = \frac{0.00696 \text{ H/m}}{4\pi \times 10^{-7} \text{ H/m}} = 5540$$

(c) The reluctance of the core is

$$\mathscr{R} = \frac{\mathscr{F}}{\phi} = \frac{63.25 \text{ A-turns}}{0.012 \text{ Wb}} = 5270 \text{ A-turns/Wb}$$ ■

Energy Losses in a Ferromagnetic Core

Instead of applying a direct current to the windings on the core, let us now apply an alternating current and observe what happens. The current to be applied is shown in Figure 1–11a. Assume that the flux in the core is initially zero. As the current increases for the first time, the flux in the core traces out path *ab* in Figure 1–11b. This is basically the saturation curve shown in Figure 1–10. However, when the current falls again, *the flux traces out a different path from the one it followed when the current increased.* As the current decreases, the flux in the core traces out path *bcd*, and later when the current increases again, the flux traces out path *deb*. Notice that the amount of flux present in the core depends not only on the amount of current applied to the windings of the core, but also on the previous history of the flux in the core. This dependence on the preceding flux history and the resulting failure to retrace flux paths is called *hysteresis*. Path *bcdeb* traced out in Figure 1–11b as the applied current changes is called a *hysteresis loop*.

Figure 1–11 I The hysteresis loop traced out by the flux in a core when the current *i(t)* is applied to it.

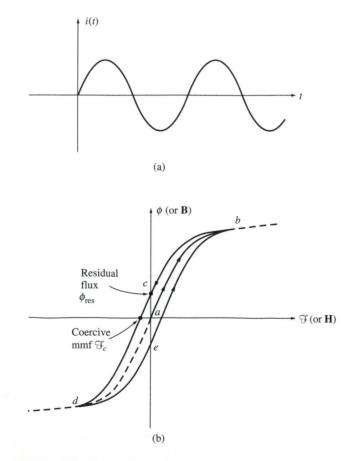

Notice that if a large magnetomotive force is first applied to the core and then removed, the flux path in the core will be *abc*. When the magnetomotive force is removed, the flux in the core *does not* go to zero. Instead, a magnetic field is left in the core. This magnetic field is called the *residual flux* in the core. It is in precisely this manner that permanent magnets are produced. To force the flux to zero, an amount of magnetomotive force known as the *coercive magnetomotive force* $\mathcal{F}_c$ must be applied to the core in the opposite direction.

Why does hysteresis occur? To understand the behavior of ferromagnetic materials, it is necessary to know something about their structure. The atoms of iron and similar metals (cobalt, nickel, and some of their alloys) tend to have their magnetic fields closely aligned with each other. Within the metal, there are many small regions called *domains*. In each domain, all the atoms are aligned with their magnetic fields pointing in the same direction, so each domain within the material acts as a small permanent magnet. The reason that a whole block of iron can appear to have no flux is that these numerous tiny domains are oriented randomly within the material. An example of the domain structure within a piece of iron is shown in Figure 1–12.

Figure 1–12 | (a) Magnetic domains oriented randomly. (b) Magnetic domains lined up in the presence of an external magnetic field.

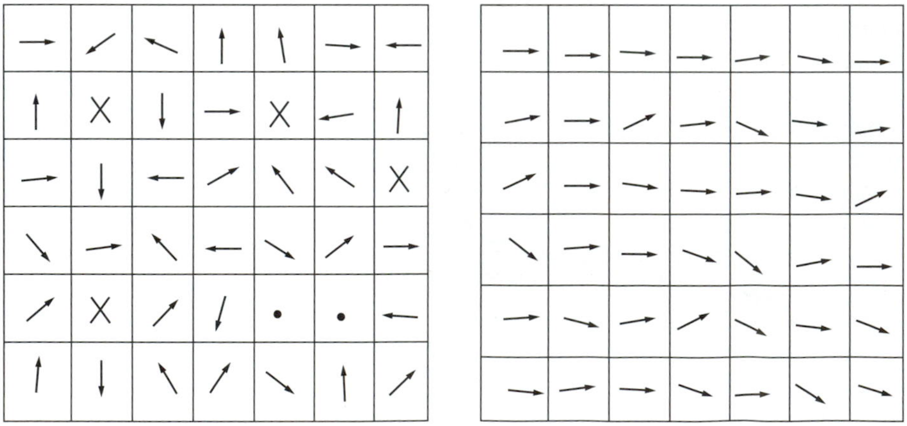

(a)   (b)

When an external magnetic field is applied to this block of iron, it causes domains that happen to point in the direction of the field to grow at the expense of domains pointed in other directions. Domains pointing in the direction of the magnetic field grow because the atoms at their boundaries physically switch orientation to align themselves with the applied magnetic field. The extra atoms aligned with the field increase the magnetic flux in the iron, which in turn causes more atoms to switch orientation, further increasing the strength of the magnetic field. It is this positive feedback effect that causes iron to have a permeability much higher than air.

As the strength of the external magnetic field continues to increase, whole domains that are aligned in the wrong direction eventually reorient themselves as a unit to line up with the field. Finally, when nearly all the atoms and domains in the iron are lined up with the external field, any further increase in the magnetomotive force can cause only the same flux increase that it would in free space. (Once everything is aligned, there can be no more feedback effect to strengthen the field.) At this point, the iron is *saturated* with flux. This is the situation in the saturated region of the magnetization curve in Figure 1–10.

The key to hysteresis is that when the external magnetic field is removed, the domains do not completely randomize again. Why do the domains remain lined up? Because turning the atoms in them requires *energy.* Originally, energy was provided by the external magnetic field to accomplish the alignment; when the field is removed, there is no source of energy to cause all the domains to rotate back. The piece of iron is now a permanent magnet.

Once the domains are aligned, some of them will remain aligned until a source of external energy is supplied to change them. Examples of sources of external energy that can change the boundaries between domains and/or the alignment of domains are magnetomotive force applied in another direction, a large mechanical shock, and heating. Any of these events can impart energy to the domains and enable them to change alignment. (It is for this reason that a permanent magnet can lose its magnetism if it is dropped, hit with a hammer, or heated.)

The fact that turning domains in the iron requires energy leads to a common type of energy loss in all machines and transformers. The *hysteresis loss* in an iron core is the energy required to accomplish the reorientation of domains during each cycle of the alternating current applied to the core. It can be shown that the area enclosed in the hysteresis loop formed by applying an alternating current to the core is directly proportional to the energy lost in a given AC cycle. The smaller the applied magnetomotive force excursions on the core, the smaller the area of the resulting hysteresis loop and so the smaller the resulting losses. Figure 1–13 illustrates this point.

Another type of loss should be mentioned at this point, since it is also caused by varying magnetic fields in an iron core. This loss is the *eddy current* loss. The mechanism of eddy current losses is explained later after Faraday's law has been introduced. Both hysteresis and eddy current losses cause heating in the core material, and both losses must be considered in the design of any machine or transformer. Since both losses occur within the metal of the core, they are usually lumped together and called *core losses.*

1.5 | FARADAY'S LAW—INDUCED VOLTAGE FROM A TIME-CHANGING MAGNETIC FIELD

So far, attention has been focused on the production of a magnetic field and on its properties. It is now time to examine the various ways in which an existing magnetic field can affect its surroundings.

The first major effect to be considered is called *Faraday's law.* It is the basis of transformer operation. Faraday's law states that if a flux passes through a turn of a

Figure 1–13 | The effect of the size of magnetomotive force excursions on the magnitude of the hysteresis loss.

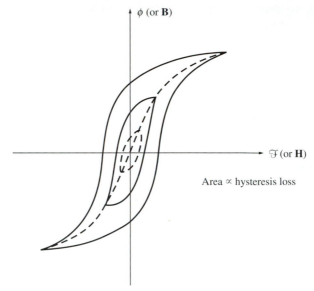

ϕ (or **B**)

$\mathcal{F}$ (or **H**)

Area $\propto$ hysteresis loss

coil of wire, a voltage will be induced in the turn of wire that is directly proportional to the *rate of change* in the flux with respect to time. In equation form,

$$e_{\text{ind}} = -\frac{d\phi}{dt} \tag{1–35}$$

where e_{ind} is the voltage induced in the turn of the coil and ϕ is the flux passing through the turn. If a coil has N turns and if the same flux passes through all of them, then the voltage induced across the whole coil is given by

$$\boxed{e_{\text{ind}} = -N\frac{d\phi}{dt}} \tag{1–36}$$

where

$\qquad e_{\text{ind}}$ = voltage induced in the coil

$\qquad N$ = number of turns of wire in coil

$\qquad \phi$ = flux passing through coil

The minus sign in the equations is an expression of *Lenz's law.* Lenz's law states that the direction of the voltage buildup in the coil is such that if the coil ends were short circuited, it would produce current that would cause a flux *opposing* the original flux change. Since the induced voltage opposes the change that causes it, a minus sign is included in Equation (1–36). To understand this concept clearly, examine Figure 1–14. If the flux shown in the figure is *increasing* in strength, then the voltage built up in the coil will tend to establish a flux that will oppose the increase. A current

Figure 1–14 | The meaning of Lenz's law: (a) A coil enclosing an increasing magnetic flux; (b) determining the resulting voltage polarity.

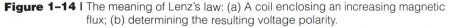

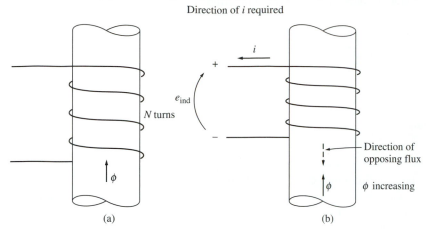

flowing as shown in Figure 1–14b would produce a flux opposing the increase, so the voltage on the coil must be built up with the polarity required to drive that current through the external circuit. Therefore, the voltage must be built up with the polarity shown in the figure. Since the polarity of the resulting voltage can be determined from physical considerations, the minus sign in Equations (1–35) and (1–36) is often left out. It is left out of Faraday's law in the remainder of this book.

There is one major difficulty involved in using Equation (1–36) in practical problems. That equation assumes that exactly the same flux is present in each turn of the coil. Unfortunately, the flux leaking out of the core into the surrounding air prevents this from being true. If the windings are tightly coupled, so that the vast majority of the flux passing through one turn of the coil does indeed pass through all of them, then Equation (1–36) will give valid answers. But if leakage is quite high or if extreme accuracy is required, a different expression that does not make that assumption will be needed. The magnitude of the voltage in the ith turn of the coil is always given by

$$e_{\text{ind}} = \frac{d(\phi_i)}{dt} \tag{1–37}$$

If there are N turns in the coil of wire, the total voltage on the coil is

$$e_{\text{ind}} = \sum_{i=1}^{N} e_i \tag{1–38}$$

$$= \sum_{i=1}^{N} \frac{d(\phi_i)}{dt} \tag{1–39}$$

$$= \frac{d}{dt} \left(\sum_{i=1}^{N} \phi_i \right) \tag{1–40}$$

The term in parentheses in Equation (1–40) is called *the flux linkage* λ of the coil, and Faraday's law can be rewritten in terms of flux linkage as

$$e_{\text{ind}} = \frac{d\lambda}{dt} \tag{1–41}$$

where

$$\lambda = \sum_{i=1}^{N} \phi_i \tag{1–42}$$

The units of flux linkage are weber-turns.

Faraday's law is the fundamental property of magnetic fields involved in transformer operation. The effect of Lenz's law in transformers is to predict the polarity of the voltages induced in transformer windings.

Faraday's law also explains the eddy current losses mentioned previously. A time-changing flux induces voltage *within* a ferromagnetic core in just the same manner as it would in a wire wrapped around that core. These voltages cause swirls of current to flow within the core, much like the eddies seen at the edges of a river. It is the shape of these currents that gives rise to the name *eddy currents*. These eddy currents are flowing in a resistive material (the iron of the core), so energy is dissipated by them. The lost energy goes into heating the iron core.

The amount of energy lost to eddy currents is proportional to the size of the paths they follow within the core. For this reason, it is customary to break up any ferromagnetic core that may be subject to alternating fluxes into many small strips, or *laminations,* and to build the core up out of these strips. An insulating oxide or resin is used between the strips, so that the current paths for eddy currents are limited to very small areas. Because the insulating layers are extremely thin, this action reduces eddy current losses with very little effect on the core's magnetic properties. Actual eddy current losses are proportional to the square of the lamination thickness, so there is a strong incentive to make the laminations as thin as economically possible.

EXAMPLE 1–6

Figure 1–15 shows a coil of wire wrapped around an iron core. If the flux in the core is given by the equation

$$\phi = 0.05 \sin 377t \qquad \text{Wb}$$

If there are 100 turns on the core, what voltage is produced at the terminals of the coil? Of what polarity is the voltage during the time when flux is *increasing* in the reference direction shown in the figure? Assume that all the magnetic flux stays within the core (i.e., assume that the flux leakage is zero).

Figure 1–15 | The core of Example 1–6. Determination of the voltage polarity at the terminals is shown.

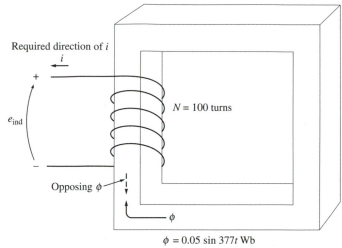

$\phi = 0.05 \sin 377t$ Wb

■ **Solution**

By the same reasoning as in the discussion on page 34, the direction of the voltage while the flux is increasing in the reference direction must be positive to negative, as shown in Figure 1–15. The *magnitude* of the voltage is given by

$$e_{ind} = N\frac{d\phi}{dt}$$

$$= (100 \text{ turns})\frac{d}{dt}(0.05 \sin 377t)$$

$$= 1885 \cos 377t$$

or alternatively,

$$e_{ind} = 1885 \sin(377t + 90°) \text{ V} \qquad ■$$

1.6 | PRODUCTION OF INDUCED FORCE ON A WIRE

A second major effect of a magnetic field on its surroundings is that it induces a force on a current-carrying wire within the field. The basic concept involved is illustrated in Figure 1–16. The figure shows a conductor present in a uniform magnetic field of flux density **B**, pointing into the page. The conductor itself is *l* meters long and contains a current of *i* amperes. The force induced on the conductor is given by

$$\mathbf{F} = i(\mathbf{l} \times \mathbf{B}) \qquad (1–43)$$

Figure 1–16 | A current-carrying wire in the presence of a magnetic field.

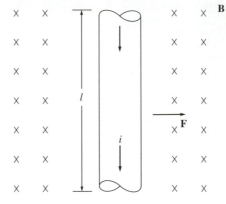

where

   i = magnitude of current in wire

   l = length of wire, with direction of **l** defined to be in the direction of
      current flow

   B = magnetic flux density vector

The direction of the force is given by the right-hand rule: If the index finger of the right hand points in the direction of the vector **l** and the middle finger points in the direction of the flux density vector **B**, then the thumb points in the direction of the resultant force on the wire. The magnitude of the force is given by the equation

$$F = ilB \sin \theta \qquad (1\text{–}44)$$

where θ is the angle between the wire and the flux density vector.

EXAMPLE 1–7

Figure 1–16 shows a wire carrying a current in the presence of a magnetic field. The magnetic flux density is 0.25 T, directed into the page. If the wire is 1.0 m long and carries 0.5 A of current in the direction from the top of the page to the bottom of the page, what are the magnitude and direction of the force induced on the wire?

■ Solution
The direction of the force is given by the right-hand rule as being to the right. The magnitude is given by

$$F = ilB \sin \theta \qquad (1\text{–}44)$$
$$= (0.5 \text{ A})(1.0 \text{ m})(0.25 \text{ T}) \sin 90° = 0.125 \text{ N}$$

Therefore,

$$\mathbf{F} = 0.125 \text{ N, directed to the right} \qquad ■$$

The induction of a force in a wire by a current in the presence of a magnetic field is the basis of *motor action.* Almost every type of motor depends on this basic principle for the forces and torques which make it move.

1.7 | INDUCED VOLTAGE ON A CONDUCTOR MOVING IN A MAGNETIC FIELD

There is a third major way in which a magnetic field interacts with its surroundings. If a wire with the proper orientation moves through a magnetic field, a voltage is induced in it. This idea is shown in Figure 1–17. The voltage induced in the wire is given by

$$e_{ind} = (\mathbf{v} \times \mathbf{B}) \cdot \mathbf{l} \tag{1–45}$$

where

$\mathbf{v}$ = velocity of the wire

$\mathbf{B}$ = magnetic flux density vector

$\mathbf{l}$ = length of conductor in the magnetic field

Vector $\mathbf{l}$ points along the direction of the wire toward the end making the smallest angle with respect to the vector $\mathbf{v} \times \mathbf{B}.$ The voltage in the wire will be built up so that the positive end is in the direction of the vector $\mathbf{v} \times \mathbf{B}.$ The following examples illustrate this concept.

Figure 1–17 | A conductor moving in the presence of a magnetic field.

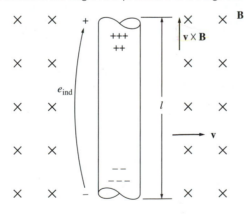

EXAMPLE 1–8

Figure 1–17 shows a conductor moving with a velocity of 5.0 m/s to the right in the presence of a magnetic field. The flux density is 0.5 T into the page, and the wire is 1.0 m in length, oriented as shown. What are the magnitude and polarity of the resulting induced voltage?

■ **Solution**

The direction of the quantity **v** × **B** in this example is up. Therefore, the voltage on the conductor will be built up positive at the top with respect to the bottom of the wire. The direction of vector **l** is up, so that it makes the smallest angle with respect to the vector **v** × **B**.

Since **v** is perpendicular to **B** and since **v** × **B** is parallel to **l,** the magnitude of the induced voltage reduces to

$$e_{\text{ind}} = (\mathbf{v} \times \mathbf{B}) \cdot \mathbf{l} \tag{1-45}$$
$$= (vB \sin 90°)\, l \cos 0°$$
$$= vBl$$
$$= (5.0 \text{ m/s})(0.5 \text{ T})(1.0 \text{ m})$$
$$= 2.5 \text{ V}$$

Thus the induced voltage is 2.5 V, positive at the top of the wire. ■

EXAMPLE 1-9

Figure 1–18 shows a conductor moving with a velocity of 10 m/s to the right in a magnetic field. The flux density is 0.5 T, out of the page, and the wire is 1.0 m in length, oriented as shown. What are the magnitude and polarity of the resulting induced voltage?

Figure 1–18 | The conductor of Example 1–9.

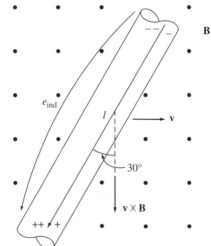

■ **Solution**

The direction of the quantity **v** × **B** is down. The wire is not oriented on an up-down line, so choose the direction of l as shown to make the smallest possible angle with the direction of $\mathbf{v} \times \mathbf{B}$. The voltage is positive at the bottom of the wire with respect to the top of the wire. The magnitude of the voltage is

$$
\begin{aligned}
e_{ind} &= (\mathbf{v} \times \mathbf{B}) \cdot \mathbf{l} \\
&= (vB \sin 90°)\, l \cos 30° \\
&= (10.0 \text{ m/s})(0.5 \text{ T})(1.0 \text{ m}) \cos 30° \\
&= 4.33 \text{ V}
\end{aligned}
\tag{1-45}
$$

■

The induction of voltages in a wire moving in a magnetic field is fundamental to the operation of all types of generators. For this reason, it is called *generator action*.

1.8 | REAL, REACTIVE, AND APPARENT POWER IN AC CIRCUITS

In a DC circuit such as the one shown in Figure 1–19a, the power supplied to the DC load is simply the product of the voltage across the load and the current flowing through it.

$$
P = VI
\tag{1-46}
$$

Unfortunately, the situation in sinusoidal AC circuits is more complex, because there can be a phase difference between the AC voltage and the AC current supplied to the load. The *instantaneous* power supplied to an AC load will still be the product

Figure 1–19 | (a) A DC voltage source supplying a load with resistance R. (b) An AC voltage source supplying a load with impedance $\mathbf{Z} = Z\angle\,\theta\,\Omega$.

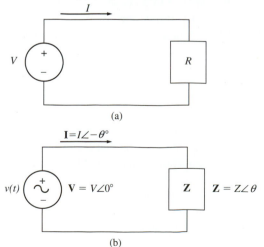

(a)

(b)

of the instantaneous voltage and the instantaneous current, but the *average* power supplied to the load will be affected by the phase angle between the voltage and the current. We will now explore the effects of this phase difference on the average power supplied to an AC load.

Figure 1–19b shows a single-phase voltage source supplying power to a single-phase load with impedance $\mathbf{Z} = Z\angle\theta\ \Omega$. If we assume that the load is inductive, then the impedance angle θ of the load will be positive, and the current will lag the voltage by θ degrees.

The voltage applied to this load is

$$v(t) = \sqrt{2}V \cos \omega t \qquad (1\text{–}47)$$

where V is the rms value of the voltage applied to the load, and the resulting current flow is

$$i(t) = \sqrt{2}I \cos(\omega t - \theta) \qquad (1\text{–}48)$$

where I is the rms value of the current flowing through the load.

The instantaneous power supplied to this load at any time t is

$$p(t) = v(t)i(t) = 2VI \cos \omega t \cos(\omega t - \theta) \qquad (1\text{–}49)$$

The angle θ in this equation is the *impedance angle* of the load. For inductive loads, the impedance angle is positive, and the current waveform lags the voltage waveform by θ degrees.

If we apply trigonometric identities to Equation (1–49), it can be manipulated into an expression of the form

$$p(t) = VI \cos \theta \,(1 + \cos 2\omega t) + VI \sin \theta \sin 2\omega t \qquad (1\text{–}50)$$

The first term of this equation represents the power supplied to the load by the component of current that is *in phase* with the voltage, while the second term represents the power supplied to the load by the component of current that is *90° out of phase* with the voltage. The components of this equation are plotted in Figure 1–20.

Note that the *first* term of the instantaneous power expression is always positive, but it produces pulses of power instead of a constant value. The average value of this term is

$$P = VI \cos \theta \qquad (1\text{–}51)$$

which is the *average* or *real* power (P) supplied to the load by term 1 of the Equation (1–50). The units of real power are watts (W), where $1\ \mathrm{W} = 1\ \mathrm{V} \times 1\ \mathrm{A}$.

Note that the *second* term of the instantaneous power expression is positive half of the time and negative half of the time, so that *the average power supplied by this term is zero*. This term represents power that is first transferred from the source to the load, and then returned from the load to the source. The power that continually bounces back and forth between the source and the load is known as *reactive power* (Q). Reactive power represents the energy that is first stored and then released in the magnetic field of an inductor, or in the electric field of a capacitor.

Figure 1–20 | The components of power supplied to a single-phase load versus time. The first component represents the power supplied by the component of current *in phase* with the voltage, while the second term represents the power supplied by the component of current *90° out of phase* with the voltage.

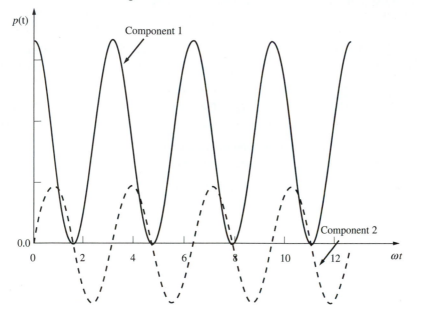

The reactive power of a load is given by

$$Q = VI \sin \theta \qquad (1–52)$$

where θ is the impedance angle of the load. By convention, Q is positive for inductive loads and negative for capacitive loads, because the impedance angle θ is positive for inductive loads and negative for capacitive loads. The units of reactive power are volt-amperes reactive (var), where 1 var = 1 V × 1 A. Even though the dimensional units are the same as for watts, reactive power is traditionally given a unique name to distinguish it from power actually supplied to a load.

The apparent power (S) supplied to a load is defined as the product of the voltage across the load and the current through the load. This is the power that "appears" to be supplied to the load if the phase angle differences between voltage and current are ignored. Therefore, the apparent power of a load is given by

$$S = VI \qquad (1–53)$$

The units of apparent power are volt-amperes (VA), where 1 VA = 1 V × 1 A. As with reactive power, apparent power is given a distinctive set of units to avoid confusing it with real and reactive power.

Alternative Forms of the Power Equations

If a load has a constant impedance, then Ohm's law can be used to derive alternative expressions for the real, reactive, and apparent powers supplied to the load. Since the magnitude of the voltage across the load is given by

$$V = IZ \tag{1–54}$$

substituting Equation (1–54) into Equations (1–51) to (1–53) produces equations for real, reactive, and apparent power expressed in terms of current and impedance:

$$P = I^2Z \cos \theta \tag{1–55}$$
$$Q = I^2Z \sin \theta \tag{1–56}$$
$$S = I^2Z \tag{1–57}$$

where Z is the magnitude of the load impedance $\mathbf{Z}$.

Since the impedance of the load $\mathbf{Z}$ can be expressed as

$$\mathbf{Z} = R + jX = Z \cos \theta + jZ \sin \theta$$

we see from this equation that $R = Z \cos \theta$ and $X = Z \sin \theta$, so the real and reactive powers of a load can also be expressed as

$$P = I^2R \tag{1–58}$$
$$Q = I^2X \tag{1–59}$$

where R is the resistance and X is the reactance of load $\mathbf{Z}$.

Complex Power

For simplicity in computer calculations, real and reactive power are sometimes represented together as a *complex power* $\mathbf{S}$, where

$$\mathbf{S} = P + jQ \tag{1–60}$$

The complex power $\mathbf{S}$ supplied to a load can be calculated from the equation

$$\mathbf{S} = \mathbf{VI}^* \tag{1–61}$$

where the asterisk represents the complex conjugate operator.

To understand this equation, let's suppose that the voltage applied to a load is $\mathbf{V} = V \angle \alpha$ and the current through the load is $\mathbf{I} = I \angle \beta$. Then the complex power supplied to the load is

$$\mathbf{S} = \mathbf{VI}^* = (V\angle\alpha)(I\angle-\beta) = VI \angle(\alpha - \beta)$$
$$= VI \cos(\alpha - \beta) + jVI \sin(\alpha - \beta)$$

The impedance angle θ is the difference between the angle of the voltage and the angle of the current ($\theta = \alpha - \beta$), so this equation reduces to

$$\mathbf{S} = VI \cos \theta + jVI \sin \theta$$
$$= P + jQ$$

The Relationships between Impedance Angle, Current Angle, and Power

As we know from basic circuit theory, an inductive load (Figure 1–21) has a positive impedance angle θ, since the reactance of an inductor is positive. If the impedance angle θ of a load is positive, the phase angle of the current flowing through the load will *lag* the phase angle of the voltage across the load by θ.

$$\mathbf{I} = \frac{\mathbf{V}}{\mathbf{Z}} = \frac{V \angle 0}{Z \angle \theta} = \frac{V}{Z} \angle -\theta$$

Also, if the impedance angle θ of a load is positive, the reactive power consumed by the load will be positive (Equation 1–56), and the load is said to be consuming both real and reactive power from the source.

Figure 1–21 | An inductive load has a *positive* impedance angle θ. This load produces a *lagging* current, and it consumes both real power P and reactive power Q from the source.

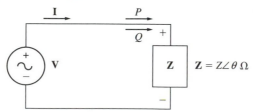

In contrast, a capacitive load (Figure 1–22) has a negative impedance angle θ, since the reactance of a capacitor is negative. If the impedance angle θ of a load is negative, the phase angle of the current flowing through the load will *lead* the phase angle of the voltage across the load by θ. Also, if the impedance angle θ of a load is negative, the reactive power Q consumed by the load will be *negative* (Equation 1–56). In this case, we say that the load is consuming real power from the source and *supplying* reactive power to the source.

Figure 1–22 | A capacitive load has a *negative* impedance angle θ. This load produces a *leading* current, and it consumes real power P from the source and while supplying reactive power Q to the source.

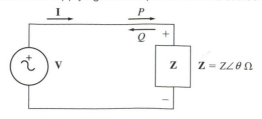

The Power Triangle

The real, reactive, and apparent powers supplied to a load are related by the *power triangle*. A power triangle is shown in Figure 1–23. The angle in the lower left corner is the impedance angle θ. The adjacent side of this triangle is the real power P supplied to the load, the opposite side of the triangle is the reactive power Q supplied to the load, and the hypotenuse of the triangle is the apparent power S of the load.

The quantity $\cos\theta$ is usually known as the *power factor* of a load. The power factor is defined as the fraction of the apparent power S that is actually supplying real power to a load. Thus,

$$\text{PF} = \cos\theta \qquad (1\text{–}62)$$

where θ is the impedance angle of the load.

Note that $\cos\theta = \cos(-\theta)$, so the power factor produced by an impedance angle of $+30°$ is exactly the same as the power factor produced by an impedance angle of $-30°$. Because we can't tell whether a load is inductive or capacitive from the power factor alone, it is customary to state whether the current is leading or lagging the voltage whenever a power factor is quoted.

The power triangle makes the relationships among real power, reactive power, apparent power, and the power factor clear, and provides a convenient way to calculate various power-related quantities if some of them are known.

Figure 1–23 | The power triangle.

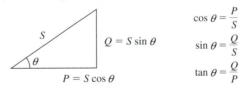

$$\cos\theta = \frac{P}{S}$$

$$\sin\theta = \frac{Q}{S}$$

$$\tan\theta = \frac{Q}{P}$$

$Q = S\sin\theta$

$P = S\cos\theta$

EXAMPLE 1–10

Figure 1–24 shows an AC voltage source supplying power to a load with impedance $\mathbf{Z} = 20 \angle -30°\ \Omega$. Calculate the current $\mathbf{I}$ supplied to the load, the power factor of the load, and the real, reactive, apparent, and complex power supplied to the load.

Figure 1–24 | The circuit of Example 1–10.

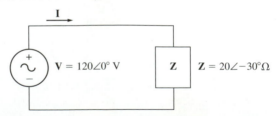

$\mathbf{I}$

$\mathbf{V} = 120\angle 0°\ \text{V}$

$\mathbf{Z}$ $\mathbf{Z} = 20\angle -30°\Omega$

■ Solution

The current supplied to this load is

$$\mathbf{I} = \frac{\mathbf{V}}{\mathbf{Z}} = \frac{120\angle 0° \text{ V}}{20\angle -30° \text{ }\Omega} = 6\angle 30° \text{ A}$$

The power factor of the load is

$$PF = \cos \theta = \cos (-30°) = 0.866 \text{ leading} \qquad (1-62)$$

(Note that this is a capacitive load, so the impedance angle θ is negative, and the current *leads* the voltage.)

The real power supplied to the load is

$$P = VI \cos \theta \qquad (1-51)$$
$$P = (120 \text{ V})(6 \text{ A}) \cos (-30°) = 623.5 \text{ W}$$

The reactive power supplied to the load is

$$Q = VI \sin \theta \qquad (1-52)$$
$$Q = (120 \text{ V})(6 \text{ A}) \sin (-30°) = -360 \text{ VAR}$$

The apparent power supplied to the load is

$$S = VI \qquad (1-53)$$
$$Q = (120 \text{ V})(6 \text{ A}) = 720 \text{ VA}$$

The complex power supplied to the load is

$$\mathbf{S} = \mathbf{VI}^* \qquad (1-61)$$
$$= (120\angle 0° \text{ V})(6\angle -30° \text{ A})^*$$
$$= (120\angle 0° \text{ V})(6\angle 30° \text{ A}) = 720\angle 30° \text{ VA}$$
$$= 623.5 - j360 \text{ VA}$$

■

1.9 | SUMMARY

This chapter has reviewed briefly the mechanics of systems rotating about a single axis and introduced the sources and effects of magnetic fields important in the understanding of transformers, motors, and generators.

Historically, the English system of units has been used to measure the mechanical quantities associated with machines in English-speaking countries. Recently, the SI units have superseded the English system almost everywhere in the world except in the United States, but rapid progress is being made even there. Since SI is becoming almost universal, most (but not all) of the examples in this book use this system of units for mechanical measurements. Electrical quantities are always measured in SI units.

In the section on mechanics, the concepts of angular position, angular velocity, angular acceleration, torque, Newton's law, work, and power were explained for the special case of rotation about a single axis. Some fundamental relationships (such as the power and speed equations) were given in both SI and English units.

The production of a magnetic field by a current was explained, and the special properties of ferromagnetic materials were explored in detail. The shape of the magnetization curve and the concept of hysteresis were explained in terms of the domain theory of ferromagnetic materials, and eddy current losses were discussed.

Faraday's law states that a voltage will be generated in a coil of wire that is proportional to the time rate of change in the flux passing through it. Faraday's law is the basis of transformer action, which is explored in detail in Chapter 3.

A current-carrying wire present in a magnetic field, if it is oriented properly, will have a force induced on it. This behavior is the basis of motor action in all real machines.

A wire moving through a magnetic field with the proper orientation will have a voltage induced in it. This behavior is the basis of generator action in all real machines.

In AC circuits, the real power P is the average power supplied by a source to a load. The reactive power Q is the component of power that is exchanged back and forth between a source and a load. By convention, positive reactive power is consumed by inductive loads ($+\theta$) and negative reactive power is consumed (or positive reactive power is supplied) by capacitive loads ($-\theta$). The apparent power S is the power that "appears" to be supplied to the load if only the magnitudes of the voltages and currents are considered.

1.10 | QUESTIONS

1–1. What is torque? What role does torque play in the rotational motion of machines?

1–2. What is Ampere's law?

1–3. What is magnetizing intensity? What is magnetic flux density? How are they related?

1–4. How does the magnetic circuit concept aid in the design of transformer and machine cores?

1–5. What is reluctance?

1–6. What is a ferromagnetic material? Why is the permeability of ferromagnetic materials so high?

1–7. How does the relative permeability of a ferromagnetic material vary with magnetomotive force?

1–8. What is hysteresis? Explain hysteresis in terms of magnetic domain theory.

1–9. What are eddy current losses? What can be done to minimize eddy current losses in a core?

1–10. Why are all cores exposed to AC flux variations laminated?

1–11. What is Faraday's law?

1–12. What conditions are necessary for a magnetic field to produce a force on a wire?

1–13. What conditions are necessary for a magnetic field to produce a voltage in a wire?

1–14. Will current be leading or lagging voltage in an inductive load? Will the reactive power of the load be positive or negative?

1–15. What are real, reactive, and apparent power? What units are they measured in? How are they related?

1–16. What is power factor?

1.11 | PROBLEMS

1–1. A motor's shaft is spinning at a speed of 1800 r/min. What is the shaft speed in radians per second?

1–2. A flywheel with a moment of inertia of 4 kg$-$m^2 is initially at rest. If a torque of 5 N$-$m (counterclockwise) is suddenly applied to the flywheel, what will be the speed of the flywheel after 5 s? Express that speed in both radians per second and revolutions per minute.

1–3. A force of 5 N is applied to a cylinder, as shown in Figure P1–1. What are the magnitude and direction of the torque produced on the cylinder? What is the angular acceleration α of the cylinder?

Figure P1–1 | The cylinder of Problem 1–3.

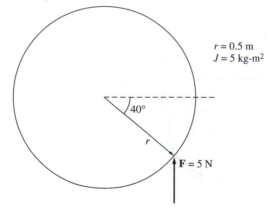

1–4. A motor is supplying 70 N$-$m of torque to its load. If the motor's shaft is turning at 1500 r/min, what is the mechanical power supplied to the load in watts? In horsepower?

1–5. A ferromagnetic core is shown in Figure P1–2. The depth of the core is 5 cm. The other dimensions of the core are as shown in the figure. Find the value of the current that will produce a flux of 0.003 Wb. With this current, what is the flux density at the top of the core? What is the flux density at the right side of the core? Assume that the relative permeability of the core is 1000.

1–6. A ferromagnetic core with a relative permeability of 2000 is shown in Figure P1–3. The dimensions are as shown in the diagram, and the depth of

Figure P1-2 I The core of Problems 1–5 and 1–16.

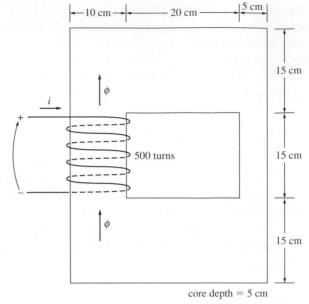

core depth = 5 cm

Figure P1-3 I The core of Problem 1–6.

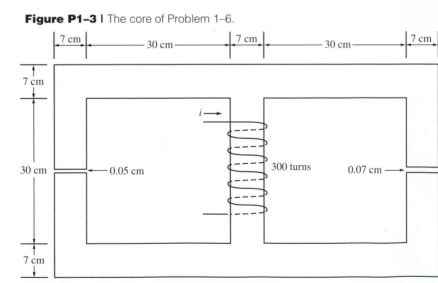

core depth = 7 cm

the core is 7 cm. The air gaps on the left and right sides of the core are 0.050 and 0.070 cm, respectively. Because of fringing effects, the effective area of the air gaps is 5 percent larger than their physical size. If there are 300 turns in the coil wrapped around the center leg of the core and if the current in the coil is 1.0 A, what is the flux in each of the left, center, and right legs of the core? What is the flux density in each air gap?

1–7. A two-legged core is shown in Figure P1–4. The winding on the left leg of the core (N_1) has 600 turns, and the winding on the right (N_2) has 200 turns. The coils are wound in the directions shown in the figure. If the dimensions are as shown, then what flux would be produced by currents $i_1 = 0.5$ A and $i_2 = 1.00$ A? Assume $\mu_r = 1000$ and constant.

Figure P1–4 | The core of Problems 1–7 and 1–12.

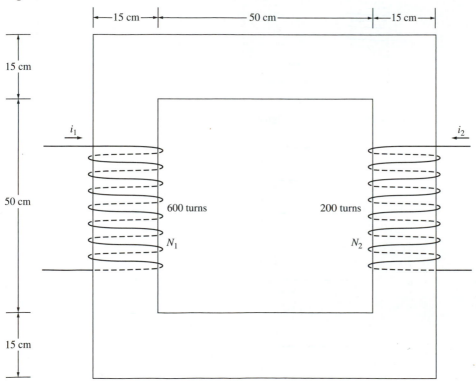

core depth = 15 cm

1–8. A core with three legs is shown in Figure P1–5. Its depth is 5 cm, and there are 200 turns on the leftmost leg. The relative permeability of the core can be assumed to be 1500 and constant. What flux exists in each of the three legs of the core? What is the flux density in each of the legs? Assume a 4 percent increase in the effective area of the air gap due to fringing effects.

1–9. The wire shown in Figure P1–6 is carrying 2.0 A in the presence of a magnetic field. Calculate the magnitude and direction of the force induced on the wire.

Figure P1–5 | The core of Problem 1–8.

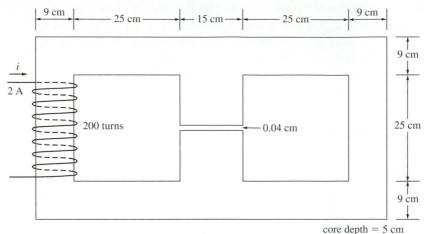

Figure P1–6 | A current-carrying wire in a magnetic field (Problem 1–9).

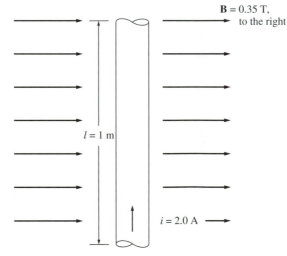

1–10. The wire shown in Figure P1–7 is moving in the presence of a magnetic field. With the information given in the figure, determine the magnitude and direction of the induced voltage in the wire.

1–11. Repeat Problem 1–10 for the wire in Figure P1–8.

1–12. The core shown in Figure P1–4 is made of a steel whose magnetization curve is shown in Figure P1–9. Repeat Problem 1–7, but this time do *not* assume a constant value of μ_r. How much flux is produced in the core by the currents specified? What is the relative permeability of this core under these conditions? Was the assumption in Problem 1–7 that the relative permeability was equal to 1000 a good assumption for these conditions? Is it a good assumption in general?

Figure P1–7 I A wire moving in a magnetic field (Problem 1–10).

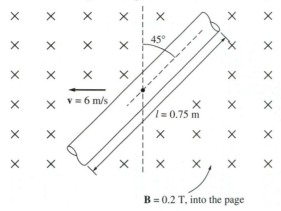

$\mathbf{B} = 0.2$ T, into the page

Figure P1–8 I A wire moving in a magnetic field (Problem 1–11).

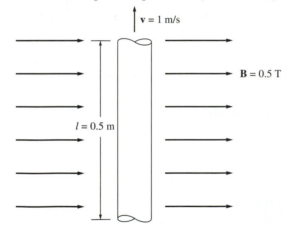

1–13. A core with three legs is shown in Figure P1–10. Its depth is 8 cm, and there are 400 turns on the center leg. The remaining dimensions are shown in the figure. The core is composed of a steel having the magnetization curve shown in Figure 1–10c. Answer the following questions about this core:

(a) What current is required to produce a flux density of 0.5 T in the central leg of the core?

(b) What current is required to produce a flux density of 1.0 T in the central leg of the core? Is it twice the current in part (a)?

(c) What are the reluctances of the central and right legs of the core under the conditions in part (a)?

Figure P1–9 | The magnetization curve for the core material of Problems 1–12 and 1–14.

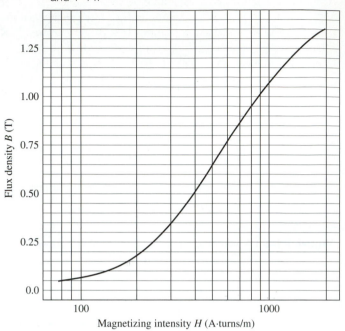

Figure P1–10 | The core of Problem 1–13.

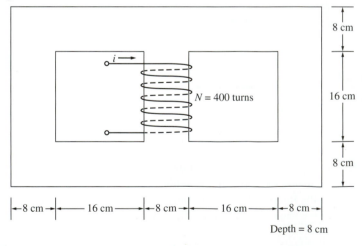

(d) What are the reluctances of the central and right legs of the core under the conditions in part *(b)*?

(e) What conclusion can you make about reluctances in real magnetic cores?

1–14. A two-legged magnetic core with an air gap is shown in Figure P1–11. The depth of the core is 5 cm, the length of the air gap in the core is 0.07 cm, and the number of turns on the coil is 500. The magnetization curve of the core material is shown in Figure P1–9. Assume a 5 percent increase in effective air-gap area to account for fringing. How much current is required to produce an air-gap flux density of 0.5 T? What are the flux densities of the four sides of the core at that current? What is the total flux present in the air gap?

Figure P1–11 | The core of Problem 1–14.

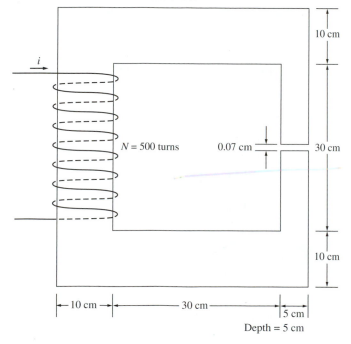

1–15. A transformer core with an effective mean path length of 10 in has a 300-turn coil wrapped around one leg. Its cross-sectional area is 0.25 in², and its magnetization curve is shown in Figure 1–10c. If current of 0.25 A is flowing in the coil, what is the total flux in the core? What is the flux density?

1–16. The core shown in Figure P1–2 has the flux ϕ shown in Figure P1–12. Sketch the voltage present at the terminals of the coil.

1–17. Figure P1–13 shows the core of a simple DC motor. The magnetization curve for the metal in this core is given by Figure 1–10c and d. Assume that the cross-sectional area of each air gap is 18 cm² and that the width of each air gap is 0.05 cm. The effective diameter of the rotor core is 4 cm.

Figure P1–12 | Plot of flux ϕ as a function of time for Problem 1–16.

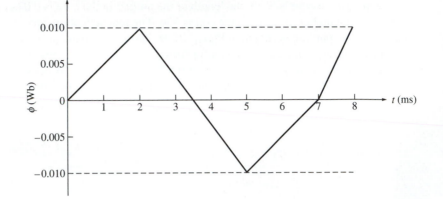

Figure P1–13 | The core of Problem 1–17.

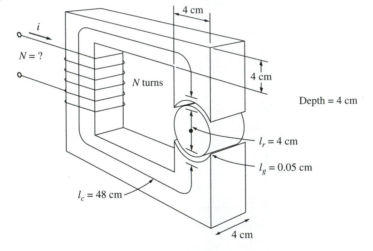

(a) It is desired to build a machine with as great a flux density as possible while avoiding excessive saturation in the core. What would be a reasonable maximum flux density for this core?

(b) What would be the total flux in the core at the flux density of part *(a)*?

(c) The maximum possible field current for this machine is 1 A. Select a reasonable number of turns of wire to provide the desired flux density while not exceeding the maximum available current.

1–18. Assume that the voltage applied to a load is $\mathbf{V} = 208\angle -30°$ V and the current flowing through the load is $\mathbf{I} = 5\angle 15°$ A.

(a) Calculate the complex power $\mathbf{S}$ consumed by this load.

(b) Is this load inductive or capacitive?

(c) Calculate the power factor of this load.

(d) Calculate the reactive power consumed or supplied by this load. Does the load consume reactive power from the source or supply it to the source?

1–19. Figure P1–14 shows a simple single-phase AC power system with three loads. The impedances of these three loads are

$$\mathbf{Z}_1 = 5\angle 30° \; \Omega \qquad \mathbf{Z}_2 = 5\angle 45° \; \Omega \qquad \mathbf{Z}_3 = 5\angle -90° \; \Omega$$

Answer the following questions about this power system.

(a) Assume that the switch shown in the figure is open, and calculate the current **I**, the power factor, and the real, reactive, and apparent power being supplied by the load.

(b) Assume that the switch shown in the figure is closed, and calculate the current **I**, the power factor, and the real, reactive, and apparent power being supplied by the load.

(c) What happened to the current flowing from the source when the switch closed? Why?

Figure P1–14 | The circuit of Problem 1–19.

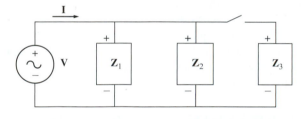

1–20. Demonstrate that Equation (1–50) can be derived from Equation (1–49) using simple trigonometric identities:

$$p(t) = v(t)i(t) = 2VI \cos \omega t \cos(\omega t - \theta) \qquad (1\text{–}49)$$
$$p(t) = VI \cos \theta \, (1 + \cos 2\omega t) + VI \sin \theta \sin 2\omega t \qquad (1\text{–}50)$$

1.12 | REFERENCES

1. Alexander, Charles K., and Matthew N. O. Sadiku: *Fundamentals of Electric Circuits,* McGraw-Hill, 2000.

2. Beer, F., and E. Johnston, Jr.: *Vector Mechanics for Engineers: Dynamics,* 6th ed., McGraw-Hill, New York, 1997.

3. Hayt, William H.: *Engineering Electromagnetics,* 5th ed., McGraw-Hill, New York, 1989.

4. Mulligan, J. F.: *Introductory College Physics,* 2nd ed., McGraw-Hill, New York, 1991.

5. Sears, Francis W., Mark W. Zemansky, and Hugh D. Young: *University Physics,* Addison-Wesley, Reading, Mass., 1982.

Three-Phase Circuits

Almost all electric power generation and most of the power transmission in the world today is in the form of three-phase AC circuits. A three-phase AC power system consists of three-phase generators, transmission lines, and loads. AC power systems have a great advantage over DC systems in that their voltage levels can be changed with transformers to reduce *transmission losses, as we will show in Chapter 3. *Three-phase* AC power systems have two major advantages over single-phase AC power systems: (1) it is possible to get more power per kilogram of metal from a three-phase machine and (2) the power delivered to a three-phase load is constant at all times, instead of pulsing as it does in single-phase systems. Three-phase systems also make the use of induction motors easier by allowing them to start without special auxiliary starting windings.

2.1 | GENERATION OF THREE-PHASE VOLTAGES AND CURRENTS

A three-phase generator consists of three single-phase generators, with voltages equal in magnitude but differing in phase angle from the others by 120°. Each of these three generators could be connected to one of three identical loads by a pair of wires, and the resulting power system would be as shown in Figure 2–1c. Such a system consists of three single-phase circuits that happen to differ in phase angle by 120°. The current flowing to each load can be found from the equation

$$\mathbf{I} = \frac{\mathbf{V}}{\mathbf{Z}} \qquad (2-1)$$

Therefore, the currents flowing in the three phases are

$$\mathbf{I}_A = \frac{V\angle 0°}{Z\angle\theta} = I\angle -\theta \qquad (2-2)$$

$$\mathbf{I}_B = \frac{V\angle -120°}{Z\angle\theta} = I\angle -120° - \theta \qquad (2-3)$$

$$\mathbf{I}_C = \frac{V\angle -240°}{Z\angle\theta} = I\angle -240° - \theta \qquad (2-4)$$

Figure 2–1 | (a) A three-phase generator, consisting of three single-phase sources equal in magnitude and 120° apart in phase. (b) The voltages in each phase of the generator. (c) The three phases of the generator connected to three identical loads. (d) Phasor diagram showing the voltages in each phase.

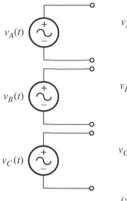
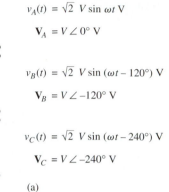

$$v_A(t) = \sqrt{2}\ V \sin \omega t \text{ V}$$

$$\mathbf{V}_A = V \angle 0° \text{ V}$$

$$v_B(t) = \sqrt{2}\ V \sin (\omega t - 120°) \text{ V}$$

$$\mathbf{V}_B = V \angle -120° \text{ V}$$

$$v_C(t) = \sqrt{2}\ V \sin (\omega t - 240°) \text{ V}$$

$$\mathbf{V}_C = V \angle -240° \text{ V}$$

(a)

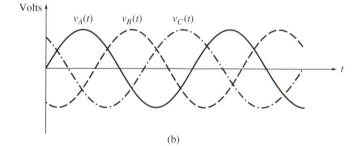

(b)

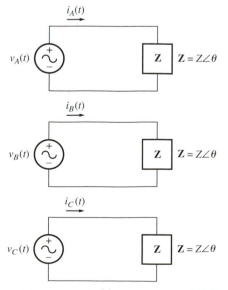

(c)

(continued)

Figure 2–1 | *(concluded)*

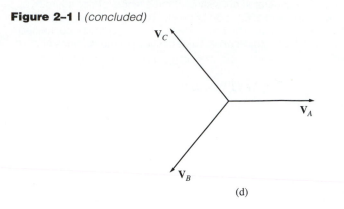

(d)

It is possible to connect the negative ends of these three single-phase generators and loads together, so that they share a common return line (called the *neutral*). The resulting system is shown in Figure 2–2; note that now only *four* wires are required to supply power from the three generators to the three loads.

Figure 2–2 | The three circuits connected together with a common neutral.

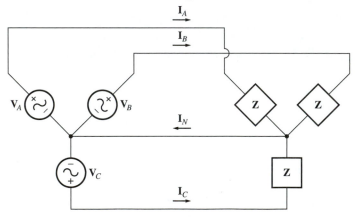

How much current is flowing in the single neutral wire shown in Figure 2–2? The return current will be the sum of the currents flowing to each individual load in the power system. This current is given by

$$\mathbf{I}_N = \mathbf{I}_A + \mathbf{I}_B + \mathbf{I}_C \qquad (2\text{–}5)$$
$$= I\angle-\theta + I\angle-\theta - 120° + I\angle-\theta - 240°$$
$$= I\cos(-\theta) + jI\sin(-\theta)$$
$$\quad + I\cos(-\theta - 120°) + jI\sin(-\theta - 120°)$$
$$\quad + I\cos(-\theta - 240°) + jI\sin(-\theta - 240°)$$
$$= I\left[\cos(-\theta) + \cos(-\theta - 120°) + \cos(-\theta - 240°)\right]$$
$$\quad + jI\left[\sin(-\theta) + \sin(-\theta - 120°) + \sin(-\theta - 240°)\right]$$

Recall the elementary trigonometric identities:

$$\cos(\alpha - \beta) = \cos\alpha\cos\beta + \sin\alpha\sin\beta \qquad (2\text{--}6)$$
$$\sin(\alpha - \beta) = \sin\alpha\cos\beta - \cos\alpha\sin\beta \qquad (2\text{--}7)$$

Applying these trigonometric identities yields

$$\mathbf{I}_N = I[\cos(-\theta) + \cos(-\theta)\cos 120° + \sin(-\theta)\sin 120° + \cos(-\theta)\cos 240°$$
$$+ \sin(-\theta)\sin 240°]$$
$$+ jI[\sin(-\theta) + \sin(-\theta)\cos 120° - \cos(-\theta)\sin 120°$$
$$+ \sin(-\theta)\cos 240° - \cos(-\theta)\sin 240°]$$

$$\mathbf{I}_N = I\left[\cos(-\theta) - \frac{1}{2}\cos(-\theta) + \frac{\sqrt{3}}{2}\sin(-\theta) - \frac{1}{2}\cos(-\theta) - \frac{\sqrt{3}}{2}\sin(-\theta)\right]$$
$$+ jI\left[\sin(-\theta) - \frac{1}{2}\sin(-\theta) - \frac{\sqrt{3}}{2}\cos(-\theta) - \frac{1}{2}\sin(-\theta) + \frac{\sqrt{3}}{2}\cos(-\theta)\right]$$

$$\mathbf{I}_N = 0 \text{ A}$$

As long as the three loads are equal, the return current in the neutral is zero! A three-phase power system in which the three generators have voltages that are exactly equal in magnitude and 120° different in phase, and in which all three loads are identical, is called a *balanced three-phase system.* In such a system, the neutral is actually unnecessary, and we could get by with only *three* wires instead of the original six.

Phase Sequence The *phase sequence* of a three-phase power system is the order in which the voltages in the individual phases peak. The three-phase power system illustrated in Figure 2–1 is said to have phase sequence *abc*, since the voltages in the three phases peak in the order *a*, *b*, *c* (see Figure 2–1b). The phasor diagram of a power system with an *abc* phase sequence is shown in Figure 2–3a.

Figure 2–3 | (a) The phase voltages in a power system with an *abc* phase sequence. (b) The phase voltages in a power system with an *acb* phase sequence.

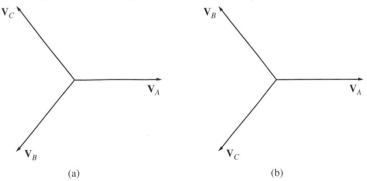

(a)   (b)

It is also possible to connect the three phases of a power system so that the voltages in the phases peak in order the order *a*, *c*, *b*. This type of power system is said to have phase sequence *acb*. The phasor diagram of a power system with an *acb* phase sequence is shown in Figure 2–3b.

The result derived on page 59 is equally valid for both *abc* and *acb* phase sequences. In either case, if the power system is balanced, the current flowing in the neutral will be 0.

2.2 | VOLTAGES AND CURRENTS IN A THREE-PHASE CIRCUIT

A connection of the sort shown in Figure 2–2 is called a wye (Y) connection because it looks like the letter Y. Another possible connection is the delta (Δ) connection, in which the three generators are connected head to tail. The Δ connection is possible because the sum of the three voltages $\mathbf{V}_A + \mathbf{V}_B + \mathbf{V}_C = 0$, so that no short-circuit currents will flow when the three sources are connected head to tail.

Each generator and each load in a three-phase power system may be either Y- or Δ-connected. Any number of Y- and Δ-connected generators and loads may be mixed on a power system.

Figure 2–4 shows three-phase generators connected in Y and in Δ. The voltages and currents in a given phase are called *phase quantities,* and the voltages between lines and currents in the lines connected to the generators are called *line quantities.* The relationship between the line quantities and phase quantities for a given generator or load depends on the type of connection used for that generator or load. These relationships will now be explored for each of the Y and Δ connections.

Figure 2–4 | (a) Y connection. (b) Δ connection.

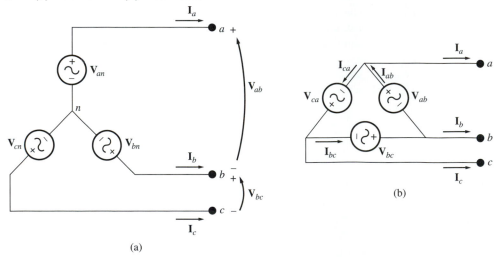

(a)

(b)

Voltages and Currents in the Wye (Y) Connection

A Y-connected three-phase generator with an *abc* phase sequence connected to a resistive load is shown in Figure 2–5. The phase voltages in this generator are given by

$$\mathbf{V}_{an} = V_\phi \angle 0°$$
$$\mathbf{V}_{bn} = V_\phi \angle -120°$$
$$\mathbf{V}_{cn} = V_\phi \angle -240°$$

$$(2\text{–}8)$$

Figure 2–5 | Y-connected generator with a resistive load.

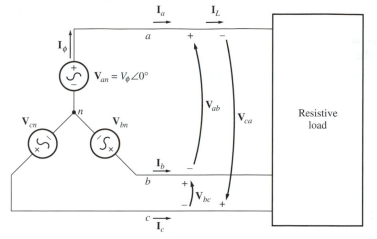

Since the load connected to this generator is assumed to be resistive, the current in each phase of the generator will be at the same angle as the voltage. Therefore, the current in each phase will be given by

$$\mathbf{I}_a = I_\phi \angle 0°$$
$$\mathbf{I}_b = I_\phi \angle -120°$$
$$\mathbf{I}_c = I_\phi \angle -240°$$

$$(2\text{–}9)$$

From Figure 2–5, it is obvious that the current in any line is the same as the current in the corresponding phase. Therefore, for a Y connection,

$$\boxed{I_L = I_\phi \qquad \text{Y connection}}$$

$$(2\text{–}10)$$

The relationship between line voltage and phase voltage is a bit more complex. By Kirchhoff's voltage law, the line-to-line voltage $\mathbf{V}_{ab}$ is given by

$$\begin{aligned}
\mathbf{V}_{ab} &= \mathbf{V}_a - \mathbf{V}_b \\
&= V_\phi \angle 0° - V_\phi \angle -120° \\
&= V_\phi - \left(-\frac{1}{2} V_\phi - j\frac{\sqrt{3}}{2} V_\phi\right) = \frac{3}{2} V_\phi + j\frac{\sqrt{3}}{2} V_\phi \\
&= \sqrt{3} V_\phi \left(\frac{\sqrt{3}}{2} + j\frac{1}{2}\right) \\
&= \sqrt{3} V_\phi \angle 30°
\end{aligned}$$

Therefore, the relationship between the magnitudes of the line-to-line voltage and the line-to-neutral (phase) voltage in a Y-connected generator or load is

$$V_{LL} = \sqrt{3}V_\phi \qquad \text{Y connection} \tag{2–11}$$

In addition, the line voltages are shifted 30° with respect to the phase voltages. A phasor diagram of the line and phase voltages for the Y connection in Figure 2–5 is shown in Figure 2–6.

Figure 2–6 | Line-to-line and phase (line-to-neutral) voltages for the Y connection in Figure 2–5.

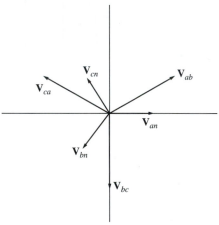

Note that for Y connections with the *abc* phase sequence such as the one in Figure 2–5, the voltage of a line *leads* the corresponding phase voltage by 30°. For Y connections with the *acb* phase sequence, the voltage of a line *lags* the corresponding phase voltage by 30°, as you will be asked to demonstrate in a problem at the end of the chapter.

Although the relationships between line and phase voltages and currents for the Y connection were derived for the assumption of a unity power factor, they are in fact valid for any power factor. The assumption of unity-power-factor loads simply made the mathematics slightly easier in this development.

Voltages and Currents in the Delta (Δ) Connection

A Δ-connected three-phase generator connected to a resistive load is shown in Figure 2–7. The phase voltages in this generator are given by

$$\begin{aligned}
\mathbf{V}_{ab} &= V_\phi \angle 0° \\
\mathbf{V}_{bc} &= V_\phi \angle -120° \\
\mathbf{V}_{ca} &= V_\phi \angle -240°
\end{aligned} \tag{2–12}$$

Because the load is resistive, the phase currents are given by

$$\mathbf{I}_{ab} = I_\phi \angle 0°$$
$$\mathbf{I}_{bc} = I_\phi \angle -120° \qquad (2-13)$$
$$\mathbf{I}_{ca} = I_\phi \angle -240°$$

Figure 2–7 | Δ-connected generator with a resistive load.

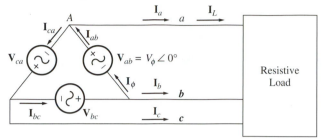

In the case of the Δ connection, it is obvious that the line-to-line voltage between any two lines will be the same as the voltage in the corresponding phase. *In a Δ connection,*

$$\boxed{V_{LL} = V_\phi \qquad \text{Δ connection}} \qquad (2-14)$$

The relationship between line current and phase current is more complex. It can be found by applying Kirchhoff's current law at a node of the Δ. Applying Kirchhoff's current law to node *A* yields the equation

$$\mathbf{I}_a = \mathbf{I}_{ab} - \mathbf{I}_{ca}$$
$$= I_\phi \angle 0° - I_\phi \angle -240°$$
$$= I_\phi - \left(-\frac{1}{2} I_\phi + j\frac{\sqrt{3}}{2} I_\phi\right) = \frac{3}{2} I_\phi - j\frac{\sqrt{3}}{2} I_\phi$$
$$= \sqrt{3} I_\phi \left(\frac{\sqrt{3}}{2} - j\frac{1}{2}\right)$$
$$= \sqrt{3} I_\phi \angle -30°$$

Therefore, the relationship between the magnitudes of the line and phase currents in a Δ-connected generator or load is

$$\boxed{I_L = \sqrt{3} I_\phi \qquad \text{Δ connection}} \qquad (2-15)$$

and the line currents are shifted 30° relative to the corresponding phase currents.

Note that for Δ connections with the *abc* phase sequence such as the one shown in Figure 2–7, the current of a line *lags* the corresponding phase current by 30° (see Figure 2–8). For Δ connections with the *acb* phase sequence, the current of a line *leads* the corresponding phase current by 30°.

Figure 2–8 | Line and phase currents for the Δ connection in Figure 2–7.

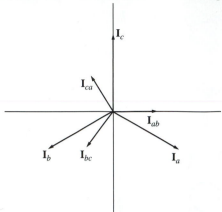

The voltage and current relationships for Y- and Δ-connected sources and loads are summarized in Table 2–1.

Table 2–1 | Summary of relationships in Y and Δ connections

	Y connection	**Δ connection**
Voltage magnitudes	$V_{LL} = \sqrt{3}\, V_\phi$	$V_{LL} = V_\phi$
Current magnitudes	$I_L = I_\phi$	$I_L = \sqrt{3}\, I_\phi$
abc phase sequence	$\mathbf{V}_{ab}$ leads $\mathbf{V}_a$ by 30°	$\mathbf{I}_a$ lags $\mathbf{I}_{ab}$ by 30°
acb phase sequence	$\mathbf{V}_{ab}$ lags $\mathbf{V}_a$ by 30°	$\mathbf{I}_a$ leads $\mathbf{I}_{ab}$ by 30°

2.3 | POWER RELATIONSHIPS IN THREE-PHASE CIRCUITS

Figure 2–9 shows a balanced Y-connected load whose phase impedance is $\mathbf{Z}_\phi = Z\angle\theta°$. If the three-phase voltages applied to this load are given by

$$\begin{aligned}
v_{an}(t) &= \sqrt{2}V \sin \omega t \\
v_{bn}(t) &= \sqrt{2}V \sin(\omega t - 120°) \\
v_{cn}(t) &= \sqrt{2}V \sin(\omega t - 240°)
\end{aligned} \tag{2–16}$$

then the three-phase currents flowing in the load are given by

$$\begin{aligned}
i_a(t) &= \sqrt{2}I \sin(\omega t - \theta) \\
i_b(t) &= \sqrt{2}I \sin(\omega t - 120° - \theta) \\
i_c(t) &= \sqrt{2}I \sin(\omega t - 240° - \theta)
\end{aligned} \tag{2–17}$$

where $I = V/Z$. How much power is being supplied to this load from the source?

Figure 2-9 | A balanced Y-connected load.

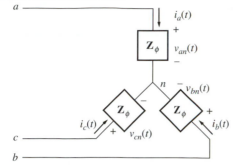

The instantaneous power supplied to one phase of the load is given by the equation

$$p(t) = v(t)i(t) \qquad (2\text{--}18)$$

Therefore, the instantaneous power supplied to each of the three phases is

$$p_a(t) = v_{an}(t)i_a(t) = 2VI\,\sin(\omega t)\,\sin(\omega t - \theta)$$
$$p_b(t) = v_{bn}(t)i_b(t) = 2VI\,\sin(\omega t - 120°)\,\sin(\omega t - 120° - \theta) \qquad (2\text{--}19)$$
$$p_c(t) = v_{cn}(t)i_c(t) = 2VI\,\sin(\omega t - 240°)\,\sin(\omega t - 240° - \theta)$$

A trigonometric identity states that

$$\sin \alpha \sin \beta = \frac{1}{2}[\cos(\alpha - \beta) - \cos(\alpha - \beta)] \qquad (2\text{--}20)$$

Applying this identity to Equations (2–19) yields new expressions for the power in each phase of the load:

$$p_a(t) = VI[\cos \theta - \cos(2\omega t - \theta)]$$
$$p_b(t) = VI[\cos \theta - \cos(2\omega t - 240° - \theta)] \qquad (2\text{--}21)$$
$$p_c(t) = VI[\cos \theta - \cos(2\omega t - 480° - \theta)]$$

The total power supplied to the entire three-phase load is the sum of the power supplied to each of the individual phases. The power supplied by each phase consists of a constant component plus a pulsing component. However, *the pulsing components in the three phases cancel each other out since they are 120° out of phase with each other*, and the final power supplied by the three-phase power system is constant. This power is given by the equation:

$$P_{tot}(t) = p_A(t) + p_B(t) + p_C(t) = 3VI \cos \theta \qquad (2\text{--}22)$$

The instantaneous power in phases *a*, *b*, and *c* are shown as a function of time in Figure 2–10. Note that *the total power supplied to a balanced three-phase load is constant at all times*. The fact that a constant power is supplied by a three-phase power system is one of its major advantages compared to single-phase sources.

Figure 2–10 | Instantaneous power in phases *a*, *b*, and *c*, plus the total power supplied to the load.

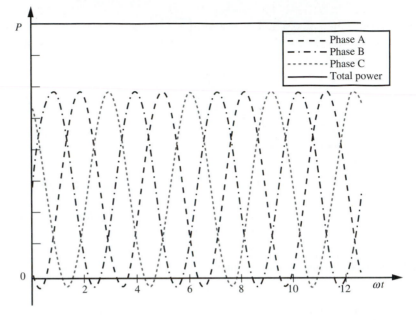

Three-Phase Power Equations Involving Phase Quantities

The single-phase power Equations (1–51) to (1–57) apply to *each phase* of a Y- or Δ-connected three-phase load, so the real, reactive, and apparent powers supplied to a balanced three-phase load are given by

$$P = 3V_\phi I_\phi \cos \theta \qquad (2\text{–}23)$$

$$Q = 3V_\phi I_\phi \sin \theta \qquad (2\text{–}24)$$

$$S = 3V_\phi I_\phi \qquad (2\text{–}25)$$

$$P = 3I_\phi^2 Z \cos \theta \qquad (2\text{–}26)$$

$$Q = 3I_\phi^2 Z \sin \theta \qquad (2\text{–}27)$$

$$S = 3I_\phi^2 Z \qquad (2\text{–}28)$$

The angle θ is again the angle between the voltage and the current in any phase of the load (it is the same in all phases), and the power factor of the load is the cosine of the impedance angle θ. The power-triangle relationships apply as well.

Three-Phase Power Equations Involving Line Quantities

It is also possible to derive expressions for the power in a balanced three-phase load in terms of line quantities. This derivation must be done separately for Y- and Δ-connected loads, since the relationships between the line and phase quantities are different for each type of connection.

For a Y-connected load, the power consumed by a load is given by

$$P = 3V_\phi I_\phi \cos\theta \qquad (2\text{--}23)$$

For this type of load, $I_L = I_\phi$ and $V_{LL} = \sqrt{3}V_\phi$, so the power consumed by the load can also be expressed as

$$P = 3\left(\frac{V_{LL}}{\sqrt{3}}\right)I_L \cos\theta$$

$$\boxed{P = \sqrt{3}V_{LL} I_L \cos\theta} \qquad (2\text{--}29)$$

For a Δ-connected load, the power consumed by a load is given by

$$P = 3V_\phi I_\phi \cos\theta \qquad (2\text{--}23)$$

For this type of load, $I_L = \sqrt{3}I_\phi$ and $V_{LL} = V_\phi$, so the power consumed by the load can also be expressed in terms of line quantities as

$$P = 3V_{LL}\left(\frac{I_L}{\sqrt{3}}\right)\cos\theta$$
$$= \sqrt{3}V_{LL}I_L \cos\theta \qquad (2\text{--}29)$$

This is exactly the same equation that was derived for a Y-connected load, so Equation (2–29) gives the power of a balanced three-phase load in terms of line quantities *regardless of the connection of the load.* The reactive and apparent powers of the load in terms of line quantities are

$$\boxed{Q = \sqrt{3}V_{LL} I_L \sin\theta} \qquad (2\text{--}30)$$

$$\boxed{S = \sqrt{3}V_{LL} I_L} \qquad (2\text{--}31)$$

It is important to realize that the cos θ and sin θ terms in Equations (2–29) and (2–30) are the cosine and sine of the angle between the *phase* voltage and the *phase* current, not the angle between the line-to-line voltage and the line current. Remember that there is a 30° phase shift between the line-to-line and phase voltage for a Y connection, and between the line and phase current for a Δ connection, so it is important not to take the cosine of the angle between the line-to-line voltage and line current.

2.4 | ANALYSIS OF BALANCED THREE-PHASE SYSTEMS

If a three-phase power system is balanced, it is possible to determine the voltages, currents, and powers at various points in the circuit with a *per-phase equivalent circuit*. This idea is illustrated in Figure 2–11. Figure 2–11a shows a Y-connected generator supplying power to a Y-connected load through a three-phase transmission line.

In such a balanced system, a neutral wire may be inserted with no effect on the system, since no current flows in that wire. This system with the extra wire inserted is shown in Figure 2–11b. Also, notice that each of the three phases is *identical* except for a 120° shift in phase angle. Therefore, it is possible to analyze a circuit consisting of *one phase and the neutral,* and the results of that analysis will be valid for the other two phases as well if the 120° phase shift is included. Such a per-phase circuit is shown in Figure 2–11c.

There is one problem associated with this approach, however. It requires that a neutral line be available (at least conceptually) to provide a return path for current flow from the loads to the generator. This is fine for Y-connected sources and loads, but no neutral can be connected to Δ-connected sources and loads.

How can Δ-connected sources and loads be included in a power system to be analyzed? The standard approach is to transform the impedances by the Y–Δ transform of elementary circuit theory. For the special case of balanced loads, the Y–Δ transformation states that a Δ-connected load consisting of three equal impedances, each of value Z, is totally equivalent to a Y-connected load consisting of three impedances, each of value Z/3 (see Figure 2–12). This equivalence means that the voltages, currents, and powers supplied to the two loads cannot be distinguished in any fashion by anything external to the load itself.

If Δ-connected sources or loads include voltage sources, then the magnitudes of the voltage sources must be scaled according to Equation (2–11), and the effect of the 30° phase shift must be included as well.

EXAMPLE 2–1

A 208-V three-phase power system is shown in Figure 2–13. It consists of an ideal 208-V Y-connected three-phase generator connected through a three-phase transmission line to a Y-connected load. The transmission line has an impedance of $0.06 + j0.12$ Ω per phase, and the load has an impedance of $12 + j9$ Ω per phase. For this simple power system, find

(a) The magnitude of the line current I_L

(b) The magnitude of the load's line and phase voltages V_{LL} and $V_{\phi L}$

(c) The real, reactive, and apparent powers consumed by the load

(d) The power factor of the load

(e) The real, reactive, and apparent powers consumed by the transmission line

(f) The real, reactive, and apparent powers supplied by the generator

(g) The generator's power factor

Figure 2–11 I (a) A Y-connected generator and load. (b) System with neutral inserted. (c) The per-phase equivalent circuit.

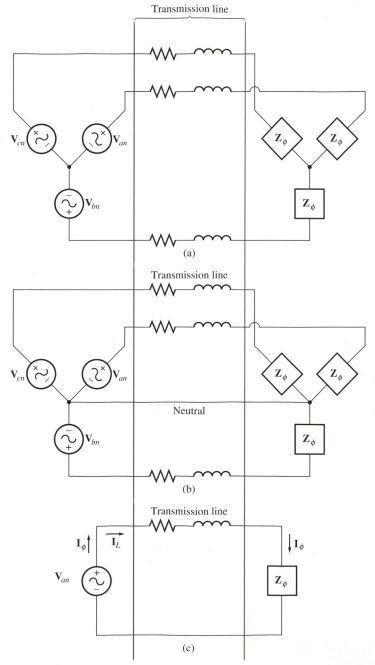

Figure 2–12 | Y-Δ transformation. A Y-connected impedance of $Z/3$ Ω is totally equivalent to a Δ-connected impedance of Z Ω to any circuit connected to the load's terminals.

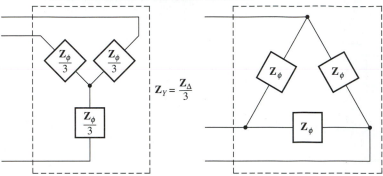

$$\mathbf{Z}_Y = \frac{\mathbf{Z}_\Delta}{3}$$

Figure 2–13 | The three-phase circuit of Example 2–1.

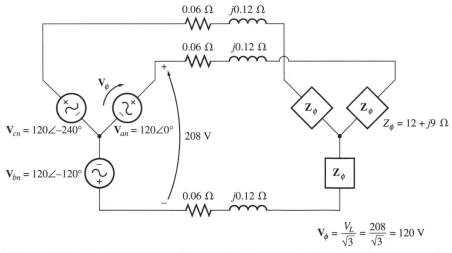

$$\mathbf{V}_\phi = \frac{V_L}{\sqrt{3}} = \frac{208}{\sqrt{3}} = 120 \text{ V}$$

■ Solution

Since both the generator and the load on this power system are Y-connected, it is very simple to construct a per-phase equivalent circuit. This circuit is shown in Figure 2–14.

(a) The line current flowing in the per-phase equivalent circuit is given by

$$\mathbf{I}_{line} = \frac{\mathbf{V}}{\mathbf{Z}_{line} + \mathbf{Z}_{load}}$$

$$= \frac{120 \angle 0° \text{ V}}{(0.06 + j0.12 \text{ } \Omega) + (12 + j9\Omega)}$$

$$= \frac{120 \angle 0°}{12.06 + j9.12} = \frac{120 \angle 0°}{15.12\angle 37.1°}$$

$$= 7.94\angle -37.1° \text{ A}$$

The magnitude of the line current is thus 7.94 A.

Figure 2–14 | Per-phase circuit in Example 2–1.

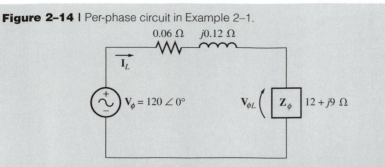

(b) The phase voltage on the load is the voltage across one phase of the load. This voltage is the product of the phase impedance and the phase current of the load:

$$\mathbf{V}_{\phi L} = \mathbf{I}_{\phi L}\mathbf{Z}_{\phi L}$$
$$= (7.94\angle -37.1° \text{ A})(12 + j9 \ \Omega)$$
$$= (7.94\angle -37.1° \text{ A})(15\angle 36.9° \ \Omega)$$
$$= 119.1\angle -0.2° \text{ V}$$

Therefore, the magnitude of the load's phase voltage is

$$V_{\phi L} = 119.1 \text{ V}$$

and the magnitude of the load's line voltage is

$$V_{LL} = \sqrt{3}V_{\phi L} = 206.3 \text{ V}$$

(c) The real power consumed by the load is

$$P_{\text{load}} = 3V_{\phi}I_{\phi} \cos \theta$$
$$= 3(119.1 \text{ V})(7.94 \text{ A}) \cos 36.9°$$
$$= 2270 \text{ W}$$

The reactive power consumed by the load is

$$Q_{\text{load}} = 3V_{\phi}I_{\phi} \sin \theta$$
$$= 3(119.1 \text{ V})(7.94 \text{ A}) \sin 36.9°$$
$$= 1702 \text{ var}$$

The apparent power consumed by the load is

$$S_{\text{load}} = 3V_{\phi}I_{\phi}$$
$$= 3(119.1 \text{ V})(7.94 \text{ A})$$
$$= 2839 \text{ VA}$$

(d) The load power factor is

$$\text{PF}_{\text{load}} = \cos \theta = \cos 36.9° = 0.8 \text{ lagging}$$

(e) The current in the transmission line is $7.94\angle-37.1$ A, and the impedance of the line is $0.06 + j0.12$ Ω or $0.134\angle63.4°$ Ω per phase. Therefore, the real, reactive, and apparent powers consumed in the line are

$$P_{line} = 3I_\phi^2 Z \cos\theta \tag{2-26}$$
$$= 3(7.94 \text{ A})^2 (0.134 \text{ }\Omega) \cos 63.4°$$
$$= 11.3 \text{ W}$$
$$Q_{line} = 3I_\phi^2 Z \sin\theta \tag{2-27}$$
$$= 3(7.94 \text{ A})^2 (0.134 \text{ }\Omega) \sin 63.4°$$
$$= 22.7 \text{ var}$$
$$S_{line} = 3I_\phi^2 Z \tag{2-28}$$
$$= 3(7.94 \text{ A})^2 (0.134 \text{ }\Omega)$$
$$= 25.3 \text{ VA}$$

(f) The real and reactive powers supplied by the generator are the sum of the powers consumed by the line and the load:

$$P_{gen} = P_{line} + P_{load}$$
$$= 11.3 \text{ W} + 2270 \text{ W} = 2281 \text{ W}$$
$$Q_{gen} = Q_{line} + Q_{load}$$
$$= 22.7 \text{ var} + 1702 \text{ var} = 1725 \text{ VAR}$$

The apparent power of the generator is the square root of the sum of the squares of the real and reactive powers:

$$S_{gen} = \sqrt{P_{gen}^2 + Q_{gen}^2} = 2860 \text{ VA}$$

(g) From the power triangle, the power-factor angle θ is

$$\theta_{gen} = \tan^{-1}\frac{Q_{gen}}{P_{gen}} = \tan^{-1}\frac{1725 \text{ VAR}}{2281 \text{ W}} = 37.1°$$

Therefore, the generator's power factor is

$$PF_{gen} = \cos 37.1° = 0.798 \text{ lagging}$$ ■

EXAMPLE 2–2

Repeat Example 2–1 for a Δ-connected load, with everything else unchanged.

■ Solution

This power system is shown in Figure 2–15. Since the load on this power system is Δ connected, it must first be converted to an equivalent Y form. The phase impedance of the Δ-connected load is $12 + j9$ Ω so the equivalent phase impedance of the corresponding Y form is

$$Z_Y = \frac{Z_\Delta}{3} = 4 + j3 \text{ }\Omega$$

Figure 2–15 | Three-phase circuit in Example 2–2.

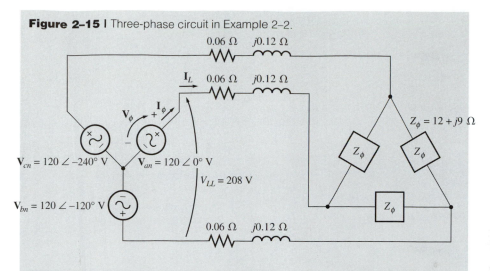

The resulting per-phase equivalent circuit of this system is shown in Figure 2–16.

Figure 2–16 | Per-phase circuit in Example 2–2.

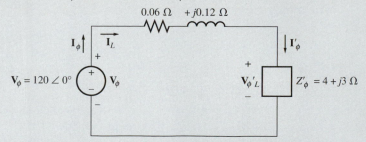

(a) The line current flowing in the per-phase equivalent circuit is given by

$$\mathbf{I}_{line} = \frac{\mathbf{V}}{\mathbf{Z}_{line} + \mathbf{Z}_{load}}$$

$$= \frac{120\angle 0°\ \text{V}}{(0.06 + j0.12\ \Omega) + (4 + j3\ \Omega)}$$

$$= \frac{120\angle 0°}{4.06 + j3.12} = \frac{120\angle 0°}{5.12\angle 37.5°}$$

$$= 23.4\angle -37.5°\ \text{A}$$

The magnitude of the line current is thus 23.4 A.

(b) The phase voltage on the equivalent Y load is the voltage across one phase of the load. This voltage is the product of the phase impedance and the phase current of the load:

$$\mathbf{V}'_{\phi L} = \mathbf{I}'_{\phi L}\mathbf{Z}'_{\phi L}$$

$$= (23.4\angle -37.5°\ \text{A})(4 + j3\ \Omega)$$

$$= (23.4\angle -37.5°\ \text{A})(5\angle 36.9°\ \Omega) = 117\angle -0.6°\ \text{V}$$

The original load was Δ connected, so the phase voltage of the *original* load is

$$\mathbf{V}_{\phi L} = \sqrt{3} \, (117 \text{ V}) = 203 \text{ V}$$

and the magnitude of the load's line voltage is

$$V_{LL} = V_{\phi L} = 203 \text{ V}$$

(c) The real power consumed by the equivalent Y load (which is the same as the power in the actual load) is

$$\begin{aligned} P_{\text{load}} &= 3V_\phi I_\phi \cos\theta \\ &= 3(117 \text{ V})(23.4 \text{ A}) \cos 36.9° \\ &= 6571 \text{ W} \end{aligned}$$

The reactive power consumed by the load is

$$\begin{aligned} Q_{\text{load}} &= 3V_\phi I_\phi \sin\theta \\ &= 3(117 \text{ V})(23.4 \text{ A}) \sin 36.9° \\ &= 4928 \text{ var} \end{aligned}$$

The apparent power consumed by the load is

$$\begin{aligned} S_{\text{load}} &= 3V_\phi I_\phi \\ &= 3(117 \text{ V})(23.4 \text{ A}) \\ &= 8213 \text{ VA} \end{aligned}$$

(d) The load power factor is

$$\text{PF}_{\text{load}} = \cos\theta = \cos 36.9° = 0.8 \text{ lagging}$$

(e) The current in the transmission is $23.4\angle{-37.5°}$ A, and the impedance of the line is $0.06 + j0.12 \ \Omega$ or $0.134\angle 63.4° \ \Omega$ per phase. Therefore, the real, reactive, and apparent powers consumed in the line are

$$\begin{aligned} P_{\text{line}} &= 3I_\phi^2 Z \cos\theta && \text{(2–26)} \\ &= 3(23.4 \text{ A})^2(0.134 \ \Omega) \cos 63.4° \\ &= 98.6 \text{ W} \end{aligned}$$

$$\begin{aligned} Q_{\text{line}} &= 3I_\phi^2 Z \sin\theta && \text{(2–27)} \\ &= 3(23.4 \text{ A})^2(0.134 \ \Omega) \sin 63.4° \\ &= 197 \text{ var} \end{aligned}$$

$$\begin{aligned} S_{\text{line}} &= 3I_\phi^2 Z && \text{(2–28)} \\ &= 3(23.4 \text{ A})^2(0.134 \ \Omega) \\ &= 220 \text{ VA} \end{aligned}$$

(f) The real and reactive powers supplied by the generator are the sums of the powers consumed by the line and the load:

$$\begin{aligned} P_{\text{gen}} &= P_{\text{line}} + P_{\text{load}} \\ &= 98.6 \text{ W} + 6571 \text{ W} = 6670 \text{ W} \\ Q_{\text{gen}} &= Q_{\text{line}} + Q_{\text{load}} \\ &= 197 \text{ VAR} + 4928 \text{ VAR} = 5125 \text{ var} \end{aligned}$$

The apparent power of the generator is the square root of the sum of the squares of the real and reactive powers:

$$S_{gen} = \sqrt{P_{gen}^2 + Q_{gen}^2} = 8411 \text{ VA}$$

(g) From the power triangle, the power-factor angle θ is

$$\theta_{gen} = \tan^{-1}\frac{Q_{gen}}{P_{gen}} = \tan^{-1}\frac{5125 \text{ VAR}}{6670 \text{ W}} = 37.6°$$

Therefore, the generator's power factor is

$$PF_{gen} = \cos 37.6° = 0.792 \text{ lagging}$$ ■

2.5 | ONE-LINE DIAGRAMS

As we have seen in this chapter, a balanced three-phase power system has three lines connecting each source with each load, one for each of the phases in the power system. The three phases are all similar, with voltages and currents equal in amplitude and shifted in phase from each other by 120°. Because the three phases are all basically the same, it is customary to sketch power systems in a simple form with a *single line* representing all three phases of the real power system. These *one-line diagrams* provide a compact way to represent the interconnections of a power system. One-line diagrams typically include all of the major components of a power system, such as generators, transformers, transmission lines, and loads with the transmission lines represented by a single line. The voltages and types of connections of each generator and load are usually shown on the diagram. A simple power system is shown in Figure 2–17, together with the corresponding one-line diagram.

2.6 | USING THE POWER TRIANGLE

If the transmission lines in a power system can be assumed to have negligible impedance, then an important simplification is possible in the calculation of three-phase currents and powers. This simplification depends on the use of the real and reactive powers of each load to determine the currents and power factors at various points in the system.

For example, consider the simple power system shown in Figure 2–17. If the transmission line in that power system is assumed to be lossless, the line voltage at the generator will be the same as the line voltage at the loads. If the generator voltage is specified, then we can find the current and power factor at any point in this power system as follows:

1. Determine the line voltage at the generator and the loads. Since the transmission line is assumed to be lossless, these two voltages will be identical.
2. Determine the real and reactive powers of each load on the power system. We can use the known load voltage to perform this calculation.

Figure 2–17 I (a) A simple power system with a Y-connected generator, a
Δ-connected load, and a Y-connected load. (b) The corresponding
one-line diagram.

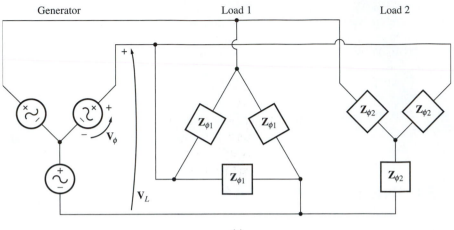

(a)

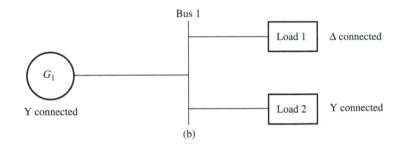

(b)

3. Find the total real and reactive powers supplied to all loads "downstream"
   from the point being examined.
4. Determine the system power factor at that point, using the power-triangle
   relationships.
5. Use Equation (2–29) to determine line currents, or Equation (2–23)
   to determine phase currents, at that point.

This approach is commonly employed by engineers estimating the currents and
power flows at various points on distribution systems within an industrial plant.
Within a single plant, the lengths of transmission lines will be quite short and their
impedances will be relatively small, and so only small errors will occur if the impedances are neglected. An engineer can treat the line voltage as constant, and use
the power triangle method to quickly calculate the effect of adding a load on the
overall system current and power factor.

EXAMPLE 2–3

Figure 2–18 shows a one-line diagram of a small 480-V industrial distribution system. The power system supplies a constant line voltage of 480 V, and the impedance of the distribution lines is negligible. Load 1 is a Δ-connected load with a phase impedance of $10\angle 30°$ Ω, and load 2 is a Y-connected load with a phase impedance of $5\angle -36.87°$ Ω.

(a) Find the overall power factor of the distribution system.

(b) Find the total line current supplied to the distribution system.

Figure 2–18 | The system in Example 2–3.

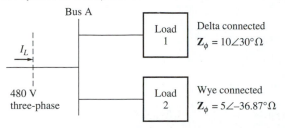

■ **Solution**

The lines in this system are assumed impedanceless, so there will be no voltage drops within the system. Since load 1 is Δ connected, its phase voltage will be 480 V. Since load 2 is Y connected, its phase voltage will be $480/\sqrt{3} = 277$ V.

The phase current in load 1 is

$$I_{\phi 1} = \frac{480\ \text{V}}{10\ \Omega} = 48\ \text{A}$$

Therefore, the real and reactive powers of load 1 are

$$P_1 = 3V_{\phi 1}I_{\phi 1}\cos\theta$$
$$= 3(480\ \text{V})(48\ \text{A})\cos 30° = 59.9\ \text{kW}$$
$$Q_1 = 3V_{\phi 1}I_{\phi 1}\sin\theta$$
$$= 3(480\ \text{V})(48\ \text{A})\sin 30° = 34.6\ \text{kvar}$$

The phase current in load 2 is

$$I_{\phi 2} = \frac{277\ \text{V}}{5\ \Omega} = 55.4\ \text{A}$$

Therefore, the real and reactive powers of load 2 are

$$P_2 = 3V_{\phi 2}I_{\phi 2}\cos\theta$$
$$= 3(277\ \text{V})(55.4\ \text{A})\cos(-36.87°) = 36.8\ \text{kW}$$
$$Q_2 = 3V_{\phi 2}I_{\phi 2}\sin\theta$$
$$= 3(277\ \text{V})(55.4\ \text{A})\sin(-36.87°) = -27.6\ \text{kvar}$$

(a) The total real and reactive powers supplied by the distribution system are

$$P_{tot} = P_1 + P_2$$
$$= 59.9 \text{ kW} + 36.8 \text{ kW} = 96.7 \text{ kW}$$
$$Q_{tot} = Q_1 + Q_2$$
$$= 34.6 \text{ kvar} - 27.6 \text{ kvar} = 7.00 \text{ kvar}$$

From the power triangle, the effective impedance angle θ is given by

$$\theta = \tan^{-1} \frac{Q}{P}$$
$$= \tan^{-1} \frac{7.00 \text{ kvar}}{96.7 \text{ kW}} = 4.14°$$

The system power factor is thus

$$PF = \cos \theta = \cos(4.14°) = 0.997 \text{ lagging}$$

(b) The total line current is given by

$$I_L = \frac{P}{\sqrt{3} V_L \cos \theta}$$
$$I_L = \frac{96.7 \text{ kW}}{\sqrt{3}(480 \text{ V})(0.997)} = 117 \text{ A} \qquad ■$$

2.7 | QUESTIONS

2–1. What types of connections are possible for three-phase generators and loads?

2–2. What is meant by the term "balanced" in a balanced three-phase system?

2–3. What is the relationship between phase and line voltages and currents for a wye (Y) connection?

2–4. What is the relationship between phase and line voltages and currents for a delta (Δ) connection?

2–5. What is phase sequence?

2–6. Write the equations for real, reactive, and apparent power in three-phase circuits, in terms of both line and phase quantities.

2–7. What is a Y–Δ transform?

2.8 | PROBLEMS

2–1. Three impedances of $4 + j3 \ \Omega$ are Δ connected and tied to a three-phase 208-V power line. Find I_ϕ, I_L, P, Q, S, and the power factor of this load.

2–2. Figure P2–1 shows a three-phase power system with two loads. The Δ-connected generator is producing a line voltage of 480 V, and the line impedance is $0.09 + j0.16$ Ω. Load 1 is Y connected, with a phase impedance of $2.5\angle36.87°$ Ω and load 2 is Δ connected, with a phase impedance of $5\angle-20°$ Ω.

Figure P2–1 | The system in Problem 2–2.

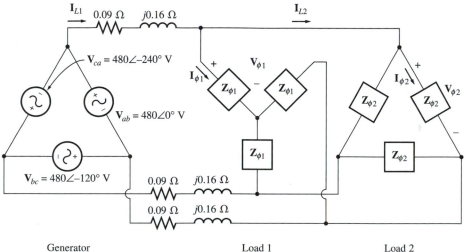

Generator   Load 1   Load 2

$$Z_{\phi1} = 2.5\angle36.87°\,\Omega$$
$$Z_{\phi2} = 5\angle-20°\,\Omega$$

(a) What is the line voltage of the two loads?

(b) What is the voltage drop on the transmission lines?

(c) Find the real and reactive powers supplied to each load.

(d) Find the real and reactive power losses in the transmission line.

(e) Find the real power, reactive power, and power factor supplied by the generator.

2–3. Figure P2–2 shows a one-line diagram of a simple power system containing a single 480-V generator and three loads. Assume that the transmission lines in this power system are lossless, and answer the following questions.

(a) Assume that Load 1 is Y connected. What are the phase voltage and currents in that load?

(b) Assume that Load 2 is Δ connected. What are the phase voltage and currents in that load?

(c) What real, reactive, and apparent power does the generator supply when the switch is open?

(d) What is the total line current I_L when the switch is open?

(e) What real, reactive, and apparent power does the generator supply when the switch is closed?

Figure P2–2 I The power system in Problem 2–3.

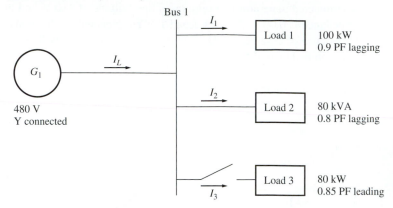

(f) What is the total line current I_L when the switch is closed?

(g) How does the total line current I_L compare to the sum of the three individual currents $I_1 + I_2 + I_3$? If they are not equal, why not?

2–4. Prove that the line voltage of a Y-connected generator with an *acb* phase sequence lags the corresponding phase voltage by 30°. Draw a phasor diagram showing the phase and line voltages for this generator.

2–5. Find the magnitudes and angles of each line and phase voltage and current on the load shown in Figure P2–3.

Figure P2–3 I The system in Problem 2–5.

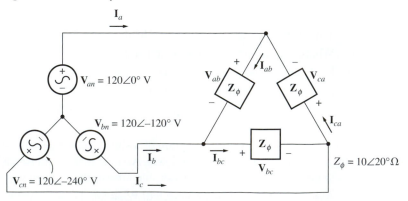

2–6. Figure P2–4 shows a one-line diagram of a small 480-V distribution system in an industrial plant. An engineer working at the plant wishes to calculate the current that will be drawn from the power utility company with and without the capacitor bank switched into the system. For the purposes of this calculation, the engineer will assume that the lines in the system have zero impedance.

Figure P2–4 | The system in Problem 2–6.

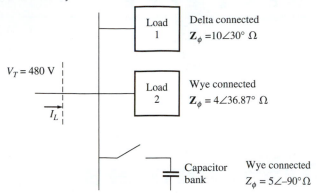

(a) If the switch shown is open, find the real, reactive, and apparent powers in the system. Find the total current supplied to the distribution system by the utility.

(b) Repeat part *(a)* with the switch closed.

(c) What happened to the total current supplied by the power system when the switch closed? Why?

2.9 | REFERENCE

1. Alexander, Charles K., and Matthew N. O. Sadiku: *Fundamentals of Electric Circuits,* McGraw-Hill, 2000.

Transformers

A *transformer* is a device that changes AC electric power at one voltage level into AC electric power at another voltage level through the action of a magnetic field. It consists of two or more coils of wire wrapped around a common ferromagnetic core. These coils are (usually) not directly connected. The only connection between the coils is the common magnetic flux present within the core.

One of the transformer windings is connected to a source of AC electric power, and the second (and perhaps third) transformer winding supplies electric power to loads. The transformer winding connected to the power source is called the *primary winding* or *input winding,* and the winding connected to the loads is called the *secondary winding* or *output winding.* If there is a third winding on the transformer, it is called the *tertiary winding.*

3.1 | WHY TRANSFORMERS ARE IMPORTANT TO MODERN LIFE

The first power distribution system in the United States was a 120-V DC system invented by Thomas A. Edison to supply power for incandescent light bulbs. Edison's first central power station went into operation in New York City in September 1882. Unfortunately, his power system generated and transmitted power at such low voltages that very large currents were necessary to supply significant amounts of power. These high currents caused huge voltage drops and power losses in the transmission lines, severely restricting the service area of a generating station. In the 1880s, central power stations were located every few city blocks to overcome this problem. The fact that power could not be transmitted far with low-voltage DC power systems meant that generating stations had to be small and localized and so were relatively inefficient.

The invention of the transformer (Figure 3–1) and the concurrent development of AC power sources eliminated forever these restrictions on the range and power level of power systems. A transformer ideally changes one AC voltage level to another voltage level without affecting the actual power supplied. If the transformer

Figure 3–1 | The first practical modern transformer, built by William Stanley in 1885. Note that the core is made up of individual sheets of metal (laminations). *(Courtesy of General Electric Company.)*

steps up the voltage level of a circuit, it must decrease the current to keep the power into the device equal to the power out of it. Therefore, AC electric power can be generated at one central location, its voltage stepped up for transmission over long distances at very low losses, and its voltage stepped down again for final use. Since the transmission losses in the lines of a power system are proportional to the square of the current in the lines, raising the transmission voltage and reducing the resulting transmission currents by a factor of 10 with transformers reduces power transmission losses by a factor of 100. Without the transformer, it would simply not be possible to use electric power in many of the ways it is used today.

In a modern power system, transformers are literally found everywhere. Electric power is generated at voltages of 12 to 25 kV. Transformers at the output of the generators step up the voltage to between 110 kV and nearly 1000 kV for transmission over long distances at very low losses. Substation transformers then step down the voltage to the 13- to 34.5-kV range for local distribution, and distribution transformers finally permit the power to be used safely in homes, offices, and factories at voltages as low as 120 V.

3.2 | TYPES AND CONSTRUCTION OF TRANSFORMERS

The principal purpose of a transformer is to convert AC power at one voltage level into AC power of the same frequency at another voltage level. Transformers are also used for a variety of other purposes (e.g., voltage sampling, current sampling, and impedance transformation), but this chapter is primarily devoted to the power transformer.

Power transformers are constructed on one of two types of cores. One type of construction consists of a simple rectangular laminated piece of steel with the transformer windings wrapped around two sides of the rectangle. This type of construction is known as *core form* and is illustrated in Figure 3–2. The other type consists of

Figure 3–2 | Core-form transformer construction.

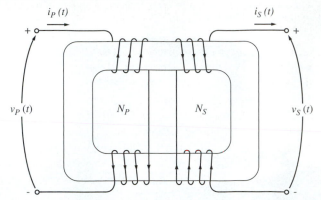

a three-legged laminated core with the windings wrapped around the center leg. This type of construction is known as *shell form* and is illustrated in Figure 3–3. In either case, the core is constructed of thin laminations electrically isolated from each other in order to minimize eddy currents.

Figure 3–3 | (a) Shell-form transformer construction. (b) A typical shell-form transformer. *(Courtesy of General Electric Company.)*

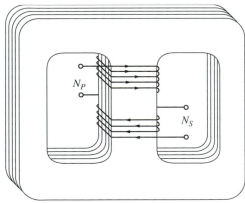

(a)

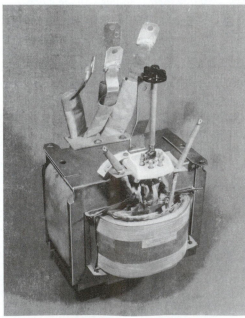

(b)

The primary and secondary windings in a physical transformer are wrapped one on top of the other, with the low-voltage winding innermost. Such an arrangement serves two purposes:

1. It simplifies the problem of insulating the high-voltage winding from the core.
2. It results in much less leakage flux than would be the case if the two windings were separated by a distance on the core.

Power transformers are given a variety of different names, depending on their use in power systems. A transformer connected to the output of a generator and used to step its voltage up to transmission levels (110+ kV) is sometimes called a *unit transformer*. The transformer at the other end of the transmission line, which steps the voltage down from transmission levels to distribution levels (from 2.3 to 34.5 kV), is called a *substation transformer*. Finally, the transformer that takes the distribution voltage and steps it down to the final voltage at which the power is actually used (110, 208, 220 V, etc.) is called a *distribution transformer*. All these devices are essentially the same—the only difference among them is their intended use.

In addition to the various power transformers, two special-purpose transformers are used with electric machinery and power systems. The first of these special transformers is a device specially designed to sample a high voltage and produce a low secondary voltage directly proportional to it. Such a transformer is called a *potential transformer*. A power transformer also produces a secondary voltage directly proportional to its primary voltage; the difference between a potential transformer and a power transformer is that the potential transformer is designed to handle only a very small current. The second type of special transformer is a device designed to provide a secondary current much smaller than but directly proportional to its primary current. This device is called a *current transformer*. Both special-purpose transformers are discussed in a later section of this chapter.

3.3 | THE IDEAL TRANSFORMER

An *ideal transformer* is a lossless device with an input winding and an output winding. In an ideal transformer, the relationships between the input voltage and the output voltage, and between the input current and the output current, are given by Equations (3–1) and (3–4) below. Figure 3–4 shows an ideal transformer.

The transformer shown in Figure 3–4 has N_P turns of wire on its primary side and N_S turns of wire on its secondary side. The relationship between the voltage $v_P(t)$ applied to the primary side of the ideal transformer and the voltage $v_S(t)$ produced on the secondary side is

$$\boxed{\frac{v_P(t)}{v_S(t)} = \frac{N_P}{N_S} = a} \qquad (3–1)$$

where a is defined to be the *turns ratio* of the transformer:

$$a = \frac{N_P}{N_S} \qquad (3–2)$$

Figure 3–4 | (a) Sketch of an ideal transformer. (b) Schematic symbols of a transformer.

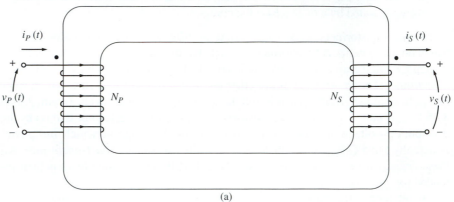

(a)

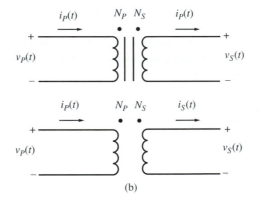

(b)

The relationship between the current $i_P(t)$ flowing into the primary side of the ideal transformer and the current $i_S(t)$ flowing out of the secondary side of the transformer is

$$N_P i_P(t) = N_S i_S(t)$$ (3–3)

or

$$\frac{i_P(t)}{i_S(t)} = \frac{1}{a}$$ (3–4)

In terms of phasor quantities, these equations are

$$\frac{\mathbf{V}_P}{\mathbf{V}_S} = a$$ (3–5)

or

$$\boxed{\frac{\mathbf{I}_P}{\mathbf{I}_S} = \frac{1}{a}}$$ (3–6)

Notice that the phase angle of $\mathbf{V}_P$ is the same as the angle of $\mathbf{V}_S$ and the phase angle of $\mathbf{I}_P$ is the same as the phase angle of $\mathbf{I}_S$. The turns ratio of the ideal transformer affects the *magnitudes* of the voltages and currents, but not their *angles*.

Equations (3–1) to (3–6) describe the relationships between the magnitudes and angles of the voltages and currents on the primary and secondary sides of the transformer, but they leave one question unanswered: Given that the primary circuit's voltage is positive at a specific end of the coil, what would the *polarity* of the secondary circuit's voltage be? In real transformers, it would be possible to tell the secondary's polarity only if the transformer were opened and its windings examined. To avoid this necessity, transformers utilize the *dot convention*. The dots appearing at one end of each winding in Figure 3–4 tell the polarity of the voltage and current on the secondary side of the transformer. The relationship is as follows:

1. If the primary *voltage* is positive at the dotted end of the winding with respect to the undotted end, then the secondary voltage will be positive at the dotted end also. Voltage polarities are the same with respect to the dots on each side of the core.

2. If the primary *current* of the transformer flows *into* the dotted end of the primary winding, the secondary current will flow *out* of the dotted end of the secondary winding.

The physical meaning of the dot convention and the reason polarities work out this way will be explained in Section 3.4, which deals with the real transformer.

Power in an Ideal Transformer

The power supplied to the transformer by the primary circuit is given by the equation

$$\boxed{P_{\text{in}} = V_P I_P \cos \theta_P}$$ (3–7)

where θ_P is the angle between the primary voltage and the primary current. The power supplied by the transformer secondary circuit to its loads is given by the equation

$$\boxed{P_{\text{out}} = V_S I_S \cos \theta_S}$$ (3–8)

where θ_S is the angle between the secondary voltage and the secondary current. Since voltage and current angles are unaffected by an ideal transformer, $\theta_P = \theta_S = \theta$. The primary and secondary windings of an ideal transformer have the *same power factor*.

How does the power going into the primary circuit of the ideal transformer compare to the power coming out of the other side? It is possible to find out through a simple application of the voltage and current equations [Equations (3–5) and (3–6)]. The power out of a transformer is

$$P_{\text{out}} = V_S I_S \cos \theta \qquad (3\text{–}9)$$

Applying the turns-ratio equations gives $V_S = V_P/a$ and $I_S = aI_P$, so

$$P_{\text{out}} = \frac{V_P}{a} (aI_P) \cos \theta \qquad (3\text{–}10)$$

$$\boxed{P_{\text{out}} = V_P I_P \cos \theta = P_{\text{in}}} \qquad (3\text{–}11)$$

Thus, *the output power of an ideal transformer is equal to its input power.*

The same relationship applies to reactive power Q and apparent power S:

$$\boxed{Q_{\text{in}} = V_P I_P \sin \theta = V_S I_S \sin \theta = Q_{\text{out}}} \qquad (3\text{–}12)$$

and

$$\boxed{S_{\text{in}} = V_P I_P = V_S I_S = S_{\text{out}}} \qquad (3\text{–}13)$$

Impedance Transformation through a Transformer

The *impedance* of a device or an element is defined as the ratio of the phasor voltage across it to the phasor current flowing through it:

$$Z_L = \frac{\mathbf{V}_L}{\mathbf{I}_L} \qquad (3\text{–}14)$$

One of the interesting properties of a transformer is that, since it changes voltage and current levels, it changes the *ratio* between voltage and current and hence the apparent impedance of an element. To understand this idea, refer to Figure 3–5. If the secondary current is called $\mathbf{I}_S$ and the secondary voltage $\mathbf{V}_S$, then the impedance of the load is given by

$$Z_L = \frac{\mathbf{V}_S}{\mathbf{I}_S} \qquad (3\text{–}15)$$

The apparent impedance of the primary circuit of the transformer is

$$Z_L' = \frac{\mathbf{V}_P}{\mathbf{I}_P} \qquad (3\text{–}16)$$

Since the primary voltage can be expressed as

$$\mathbf{V}_P = a\mathbf{V}_S$$

and the primary current can be expressed as

$$\mathbf{I}_P = \frac{\mathbf{I}_S}{a}$$

the apparent impedance of the primary is

$$Z'_L = \frac{\mathbf{V}_P}{\mathbf{I}_P} = \frac{a\mathbf{V}_S}{\mathbf{I}_S/a} = a^2 \frac{\mathbf{V}_S}{\mathbf{I}_S}$$

$$\boxed{Z'_L = a^2 Z_L}$$

(3–17)

Figure 3–5 | (a) Definition of impedance. (b) Impedance scaling through a transformer.

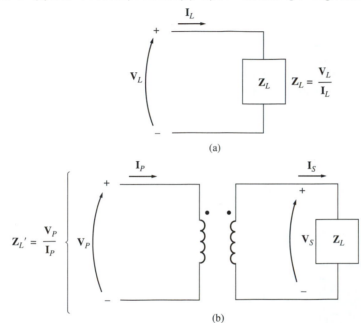

(a)

(b)

With a transformer, it is possible to match the magnitude of a load impedance to a source impedance simply by picking the proper turns ratio.

Analysis of Circuits Containing Ideal Transformers

If a circuit contains an ideal transformer, then the easiest way to analyze the circuit for its voltages and currents is to replace the portion of the circuit on one side of the transformer by an equivalent circuit with the same terminal characteristics. After the equivalent circuit has been substituted for one side, then the new circuit (without a transformer present) can be solved for its voltages and currents. In the portion of the circuit that was not replaced, the solutions obtained will be the correct values of voltage and current for the original circuit. Then the turns ratio of the transformer can be used to determine the voltages and currents on the other side of the transformer. The process of replacing one side of a transformer by its equivalent at the other side's voltage level is known as *referring* the first side of the transformer to the second side.

How is the equivalent circuit formed? Its shape is exactly the same as the shape of the original circuit. The values of voltages on the side being replaced are scaled by Equation (3–5), and the values of the impedances are scaled by Equation (3–17). The polarities of voltage sources in the equivalent circuit will be reversed from their direction in the original circuit if the dots on one side of the transformer windings are reversed compared to the dots on the other side of the transformer windings.

The solution for circuits containing ideal transformers is illustrated in the following example.

EXAMPLE 3–1

A single-phase power system consists of a 480-V 60-Hz generator supplying a load $\mathbf{Z}_{load} = 4 + j3 \; \Omega$ through a transmission line of impedance $\mathbf{Z}_{line} = 0.18 + j0.24 \; \Omega$. Answer the following questions about this system.

a. If the power system is exactly as described above (Figure 3–6a), what will the voltage at the load be? What will the transmission line losses be?

b. Suppose a 1:10 step-up transformer is placed at the generator end of the transmission line and a 10:1 step-down transformer is placed at the load end of the line (Figure 3–6b). What will the load voltage be now? What will the transmission line losses be now?

Figure 3–6 I The power system of Example 3–1 (a) without and (b) with transformers at the ends of the transmission line.

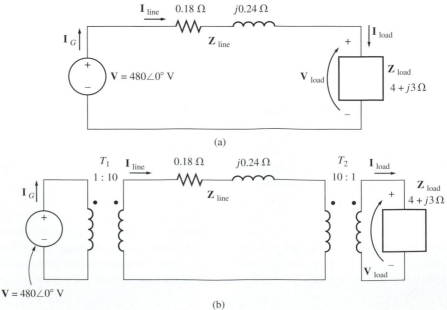

Solution

a. Figure 3–6a shows the power system without transformers. Here $I_G = I_{line} = I_{load}$. The line current in this system is given by

$$I_{line} = \frac{V}{Z_{line} + Z_{load}}$$

$$= \frac{480\angle 0° \text{ V}}{(0.18 \ \Omega + j0.24 \ \Omega) + (4\Omega + j3\Omega)}$$

$$= \frac{480\angle 0°}{4.18 + j3.24} = \frac{480\angle 0°}{5.29\angle 37.8°}$$

$$= 90.8\angle -37.8° \text{ A}$$

Therefore the load voltage is

$$\mathbf{V}_{load} = \mathbf{I}_{line} \ \mathbf{Z}_{load}$$

$$= (90.8\angle -37.8° \text{ A})(4 \ \Omega + j3 \ \Omega)$$

$$= (90.8\angle -37.8° \text{ A})(5\angle 36.9° \ \Omega)$$

$$= 454\angle -0.9° \text{ V}$$

and the line losses are

$$P_{loss} = (I_{line})^2 \ R_{line}$$

$$P_{loss} = (90.8 \text{ A})^2 \ (0.18 \ \Omega) = 1484 \text{ W}$$

b. Figure 3–6b shows the power system with the transformers. To analyze this system, it is necessary to convert it to a common voltage level. This is done in two steps:

1. Eliminate transformer T_2 by referring the load over to the transmission line's voltage level.
2. Eliminate transformer T_1 by referring the transmission line's elements and the equivalent load at the transmission line's voltage over to the source side.

The value of the load's impedance when referred to the transmission system's voltage is

$$\mathbf{Z}'_{load} = a^2 \mathbf{Z}_{load}$$

$$= \left(\frac{10}{1}\right)^2 (4 \ \Omega + j3 \ \Omega)$$

$$= 400 \ \Omega + j300 \ \Omega$$

The total impedance at the transmission line level is now

$$\mathbf{Z}_{eq} = \mathbf{Z}_{line} + \mathbf{Z}'_{load}$$

$$= 400.18 + j300.24 \ \Omega = 500.3\angle 36.88° \ \Omega$$

This equivalent circuit is shown in Figure 3–7a. The total impedance at the transmission line level ($\mathbf{Z}_{line} + \mathbf{Z}'_{load}$) is now reflected across T_1 to the source's voltage level:

$$\mathbf{Z}'_{eq} = a^2 \mathbf{Z}_{eq}$$
$$= a^2 (\mathbf{Z}_{line} + \mathbf{Z}'_{load})$$

$$= \left(\frac{1}{10}\right)^2 (0.18 \ \Omega + j0.24 \ \Omega + 400 \ \Omega + j300 \ \Omega)$$
$$= (0.0018 \ \Omega + j0.0024 \ \Omega + 4 \ \Omega + j3 \ \Omega)$$
$$= 5.003 \angle 36.88° \ \Omega$$

Notice that $\mathbf{Z}''_{load} = 4 + j3 \ \Omega$ and $\mathbf{Z}'_{line} = 0.0018 + j0.0024 \ \Omega$. The resulting equivalent circuit is shown in Figure 3–7b. The generator's current is

$$\mathbf{I}_G = \frac{480 \angle 0° \ \text{V}}{5.003 \angle 36.88° \ \Omega} = 95.94 \angle -36.88° \ \text{A}$$

Knowing the current $\mathbf{I}_G$, we can now work back and find $\mathbf{I}_{ine}$ and $\mathbf{I}_{load}$. Working back through T_1, we get

$$N_{P1} \mathbf{I}_G = N_{S1} \mathbf{I}_{line}$$
$$\mathbf{I}_{line} = \frac{N_{P1}}{N_{S1}} \mathbf{I}_G$$

$$= \frac{1}{10} (95.94 \angle -36.88° \ \text{A}) = 9.594 \angle -36.88° \ \text{A}$$

Working back through T_2 gives

$$N_{P2} \mathbf{I}_{line} = N_{S2} \mathbf{I}_{load}$$
$$\mathbf{I}_{load} = \frac{N_{P2}}{N_{S2}} \mathbf{I}_{line}$$

$$= \frac{10}{1} (9.594 \angle -36.88° \ \text{A}) = 95.94 \angle -36.88° \ \text{A}$$

It is now possible to answer the questions originally asked. The load voltage is given by

$$\mathbf{V}_{load} = \mathbf{I}_{load} \mathbf{Z}_{load}$$
$$= (95.94 \angle -36.88° \ \text{A})(5 \angle 36.87° \ \Omega)$$
$$= 479.7 \angle -0.01° \ \text{V}$$

and the line losses are given by

$$P_{loss} = (I_{line})^2 R_{line}$$
$$= (9.594 \ \text{A})^2 (0.18 \ \Omega) = 16.7 \ \text{W}$$ ∎

Notice that raising the transmission voltage of the power system reduced transmission losses by a factor of nearly 90. Also, the voltage at the load dropped much less in the system with transformers compared to the system without transformers. This simple example dramatically illustrates the advantages of using higher-voltage transmission lines as well as the extreme importance of transformers in modern power systems.

Figure 3–7 | (a) System with the load referred to the transmission system voltage level. (b) System with the load and transmission line referred to the generator's voltage level.

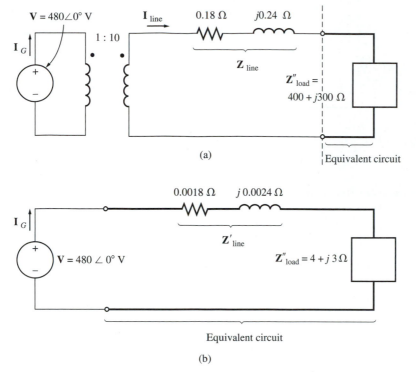

(a)

(b)

3.4 | THEORY OF OPERATION OF REAL SINGLE-PHASE TRANSFORMERS

The ideal transformers described in Section 3.3 can of course never actually be made. What can be produced are real transformers—two or more coils of wire physically wrapped around a ferromagnetic core. The characteristics of a real transformer approximate the characteristics of an ideal transformer, but only to a degree. This section deals with the behavior of real transformers.

To understand the operation of a real transformer, refer to Figure 3–8. Figure 3–8 shows a transformer consisting of two coils of wire wrapped around a transformer core. The primary of the transformer is connected to an AC power source, and the secondary winding is open-circuited. The hysteresis curve of the transformer is shown in Figure 3–9.

The basis of transformer operation can be derived from Faraday's law

$$e_{\text{ind}} = \frac{d\lambda}{dt} \tag{1–41}$$

Figure 3–8 I Sketch of a real transformer with no load attached to its secondary.

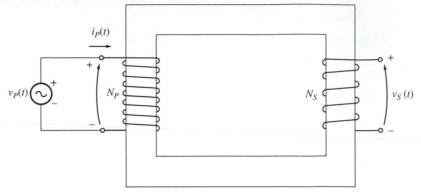

Figure 3–9 I The hysteresis curve of the transformer.

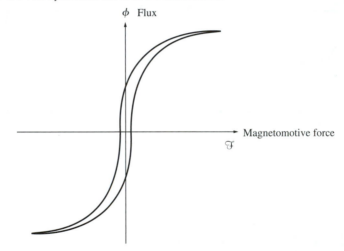

where λ is the flux linkage in the coil across which the voltage is being induced. The flux linkage λ is the sum of the flux passing through each turn in the coil added over all the turns of the coil:

$$\lambda = \sum_{i=1}^{N} \phi_i \qquad (1\text{–}42)$$

The total flux linkage through a coil is not just $N\phi$, where N is the number of turns in the coil, because the flux passing through each turn of a coil is slightly different from the flux in the other turns, depending on the position of the turn within the coil.

However, it is possible to define an *average* flux per turn in a coil. If the total flux linkage in all the turns of the coils is λ and if there are N turns, then the *average flux per turn* is given by

$$\bar{\phi} = \frac{\lambda}{N} \qquad (3\text{-}18)$$

and Faraday's law can be written as

$$e_{\text{ind}} = N \frac{d\bar{\phi}}{dt} \qquad (3\text{-}19)$$

The Voltage Ratio across a Transformer

If the voltage of the source in Figure 3–8 is $v_P(t)$, then that voltage is placed directly across the coils of the primary winding of the transformer. How will the transformer react to this applied voltage? Faraday's law explains what will happen. When Equation (3–19) is solved for the average flux present in the primary winding of the transformer, the result is

$$\bar{\phi} = \frac{1}{N_P} \int v_P(t)\,dt \qquad (3\text{-}20)$$

This equation states that the average flux in the winding is proportional to the integral of the voltage applied to the winding, and the constant of proportionality is the reciprocal of the number of turns in the primary winding $1/N_P$.

This flux is present in the *primary coil* of the transformer. What effect does it have on the secondary coil of the transformer? The effect depends on how much of the flux reaches the secondary coil. Not all the flux produced in the primary coil also passes through the secondary coil—some of the flux lines leave the iron core and pass through the air instead (see Figure 3–10). The portion of the flux that goes through one of the transformer coils but not the other one is called *leakage flux*. The flux in the primary coil of the transformer can thus be divided into two components: a *mutual flux*, which remains in the core and links both windings, and a small *leakage flux*, which passes through the primary winding but returns through the air, bypassing the secondary winding:

$$\boxed{\bar{\phi}_P = \phi_M + \phi_{LP}} \qquad (3\text{-}21)$$

where

$\bar{\phi}_P$ = total average primary flux
ϕ_M = flux component linking both primary and secondary coils
ϕ_{LP} = primary leakage flux

There is a similar division of flux in the secondary winding between mutual flux and leakage flux that passes through the secondary winding but returns through the air, bypassing the primary winding:

$$\boxed{\bar{\phi}_S = \phi_M + \phi_{LS}} \qquad (3\text{-}22)$$

where

$\bar{\phi}_S$ = total average secondary flux

ϕ_M = flux component linking both primary and secondary coils

ϕ_{LS} = secondary leakage flux

With the division of the average primary flux into mutual and leakage components, Faraday's law for the primary circuit can be reexpressed as

$$v_P(t) = N_P \frac{d\bar{\phi}_P}{dt}$$

$$= N_P \frac{d\phi_M}{dt} + N_P \frac{d\phi_{LP}}{dt} \qquad (3\text{--}23)$$

The first term of this expression can be called $e_P(t)$, and the second term can be called $e_{LP}(t)$. If this is done, then Equation (3–23) can be rewritten as

$$v_P(t) = e_P(t) + e_{LP}(t) \qquad (3\text{--}24)$$

Figure 3–10 | Mutual and leakage fluxes in a transformer core.

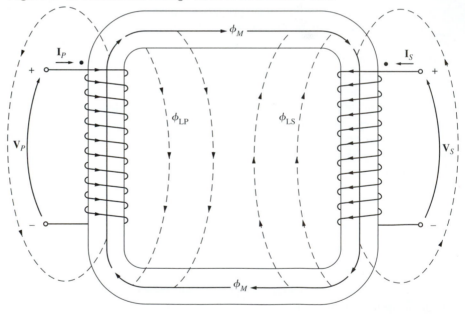

The voltage on the secondary coil of the transformer can also be expressed in terms of Faraday's law as

$$v_S(t) = N_S \frac{d\bar{\phi}_S}{dt}$$

$$= N_S \frac{d\phi_M}{dt} + N_S \frac{d\phi_{LS}}{dt} \qquad (3\text{--}25)$$

$$= e_S(t) + e_{LS}(t) \qquad (3\text{--}26)$$

The primary voltage *due to the mutual flux* is given by

$$e_P(t) = N_P \frac{d\phi_M}{dt} \tag{3-27}$$

and the secondary voltage *due to the mutual flux* is given by

$$e_S(t) = N_S \frac{d\phi_M}{dt} \tag{3-28}$$

Notice from these two relationships that

$$\frac{e_P(t)}{N_P} = \frac{d\phi_M}{dt} = \frac{e_S(t)}{N_S}$$

Therefore,

$$\boxed{\frac{e_P(t)}{e_S(t)} = \frac{N_P}{N_S} = a} \tag{3-29}$$

This equation means that *the ratio of the primary voltage caused by the mutual flux to the secondary voltage caused by the mutual flux is equal to the turns ratio of the transformer.* Since in a well-designed transformer $\phi_M \gg \phi_{LP}$ and $\phi_M \gg \phi_{LS}$, the ratio of the total voltage on the primary of a transformer to the total voltage on the secondary of a transformer is approximately

$$\frac{v_P(t)}{v_S(t)} = \frac{N_P}{N_S} = a \tag{3-30}$$

The smaller the leakage fluxes of the transformer are, the closer the total transformer voltage ratio approximates that of the ideal transformer discussed in Section 3.3.

The Magnetization Current in a Real Transformer

When an AC power source is connected to a transformer as shown in Figure 3–8, a current flows in its primary circuit, *even when the secondary circuit is open circuited.* This current is the current required to produce flux in a real ferromagnetic core, as explained in Chapter 1. It consists of two components:

1. The *magnetization current* i_M, which is the current required to produce the flux in the transformer core
2. The *core-loss current* i_{h+e}, which is the current required to make up for hysteresis and eddy current losses

Figure 3–11 shows the magnetization curve of a typical transformer core. If the flux in the transformer core is known, then the magnitude of the magnetization current can be found directly from Figure 3–11.

Figure 3–11 | (a) The magnetization curve of the transformer core; (b) the magnetization current caused by the flux in the transformer core.

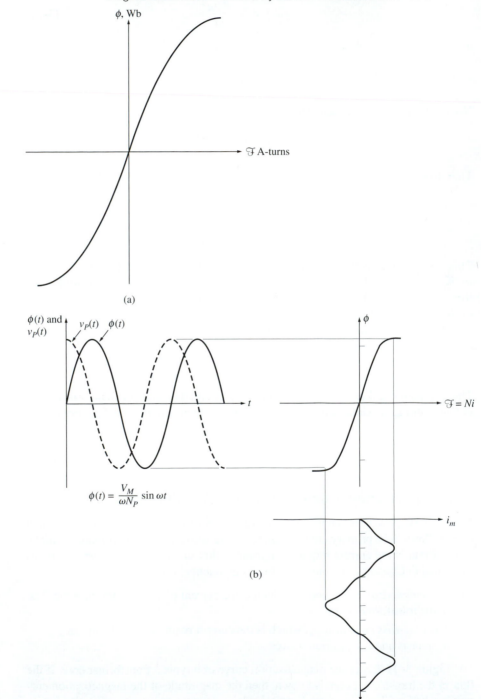

$$\phi(t) = \frac{V_M}{\omega N_P} \sin \omega t$$

(a)

(b)

Ignoring for the moment the effects of leakage flux, we see that the average flux in the core is given by

$$\bar{\phi} = \frac{1}{N_P} \int v_P(t)dt \qquad (3\text{--}20)$$

If the primary voltage is given by the expression $v_P(t) = V_M \cos \omega t$ V, then the resulting flux must be

$$\bar{\phi} = \frac{1}{N_P} \int V_M \cos \omega t \, dt$$

$$= \frac{V_M}{\omega N_P} \sin \omega t \qquad \text{Wb} \qquad (3\text{--}31)$$

If the values of current required to produce a given flux (Figure 3–11a) are compared to the flux in the core at different times, it is possible to construct a sketch of the magnetization current in the winding on the core. Such a sketch is shown in Figure 3–11b. Notice the following points about the magnetization current:

1. The magnetization current in the transformer is not sinusoidal. The higher-frequency components in the magnetization current are due to magnetic saturation in the transformer core.

2. Once the peak flux reaches the saturation point in the core, a small increase in peak flux requires a very large increase in the peak magnetization current.

3. The fundamental component of the magnetization current lags the voltage applied to the core by 90°.

4. The higher-frequency components in the magnetization current can be quite large compared to the fundamental component. In general, the further a transformer core is driven into saturation, the larger the harmonic components will become.

The other component of the no-load current in the transformer is the current required to supply power to make up the hysteresis and eddy current losses in the core. This is the core-loss current. Assume that the flux in the core is sinusoidal. Since the eddy currents in the core are proportional to $d\phi/dt$, the eddy currents are largest when the flux in the core is passing through 0 Wb. Therefore, the core-loss current is greatest as the flux passes through zero. The total current required to make up for core losses is shown in Figure 3–12.

Notice the following points about the core-loss current:

1. The core-loss current is nonlinear because of the nonlinear effects of hysteresis.

2. The fundamental component of the core-loss current is in phase with the voltage applied to the core.

The total no-load current in the core is called the *excitation current* of the transformer. It is just the sum of the magnetization current and the core-loss current in the core:

$$i_{ex} = i_m + i_{h+e} \qquad (3\text{--}32)$$

Figure 3–12 | The core-loss current in a transformer.

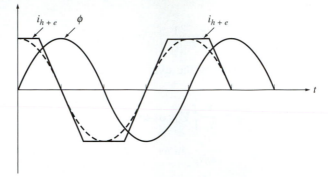

The total excitation current in a typical transformer core is shown in Figure 3–13.

Figure 3–13 | The total excitation current in a transformer.

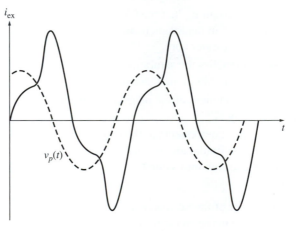

The Current Ratio on a Transformer and the Dot Convention

Now suppose that a load is connected to the secondary of the transformer. The resulting circuit is shown in Figure 3–14. Notice the dots on the windings of the transformer. As in the ideal transformer previously described, the dots help determine the polarity of the voltages and currents in the core without having physically to examine its windings. The physical significance of the dot convention is that *a current flowing into the dotted end of a winding produces a positive magnetomotive force* $\mathcal{F}$, while a current flowing into the undotted end of a winding produces a negative magnetomotive force. Therefore, two currents flowing into the dotted ends of their respective windings produce magnetomotive forces that add. If one current flows into a dotted end of a winding and one flows out of a dotted end, then the magnetomotive forces will subtract from each other.

Figure 3–14 | A real transformer with a load connected to its secondary.

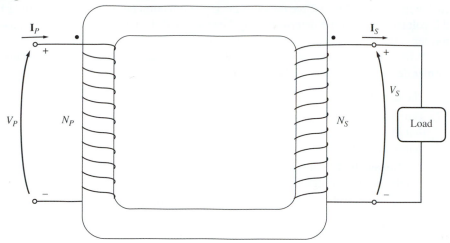

In the situation shown in Figure 3–14, the primary current produces a positive magnetomotive force $\mathcal{F}_P = N_P i_P$, and the secondary current produces a negative magnetomotive force $\mathcal{F}_S = N_S i_S$. Therefore, the net magnetomotive force on the core must be

$$\mathcal{F}_{net} = N_P i_P - N_S i_S \tag{3–33}$$

This net magnetomotive force must produce the net flux in the core, so the net magnetomotive force must be equal to

$$\boxed{\mathcal{F}_{net} = N_P i_P - N_S i_S = \phi \mathcal{R}} \tag{3–34}$$

where $\mathcal{R}$ is the reluctance of the transformer core. Because the reluctance of a well-designed transformer core is very small (nearly zero) until the core is saturated, the relationship between the primary and secondary currents is approximately

$$\mathcal{F}_{net} = N_P i_P - N_S i_S \approx 0 \tag{3–35}$$

as long as the core is unsaturated. Therefore,

$$\boxed{N_P i_P \approx N_S i_S} \tag{3–36}$$

or
$$\boxed{\frac{i_P}{i_S} \approx \frac{N_S}{N_P} = \frac{1}{a}} \tag{3–37}$$

It is the fact that the magnetomotive force in the core is nearly zero that gives the dot convention the meaning described in Section 3.3. In order for the magnetomotive force to be nearly zero, current must flow *into one dotted end* and *out of the other dotted end.* The voltages must be built up in the same way with respect to the dots on each winding in order to drive the currents in the direction required. (The polarity of the voltages can also be determined by Lenz's law if the construction of the transformer coils is visible.)

What assumptions are required to convert a real transformer into the ideal transformer described previously? They are as follows:

1. The core must have no hysteresis or eddy currents.
2. The magnetization curve must have the shape shown in Figure 3–15. Notice that for an unsaturated core the net magnetomotive force $\mathcal{F}_{net} = 0$, implying that $N_P i_P = N_S i_S$.
3. The leakage flux in the core must be zero, implying that all the flux in the core couples both windings.
4. The resistance of the transformer windings must be zero.

While these conditions are never exactly met, well-designed power transformers can come quite close.

Figure 3–15 | The magnetization curve of an ideal transformer.

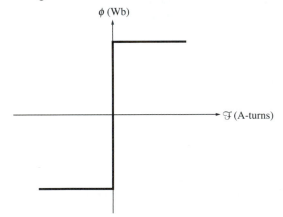

3.5 | THE EQUIVALENT CIRCUIT OF A TRANSFORMER

The losses that occur in real transformers have to be accounted for in any accurate model of transformer behavior. The major items to be considered in the construction of such a model are

1. *Copper (I^2R) losses.* Copper losses are the resistive heating losses *in the primary and secondary windings* of the transformer. They are proportional to the square of the current in the windings.
2. *Eddy current losses.* Eddy current losses are resistive heating losses *in the core* of the transformer. They are proportional to the square of the voltage applied to the transformer.

3. *Hysteresis losses.* Hysteresis losses are associated with the rearrangement of the magnetic domains in the core during each half-cycle, as explained in Chapter 1. They are a complex, nonlinear function of the voltage applied to the transformer.

4. *Leakage flux.* The fluxes ϕ_{LP} and ϕ_{LS} that escape the core and pass through only one of the transformer windings are leakage fluxes. These escaped fluxes produce a *self-inductance* in the primary and secondary coils, and the effects of this inductance must be accounted for.

The Exact Equivalent Circuit of a Real Transformer

It is possible to construct an equivalent circuit that takes into account all the major imperfections in real transformers. Each major imperfection is considered in turn, and its effect is included in the transformer model.

The easiest effect to model is the copper losses. Copper losses are resistive losses in the primary and secondary windings of the transformer core. They are modeled by placing a resistor R_P in the primary circuit of the transformer and a resistor R_S in the secondary circuit.

As explained in Section 3.4, the leakage flux in the primary windings ϕ_{LP} produces a voltage e_{LP} given by

$$e_{LP}(t) = N_P \frac{d\phi_{LP}}{dt} \tag{3–38}$$

and the leakage flux in the secondary windings ϕ_{LS} produces a voltage e_{LS} given by

$$e_{LS}(t) = N_S \frac{d\phi_{LS}}{dt} \tag{3–39}$$

Since much of the leakage flux path is through air, and since air has a *constant* reluctance much higher than the core reluctance, the flux ϕ_{LP} is directly proportional to the primary circuit current i_P and the flux ϕ_{LS} is directly proportional to the secondary current i_S:

$$\phi_{LP} = (\mathcal{P}N_P)i_P \tag{3–40}$$
$$\phi_{LS} = (\mathcal{P}N_S)i_S \tag{3–41}$$

where

$\mathcal{P}$ = permeance of flux path
N_P = number of turns on primary coil
N_S = number of turns on secondary coil

Substitute Equations (3–40) and (3–41) into Equations (3–38) and (3–39). The result is

$$e_{LP}(t) = N_P \frac{d}{dt}(\mathcal{P}N_P)i_P = N_P^2\mathcal{P}\frac{di_P}{dt} \tag{3–42}$$

$$e_{LS}(t) = N_S \frac{d}{dt}(\mathcal{P}N_S)i_S = N_S^2\mathcal{P}\frac{di_S}{dt} \tag{3–43}$$

The constants in these equations can be lumped together. Then

$$e_{LP}(t) = L_P \frac{di_P}{dt} \tag{3-44}$$

$$e_{LS}(t) = L_S \frac{di_S}{dt} \tag{3-45}$$

where $L_P = N_P^2 \mathcal{P}$ is the self-inductance of the primary coil and $L_S = N_S^2 \mathcal{P}$ is the self-inductance of the secondary coil. Therefore, the leakage flux will be modeled by primary and secondary inductors.

How can the core excitation effects be modeled? The magnetization current i_m is a current proportional (in the unsaturated region) to the voltage applied to the core and *lagging the applied voltage by* 90°, so it can be modeled by a reactance X_M connected across the primary voltage source. The core-loss current i_{h+e} is a current proportional to the voltage applied to the core that is *in phase with the applied voltage,* so it can be modeled by a resistance R_C connected across the primary voltage source. (Remember that both these currents are really nonlinear, so the inductance X_M and the resistance R_C are, at best, approximations of the real excitation effects.)

The resulting equivalent circuit is shown in Figure 3–16. Notice that the elements forming the excitation branch are placed inside the primary resistance R_P and the primary inductance L_P. This is because the voltage actually applied to the core is really equal to the input voltage less the internal voltage drops of the winding.

Figure 3–16 | The model of a real transformer.

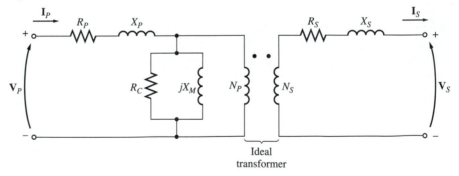

Although Figure 3–16 is an accurate model of a transformer, it is not a very useful one. To analyze practical circuits containing transformers, it is normally necessary to convert the entire circuit to an equivalent circuit at a single voltage level. (Such a conversion was done in Example 3–1.) Therefore, the equivalent circuit must be referred either to its primary side or to its secondary side in problem solutions. Figure 3–17a is the equivalent circuit of the transformer referred to its primary side, and Figure 3–17b is the equivalent circuit referred to its secondary side.

Figure 3–17 | (a) The transformer model referred to its primary voltage level.
(b) The transformer model referred to its secondary voltage level.

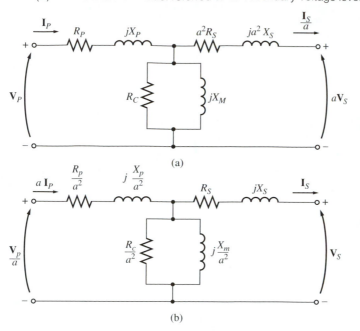

(a)

(b)

Approximate Equivalent Circuits of a Transformer

The transformer models shown above are often more complex than necessary in order to get good results in practical engineering applications. One of the principal complaints about them is that the excitation branch of the model adds another node to the circuit being analyzed, making the circuit solution more complex than necessary. The excitation branch has a very small current compared to the load current of the transformers. In fact, it is so small that under normal circumstances it causes a completely negligible voltage drop in R_P and X_P. Because this is true, a simplified equivalent circuit can be produced that works almost as well as the original model. The excitation branch is simply moved to the front of the transformer, and the primary and secondary impedances are left in series with each other. These impedances are just added, creating the approximate equivalent circuits in Figure 3–18a and b.

In some applications, the excitation branch may be neglected entirely without causing serious error. In these cases, the equivalent circuit of the transformer reduces to the simple circuits in Figure 3–18c and d.

Determining the Values of Components in the Transformer Model

It is possible to experimentally determine the values of the reactances and resistances in the transformer model. An adequate approximation of these values can be obtained with only two tests, the open-circuit test and the short-circuit test.

Figure 3–18 I Approximate transformer models: (a) Referred to the primary side; (b) referred to the secondary side; (c) with no excitation branch, referred to the primary side; (d) with no excitation branch, referred to the secondary side.

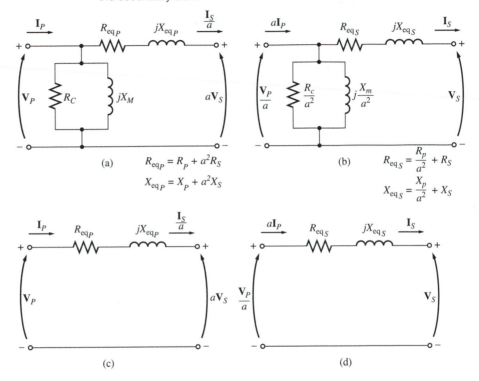

(a) $R_{eqP} = R_P + a^2 R_S$

$X_{eqP} = X_P + a^2 X_S$

(b) $R_{eqS} = \dfrac{R_P}{a^2} + R_S$

$X_{eqS} = \dfrac{X_P}{a^2} + X_S$

In the *open-circuit test,* a transformer's secondary winding is open-circuited, and its primary winding is connected to a full-rated line voltage. Look at the equivalent circuit in Figure 3–17. Under the conditions described, all the input current must be flowing through the excitation branch of the transformer. The series elements R_P and X_P are too small in comparison to R_C and X_M to cause a significant voltage drop, so essentially all the input voltage is dropped across the excitation branch.

The open-circuit test connections are shown in Figure 3–19. Full line voltage is applied to the primary of the transformer, and the input voltage, input current, and input power to the transformer are measured. From this information, it is possible to determine the power factor of the input current and therefore both the *magnitude* and the *angle* of the excitation impedance.

The easiest way to calculate the values of R_C and X_M is to look first at the *admittance* of the excitation branch. The conductance of the core-loss resistor is given by

$$G_C = \frac{1}{R_C} \tag{3–46}$$

Figure 3–19 | Connection for transformer open-circuit test.

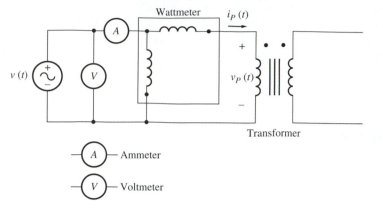

and the susceptance of the magnetizing inductor is given by

$$B_M = \frac{1}{X_M} \tag{3–47}$$

Since these two elements are in parallel, their admittances add, and the total excitation admittance is

$$Y_E = G_C - jB_M \tag{3–48}$$

$$Y_E = \frac{1}{R_C} - j\frac{1}{X_M} \tag{3–49}$$

The *magnitude* of the excitation admittance (referred to the primary circuit) can be found from the open-circuit test voltage and current:

$$|Y_E| = \frac{I_{OC}}{V_{OC}} \tag{3–50}$$

The *angle* of the admittance can be found from a knowledge of the circuit power factor. The open-circuit power factor (PF) is given by

$$\text{PF} = \cos\theta = \frac{P_{OC}}{V_{OC}I_{OC}} \tag{3–51}$$

and the power-factor angle θ is given by

$$\theta = \cos^{-1}\frac{P_{OC}}{V_{OC}I_{OC}} \tag{3–52}$$

The power factor is always lagging for a real transformer, so the angle of the current always lags the angle of the voltage by θ degrees. Therefore, the admittance Y_E is

$$Y_E = \frac{I_{OC}}{V_{OC}} \angle -\theta$$

$$Y_E = \frac{I_{OC}}{V_{OC}} \angle -\cos^{-1}\text{PF} \tag{3–53}$$

By comparing Equations (3–49 and (3–53), it is possible to determine the values of R_C and X_M directly from the open-circuit test data.

In the *short-circuit test,* the secondary terminals of the transformer are short-circuited, and the primary terminals are connected to a fairly low voltage source, as shown in Figure 3–20. The input voltage is adjusted until the current in the short-circuited windings is equal to its rated value. (Be sure to keep the primary voltage at a safe level. It would not be a good idea to burn out the transformer's windings while trying to test it.) The input voltage, current, and power are again measured.

Figure 3–20 | Connection for transformer short-circuit test.

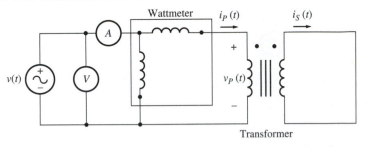

Since the input voltage is so low during the short-circuit test, negligible current flows through the excitation branch. If the excitation current is ignored, then all the voltage drop in the transformer can be attributed to the series elements in the circuit. The magnitude of the series impedances referred to the primary side of the transformer is

$$|Z_{SE}| = \frac{V_{SC}}{I_{SC}} \tag{3–54}$$

The power factor of the current is given by

$$PF = \cos\theta = \frac{P_{SC}}{V_{SC}I_{SC}} \tag{3–55}$$

and is lagging. The current angle is thus negative, and the overall impedance angle θ is positive:

$$\theta = \cos^{-1}\frac{P_{SC}}{V_{SC}I_{SC}} \tag{3–56}$$

Therefore,

$$Z_{SE} = \frac{V_{SC}\angle 0°}{I_{SC}\angle -\theta°} = \frac{V_{SC}}{I_{SC}} \angle \theta° \tag{3–57}$$

The series impedance Z_{SE} is equal to

$$Z_{SE} = R_{eq} + jX_{eq}$$
$$Z_{SE} = (R_P + a^2 R_S) + j(X_P + a^2 X_S) \tag{3–58}$$

It is possible to determine the total series impedance referred to the primary side by using this technique, but there is no easy way to split the series impedance into primary and secondary components. Fortunately, such separation is not necessary to solve normal problems.

Also these same tests may be performed on the *secondary* side of the transformer if it is more convenient to do so because of voltage levels or other reasons. If the tests are performed on the secondary side, the results will naturally yield the equivalent circuit impedances referred to the secondary side of the transformer instead of to the primary side.

EXAMPLE 3–2

The equivalent circuit impedances of a 20-kVA, 8000/240-V, 60-Hz transformer are to be determined. The open-circuit test and the short-circuit test were performed on the primary side of the transformer, and the following data were taken:

Open-circuit test (on primary)	Short-circuit test (on primary)
V_{OC} = 8000 V	V_{SC} = 489 V
I_{OC} = 0.214 A	I_{SC} = 2.5 A
V_{OC} = 400 W	P_{SC} = 240 W

Find the impedances of the approximate equivalent circuit referred to the primary side, and sketch that circuit.

Solution
The power factor during the *open-circuit* test is

$$PF = \cos\theta = \frac{P_{OC}}{V_{OC}I_{OC}} \tag{3–51}$$

$$= \frac{400\ W}{(8000\ V)(0.214\ A)}$$

$$= 0.234\ lagging$$

The excitation admittance is given by

$$Y_E = \frac{I_{OC}}{V_{OC}} \angle -\cos^{-1} PF \tag{3–53}$$

$$= \frac{0.214\ A}{8000\ V}\ V \angle -\cos^{-1} 0.234$$

$$= 0.0000268 \angle -76.5° \text{ S}$$

$$= 0.0000063 - j0.0000261 = \frac{1}{R_C} - j\frac{1}{X_M}$$

Therefore,

$$R_C = \frac{1}{0.0000063} = 159 \text{ k}\Omega$$

$$X_M = \frac{1}{0.0000261} = 38.4 \text{ k}\Omega$$

The power factor during the *short-circuit* test is

$$PF = \cos\theta = \frac{P_{SC}}{V_{SC}I_{SC}} \qquad (3\text{-}55)$$

$$= \frac{240 \text{ W}}{(489 \text{ V})(2.5 \text{ A})} = 0.196 \text{ lagging}$$

The series impedance is given by

$$Z_{SE} = \frac{V_{SC}}{I_{SC}} \angle \cos^{-1} PF$$

$$Z_{SE} = \frac{489 \text{ V}}{2.5 \text{ A}} \angle 78.7°$$

$$Z_{SE} = 195.6\angle 78.7° = 38.3 + j192 \; \Omega$$

Therefore, the equivalent resistance and reactance are

$$R_{eq} = 38.3 \; \Omega \qquad X_{eq} = 192 \; \Omega$$

The resulting simplified equivalent circuit referred to the primary side is shown in Figure 3–21. ∎

Figure 3–21 | The equivalent circuit of Example 3–2, referred to the primary side of the transformer.

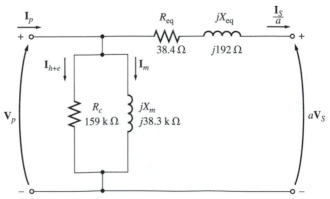

3.6 | THE PER-UNIT SYSTEM

As the relatively simple Example 3–1 showed, solving circuits containing transformers can be quite a tedious operation because of the need to refer all the different voltage levels on different sides of the transformers in the system to a common level. Only after this step has been taken can the system be solved for its voltages and currents.

There is another approach to solving circuits containing transformers which eliminates the need for explicit voltage-level conversions at every transformer in the system. Instead, the required conversions are handled automatically by the method itself, without ever requiring the user to worry about impedance transformations. Because such impedance transformations can be avoided, circuits containing many transformers can be solved easily with less chance of error. This method of calculation is known as the *per-unit (pu) system.*

There is yet another advantage to the per-unit system that is quite significant for electric machinery and transformers. As the size of a machine or transformer varies, its internal impedances vary widely. Thus, a primary circuit reactance of $0.1 \ \Omega$ might be an atrociously high number for one transformer and a ridiculously low number for another—it all depends on the device's voltage and power ratings. However, it turns out that in a per-unit system related to the device's ratings, *machine and transformer impedances fall within fairly narrow ranges* for each type and construction of device. This fact can serve as a useful check in problem solutions.

In the per-unit system, the voltages, currents, powers, impedances, and other electrical quantities are not measured in their usual SI units (volts, amperes, watts, ohms, etc.). Instead, *each electrical quantity is measured as a decimal fraction* of some base level. Any quantity can be expressed on a per-unit basis by the equation

$$\text{Quantity per unit} = \frac{\text{actual value}}{\text{base value of quantity}} \qquad (3\text{--}59)$$

where "actual quantity" is a value in volts, amperes, ohms, etc.

It is customary to select two base quantities to define a given per-unit system. The ones usually selected are voltage and power (or apparent power). Once these base quantities have been selected, all the other base values are related to them by the usual electrical laws. In a single-phase system, these relationships are

$$P_{\text{base}}, Q_{\text{base}}, \text{ or } S_{\text{base}} = V_{\text{base}} I_{\text{base}} \qquad (3\text{--}60)$$

$$Z_{\text{base}} = \frac{V_{\text{base}}}{I_{\text{base}}} \qquad (3\text{--}61)$$

$$Y_{\text{base}} = \frac{I_{\text{base}}}{V_{\text{base}}} \qquad (3\text{--}62)$$

and

$$Z_{\text{base}} = \frac{(V_{\text{base}})^2}{S_{\text{base}}} \qquad (3\text{--}63)$$

Once the base values of S (or P) and V have been selected, all other base values can be computed easily from Equations (3–60) to (3–63).

In a power system, a base apparent power and voltage are selected *at a specific point in the system.* A transformer has no effect on the base apparent power of the system, since the apparent power into a transformer equals the apparent power out of the transformer [Equation (3–11)], so *the base apparent power remains constant everywhere in the power system.* On the other hand, voltage changes when it goes through a transformer, so the value of V_{base} changes at every transformer in the system according to its turns ratio. Because the *base quantities* change in passing through a transformer, the process of referring quantities to a common voltage level is automatically taken care of during per-unit conversion.

EXAMPLE 3–3

A simple power system is shown in Figure 3–22. This system contains a 480-V generator connected to an ideal 1:10 step-up transformer, a transmission line, an ideal 20:1 step-down transformer, and a load. The impedance of the transmission line is $20 + j60\ \Omega$, and the impedance of the load is $10\angle30°\ \Omega$. The base values for this system are chosen to be 480 V and 10 kVA at the generator.

a.   Find the base voltage, current, impedance, and apparent power at every point in the power system.

b.   Convert this system to its per-unit equivalent circuit.

c.   Find the power supplied to the load in this system.

d.   Find the power lost in the transmission line.

Figure 3–22 | The power system of Example 3–3.

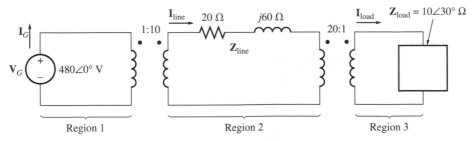

Region 1   Region 2   Region 3

Solution

a.   *In the generator region, $V_{\text{base 1}}$ = 480 V and S_{base} = 10 kVA, so*

$$I_{\text{base 1}} = \frac{S_{\text{base}}}{V_{\text{base 1}}} = \frac{10{,}000\ \text{VA}}{480\ \text{V}} = 20.83\ \text{A}$$

$$Z_{\text{base 1}} = \frac{V_{\text{base 1}}}{I_{\text{base 1}}} = \frac{480\ \text{V}}{20.83\ \text{A}} = 23.04\ \Omega$$

The turns ratio of transformer T_1 is a = 1/10 = 0.1, so the base voltage *in the transmission line region* is

$$V_{\text{base 2}} = \frac{V_{\text{base 1}}}{a} = \frac{480\ \text{V}}{0.1} = 4800\ \text{V}$$

The other base quantities are

$$S_{base\,2} = 10 \text{ kVA}$$

$$I_{base\,2} = \frac{10,000 \text{ VA}}{4800 \text{ V}} = 2.083 \text{ A}$$

$$Z_{base\,2} = \frac{4800 \text{ V}}{2.083 \text{ A}} = 2304 \text{ } \Omega$$

The turns ratio of transformer T_2 is $a = 20/1 = 20$, so the base voltage *in the load region* is

$$V_{base\,3} = \frac{V_{base\,2}}{a} = \frac{4800 \text{ V}}{20} = 240 \text{ V}$$

The other base quantities are

$$S_{base\,3} = 10 \text{ kVA}$$

$$I_{base\,3} = \frac{10,000 \text{ VA}}{240 \text{ V}} = 41.67 \text{ A}$$

$$Z_{base\,3} = \frac{240 \text{ V}}{41.67 \text{ A}} = 5.76 \text{ } \Omega$$

b. To convert a power system to a per-unit system, each component must be divided by its base value in its region of the system. The *generator's* per-unit voltage is its actual value divided by its base value:

$$V_{G,\,pu} = \frac{480\angle 0° \text{ V}}{480 \text{ V}} = 1.0\angle 0° \text{ pu}$$

The *transmission line's* per-unit impedance is its actual value divided by its base value:

$$\mathbf{Z}_{line,\,pu} = \frac{20 + j60 \text{ } \Omega}{2304 \text{ } \Omega} = 0.0087 + j0.0260 \text{ pu}$$

The *load's* per-unit impedance is also given by actual value divided by its base value:

$$\mathbf{Z}_{load,\,pu} = \frac{10\angle 30° \text{ } \Omega}{5.76 \text{ } \Omega} = 1.736\angle 30° \text{ pu}$$

The per-unit equivalent circuit of the power system is shown in Figure 3–23.

c. The current flowing in this per-unit power system is

$$\mathbf{I}_{pu} = \frac{\mathbf{V}_{pu}}{\mathbf{Z}_{tot,\,pu}}$$

$$\mathbf{I}_{pu} = \frac{1\angle 0°}{(0.0087 + j0.0260) + (1.736\angle 30°)}$$

$$= \frac{1\angle 0°}{(0.0087 + j0.0260) + (1.503 + j0.868)}$$

$$= \frac{1\angle 0°}{1.512 + j0.894} = \frac{1\angle 0°}{1.757\angle 30.6°}$$

$$= 0.569\angle -30.6° \text{ pu}$$

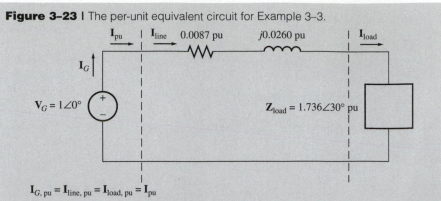

Figure 3–23 | The per-unit equivalent circuit for Example 3–3.

$$\mathbf{I}_{G,\,pu} = \mathbf{I}_{line,\,pu} = \mathbf{I}_{load,\,pu} = \mathbf{I}_{pu}$$

Therefore, the per-unit power of the load is

$$P_{load,\,pu} = I_{pu}^2 R_{pu} = (0.569)^2(1.503) = 0.487$$

and the actual power supplied to the load is

$$P_{load} = P_{load,\,pu}S_{base} = (0.487)(10{,}000\text{ VA})$$
$$= 4870\text{ W}$$

d.   The per-unit power lost in the transmission line is

$$P_{line,pu} = I_{pu}^2 R_{line,\,pu} = (0.569)^2(0.0087) = 0.00282$$

and the actual power lost in the transmission line is

$$P_{line} = P_{line,\,pu}S_{base} = (0.00282)(10{,}000\text{ VA})$$
$$= 28.2\text{ W}$$

When only one device (transformer or motor) is being analyzed, its own ratings are usually used as the base for the per-unit system. If a per-unit system based on the transformer's own ratings is used, a power or distribution transformer's characteristics will not vary much over a wide range of voltage and power ratings. For example, the series resistance of a transformer is usually about 0.01 per unit, and the series reactance is usually between 0.02 and 0.10 per unit. In general, the larger the transformer, the smaller the series impedances. The magnetizing reactance is usually between about 10 and 40 per unit, while the core-loss resistance is usually between about 50 and 200 per unit. Because per-unit values provide a convenient and meaningful way to compare transformer characteristics when they are of different sizes, transformer impedances are normally given in per-unit or as a percentage on the transformer's nameplate (a percentage impedance is the per-unit impedance multiplied by 100; see Figure 3–42, later in this chapter).

The same idea applies to synchronous and induction machines as well: Their per-unit impedances fall within relatively narrow ranges over quite large size ranges.

If more than one machine and one transformer are included in a single power system, the system base voltage and power may be chosen arbitrarily. However, the *entire system must have the same base power, and the base voltages at various points in the system must be related by the voltage ratios of the transformers.* One common procedure is to choose the system base quantities to be equal to the base of the largest component in the system. Per-unit values given to another base can be converted to the new base by converting them to their actual values (volts, amperes, ohms, etc.) as an in-between step. Alternatively, they can be converted directly by the equations

$$(P, Q, S)_{\text{pu on base 2}} = (P, Q, S)_{\text{pu on base 1}} \frac{S_{\text{base 1}}}{S_{\text{base 2}}} \tag{3–64}$$

$$V_{\text{pu on base 2}} = V_{\text{pu on base 1}} \frac{V_{\text{base 1}}}{V_{\text{base 2}}} \tag{3–65}$$

$$(R, X, Z)_{\text{pu on base 2}} = (R, X, Z)_{\text{pu on base 1}} \frac{(V_{\text{base 1}})^2 (S_{\text{base 2}})}{(V_{\text{base 2}})^2 (S_{\text{base 1}})} \tag{3–66}$$

Figure 3–24 | (a) A typical 13.2-kV to 120/240-V distribution transformer. *(Courtesy of General Electric Company.)* (b) A cutaway view of the distribution transformer showing the shell-form transformer inside it. *(Courtesy of General Electric Company.)*

EXAMPLE 3–4

Sketch the approximate per-unit equivalent circuit for the transformer in Example 3–2. Use the transformer's ratings as the system base.

Solution
The transformer in Example 3–2 is rated at 20 kVA, 8000/240 V. The approximate equivalent circuit (Figure 3–21) developed in the example was referred to the high-voltage side of the transformer, so to convert it to per-unit, the primary circuit base impedance must be found. On the primary,

$$V_{\text{base 1}} = 8000 \text{ V}$$
$$S_{\text{base 1}} = 20,000 \text{ VA}$$
$$Z_{\text{base 1}} = \frac{(V_{\text{base 1}})^2}{S_{\text{base 1}}} = \frac{(8000 \text{ V})^2}{20,000 \text{ VA}} = 3200 \text{ } \Omega$$

Therefore,

$$Z_{\text{SE, pu}} = \frac{38.4 + j192 \text{ } \Omega}{3200 \text{ } \Omega} = 0.012 + j0.06 \text{ pu}$$
$$R_{C, \text{pu}} = \frac{159 \text{ k}\Omega}{3200 \text{ } \Omega} = 49.7 \text{ pu}$$
$$X_{m, \text{pu}} = \frac{38.4 \text{ k}\Omega}{3200 \text{ } \Omega} = 12 \text{ pu}$$

The per-unit approximate equivalent circuit, expressed to the transformer's own base, is shown in Figure 3–25. ∎

Figure 3–25 | The per–unit equivalent circuit of Example 3–4.

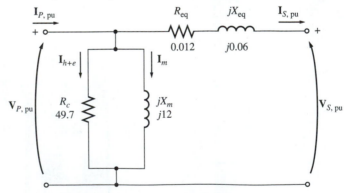

3.7 | TRANSFORMER VOLTAGE REGULATION AND EFFICIENCY

Because a real transformer has series impedances within it, the output voltage of a transformer varies with the load even if the input voltage remains constant. To conveniently compare transformers in this respect, it is customary to define a quantity called *voltage regulation* (VR). *Full-load voltage regulation* is a quantity that compares the output voltage of the transformer at no load with the output voltage at full load. It is defined by the equation

$$\text{VR} = \frac{V_{S,\,\text{nl}} - V_{S,\,\text{fl}}}{V_{S,\,\text{fl}}} \times 100\% \tag{3–67}$$

Since at no load, $V_S = V_P/a$, the voltage regulation can also be expressed as

$$\text{VR} = \frac{V_P/a - V_{S,\,\text{fl}}}{V_{S,\,\text{fl}}} \times 100\% \tag{3–68}$$

If the transformer equivalent circuit is in the per-unit system, then voltage regulation can be expressed as

$$\text{VR} = \frac{V_{P,\,\text{pu}} - V_{S,\,\text{fl,\,pu}}}{V_{S,\,\text{fl,\,pu}}} \times 100\% \tag{3–69}$$

Usually it is a good practice to have as small a voltage regulation as possible. For an ideal transformer, VR = 0 percent. It is not always a good idea to have a low-voltage regulation, though—sometimes high-impedance and high-voltage-regulation transformers are deliberately used to reduce the fault currents in a circuit.

How can the voltage regulation of a transformer be determined?

The Transformer Phasor Diagram

To determine the voltage regulation of a transformer, it is necessary to understand the voltage drops within it. Consider the simplified transformer equivalent circuit in Figure 3–18b. The effects of the excitation branch on transformer voltage regulation can be ignored, so only the series impedances need be considered. The voltage regulation of a transformer depends both on the magnitude of these series impedances and on the phase angle of the current flowing through the transformer. The easiest way to determine the effect of the impedances and the current phase angles on the transformer voltage regulation is to examine a *phasor diagram*, a sketch of the phasor voltages and currents in the transformer.

In all the following phasor diagrams, the phasor voltage $\mathbf{V}_S$ is assumed to be at an angle of 0°, and all other voltages and currents are compared to that reference. By applying Kirchhoff's voltage law to the equivalent circuit in Figure 3–18b, the primary voltage can be found as

$$\frac{\mathbf{V}_P}{a} = \mathbf{V}_S + R_{\text{eq}}\mathbf{I}_S + jX_{\text{eq}}\mathbf{I}_S \tag{3–70}$$

A transformer phasor diagram is just a visual representation of this equation.

Figure 3–26 shows a phasor diagram of a transformer operating at a lagging power factor. It is easy to see that $V_P/a > V_S$ for lagging loads, so the voltage regulation of a transformer with lagging loads must be greater than zero.

Figure 3–26 I Phasor diagram of a transformer operating at a lagging power factor.

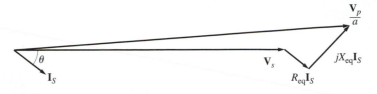

A phasor diagram at unity power factor is shown in Figure 3–27a. Here again, the voltage at the secondary is lower than the voltage at the primary, so VR > 0. However, this time the voltage regulation is a smaller number than it was with a lagging current. If the secondary current is leading, the secondary voltage can actually be *higher* than the referred primary voltage. If this happens, the transformer actually has a *negative* voltage regulation (see Figure 3–27b).

Figure 3–27 I Phasor diagram of a transformer operating at (a) unity and (b) leading power factor.

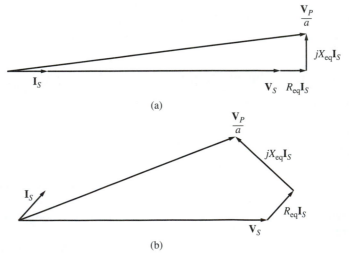

Transformer Efficiency

Transformers are also compared and judged on their efficiencies. The efficiency of a device is defined by the equation

$$\eta = \frac{P_{out}}{P_{in}} \times 100\% \tag{3-71}$$

$$\eta = \frac{P_{out}}{P_{out} + P_{loss}} \times 100\% \tag{3-72}$$

These equations apply to motors and generators as well as to transformers.

The transformer equivalent circuits make efficiency calculations easy. There are three types of losses present in transformers:

1. *Copper (I^2R) losses.* These losses are accounted for by the series resistance in the equivalent circuit.
2. *Hysteresis losses.* These losses were explained in Chapter 1 and are accounted for by resistor R_C.
3. *Eddy current losses.* These losses were explained in Chapter 1 and are accounted for by resistor R_C.

To calculate the efficiency of a transformer at a given load, just add the losses from each resistor and apply Equation (3–72). Since the output power is given by

$$P_{out} = V_S I_S \cos \theta_S \tag{3-73}$$

the efficiency of the transformer can be expressed by

$$\eta = \frac{V_S I_S \cos \theta}{P_{Cu} + P_{core} + V_S I_S \cos \theta} \times 100\% \tag{3-74}$$

EXAMPLE 3–5

A 15-kVA, 2300/230-V transformer is to be tested to determine its excitation branch components, its series impedances, and its voltage regulation. The following test data have been taken from the primary side of the transformer:

Open-circuit test	Short-circuit test
$V_{OC} = 2300$ V	$V_{SC} = 47$ V
$I_{OC} = 0.21$ A	$I_{SC} = 6.0$ A
$P_{OC} = 50$ W	$P_{SC} = 160$ W

The data have been taken using the connections shown in Figure 3–19 and Figure 3–20.

a. Find the equivalent circuit of this transformer referred to the high-voltage side.
b. Find the equivalent circuit of this transformer referred to the low-voltage side.
c. Calculate the full-load voltage regulation at 0.8 lagging power factor, 1.0 power factor, and at 0.8 leading power factor.

d. Plot the voltage regulation as load is increased from no load to full load at power factors of 0.8 lagging, 1.0, and 0.8 leading.

e. What is the efficiency of the transformer at full load with a power factor of 0.8 lagging?

Solution

a. The excitation branch values of the transformer equivalent circuit can be calculated from the *open-circuit test* data, and the series elements can be calculated from the *short-circuit test* data. From the open-circuit test data, the open-circuit impedance angle is

$$\theta_{OC} = \cos^{-1} \frac{P_{OC}}{V_{OC} I_{OC}}$$

$$\theta_{OC} = \cos^{-1} \frac{50 \text{ W}}{(2300 \text{ V})(0.21 \text{ A})} = 84°$$

The excitation admittance is thus

$$Y_E = \frac{I_{OC}}{V_{OC}} \angle -84°$$

$$Y_E = \frac{0.21 \text{ A}}{2300 \text{ V}} \angle -84°$$

$$Y_E = 9.13 \times 10^{-5} \angle -84° \text{ S} = 0.0000095 - j0.0000908 \text{ S}$$

The elements of the excitation branch referred to the primary are

$$R_C = \frac{1}{0.0000095} = 105 \text{ k}\Omega$$

$$X_M = \frac{1}{0.0000908} = 11 \text{ k}\Omega$$

From the short-circuit test data, the short-circuit impedance angle is

$$\theta_{SC} = \cos^{-1} \frac{P_{SC}}{V_{SC} I_{SC}}$$

$$\theta_{SC} = \cos^{-1} \frac{160 \text{ W}}{(47 \text{ V})(6 \text{ A})} = 55.4°$$

The equivalent series impedance is thus

$$\mathbf{Z}_{SE} = \frac{V_{SC}}{I_{SC}} \angle \theta_{SC}$$

$$\mathbf{Z}_{SE} = \frac{47 \text{ V}}{6 \text{ A}} \angle 55.4° \; \Omega$$

$$\mathbf{Z}_{SE} = 7.833 \angle 55.4° = 4.45 + j6.45 \; \Omega$$

The series elements referred to the primary are

$$R_{eq} = 4.45 \; \Omega \qquad X_{eq} = 6.45 \; \Omega$$

This equivalent circuit is shown in Figure 3–28a.

b. To find the equivalent circuit referred to the low-voltage side, it is simply necessary to divide the impedance by a^2. Since $a = N_P/N_S = 10$, the resulting values are

$$R_C = 1050 \ \Omega \qquad R_{eq} = 0.0445 \ \Omega$$
$$X_M = 110 \ \Omega \qquad X_{eq} = 0.0645 \ \Omega$$

The resulting equivalent circuit is shown in Figure 3–28b.

Figure 3–28 | The transformer equivalent circuit for Example 3–5 referred to (a) its primary side and (b) its secondary side.

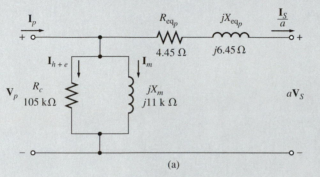

(a)

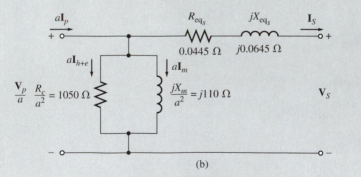

(b)

c. The full-load current on the secondary side of this transformer is

$$I_{S,\ rated} = \frac{S_{rated}}{V_{S,\ rated}} = \frac{15{,}000 \ \text{VA}}{230 \ \text{V}} = 65.2 \ \text{A}$$

To calculate V_P/a, use Equation (3–70):

$$\frac{\mathbf{V}_P}{a} = \mathbf{V}_S + R_{eq}\mathbf{I}_S + jX_{eq}\mathbf{I}_S \qquad (3\text{–}70)$$

At PF = 0.8 lagging, current $\mathbf{I}_S = 65.2\angle -36.9°$ A. Therefore,

$$\frac{\mathbf{V}_P}{a} = 230\angle 0° \text{ V} + (0.0445 \ \Omega)(65.2\angle -36.9° \text{ A}) + j(0.0645 \ \Omega)(65.2\angle -36.9° \text{ A})$$
$$= 230\angle 0° \text{ V} + 2.90\angle -36.9° \text{ V} + 4.21\angle 53.1° \text{ V}$$
$$= 230 + 2.32 - j1.74 + 2.52 + j3.36$$
$$= 234.84 + j1.62 = 234.85\angle 0.40° \text{ V}$$

The resulting voltage regulation is

$$\text{VR} = \frac{V_P/a - V_{S,\text{fl}}}{V_{S,\text{fl}}} \times 100\% \tag{3-68}$$

$$= \frac{234.85 \text{ V} - 230 \text{ V}}{230 \text{ V}} \times 100\% = 2.1\%$$

At PF = 1.0, current $\mathbf{I}_S = 65.2\angle 0°$ A. Therefore,

$$\frac{\mathbf{V}_P}{a} = 230\angle 0° \text{ V} + (0.0445 \ \Omega)(65.2\angle 0° \text{ A}) + j(0.0645 \ \Omega)(65.2\angle 0° \text{ A})$$
$$= 230\angle 0° \text{ V} + 2.90\angle 0° \text{ V} + 4.21\angle 90° \text{ V}$$
$$= 230 + 2.90 - j4.21$$
$$= 232.9 + j4.21 = 232.94\angle 1.04° \text{ V}$$

The resulting voltage regulation is

$$\text{VR} = \frac{232.94 \text{ V} - 230 \text{ V}}{230 \text{ V}} \times 100\% = 1.28\%$$

At PF = 0.8 leading, current $\mathbf{I}_S = 65.2\angle 36.9°$ A. Therefore,

$$\frac{\mathbf{V}_P}{a} = 230\angle 0° \text{ V} + (0.0445 \ \Omega)(65.2\angle 36.9° \text{ A}) + j(0.0645 \ \Omega)(65.2\angle 36.9° \text{ A})$$
$$= 230\angle 0° \text{ V} + 2.90\angle 36.9° \text{ V} + 4.21\angle 126.9° \text{ V}$$
$$= 230 + 2.32 + j1.74 - 2.52 + j3.36$$
$$= 229.80 + j5.10 = 229.85\angle 1.27° \text{ V}$$

The resulting voltage regulation is

$$\text{VR} = \frac{229.85 \text{ V} - 230 \text{ V}}{230 \text{ V}} \times 100\% = -0.062\%$$

Each of these three phasor diagrams is shown in Figure 3–29.

d.   The best way to plot the voltage regulation as a function of load is to repeat the calculations in part c for many different loads using MATLAB. A program to do this is shown below.

```
% M-file: trans_vr.m
% M-file to calculate and plot the voltage regulation
% of a transformer as a function of load for power
% factors of 0.8 lagging, 1.0, and 0.8 leading.
```

Figure 3–29 | Transformer phasor diagrams for Example 3–5.

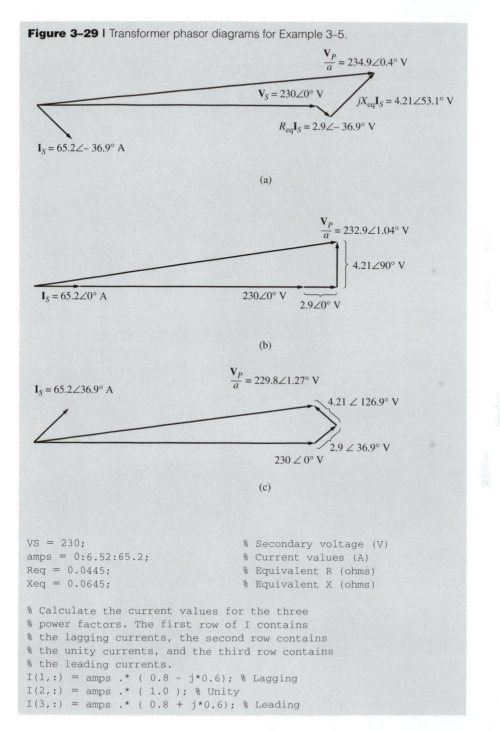

(a)

(b)

(c)

```
VS = 230;                        % Secondary voltage (V)
amps = 0:6.52:65.2;              % Current values (A)
Req = 0.0445;                    % Equivalent R (ohms)
Xeq = 0.0645;                    % Equivalent X (ohms)

% Calculate the current values for the three
% power factors. The first row of I contains
% the lagging currents, the second row contains
% the unity currents, and the third row contains
% the leading currents.
I(1,:) = amps .* ( 0.8 - j*0.6); % Lagging
I(2,:) = amps .* ( 1.0 ); % Unity
I(3,:) = amps .* ( 0.8 + j*0.6); % Leading
```

```
% Calculate VP/a.
VPa = VS + Req.*I + j.*Xeq.*I;
% Calculate voltage regulation
VR = (abs(VPa) - VS) ./ VS .* 100;

% Plot the voltage regulation
plot(amps,VR(1,:),'b-');
hold on;
plot(amps,VR(2,:),'k--');
plot(amps,VR(3,:),'r-.');
title ('Voltage Regulation Versus Load');
xlabel ('Load (A)');
ylabel ('Voltage Regulation (%)');
legend('0.8 PF lagging','1.0 PF','0.8 PF leading');
hold off;
```

The plot produced by this program is shown in Figure 3–30.

Figure 3–30 | Plot of voltage regulation versus load for the transformer of Example 3–5.

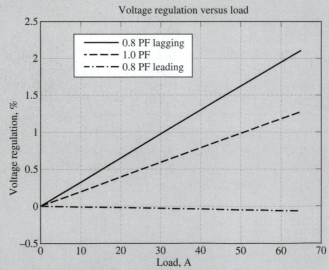

e.   To find the efficiency of the transformer, first calculate its losses. The copper losses are

$$P_{Cu} = (I_S)^2 \, R_{eq} = (65.2 \text{ A})^2 (0.0445 \ \Omega) = 189 \text{ W}$$

The core losses are given by

$$P_{core} = \frac{(V_P/a)^2}{R_C} = \frac{(234.85 \text{ V})^2}{1050 \ \Omega} = 52.5 \text{ W}$$

The output power of the transformer at this power factor is

$$P_{out} = V_S I_S \cos \theta$$
$$= (230 \text{ V})(65.2 \text{ A}) \cos 36.9° = 12{,}000 \text{ W}$$

Therefore, the efficiency of the transformer at this condition is

$$\eta = \frac{V_S I_S \cos \theta}{P_{Cu} + P_{core} + V_S I_S \cos \theta} \times 100\% \qquad (3\text{--}74)$$
$$= \frac{12{,}000 \text{ W}}{189 \text{ W} + 525 \text{ W} + 12{,}000 \text{ W}} \times 100\%$$
$$= 98.03\% \qquad \blacksquare$$

3.8 | TRANSFORMER TAPS AND VOLTAGE REGULATION

In previous sections of this chapter, transformers were described by their turns ratios or by their primary-to-secondary-voltage ratios. Throughout those sections, the turns ratio of a given transformer was treated as though it were completely fixed. In almost all real distribution transformers, this is not quite true. Distribution transformers have a series of *taps* in the windings to permit small changes in the turns ratio of the transformer after it has left the factory. A typical installation might have four taps in addition to the nominal setting with spacings of 2.5 percent of full-load voltage between them. Such an arrangement provides for adjustments up to 5 percent above or below the nominal voltage rating of the transformer.

EXAMPLE 3–6

A 500-kVA, 13,200/480-V distribution transformer has four 2.5 percent taps on its primary winding. What are the voltage ratios of this transformer at each tap setting?

Solution

The five possible voltage ratings of this transformer are

+5.0% tap	13,860/480 V
+2.5% tap	13,530/480 V
Nominal rating	13,200/480 V
−2.5% tap	12,870/480 V
−5.0% tap	12,540/480 V

$\blacksquare$

The taps on a transformer permit the transformer to be adjusted in the field to accommodate variations in local voltages. However, these taps normally cannot be changed while power is being applied to the transformer. They must be set once and left alone.

Sometimes a transformer is used on a power line whose voltage varies widely with the load. Such voltage variations might be due to a high line impedance between the generators on the power system and that particular load (perhaps it is located far out in the country). Normal loads need to be supplied an essentially constant voltage. How can a power company supply a controlled voltage through high-impedance lines to loads which are constantly changing?

One solution to this problem is to use a special transformer called a *tap changing under load* (TCUL) *transformer* or *voltage regulator.* Basically, a TCUL transformer is a transformer with the ability to change taps while power is connected to it. A voltage regulator is a TCUL transformer with built-in voltage sensing circuitry that automatically changes taps to keep the system voltage constant. Such special transformers are very common in modern power systems.

3.9 | THE AUTOTRANSFORMER

On some occasions it is desirable to change voltage levels by only a small amount. For example, it may be necessary to increase a voltage from 110 to 120 V or from 13.2 to 13.8 kV. These small rises may be made necessary by voltage drops that occur in power systems a long way from the generators. In such circumstances, it is wasteful and excessively expensive to wind a transformer with two full windings, each rated at about the same voltage. A special-purpose transformer, called an *autotransformer,* is used instead.

A diagram of a step-up autotransformer is shown in Figure 3–31. In Figure 3–31a, the two coils of the transformer are shown in the conventional manner. In Figure 3–31b, the first winding is shown connected in an additive manner to the second winding. Now, the relationship between the voltage on the first winding and the voltage on the second winding is given by the turns ratio of the transformer. However, *the voltage at the output of the whole transformer is the sum of the voltage on the first winding and the voltage on the second winding.* The first winding here is called the *common winding*, because its voltage appears on both sides of the transformer. The smaller winding is called the *series winding*, because it is connected in series with the common winding.

A diagram of a step-down autotransformer is shown in Figure 3–32. Here the voltage at the input is the sum of the voltages on the series winding and the common winding, while the voltage at the output is just the voltage on the common winding.

Because the transformer coils are physically connected, a different terminology is used for the autotransformer than for other types of transformers. The voltage on the common coil is called the *common voltage* V_C, and the current in that coil is called the *common current* I_C. The voltage on the series coil is called the *series voltage* V_{SE}, and the current in that coil is called the *series current* I_{SE}. The voltage and current on the low-voltage side of the transformer are called V_L and I_L, respectively, while the corresponding quantities on the high-voltage side of the transformer are

Figure 3–31 | A transformer with its windings (a) connected in the conventional manner and (b) reconnected as an autotransformer.

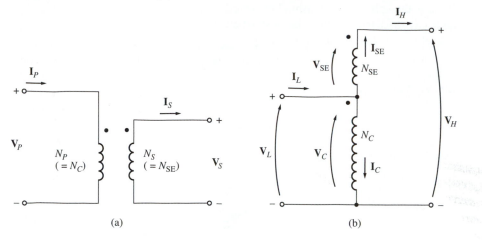

(a)   (b)

Figure 3–32 | A step-down autotransformer connection.

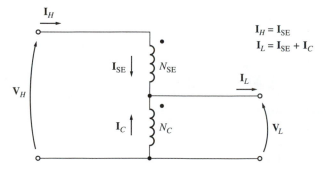

$$I_H = I_{SE}$$
$$I_L = I_{SE} + I_C$$

called $\mathbf{V}_H$ and $\mathbf{I}_H$. The primary side of the autotransformer (the side with power into it) can be either the high-voltage side or the low-voltage side, depending on whether the autotransformer is acting as a step-down or a step-up transformer. From Figure 3–31b the voltages and currents in the coils are related by the equations

$$\frac{\mathbf{V}_C}{\mathbf{V}_{SE}} = \frac{N_C}{N_{SE}} \tag{3–75}$$

$$N_C \mathbf{I}_C = N_{SE} \mathbf{I}_{SE} \tag{3–76}$$

The voltages in the coils are related to the voltages at the terminals by the equations

$$\mathbf{V}_L = \mathbf{V}_C \tag{3–77}$$
$$\mathbf{V}_H = \mathbf{V}_C + \mathbf{V}_{SE} \tag{3–78}$$

and the currents in the coils are related to the currents at the terminals by the equations

$$\mathbf{I}_L = \mathbf{I}_C + \mathbf{I}_{SE} \tag{3–79}$$
$$\mathbf{I}_H = \mathbf{I}_{SE} \tag{3–80}$$

Voltage and Current Relationships in an Autotransformer

What is the voltage relationship between the two sides of an autotransformer? It is quite easy to determine the relationship between $\mathbf{V}_H$ and $\mathbf{V}_L$. The voltage on the high side of the autotransformer is given by

$$\mathbf{V}_H = \mathbf{V}_C + \mathbf{V}_{SE} \tag{3-78}$$

But $\mathbf{V}_C/\mathbf{V}_{SE} = N_C/N_{SE}$, so

$$\mathbf{V}_H = \mathbf{V}_C + \frac{N_{SE}}{N_C}\mathbf{V}_C \tag{3-81}$$

Finally, noting that $\mathbf{V}_L = \mathbf{V}_C$, we get

$$\mathbf{V}_H = \mathbf{V}_L + \frac{N_{SE}}{N_C}\mathbf{V}_L$$
$$= \frac{N_{SE} + N_C}{N_C}\mathbf{V}_L \tag{3-82}$$

or

$$\boxed{\frac{\mathbf{V}_L}{\mathbf{V}_H} = \frac{N_C}{N_{SE} + N_C}} \tag{3-83}$$

The current relationship between the two sides of the transformer can be found by noting that

$$\mathbf{I}_L = \mathbf{I}_C + \mathbf{I}_{SE} \tag{3-79}$$

From Equation (3–76), $\mathbf{I}_C = (N_{SE}/N_C)\mathbf{I}_{SE}$, so

$$\mathbf{I}_L = N_{SE}/N_C\,\mathbf{I}_{SE} + \mathbf{I}_{SE} \tag{3-84}$$

Finally, noting that $\mathbf{I}_H = \mathbf{I}_{SE}$, we find

$$\mathbf{I}_L = \frac{N_{SE}}{N_C}\mathbf{I}_H + \mathbf{I}_H$$
$$= \frac{N_{SE} + N_C}{N_C}\mathbf{I}_H \tag{3-85}$$

or

$$\boxed{\frac{\mathbf{I}_L}{\mathbf{I}_H} = \frac{N_{SE} + N_C}{N_C}} \tag{3-86}$$

The Apparent Power Rating Advantage of Autotransformers

It is interesting to note that not all the power traveling from the primary to the secondary in the autotransformer goes through the windings. As a result, if a conventional transformer is reconnected as an autotransformer, it can handle much more power than it was originally rated for.

To understand this idea, refer again to Figure 3–31b. Notice that the input apparent power to the autotransformer is given by

$$S_{in} = V_L I_L \tag{3–87}$$

and the output apparent power is given by

$$S_{out} = V_H I_H \tag{3–88}$$

It is easy to show, by using the voltage and current equations [Equations (3–83) and (3–86)], that the input apparent power is again equal to the output apparent power:

$$S_{in} = S_{out} = S_{IO} \tag{3–89}$$

where S_{IO} is defined to be the input and output apparent powers of the transformer. However, *the apparent power in the transformer windings is*

$$S_W = V_C I_C = V_{SE} I_{SE} \tag{3–90}$$

The relationship between the power going into the primary (and out the secondary) of the transformer and the power in the transformer's actual windings can be found as follows:

$$\begin{aligned} S_W &= V_C I_C \\ &= V_L(I_L - I_H) \\ &= V_L I_L - V_L I_H \end{aligned}$$

Using Equation (3–86), we get

$$\begin{aligned} S_W &= V_L I_L - V_L I_L \frac{N_C}{N_{SE} + N_C} \\ &= V_L I_L \frac{(N_{SE} + N_C) - N_C}{N_{SE} + N_C} \tag{3–91} \\ &= S_{IO} \frac{N_{SE}}{N_{SE} + N_C} \tag{3–92} \end{aligned}$$

Therefore, the ratio of the apparent power in the primary and secondary of the autotransformer to the apparent power actually traveling through its windings is

$$\boxed{\frac{S_{IO}}{S_W} = \frac{N_{SE} + N_C}{N_{SE}}} \tag{3–93}$$

Equation (3–93) describes the *apparent power rating advantage* of an autotransformer over a conventional transformer. Here S_{IO} is the apparent power entering the primary and leaving the secondary of the transformer, while S_W is the apparent power actually traveling through the transformer's windings (the rest passes from primary to secondary without being coupled through the transformer's windings). Note that the smaller the series winding, the greater the advantage.

For example, a 5000-kVA autotransformer connecting a 110-kV system to a 138-kV system would have an N_C/N_{SE} turns ratio of 110:28. Such an autotransformer would actually have windings rated at

$$S_W = S_{IO} \frac{N_{SE}}{N_{SE} + N_C} \tag{3-92}$$

$$= (5000 \text{ kVA}) \frac{28}{28 + 110} = 1015 \text{ kVA}$$

The autotransformer would have windings rated at only about 1015 kVA, while a conventional transformer doing the same job would need windings rated at 5000 kVA. The autotransformer could be 5 times smaller than the conventional transformer and also would be much less expensive. For this reason, it is very advantageous to build transformers between two nearly equal voltages as autotransformers.[1]

The following example illustrates autotransformer analysis and the rating advantage of autotransformers.

EXAMPLE 3-7

A 100-VA, 120/12-V transformer is to be connected so as to form a step-up autotransformer (see Figure 3–33). A primary voltage of 120 V is applied to the transformer.

a. What is the secondary voltage of the transformer?
b. What is its maximum voltampere rating in this mode of operation?
c. Calculate the rating advantage of this autotransformer connection over the transformer's rating in conventional 120/12-V operation.

Figure 3–33 | The autotransformer of Example 3–7.

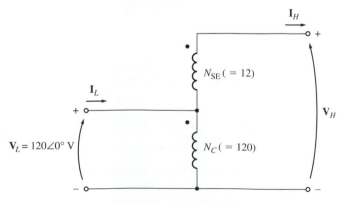

Solution

To accomplish a step-up transformation with a 120-V primary, the ratio of the turns on the common winding N_C to the turns on the series winding N_{SE} in this transformer must be 120:12 (or 10:1).

a. This transformer is being used as a step-up transformer. The secondary voltage is V_H, and from Equation (3–76),

$$V_H = \frac{N_{SE} + N_C}{N_C} V_L \qquad (3\text{–}82)$$

$$= \frac{12 + 120}{120} \, 120 \text{ V} = 132 \text{ V}$$

b. The maximum voltampere rating in either winding of this transformer is 100 VA. How much input or output apparent power can this provide? To find out, examine the series winding. The voltage V_{SE} on the winding is 12 V, and the volt-ampere rating of the winding is 100 VA. Therefore, the *maximum* series winding current is

$$I_{SE,\,max} = \frac{S_{max}}{V_{SE}} = \frac{100 \text{ VA}}{12 \text{V}} = 8.33 \text{ A}$$

Since I_{SE} is equal to the secondary current I_S (or I_H) and since the secondary voltage $V_S = V_H = 132$ V, the secondary apparent power is

$$S_{out} = V_S I_S = V_H I_H$$
$$= (132 \text{ V})(8.33 \text{ A}) = 1100 \text{ VA} = S_{in}$$

c. The rating advantage can be calculated from part *(b)* or separately from Equation (3–93). From part *(b)*,

$$\frac{S_{IO}}{S_W} = \frac{1100 \text{ VA}}{100 \text{ VA}} = 11$$

From Equation (3–93),

$$\frac{S_{IO}}{S_W} = \frac{N_{SE} + N_C}{N_{SE}} \qquad (3\text{–}93)$$

$$= \frac{12 + 120}{12} = \frac{132}{12} = 11$$

By either equation, the apparent power rating is increased by a factor of 11. ■

It is not normally possible to just reconnect an ordinary transformer as an autotransformer and use it in the manner of Example 3–7, because the insulation on the low-voltage side of the ordinary transformer may not be strong enough to withstand the full output voltage of the autotransformer connection. In transformers built specifically as autotransformers, the insulation on the smaller coil (the series winding) is made just as strong as the insulation on the larger coil.

It is common practice in power systems to use autotransformers whenever two voltages fairly close to each other in level need to be transformed, because the closer the two voltages are, the greater the autotransformer power advantage becomes. They are also used as variable transformers, where the low-voltage tap moves up and down the winding. This is a very convenient way to get a variable AC voltage. Such a variable autotransformer is shown in Figure 3–34.

The principal disadvantage of autotransformers is that, unlike ordinary transformers, *there is a direct physical connection between the primary and the secondary circuits,* so the *electrical isolation* of the two sides is lost. If a particular application does not require electrical isolation, then the autotransformer is a convenient and *inexpensive* way to tie nearly equal voltages together.

Figure 3–34 | (a) A variable-voltage autotransformer; (b) cutaway view of the autotransformer. *(Courtesy of Superior Electric Company.)*

The Internal Impedance of an Autotransformer

Autotransformers have one additional disadvantage compared to conventional transformers. It turns out that, compared to a given transformer connected in the conventional manner, the effective per-unit impedance of an autotransformer is smaller by a factor equal to the reciprocal of the power advantage of the autotransformer connection.

The proof of this statement is left as a problem at the end of the chapter.

The reduced internal impedance of an autotransformer compared to a conventional two-winding transformer can be a serious problem in some applications where the series impedance is needed to limit current flows during power system faults (short circuits). The effect of the smaller internal impedance provided by an autotransformer must be taken into account in practical applications before autotransformers are selected.

EXAMPLE 3–8

A transformer is rated at 1000 kVA, 12/1.2 kV, 60 Hz when it is operated as a conventional two-winding transformer. Under these conditions, its series resistance and reactance are given as 1 and 8 percent per unit, respectively. This transformer is to be used as a 13.2/12-kV step-down autotransformer in a power distribution system. In the autotransformer connection

a.   What is the transformer's rating when used in this manner?

b.   What is the transformer's series impedance in per-unit?

Solution

a.   The N_C/N_{SE} turns ratio must be 12:1.2 or 10:1. The voltage rating of this transformer will be 13.2/12 kV, and the apparent power (volt-ampere) rating will be

$$S_{IO} = \frac{N_{SE} + N_C}{N_{SE}} S_W$$

$$= \frac{1 + 10}{1} \, 1000 \text{ kVA} = 11{,}000 \text{ kVA}$$

b.   The transformer's impedance in a per-unit system when connected in the conventional manner is

$$Z_{eq} = 0.01 + j0.08 \text{ pu} \qquad \text{separate windings}$$

The apparent power advantage of this autotransformer is 11, so the per-unit impedance of the autotransformer connected as described is

$$Z_{eq} = \frac{0.01 + j0.08}{11}$$

$$= 0.00091 + j0.00727 \text{ pu} \qquad \text{autotransformer} \qquad ■$$

3.10 | THREE-PHASE TRANSFORMERS

Almost all the major power generation and distribution systems in the world today are three-phase AC systems. Since three-phase systems play such an important role in modern life, it is necessary to understand how transformers are used in them.

Transformers for three-phase circuits can be constructed in one of two ways. One approach is simply to take three single-phase transformers and connect them in a three-phase bank. An alternative approach is to make a three-phase transformer consisting of three sets of windings wrapped on a common core. These two possible types of transformer construction are shown in Figures 3–35 and 3–36. The

Figure 3–35 I A three-phase transformer bank composed of independent transformers.

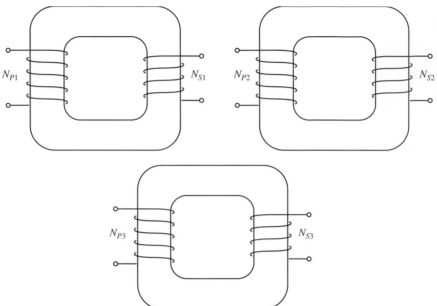

Figure 3–36 I A three-phase transformer wound on a single three-legged core.

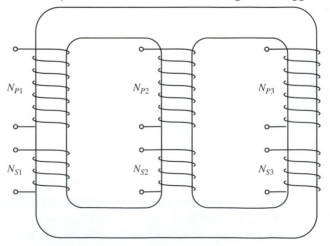

construction of a single three-phase transformer is the preferred practice today, since it is lighter, smaller, cheaper, and slightly more efficient. The older construction approach was to use three separate transformers. That approach had the advantage that each unit in the bank could be replaced individually in the event of trouble, but that does not outweigh the advantages of a combined three-phase unit for most applications. However, there are still a great many installations consisting of three single-phase units in service.

Three-Phase Transformer Connections

A three-phase transformer consists of three transformers, either separate or combined on one core. The primaries and secondaries of any three-phase transformer can be independently connected in either a wye (Y) or a delta (Δ). This gives a total of four possible connections for a three-phase transformer bank:

1. Wye–wye (Y–Y)
2. Wye–delta (Y–Δ)
3. Delta–wye (Δ–Y)
4. Delta–delta (Δ–Δ)

These connections are shown in Figure 3–37.

The key to analyzing any three-phase transformer bank is to look at a single transformer in the bank. *Any single transformer in the bank behaves exactly like the single-phase transformers already studied.* The impedance, voltage regulation, efficiency, and similar calculations for three-phase transformers are done on *a per-phase basis,* using exactly the same techniques already developed for single-phase transformers.

The advantages and disadvantages of each type of three-phase transformer connection are discussed below.

Wye–Wye Connection The Y–Y connection of three-phase transformers is shown in Figure 3–37a. In a Y–Y connection, the primary voltage on each phase of the transformer is given by $V_{\phi P} = V_{LP}/\sqrt{3}$. The primary-phase voltage is related to the secondary-phase voltage by the turns ratio of the transformer. The phase voltage on the secondary is then related to the line voltage on the secondary by $V_{LS} = \sqrt{3}\,V_{\phi S}$. Therefore, overall the voltage ratio on the transformer is

$$\boxed{\frac{V_{LP}}{V_{LS}} = \frac{\sqrt{3}\,V_{\phi P}}{\sqrt{3}\,V_{\phi S}} = a \quad \text{Y–Y}}$$

(3–94)

The Y–Y connection has two very serious problems:

1. If loads on the transformer circuit are unbalanced, then the voltages on the phases of the transformer can become severely unbalanced.
2. There is a serious problem with third-harmonic voltages.

Figure 3–37 I Three-phase transformer connections and wiring diagrams:
(a) Y–Y; (b) Y–Δ; (c) Δ–Y; (d) Δ–Δ.

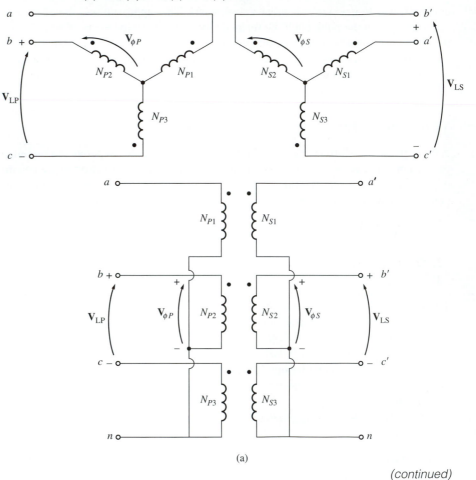

(a)

(continued)

If a three-phase set of voltages is applied to a Y–Y transformer, the voltages in any phase will be 120° apart from the voltages in any other phase. However, *the third-harmonic components of each of the three phases will be in phase with each other,* since there are three cycles in the third harmonic for each cycle of the fundamental frequency. There are always some third-harmonic components in a transformer because of the nonlinearity of the core, and these components add up. The result is a very large third-harmonic component of voltage on top of the 50- or 60-Hz fundamental voltage. This third-harmonic voltage can be larger than the fundamental voltage itself.

Both the unbalance problem and the third-harmonic problem can be solved by one of two techniques:

1. *Solidly ground the neutrals of the transformers,* especially the primary winding's neutral. This connection permits the additive third-harmonic components to cause a current flow in the neutral instead of building up large voltages. The neutral also provides a return path for any current imbalances in the load.

2. *Add a third (tertiary) winding connected in* Δ *to the transformer bank.* If a third Δ-connected winding is added to the transformer, then the third-harmonic components of voltage in the Δ will add up, causing a circulating current flow within the winding. This suppresses the third-harmonic components of voltage in the same manner as grounding the transformer neutrals.

   The Δ-connected tertiary windings need not even be brought out of the transformer case, but they often are used to supply lights and auxiliary power within the substation where it is located. The tertiary windings must be large enough to handle the circulating currents, so they are usually made about one-third the power rating of the two main windings.

One or the other of these correction techniques *must* be used any time a Y–Y transformer is installed. In practice, very few Y–Y transformers are used, since the same jobs can be done by one of the other types of three-phase transformers.

Wye–Delta Connection   The Y–Δ connection of three-phase transformers is shown in Figure 3–37b. In this connection, the primary line voltage is related to the primary phase voltage by $V_{LP} = \sqrt{3}\, V_{\phi P}$, while the secondary line voltage is equal to the secondary phase voltage $V_{LS} = V_{\phi S}$. The voltage ratio of each phase is

$$\frac{V_{\phi P}}{V_{\phi S}} = a$$

so the overall relationship between the line voltage on the primary side of the bank and the line voltage on the secondary side of the bank is

$$\frac{V_{LP}}{V_{LS}} = \frac{\sqrt{3}\, V_{\phi P}}{V_{\phi S}}$$

$$\boxed{\frac{V_{LP}}{V_{LS}} = \sqrt{3}\, a \qquad \text{Y–Δ}} \tag{3–95}$$

   The Y–Δ connection has no problem with third-harmonic components in its voltages, since they are consumed in a circulating current on the Δ side. This connection is also more stable with respect to unbalanced loads, since the Δ partially redistributes any imbalance that occurs.

   This arrangement does have one problem, though. Because of the Δ connection, the secondary voltage is shifted 30° relative to the primary voltage of the transformer. The fact that a phase shift has occurred can cause problems in paralleling the secondaries of two transformer banks together. The phase angles of transformer secondaries must be equal if they are to be paralleled, which means that attention must be

Figure 3–37 | *(continued)*

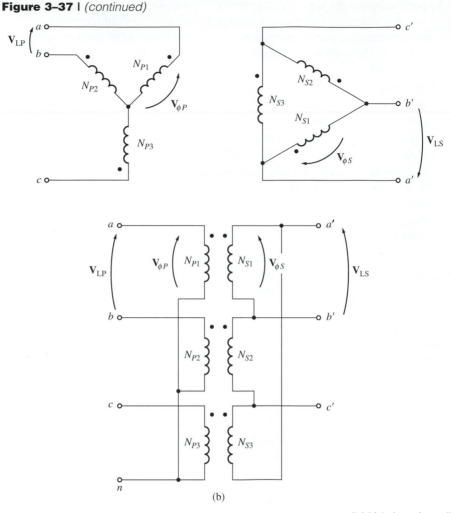

(b)

(b) Y-Δ *(continued)*

paid to the direction of the 30° phase shift occurring in each transformer bank to be paralleled together.

In the United States, it is customary to make the secondary voltage lag the primary voltage by 30°. Although this is the standard, it has not always been observed, and older installations must be checked very carefully before a new transformer is paralleled with them, to make sure that their phase angles match.

The connection shown in Figure 3–37b will cause the secondary voltage to be lagging if the system phase sequence is *abc*. If the system phase sequence is *acb*, then the connection shown in Figure 3–37b will cause the secondary voltage to be leading the primary voltage by 30°.

Figure 3–37 I *(continued)*

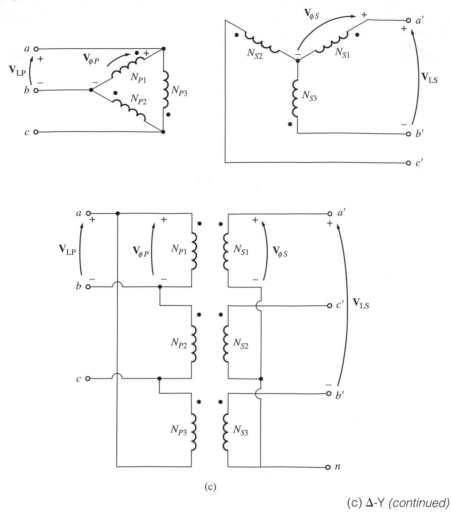

(c)

(c) Δ-Y *(continued)*

Delta–Wye Connection A Δ–Y connection of three-phase transformers is shown in Figure 3–37c. In a Δ–Y connection, the primary line voltage is equal to the primary-phase voltage $V_{LP} = V_{\phi P}$, while the secondary voltages are related by $V_{LS} = \sqrt{3}\, V_{\phi S}$. Therefore, the line-to-line voltage ratio of this transformer connection is

$$\frac{V_{LP}}{V_{LS}} = \frac{V_{\phi P}}{\sqrt{3}\, V_{\phi S}}$$

$$\boxed{\frac{V_{LP}}{V_{LS}} = \frac{a}{\sqrt{3}}} \qquad \Delta\!-\!Y \tag{3–96}$$

This connection has the same advantages and the same phase shift as the Y–Δ transformer. The connection shown in Figure 3–37c makes the secondary voltage lag the primary voltage by 30°, as before.

Delta–Delta Connection   The Δ–Δ connection is shown in Figure 3–37d. In a Δ–Δ connection, $V_{LP} = V_{\phi P}$ and $V_{LS} = V_{\phi S}$, so the relationship between primary and secondary line voltages is

$$\frac{V_{LP}}{V_{LS}} = \frac{V_{\phi P}}{V_{\phi S}} = a \qquad \Delta-\Delta \tag{3-97}$$

This transformer has no phase shift associated with it and no problems with unbalanced loads or harmonics.

Figure 3–37 | *(concluded)*

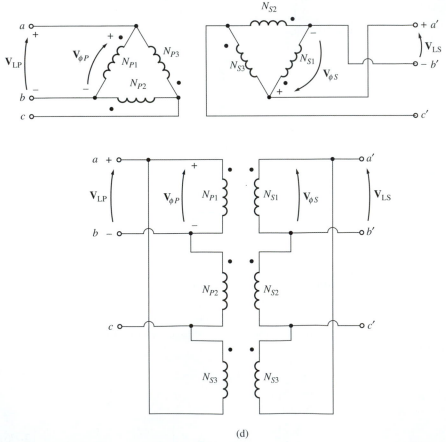

(d)

(d) Δ-Δ

The Per-Unit System for Three-Phase Transformers

The per-unit system applies just as well to three-phase transformers as to single-phase transformers. The single-phase base Equations (3–60) to (3–63) apply to three-phase systems on a *per-phase* basis. If the total base voltampere value of the transformer bank is called S_{base}, then the base voltampere value of one of the transformers, $S_{1\phi, base}$, is

$$S_{1\phi, base} = \frac{S_{base}}{3} \tag{3–98}$$

and the base phase current and impedance of the transformer are

$$I_{\phi, base} = \frac{S_{1\phi, base}}{V_{\phi, base}} \tag{3–99}$$

$$\boxed{I_{\phi, base} = \frac{S_{base}}{3\,V_{\phi, base}}} \tag{3–100}$$

$$Z_{base} = \frac{(V_{\phi, base})^2}{S_{1\phi, base}} \tag{3–101}$$

$$\boxed{Z_{base} = \frac{3(V_{\phi, base})^2}{S_{base}}} \tag{3–102}$$

Line quantities on three-phase transformer banks can also be represented in per-unit. The relationship between the base line voltage and the base phase voltage of the transformer depends on the connection of windings. If the windings are connected in delta, $V_{L, base} = V_{\phi, base}$, while if the windings are connected in wye, $V_{L, base} = \sqrt{3}\,V_{\phi, base}$. The base line current in a three-phase transformer bank is given by

$$I_{L, base} = \frac{S_{base}}{\sqrt{3}\,V_{L, base}} \tag{3–103}$$

The application of the per-unit system to three-phase transformer problems is similar to its application in the single-phase examples already given.

EXAMPLE 3–9

A 50-kVA, 13,800/208-V Δ–Y distribution transformer has a resistance of 1 percent and a reactance of 7 percent per unit.

a. What is the transformer's phase impedance referred to the high-voltage side?
b. Calculate this transformer's voltage regulation at full load and 0.8 PF lagging, using the calculated high-side impedance.
c. Calculate this transformer's voltage regulation under the same conditions, using the per-unit system.

Solution

a.   The high-voltage side of this transformer has a base line voltage of 13,800 V and a base apparent power of 50 kVA. Since the primary is Δ-connected, its phase voltage is equal to its line voltage. Therefore, its base impedance is

$$Z_{base} = \frac{3(V_{\phi, base})^2}{S_{base}} \tag{3–102}$$

$$= \frac{3(13,800\ V)^2}{50,000\ VA} = 11,426\ \Omega$$

The per-unit impedance of the transformer is

$$\mathbf{Z}_{eq} = 0.01 + j0.07\ pu$$

so the high-side impedance in ohms is

$$\mathbf{Z}_{eq} = \mathbf{Z}_{eq, pu}Z_{base}$$
$$= (0.01 + j0.07\ pu)(11,426\ \Omega) = 114.2 + j800\ \Omega$$

b.   To calculate the voltage regulation of a three-phase transformer bank, determine the voltage regulation of any single transformer in the bank. The voltages on a single transformer are phase voltages, so

$$VR = \frac{V_{\phi P} - aV_{\phi S}}{aV_{\phi S}} \times 100\%$$

The rated transformer phase voltage on the primary is 13,800 V, so the rated phase current on the primary is given by

$$I_\phi = \frac{S}{3V_\phi}$$

The rated apparent power $S = 50$ kVA, so

$$I_\phi = \frac{50,000\ VA}{3(13,800\ V)} = 1.208\ A$$

The rated phase voltage on the secondary of the transformer is 208 V/$\sqrt{3}$ = 120 V. When referred to the high-voltage side of the transformer, this voltage becomes $V'_{\phi S} = aV_{\phi S} = 13,800$ V. Assume that the transformer secondary is operating at the rated voltage and current, and find the resulting primary phase voltage:

$$\mathbf{V}_{\phi P} = a\mathbf{V}_{\phi S} + R_{eq}\mathbf{I}_\phi + jX_{eq}\mathbf{I}_\phi$$
$$= 13,800\angle 0°\ V + (114.2\ \Omega)(1.208\angle -36.87°\ A) + (j800\ \Omega)(1.208\angle -36.87°\ A)$$
$$= 13,800 + 138\angle -36.87° + 966.4\angle -53.13°$$
$$= 13,800 + 110.4 - j82.8 + 579.8 + j773.1$$
$$= 14,490 + j690.3 = 14,506\angle 2.73°\ V$$

Therefore,

$$VR = \frac{V_{\phi P} - aV_{\phi S}}{aV_{\phi S}} \times 100\%$$

$$= \frac{14{,}506 - 13{,}800}{13{,}800} \times 100\% = 5.1\%$$

c. In the per-unit system, the output voltage is $1\angle 0°$, and the current is $1\angle -36.87°$. Therefore, the input voltage is

$$\mathbf{V}_{\phi P} = 1\angle 0° + (0.01)(1\angle -36.87°) + (j0.07)(1\angle -36.87°)$$
$$= 1 + 0.008 - j0.006 + 0.042 + j0.056$$
$$= 1.05 + j0.05 = 1.051\angle 2.73°$$

The voltage regulation is

$$VR = \frac{1.051 - 1.0}{1.0} \times 100\% = 5.1\%$$ ∎

Of course, the voltage regulation of the transformer bank is the same whether the calculations are done in actual ohms or in the per-unit system.

3.11 | TRANSFORMER RATINGS AND RELATED PROBLEMS

Transformers have four major ratings: apparent power, voltage, current, and frequency. This section examines the ratings of a transformer and explains why they are chosen the way they are. It also considers the related question of the current inrush that occurs when a transformer is first connected to the line.

The Voltage and Frequency Ratings of a Transformer

The voltage rating of a transformer serves two functions. One is to protect the winding insulation from breakdown due to an excessive voltage applied to it. This is not the most serious limitation in practical transformers. The second function is related to the magnetization curve and magnetization current of the transformer. Figure 3–11 shows a magnetization curve for a transformer. If a steady-state voltage

$$v(t) = V_M \sin \omega t \quad \text{V}$$

is applied to a transformer's primary winding, the flux of the transformer is given by

$$\phi(t) = \frac{1}{N_P} \int v(t)dt$$

$$= \frac{1}{N_P} \int V_M \sin \omega t \, dt$$

$$\boxed{\phi(t) = -\frac{V_M}{\omega N_P} \cos \omega t} \tag{3–104}$$

If the applied voltage *v(t)* is increased by 10 percent, the resulting maximum flux in the core also increases by 10 percent. Above a certain point on the magnetization curve, though, a 10 percent increase in flux requires an increase in magnetization current *much* larger than 10 percent. This concept is illustrated in Figure 3–38. As the voltage increases, the high magnetization currents soon become unacceptable. The maximum applied voltage (and therefore the rated voltage) is set by the maximum acceptable magnetization current in the core.

Figure 3–38 | The effect of the peak flux in a transformer core on the required magnetization current.

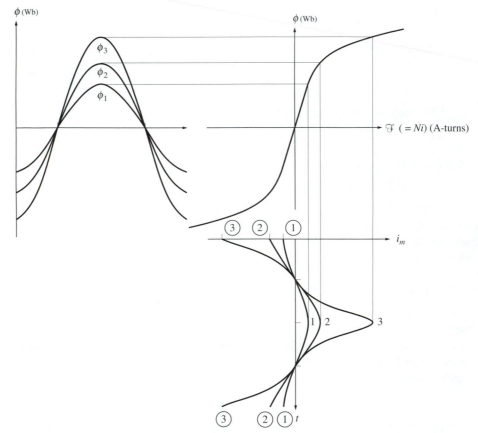

Notice that voltage and frequency are related in a reciprocal fashion if the maximum flux is to be held constant:

$$\phi_{\text{max}} = \frac{V_{\text{max}}}{\omega N_P} \tag{3–105}$$

Thus, *if a 60-Hz transformer is to be operated on 50 Hz, its applied voltage must also be reduced by one-sixth or the peak flux in the core will be too high.* This reduction in applied voltage with frequency is called *derating*. Similarly, a 50-Hz transformer may be operated at a 20 percent higher voltage on 60 Hz if this action does not cause insulation problems.

EXAMPLE 3–10

A 1-kVA, 230/115-V 60 Hz single-phase transformer has 850 turns on the primary winding and 425 turns on secondary winding. The magnetization curve for this transformer is shown in Figure 3–39.

a. Calculate and plot the magnetization current of this transformer when it is run at 230 V on a 60-Hz power source. What is the rms value of the magnetization current?

b. Calculate and plot the magnetization current of this transformer when it is run at 230 V on a 50-Hz power source. What is the rms value of the magnetization current? How does this current compare to the magnetization current at 60 Hz.

Figure 3–39 | Magnetization curve for the 230/115-V transformer of Example 3–10.

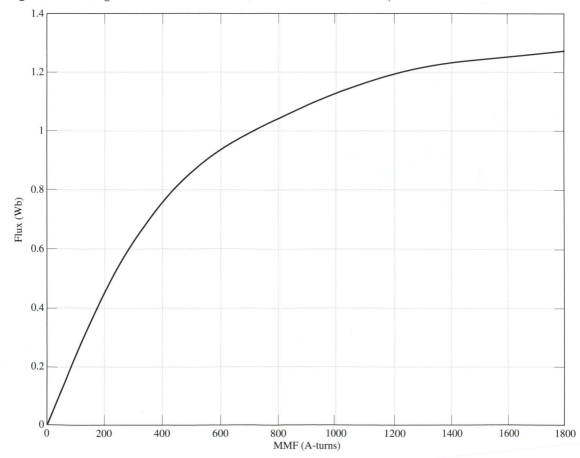

Solution

The best way to solve this problem is to calculate the flux as a function of time for this core, and then use the magnetization curve to transform each flux value to a corresponding magnetomotive force. The magnetizing current can then be determined from the equation

$$\frac{\mathcal{F}}{N_P} \tag{3–106}$$

Assuming that the voltage applied to the core is $v(t) = V_M \sin \omega t$ volts, the flux in the core as a function of time is given by Equation (3–104):

$$\phi(t) = -\frac{V_M}{\omega N_P} \cos \omega t \tag{3–104}$$

The magnetization curve for this transformer is available electronically in a file called **mag_curve_1.dat.** This file can be used by MATLAB to translate these flux values into corresponding MMF values, and Equation (3–106) can be used to find the required magnetization current values. Finally, the rms value of the magnetization current can be calculated from the equation

$$I_{\text{rms}} = \sqrt{\frac{1}{T} \int_0^T i^2 \, dt} \tag{3–107}$$

A MATLAB program to perform these calculations is shown below.

```
% M-file: mag_current.m
% M-file to calculate and plot the magnetization
% current of a 230/115 transformer operating at
% 230 volts and 50/60 Hz. This program also
% calculates the rms value of the mag. current.

% Load the magnetization curve. It is in two
% columns, with the first column being mmf and
% the second column being flux.
load mag_curve_1.dat;
mmf_data = mag_curve_1(:,1);
flux_data = mag_curve_1(:,2);

% Initialize values
VM = 325;                        % maximum voltage (V)
NP = 850;                        % Primary turns

% Calculate angular velocity for 60 Hz
freq = 60;                       % Freq (Hz)
w = 2 * pi * freq;
```

```
% Calculate flux versus time
time = 0:1/3000:1/30;                % 0 to 1/30 sec
flux = -VM/(w*NP) * cos(w .* time);


% Calculate the mmf corresponding to a given flux
% using the MATLAB interpolation function.
mmf = interp1(flux_data,mmf_data,flux);


% Calculate the magnetization current
im = mmf / NP;


% Calculate the rms value of the current
irms = sqrt(sum(im.^2)/length(im));
disp(['The rms current at 60 Hz is ', num2str(irms)]);


% Plot the magnetization current.
figure(1)
subplot(2,1,1);
plot(time,im);
title ('\bfMagnetization Current at 60 Hz');
xlabel ('\bfTime (s)');
ylabel ('\bf\itI_{m} \rm(A)');
axis([0 0.04 -2 2]);
grid on;


% Calculate angular velocity for 50 Hz
freq = 50;                           % Freq (Hz)
w = 2 * pi * freq;


% Calculate flux versus time
time = 0:1/2500:1/25;                % 0 to 1/25 sec
flux = -VM/(w*NP) * cos(w .* time);


% Calculate the mmf corresponding to a given flux
% using the Matlab interpolation function.
mmf = interp1(flux_data,mmf_data,flux);


% Calculate the magnetization current
im = mmf / NP;


% Calculate the rms value of the current
irms = sqrt(sum(im.^2)/length(im));
disp(['The rms current at 50 Hz is ', num2str(irms)]);
```

```
% Plot the magnetization current.
subplot(2,1,2);
plot(time,im);
title ('\bfMagnetization Current at 50 Hz');
xlabel ('\bfTime (s)');
ylabel ('\bf\itI_{m} \rm(A)');
axis([0 0.04 -2 2]);
grid on;
```

When this program executes, the results are:

```
» mag_current
The rms current at 60 Hz is 0.4894
The rms current at 50 Hz is 0.79252
```

The resulting magnetization currents are shown in Figure 3–40. Note that the rms magnetization current increases by more than 60% when the frequency changes from 60 Hz to 50 Hz. ∎

Figure 3–40 | (a) Magnetization current for the transformer operating at 60 Hz.
(a) Magnetization current for the transformer operating at 50 Hz.

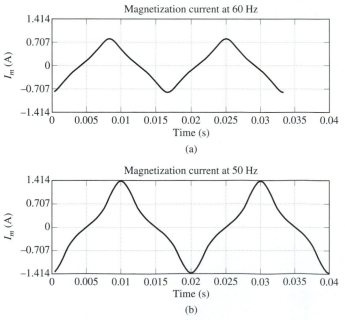

The Apparent Power Rating of a Transformer

The principal purpose of the apparent power rating of a transformer is that, together with the voltage rating, it sets the current flow through the transformer windings. The current flow is important because it controls the i^2R losses in transformer, which in turn control the heating of the transformer coils. It is the heating that is critical, since overheating the coils of a transformer *drastically* shortens the life of its insulation.

The actual volt-ampere rating of a transformer may be more than a single value. In real transformers, there may be a volt-ampere rating for the transformer by itself, and another (higher) rating for the transformer with forced cooling. The key idea behind the power rating is that the hot-spot temperature in the transformer windings *must* be limited to protect the life of the transformer.

If a transformer's voltage is reduced for any reason (e.g., if it is operated at a lower frequency than normal), then the transformer's volt-ampere rating must be reduced by an equal amount. If this is not done, then the current in the transformer's windings will exceed the maximum permissible level and cause overheating.

The Problem of Current Inrush

A problem related to the voltage level in the transformer is the problem of current inrush at starting. Suppose that the voltage

$$v(t) = V_M \sin(\omega t + \theta) \qquad \text{V} \tag{3-108}$$

is applied at the moment the transformer is first connected to the power line. The maximum flux height reached on the first half-cycle of the applied voltage depends on the phase of the voltage at the instant the voltage is applied to the transformer. If the initial voltage is

$$v(t) = V_M \sin(\omega t + 90°) = V_M \cos \omega t \qquad \text{V} \tag{3-109}$$

and if the initial flux in the core is zero, then the maximum flux during the first half-cycle will just equal the maximum flux at steady state:

$$\phi_{max} = \frac{V_{max}}{\omega N_P} \tag{3-110}$$

This flux level is just the steady-state flux, so it causes no special problems. But if the applied voltage happens to be

$$v(t) = V_M \sin \omega t \qquad \text{V}$$

the maximum flux during the first half-cycle is given by

$$\begin{aligned}
\phi(t) &= \frac{1}{N_P} \int_0^{\pi/\omega} V_M \sin \omega t \, dt \\
&= -\frac{V_M}{\omega N_P} \cos \omega t \Big|_0^{\pi/\omega} \\
&= -\frac{V_M}{\omega N_P} [(-1) - (1)]
\end{aligned}$$

$$\boxed{\phi_{max} = \frac{2V_{max}}{\omega N_P}} \tag{3-111}$$

This maximum flux is twice as high as the normal steady-state flux. If the magnetization curve in Figure 3–11 is examined, it is easy to see that doubling the maximum flux in the core results in an *enormous* magnetization current. In fact, for part of the cycle, the transformer looks like a short circuit, and a very large current flows (see Figure 3–41).

For any other phase angle of the applied voltage between 90°, which is no problem, and 0°, which is the worst case, there is some excess current flow. The applied phase angle of the voltage is not normally controlled on starting, so there can be huge inrush currents during the first several cycles after the transformer is connected to the line. The transformer and the power system to which it is connected must be able to withstand these currents.

Figure 3–41 I The current inrush due to a transformer's magnetization current on starting.

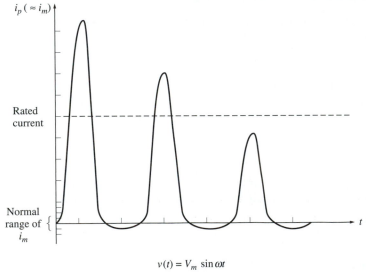

$$v(t) = V_m \sin \omega t$$

The Transformer Nameplate

A typical nameplate from a distribution transformer is shown in Figure 3–42. The information on such a nameplate includes rated voltage, rated kilovoltamperes, rated frequency, and the transformer per-unit series impedance. It also shows the voltage ratings for each tap on the transformer and the wiring schematic of the transformer.

Nameplates such as the one shown also typically include the transformer type designation and references to its operating instructions.

3.12 I INSTRUMENT TRANSFORMERS

Two special-purpose transformers are used with power systems for taking measurements. One is the potential transformer, and the other is the current transformer.

Figure 3–42 I A sample distribution transformer nameplate. Note the ratings listed: voltage, frequency, apparent power, and tap settings. *(Courtesy of General Electric Company.)*

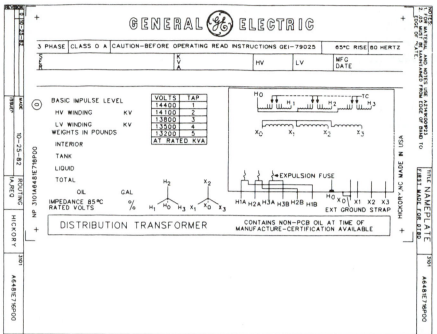

A *potential transformer* is a specially wound transformer with a high-voltage primary and a low-voltage secondary. It has a very low power rating, and its sole purpose is to provide a *sample* of the power system's voltage to the instruments monitoring it. Since the principal purpose of the transformer is voltage sampling, it must be very accurate so as not to distort the true voltage values too badly. Potential transformers of several *accuracy classes* may be purchased, depending on how accurate the readings must be for a given application.

Current transformers sample the current in a line and reduce it to a safe and measurable level. A diagram of a typical current transformer is shown in Figure 3–43. The current transformer consists of a secondary winding wrapped around a ferromagnetic ring, with the single primary line running through the center of the ring. The ferromagnetic ring holds and concentrates a small sample of the flux from the primary line. That flux then induces a voltage and current in the secondary winding.

A current transformer differs from the other transformers described in this chapter in that its windings are *loosely coupled*. Unlike all the other transformers, the mutual flux ϕ_M in the current transformer is smaller than the leakage flux ϕ_L. Because of the loose coupling, the voltage and current ratios of Equations (3–1) to (3–6) do not apply to a current transformer. Nevertheless, the secondary current in a current transformer is directly proportional to the much larger primary current, and the device can provide an accurate sample of a line's current for measurement purposes.

Figure 3–43 | Sketch of a current transformer.

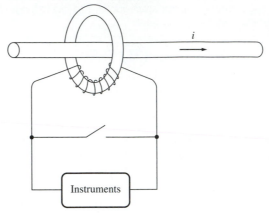

Current transformer ratings are given as ratios of primary to secondary current. A typical current transformer ratio might be 600:5, 800:5, or 1000:5. A 5-A rating is standard on the secondary of a current transformer.

It is important to keep a current transformer short-circuited at all times, since extremely high voltages can appear across its open secondary terminals. In fact, most relays and other devices using the current from a current transformer have a *shorting interlock,* which must be shut before the relay can be removed for inspection or adjustment. Without this interlock, very dangerous high voltages will appear at the secondary terminals as the relay is removed from its socket.

A much more detailed discussion of instrument transformers and their uses can be found in Reference 6.

3.13 | SUMMARY

A transformer is a device for converting electric energy at one voltage level to electric energy at another voltage level through the action of a magnetic field. It plays an extremely important role in modern life by making possible the economical long-distance transmission of electric power.

When a voltage is applied to the primary of a transformer, a flux is produced in the core as given by Faraday's law. The changing flux in the core then induces a voltage in the secondary winding of the transformer. Because transformer cores have very high permeability, the net magnetomotive force required in the core to produce its flux is very small. Since the net magnetomotive force is very small, the primary circuit's magnetomotive force must be approximately equal and opposite to the secondary circuit's magnetomotive force. This fact yields the transformer current ratio.

A real transformer has leakage fluxes that pass through either the primary or the secondary winding, but not both. In addition there are hysteresis, eddy current, and copper losses. These effects are accounted for in the equivalent circuit of the transformer. Transformer imperfections are measured in a real transformer by its voltage regulation and its efficiency.

The per-unit system is a convenient way to study systems containing transformers, because in this system the different system voltage levels disappear. In addition, the per-unit impedances of a transformer expressed to its own ratings base fall within a relatively narrow range, providing a convenient check for reasonableness in problem solutions.

An autotransformer differs from a regular transformer in that the two windings of the autotransformer are connected. The voltage on one side of the transformer is the voltage across a single winding, while the voltage on the other side of the transformer is the sum of the voltages across *both* windings. Because only a portion of the power in an autotransformer actually passes through the windings, an autotransformer has a power rating advantage compared to a regular transformer of equal size. However, the connection destroys the electrical isolation between a transformer's primary and secondary sides.

The voltage levels of three-phase circuits can be transformed by a proper combination of three transformers. Potential transformers and current transformers can sample the voltages and currents present in a circuit. Both devices are very common in large power distribution systems.

3.14 | QUESTIONS

3–1.   Is the turns ratio of a transformer the same as the ratio of voltages across the transformer? Why or why not?

3–2.   Why does the magnetization current impose an upper limit on the voltage applied to a transformer core?

3–3.   What components compose the excitation current of a transformer? How are they modeled in the transformer's equivalent circuit?

3–4.   What is the leakage flux in a transformer? Why is it modeled in a transformer equivalent circuit as an inductor?

3–5.   List and describe the types of losses that occur in a transformer.

3–6.   Why does the power factor of a load affect the voltage regulation of a transformer?

3–7.   Why does the short-circuit test essentially show only i^2R losses and not excitation losses in a transformer?

3–8.   Why does the open-circuit test essentially show only excitation losses and not i^2R losses?

3–9.   How does the per-unit system eliminate the problem of different voltage levels in a power system?

3–10.   Why can autotransformers handle more power than conventional transformers of the same size?

3–11.   What are transformer taps? Why are they used?

3–12.   What are the problems associated with the Y–Y three-phase transformer connection?

3–13.   What is a TCUL transformer?

3–14. Can a 60-Hz transformer be operated on a 50-Hz system? What actions are necessary to enable this operation?

3–15. What happens to a transformer when it is first connected to a power line? Can anything be done to mitigate this problem?

3–16. What is a potential transformer? How is it used?

3–17. What is a current transformer? How is it used?

3–18. A distribution transformer is rated at 18 kVA, 20,000/480 V, and 60 Hz. Can this transformer safely supply 15 kVA to a 415-V load at 50 Hz? Why or why not?

3–19. Why does one hear a hum when standing near a large power transformer?

3.15 PROBLEMS

3–1. The secondary winding of a transformer has a terminal voltage of $v_s(t) = 282.8 \sin 377t$ V. The turns ratio of the transformer is 50:200 ($a = 0.25$). If the secondary current of the transformer is $i_s(t) = 7.07 \sin (377t - 36.87°)$ A, what is the primary current of this transformer? What are its voltage regulation and efficiency? The impedances of this transformer referred to the primary side are

$$R_{eq} = 0.05 \ \Omega \qquad R_C = 75 \ \Omega$$
$$X_{eq} = 0.225 \ \Omega \qquad X_M = 20 \ \Omega$$

3–2. A 20-kVA, 8000/277-V distribution transformer has the following resistances and reactances:

$$R_P = 32 \ \Omega \qquad R_S = 0.05 \ \Omega$$
$$X_P = 45 \ \Omega \qquad X_S = 0.06 \ \Omega$$
$$R_C = 250 \ k\Omega \qquad X_M = 30 \ k\Omega$$

The excitation branch impedances are given referred to the high-voltage side of the transformer.

a. Find the equivalent circuit of this transformer referred to the high-voltage side.

b. Find the per-unit equivalent circuit of this transformer.

c. Assume that this transformer is supplying rated load at 277 V and 0.8 PF lagging. What is this transformer's input voltage? What is its voltage regulation?

d. What is the transformer's efficiency under the conditions of part (c)?

3–3. A 1000-VA, 230/115-V transformer has been tested to determine its equivalent circuit. The results of the tests are shown on page 155.

Open-circuit test	Short-circuit test
$V_{OC} = 230$ V	$V_{SC} = 11.2$ V
$I_{OC} = 0.95$ A	$I_{SC} = 5.0$ A
$P_{OC} = 80$ W	$P_{SC} = 20.1$ W

All data given were taken from the primary side of the transformer.

a. Find the equivalent circuit of this transformer referred to the low-voltage side of the transformer.

b. Find the transformer's voltage regulation at rated conditions and (1) 0.8 PF lagging, (2) 1.0 PF, (3) 0.8 PF leading.

c. Determine the transformer's efficiency at rated conditions and 0.8 PF lagging.

3–4. A single-phase power system is shown in Figure P3–1. The power source feeds a 100-kVA, 14/2.4-kV transformer through a feeder impedance of $38.2 + j\,140\ \Omega$. The transformer's equivalent series impedance referred to its low-voltage side is $0.12 + j\,0.5\ \Omega$. The load on the transformer is 90 kW at 0.85 PF lagging and 2300 V.

a. What is the voltage at the power source of the system?

b. What is the voltage regulation of the transformer?

c. How efficient is the overall power system?

Figure P3–1 | The circuit of Problem 3–4.

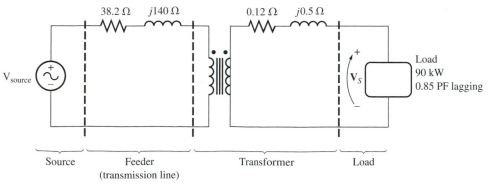

| Source | Feeder (transmission line) | Transformer | Load |

3–5. When travelers from the United States and Canada visit Europe, they encounter a different power distribution system. Wall voltages in North America are 120 V rms at 60 Hz, while typical wall voltages in Europe are 220 to 240 V at 50 Hz. Many travelers carry small step-up/step-down transformers so that they can use their appliances in the countries that they are visiting. A typical transformer might be rated at 1 kVA and 120/240 V.

It has 1000 turns of wire on the 120-V side and 500 turns of wire on the 240-V side. The magnetization curve for this transformer is shown in Figure P3–2, and can be found in file p22.mag at this book's website.

Figure P3–2 | Magnetization curve for the transformer of Problem 3–5.

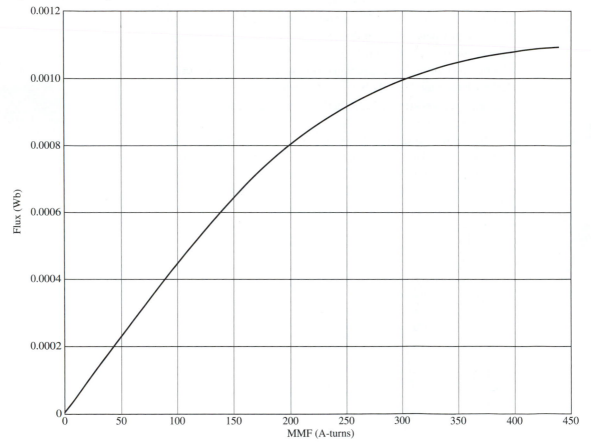

a. Suppose that this transformer is connected to a 120-V, 60-Hz power source with no load connected to the 240-V side. Sketch the magnetization current that would flow in the transformer. (Use MATLAB to plot the current accurately, if it is available.) What is the rms amplitude of the magnetization current? What percentage of full-load current is the magnetization current?

b. Now suppose that this transformer is connected to a 240-V, 50-Hz power source with no load connected to the 120-V side. Sketch the magnetization current that would flow in the transformer. (Use MATLAB to plot the current accurately, if it is available.) What is the rms amplitude of the magnetization current? What percentage of full-load current is the magnetization current?

c.   In which case is the magnetization current a higher percentage of full-load current? Why?

3–6. A 15-kVA, 8000/230-V distribution transformer has an impedance referred to the primary of $80 + j300$ Ω. The components of the excitation branch referred to the primary side are $R_C = 350$ kΩ and $X_M = 70$ kΩ.

a.   If the primary voltage is 7967 V and the load impedance is $Z_L = 3.2 + j1.5$ Ω, what is the secondary voltage of the transformer? What is the voltage regulation of the transformer?

b.   If the load is disconnected and a capacitor of $-j3.5$ Ω is connected in its place, what is the secondary voltage of the transformer? What is its voltage regulation under these conditions?

3–7. A 5000-kVA, 230/13.8-kV single-phase power transformer has a per-unit resistance of 1 percent and a per-unit reactance of 5 percent (data taken from the transformer's nameplate). The open-circuit test performed on the low-voltage side of the transformer yielded the following data:

$$V_{OC} = 13.8 \text{ kV} \qquad I_{OC} = 15.1 \text{ A} \qquad P_{OC} = 44.9 \text{ kW}$$

a.   Find the equivalent circuit referred to the low-voltage side of this transformer.

b.   If the voltage on the secondary side is 13.8 kV and the power supplied is 4000 kW at 0.8 PF lagging, find the voltage regulation of the transformer. Find its efficiency.

3–8. A 150-MVA, 15/200-kV single-phase power transformer has a per-unit resistance of 1.2 percent and a per-unit reactance of 5 percent (data taken from the transformer's nameplate). The magnetizing impedance is $j100$ per unit.

a.   Find the equivalent circuit referred to the low-voltage side of this transformer.

b.   Calculate the voltage regulation of this transformer for a full-load current at power factor of 0.8 lagging.

c.   Assume that the primary voltage of this transformer is a constant 15 kV, and plot the secondary voltage as a function of load current for currents from no-load to full-load. Repeat this process for power factors of 0.8 lagging, 1.0, and 0.8 leading.

3–9. A three-phase transformer bank is to handle 400 kVA and have a 34.5/13.8-kV voltage ratio. Find the rating of each individual transformer in the bank (high voltage, low voltage, turns ratio, and apparent power) if the transformer bank is connected to (*a*) Y–Y, (*b*) Y–Δ, (*c*) Δ–Y, (*d*) Δ–Δ.

3–10. A Y-connected bank of three identical 100-kVA, 7967/480-V transformers is supplied with power directly from a large constant-voltage bus. In the short-circuit test, the recorded values on the high-voltage side for one of these transformers are

$$V_{SC} = 560 \text{ V} \qquad I_{SC} = 12.6 \text{ A} \qquad P_{SC} = 1700 \text{ W}$$

a. If this bank delivers a rated load at 0.88 PF lagging and rated voltage, what is the line-to-line voltage on the primary of the transformer bank?

b. What is the voltage regulation under these conditions?

c. Assume that the primary voltage of this transformer is a constant 7967 V, and plot the secondary voltage as a function of load current for currents from no-load to full-load. Repeat this process for power factors of 0.85 lagging, 1.0, and 0.85 leading.

d. Plot the voltage regulation of this transformer as a function of load current for currents from no-load to full-load. Repeat this process for power factors of 0.85 lagging, 1.0, and 0.85 leading.

3–11. A 100,000-kVA, 230/115-kV Δ–Δ three-phase power transformer has a per-unit resistance of 0.02 pu and a per-unit reactance of 0.055 pu. The excitation branch elements are $R_C = 110$ pu and $X_M = 20$ pu.

a. If this transformer supplies a load of 80 MVA at 0.85 PF lagging, draw the phasor diagram of one phase of the transformer.

b. What is the voltage regulation of the transformer bank under these conditions?

c. Sketch the equivalent circuit referred to the low-voltage side of one phase of this transformer. Calculate all the transformer impedances referred to the low-voltage side.

3–12. An autotransformer is used to connect a 12.6-kV distribution line to a 13.8-kV distribution line. It must be capable of handling 2000 kVA. There are three separate autotransformers, connected Y–Y with their neutrals solidly grounded.

a. What must the N_C/N_{SE} turns ratio be to accomplish this connection?

b. How much apparent power must the windings of each autotransformer handle?

c. If one of the autotransformers were reconnected as an ordinary transformer, what would its ratings be?

3–13. A 12.4-kV single-phase generator supplies power to a load through a transmission line. The load's impedance is $Z_{load} = 500\angle36.87° \ \Omega$, and the transmission line's impedance is $Z_{line} = 60\angle60° \ \Omega$.

a. If the generator is directly connected to the load (Figure P3–3a), what is the ratio of the load voltage to the generated voltage? What are the transmission losses of the system?

b. If a 1:10 step-up transformer is placed at the output of the generator and a 10:1 transformer is placed at the load end of the transmission line, what is the new ratio of the load voltage to the generated voltage (Figure P3–3b)? What are the transmission losses of the system now? (*Note*: The transformers may be assumed to be ideal.)

3–14. A 5000-VA, 480/120-V conventional transformer is to be used to supply power from a 600-V source to a 120-V load. Consider the transformer to be ideal, and assume that all insulation can handle 600 V.

Figure P3–3 I Circuits for Problem 3–13: (a) without transformers and (b) with transformers.

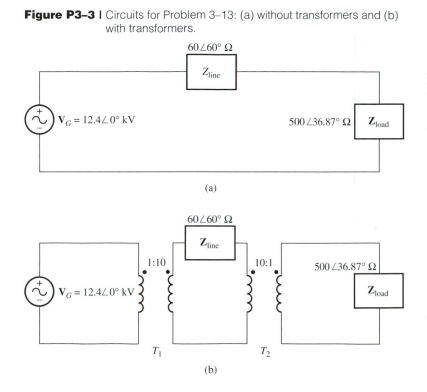

(a)

(b)

a. Sketch the transformer connection that will do the required job.
b. Find the kilovolt-ampere rating of the transformer in the configuration.
c. Find the maximum primary and secondary currents under these conditions.

3–15. A 5000-VA, 480/120-V conventional transformer is to be used to supply power from a 600-V source to a 480-V load. Consider the transformer to be ideal, and assume that all insulation can handle 600 V. Answer the questions of Problem 3–14 for this transformer.

3–16. Prove the following statement: If a transformer having a series impedance Z_{eq} is connected as an autotransformer, its per-unit series impedance Z'_{eq} as an autotransformer will be

$$Z'_{eq} = \frac{N_{SE}}{N_{SE} + N_C} = Z_{eq}$$

Note that this expression is the reciprocal of the autotransformer power advantage.

3–17. Three 25-kVA 24,000/277-V distribution transformers are connected in Δ–Y. The open-circuit test was performed on the low-voltage side of this transformer bank, and the following data were recorded:

$$V_{line,\,OC} = 480 \text{ V} \qquad I_{line,\,OC} = 4.10 \text{ A} \qquad P_{3\phi,\,OC} = 945 \text{ W}$$

The short-circuit test was performed on the high-voltage side of this transformer bank, and the following data were recorded:

$$V_{line,\,SC} = 1400 \text{ V} \qquad I_{line,\,SC} = 1.80 \text{ A} \qquad P_{3\phi,\,SC} = 912 \text{ W}$$

  a. Find the per-unit equivalent circuit of this transformer bank.
  b. Find the voltage regulation of this transformer bank at the rated load and 0.90 PF lagging.
  c. What is the transformer bank's efficiency under these conditions?

3–18. A 20-kVA, 20,000/480-V 60-Hz distribution transformer is tested with the following results:

Open-circuit test (measured from secondary side)	Short-circuit test (measured from primary side)
$V_{OC} = 480$ V	$V_{SC} = 1130$ V
$I_{OC} = 1.51$ A	$I_{SC} = 1.00$ A
$P_{OC} = 271$ W	$P_{SC} = 260$ W

  a. Find the per-unit equivalent circuit for this transformer at 60 Hz.
  b. What would the rating of this transformer be if it were operated on a 50-Hz power system?
  c. Sketch the equivalent circuit of this transformer referred to the primary side *if it is operating at 50 Hz.*

3–19. Prove that the three-phase system of voltages on the secondary of the Y–Δ transformer shown in Figure 3–37b lags the three-phase system of voltages on the primary of the transformer by 30°.

3–20. Prove that the three-phase system of voltages on the secondary of the Δ–Y transformer shown in Figure 3–37c lags the three-phase system of voltages on the primary of the transformer by 30°.

3–21. A single-phase 10-kVA, 480/120-V transformer is to be used as an autotransformer tying a 600-V distribution line to a 480-V load. When it is tested as a conventional transformer, the following values are measured on the primary (480-V) side of the transformer:

Open-circuit test	Short-circuit test
$V_{OC} = 480$ V	$V_{SC} = 10.0$ V
$I_{OC} = 0.41$ A	$I_{SC} = 10.6$ A
$P_{OC} = 38$ W	$P_{SC} = 26$ W

a. Find the per-unit equivalent circuit of this transformer when it is connected in the conventional manner. What is the efficiency of the transformer at rated conditions and unity power factor? What is the voltage regulation at those conditions?

b. Sketch the transformer connections when it is used as a 600/480-V step-down autotransformer.

c. What is the kilovolt-ampere rating of this transformer when it is used in the autotransformer connection?

d. Answer the questions in (a) for the autotransformer connection.

3–22. Figure P3–4 shows a power system consisting of a three-phase 480-V, 60-Hz generator supplying two loads through a transmission line with a pair of transformers at either end.

a. Sketch the per-phase equivalent circuit of this power system.

b. With the switch opened, find the real power P, reactive power Q, and apparent power S supplied by the generator. What is the power factor of the generator?

c. With the switch closed, find the real power P, reactive power Q, and apparent power S supplied by the generator. What is the power factor of the generator?

d. What are the transmission losses (transformer plus transmission line losses) in this system with the switch open? With the switch closed? What is the effect of adding load 2 to the system?

Figure P3–4 I A one-line diagram of the power system of Problem 3–22. Note that some impedance values are given in the per-unit system, while others are given in ohms.

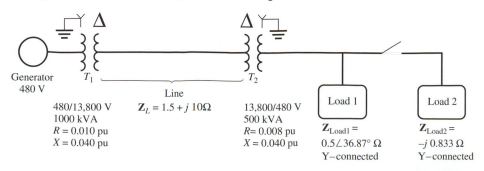

3.16 | REFERENCES

1. Beeman, Donald: *Industrial Power Systems Handbook,* McGraw-Hill, New York, 1955.

2. Chapman, Stephen J.: *Electric Machinery Fundamentals*, 3rd ed., McGraw-Hill, Burr Ridge, Ill., 1999.

3. Del Toro, V.: *Electric Machines and Power Systems,* Prentice-Hall, Englewood Cliffs, N.J., 1985.

4. Feinberg, R.: *Modern Power Transformer Practice,* Wiley, New York, 1979.

5. Fitzgerald, A. E., C. Kingsley, Jr., and S. D. Umans: *Electric Machinery,* 5th ed., McGraw-Hill, New York, 1990.

6. Horowitz, S. H., and A. G. Phadke: *Protective System Relaying*, Research Studies Press, 1992.

7. M.I.T. Staff: *Magnetic Circuits and Transformers,* Wiley, New York, 1943.

8. *Electrical Transmission and Distribution Reference Book*, Westinghouse Electric Corp., East Pittsburgh, 1964.

AC Machinery Fundamentals

AC machines are generators that convert mechanical energy to AC electric energy and motors that convert AC electric energy to mechanical energy. The fundamental principles of AC machines are very simple, but unfortunately, they are somewhat obscured by the complicated construction of real machines. This chapter will first explain the principles of AC machine operation using simple examples, and then consider some of the complications that occur in real AC machines.

There are two major classes of AC machines—synchronous machines and induction machines. *Synchronous machines* are motors and generators whose magnetic field current is supplied by a separate DC power source, while *induction machines* are motors and generators whose field current is supplied by magnetic induction (transformer action) into their field windings. The field circuits of most synchronous and induction machines are located on their rotors. This chapter covers some of the fundamentals common to both types of three-phase AC machines; synchronous machines will be covered in detail in Chapter 5, and induction machines will be covered in Chapter 6.

4.1 | THE ROTATING MAGNETIC FIELD

As any child who has ever played with magnets knows, if you place two magnets beside each other, they will twist around to line up. If there were some way to create two magnetic fields in the rotor (the rotating part) and stator (the stationary part) of a machine, then the rotor would rotate so that its magnetic field is aligned with the stator's magnetic field. If we further had some way to make the stator magnetic field rotate, the rotor would constantly "chase" the stator magnetic field around in a circle, trying to catch up and align itself with it. This, in a nutshell, is the basic principle of all AC motor operation.

How can a stator magnetic field be created and made to rotate? The fundamental principle of AC machine operation is that *if a three-phase set of currents, each of equal magnitude and differing in phase by 120°, flows in a three-phase winding, then it will produce a rotating magnetic field of constant magnitude.* The three-phase winding consists of three separate windings spaced 120 electrical degrees apart around the surface of the machine.

The rotating magnetic field concept is illustrated in the simplest case by an empty stator containing just three coils, each 120° apart (see Figure 4–1a). Since such a winding produces only one north and one south magnetic pole, it is a two-pole winding.

Figure 4–1 | (a) A simple three-phase stator. Currents in this stator are assumed positive if they flow into the unprimed end and out the primed end of the coils. The magnetic field intensity (**H**) and flux density (**B**) produced by each coil are also shown. (b) The magnetic field intensity vector **H**$_{aa'}$(t) produced by a current flowing in coil aa'.

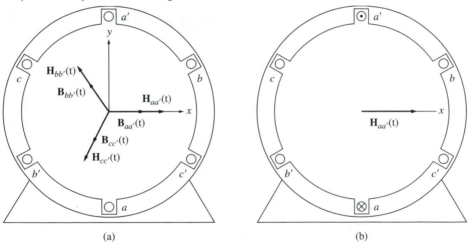

(a)   (b)

To understand the concept of the rotating magnetic field, we will apply a set of currents to the stator of Figure 4–1 and see what happens at specific instants of time. Assume that the currents in the three coils are given by the equations

$$i_{aa'}(t) = I_M \sin \omega t \quad \text{A} \tag{4–1}$$
$$i_{bb'}(t) = I_M \sin(\omega t - 120°) \quad \text{A} \tag{4–2}$$
$$i_{cc'}(t) = I_M \sin(\omega t - 240°) \quad \text{A} \tag{4–3}$$

The current in coil aa' flows into the a end of the coil and out the a' end of the coil. It produces the magnetic field intensity

$$\mathbf{H}_{aa'}(t) = H_M \sin \omega t \angle 0° \quad \text{A-turns/m} \tag{4–4}$$

where $H_M \sin \omega t$ is the magnitude of the magnetic field intensity as a function of time, while 0° is the *spatial angle* of the magnetic field intensity vector, as shown in Figure 4–1b. The direction of the magnetic field intensity vector **H**$_{aa'}$(t) can be determined from the current flowing in coil aa' by the right-hand rule. If the fingers of the right hand curl in the direction of the current flow in the coil, then the resulting magnetic field is in the direction that the thumb points. Notice that the magnitude of the magnetic field intensity vector $\mathbf{H}_{aa'}(t)$ varies sinusoidally in time, but the direction of $\mathbf{H}_{aa'}(t)$ is always constant. Similarly, the magnetic field intensity vectors $\mathbf{H}_{bb'}(t)$ and $\mathbf{H}_{cc'}(t)$ are

$$\mathbf{H}_{bb'}(t) = H_M \sin(\omega t - 120°) \angle 120° \qquad \text{A-turns/m} \qquad (4\text{--}5)$$
$$\mathbf{H}_{cc'}(t) = H_M \sin(\omega t - 240°) \angle 240° \qquad \text{A-turns/m} \qquad (4\text{--}6)$$

The magnetic flux densities resulting from these magnetic field intensities are given by Equation (1-21):

$$\mathbf{B} = \mu\mathbf{H} \qquad (1\text{--}21)$$

They are

$$\mathbf{B}_{aa'}(t) = B_M \sin \omega t \angle 0° \qquad \text{T} \qquad (4\text{--}7)$$
$$\mathbf{B}_{bb'}(t) = B_M \sin(\omega t - 120°) \angle 120° \qquad \text{T} \qquad (4\text{--}8)$$
$$\mathbf{B}_{cc'}(t) = B_M \sin(\omega t - 240°) \angle 240° \qquad \text{T} \qquad (4\text{--}9)$$

where $B_M = \mu H_M$. The currents and their corresponding flux densities can be examined at specific times to determine the resulting net magnetic field in the stator.

For example, at time $\omega t = 0°$, the magnetic flux density from coil aa' will be

$$\mathbf{B}_{aa'} = 0 \qquad (4\text{--}10)$$

The magnetic flux density from coil bb' will be

$$\mathbf{B}_{bb'} = B_M \sin(-120°) \angle 120° \qquad (4\text{--}11)$$

and the magnetic flux density from coil cc' will be

$$\mathbf{B}_{cc'} = B_M \sin(-240°) \angle 240° \qquad (4\text{--}12)$$

The total magnetic field from all three coils added together will be

$$\mathbf{B}_{net} = \mathbf{B}_{aa'} + \mathbf{B}_{bb'} + \mathbf{B}_{cc'}$$
$$= 0 + \left(-\frac{\sqrt{3}}{2}B_M\right) \angle 120° + \left(\frac{\sqrt{3}}{2}B_M\right) \angle 240°$$
$$= 1.5 B_M \angle -90° \qquad (4\text{--}13)$$

The resulting net magnetic field is shown in Figure 4–2a.

As another example, look at the magnetic field at time $\omega t = 90°$. At that time, the currents are

$$i_{aa'} = I_M \sin 90° \qquad \text{A}$$
$$i_{bb'} = I_M \sin(-30°) \qquad \text{A}$$
$$i_{cc'} = I_M \sin(-150°) \qquad \text{A}$$

and the magnetic flux densities are

$$\mathbf{B}_{aa'} = B_M \angle 0°$$
$$\mathbf{B}_{bb'} = -0.5 B_M \angle 120°$$
$$\mathbf{B}_{cc'} = -0.5 B_M \angle 240°$$

Figure 4–2 | (a) The vector magnetic field in a stator at time $\omega t = 0°$. (b) The vector magnetic field in a stator at time $\omega t = 90°$.

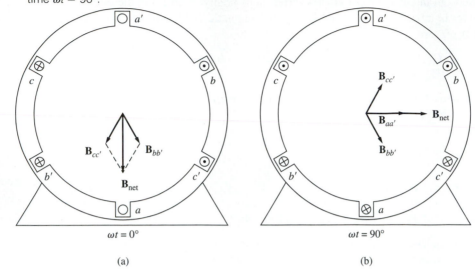

(a)   (b)

The resulting net magnetic flux density is

$$\begin{aligned}
\mathbf{B}_{net} &= \mathbf{B}_{aa'} + \mathbf{B}_{bb'} + \mathbf{B}_{cc'} \\
&= B_M\angle 0° + (-0.5B_M)\angle 120° + (-0.5B_M)\angle 240° \\
&= 1.5B_M\angle 0°
\end{aligned}$$

The resulting magnetic field is shown in Figure 4–2b. Notice that although the *direction* of the magnetic field has changed, the *magnitude* is constant. The magnetic field is maintaining a constant magnitude while rotating in a counterclockwise direction.

Proof of the Rotating Magnetic Field Concept

At any time t, the magnetic field will have the same magnitude $1.5B_M$, and it will continue to rotate at angular velocity ω. A proof of this statement for all time t is now given.

Refer again to the stator shown in Figure 4–1. By the coordinate system shown in the figure, the x direction is to the right and the y direction is upward. The vector $\hat{\mathbf{x}}$ is the unit vector in the horizontal direction, and the vector $\hat{\mathbf{y}}$ is the unit vector in the vertical direction. To find the total magnetic flux density in the stator, simply add vectorially the three component magnetic fields and determine their sum.

The net magnetic flux density in the stator is given by

$$\begin{aligned}
\mathbf{B}_{net}(t) &= \mathbf{B}_{aa'}(t) + \mathbf{B}_{bb'}(t) + \mathbf{B}_{cc'}(t) \\
&= B_M \sin \omega t \angle 0° + B_M \sin(\omega t - 120°)\angle 120° \\
&\quad + B_M \sin(\omega t - 240°)\angle 240° \quad \text{T}
\end{aligned}$$

Each of the three component magnetic fields can now be broken down into its x and y components.

$$\mathbf{B}_{net}(t) = B_M \sin \omega t \, \hat{\mathbf{x}}$$
$$-[0.5 B_M \sin(\omega t - 120°) \hat{\mathbf{x}} + \left[\frac{\sqrt{3}}{2} B_M \sin(\omega t - 120°)\right] \hat{\mathbf{y}}$$
$$-[0.5 B_M \sin(\omega t - 240°) \hat{\mathbf{x}} - \left[\frac{\sqrt{3}}{2} B_M \sin(\omega t - 240°)\right] \hat{\mathbf{y}}$$

Combining x and y components yields

$$\mathbf{B}_{net}(t) = [B_M \sin \omega t - 0.5 B_M \sin(\omega t - 120°) - 0.5 B_M \sin(\omega t - 240°)] \hat{\mathbf{x}}$$
$$+ \left[\frac{\sqrt{3}}{2} B_M \sin(\omega t - 120°)] - \frac{\sqrt{3}}{2} B_M \sin(\omega t - 240°)\right] \hat{\mathbf{y}}$$

By the angle-addition trigonometric identities,

$$\mathbf{B}_{net}(t) = \left[B_M \sin \omega t + \frac{1}{4} B_M \sin \omega t + \frac{\sqrt{3}}{4} B_M \cos \omega t + \frac{1}{4} B_M \sin \omega t - \frac{\sqrt{3}}{4} B_M \cos \omega t\right] \hat{\mathbf{x}}$$
$$+ \left[-\frac{\sqrt{3}}{4} B_M \sin \omega t - \frac{3}{4} B_M \cos \omega t + \frac{\sqrt{3}}{4} B_M \sin \omega t - \frac{3}{4} B_M \cos \omega t\right] \hat{\mathbf{y}}$$

$$\boxed{\mathbf{B}_{net}(t) = (1.5 B_M \sin \omega t)\hat{\mathbf{x}} - (1.5 B_M \cos \omega t)\hat{\mathbf{y}}} \qquad (4\text{–}14)$$

Equation (4–14) is the final expression for the net magnetic flux density. Notice that the magnitude of the field is a constant $1.5 B_M$ and that the angle changes continually in a counterclockwise direction at angular velocity ω. Notice also that at $\omega t = 0°$, $\mathbf{B}_{net} = 1.5 B_M \angle -90°$, and that at $\omega t = 90°$, $\mathbf{B}_{net} = 1.5 B_M \angle 0°$. These results agree with the specific examples examined previously.

The Relationship between Electrical Frequency and the Speed of Magnetic Field Rotation

Figure 4–3 shows that the rotating magnetic field in this stator can be represented as a north pole (where the flux leaves the stator) and a south pole (where the flux enters the stator). These magnetic poles complete one mechanical rotation around the stator surface for each electrical cycle of the applied current. Therefore, the mechanical speed of rotation of the magnetic field in revolutions per second is equal to the electric frequency in hertz:

$$f_e = f_m \qquad \text{two poles} \qquad (4\text{–}15)$$
$$\omega_e = \omega_m \qquad \text{two poles} \qquad (4\text{–}16)$$

Here f_m and ω_m are the mechanical speed in revolutions per second and radians per second, while f_e and ω_e are the electrical speed in hertz and radians per second.

Notice that the windings on the two-pole stator in Figure 4–3 occur in the order (taken counterclockwise)

$$a\text{-}c'\text{-}b\text{-}a'\text{-}c\text{-}b'$$

What would happen in a stator if this pattern were repeated twice within it? Figure 4–4a shows such a stator. There, the pattern of windings (taken counterclockwise) is

$$a\text{-}c'\text{-}b\text{-}a'\text{-}c\text{-}b'\text{-}a\text{-}c'\text{-}b\text{-}a'\text{-}c\text{-}b'$$

Figure 4–3 | The rotating magnetic field in a stator represented as moving north and south stator poles.

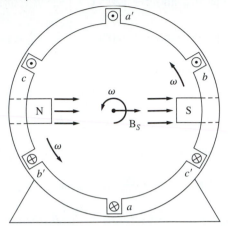

which is just the pattern of the previous stator repeated twice. When a three-phase set of currents is applied to this stator, *two* north poles and *two* south poles are produced in the stator winding, as shown in Figure 4–4b. In this winding, a pole moves only halfway around the stator surface in one electrical cycle. Since one electrical cycle is 360 electrical degrees, and since the mechanical motion is 180 mechanical degrees, the relationship between the electrical angle θ_e and the mechanical angle θ_m, in this stator is

$$\theta_e = 2\theta_m \qquad (4\text{–}17)$$

Thus for the four-pole winding, the electrical frequency of the current is twice the mechanical frequency of rotation:

$$f_e = 2f_m \qquad \text{four poles} \qquad (4\text{–}18)$$
$$\omega_e = 2\omega_m \qquad \text{four poles} \qquad (4\text{–}19)$$

In general, if the number of magnetic poles on an AC machine stator is P, then there are $P/2$ repetitions of the winding sequence a-c'-b-a'-c-b' around its inner surface, and the electrical and mechanical quantities on the stator are related by

$$\theta_e = \frac{P}{2}\,\theta_m \qquad (4\text{–}20)$$

$$f_e = \frac{P}{2}f_m \qquad (4\text{–}21)$$

$$\omega_e = \frac{P}{2}\,\omega_m \qquad (4\text{–}22)$$

Figure 4–4 | (a) A simple four-pole stator winding. (b) The resulting stator magnetic poles. Notice that there are moving poles of alternating polarity every 90° around the stator surface. (c) A winding diagram of the stator as seen from its inner surface, showing how the stator currents produce north and south magnetic poles.

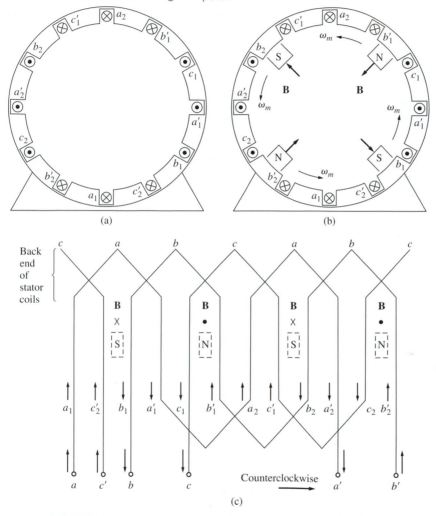

Also, noting that $f_m = n_m/60$, it is possible to relate the electrical frequency in hertz to the resulting mechanical speed of the magnetic fields in revolutions per minute. This relationship is

$$f_e = \frac{n_m P}{120} \qquad (4\text{–}23)$$

Reversing the Direction of Magnetic Field Rotation

Another interesting fact can be observed about the resulting magnetic field. *If the current in any two of the three coils is swapped, the direction of the magnetic field's rotation will be reversed.* This means that it is possible to reverse the direction of rotation of an AC motor just by switching the connections on any two of the three coils. This result is verified below.

To prove that the direction of rotation is reversed, phases *bb′* and *cc′* in Figure 4–1 are switched and the resulting flux density $\mathbf{B}_{net}$ is calculated.

The net magnetic flux density in the stator is given by

$$\mathbf{B}_{net}(t) = \mathbf{B}_{aa'}(t) + \mathbf{B}_{bb'}(t) + \mathbf{B}_{cc'}(t)$$
$$= B_M \sin \omega t \angle 0° + B_M \sin(\omega t - 240°) \angle 120°$$
$$+ B_M \sin(\omega t - 120°) \angle 240° \quad \text{T}$$

Each of the three component magnetic fields can now be broken down into its *x* and *y* components:

$$\mathbf{B}_{net}(t) = B_M \sin \omega t \, \hat{\mathbf{x}}$$
$$-[0.5B_M \sin(\omega t - 240°)\hat{\mathbf{x}} + \left[\frac{\sqrt{3}}{2} B_M \sin(\omega t - 240°)\right]\hat{\mathbf{y}}$$
$$-[0.5B_M \sin(\omega t - 120°)\hat{\mathbf{x}} - \left[\frac{\sqrt{3}}{2} B_M \sin(\omega t - 120°)\right]\hat{\mathbf{y}}$$

Combining *x* and *y* components yields

$$\mathbf{B}_{net}(t) = [B_M \sin \omega t - 0.5B_M \sin(\omega t - 240°) - 0.5B_M \sin(\omega t - 120°)]\hat{\mathbf{x}}$$
$$+ \left[\frac{\sqrt{3}}{2} B_M \sin(\omega t - 240°)] - \frac{\sqrt{3}}{2} B_M \sin(\omega t - 120°)\right]\hat{\mathbf{y}}$$

By the angle-addition trigonometric identities,

$$\mathbf{B}_{net}(t) = \left[B_M \sin \omega t + \frac{1}{4} B_M \sin \omega t - \frac{\sqrt{3}}{4} B_M \cos \omega t + \frac{1}{4} B_M \sin \omega t + \frac{\sqrt{3}}{4} B_M \cos \omega t\right]\hat{\mathbf{x}}$$
$$+ \left[-\frac{\sqrt{3}}{4} B_M \sin \omega t + \frac{3}{4} B_M \cos \omega t + \frac{\sqrt{3}}{4} B_M \sin \omega t + \frac{3}{4} B_M \cos \omega t\right]\hat{\mathbf{y}}$$

$$\boxed{\mathbf{B}_{net}(t) = (1.5B_M \sin \omega t)\hat{\mathbf{x}} + (1.5B_M \cos \omega t)\hat{\mathbf{y}}} \qquad (4\text{–}24)$$

This time the magnetic field has the same magnitude but rotates in a clockwise direction. Therefore, *switching the currents in two stator phases reverses the direction of magnetic field rotation in an AC machine.*

EXAMPLE 4–1

Create a MATLAB program that models the behavior of a rotating magnetic field in the three-phase stator shown in Figure 4–2.

■ Solution

The geometry of the loops in this stator is fixed as shown in Figure 4–2. The currents in the loops are

$$i_{aa'}(t) = I_M \sin \omega t \quad \text{A} \tag{4–1}$$
$$i_{bb'}(t) = I_M \sin(\omega t - 120°) \quad \text{A} \tag{4–2}$$
$$i_{cc'}(t) = I_M \sin(\omega t - 240°) \quad \text{A} \tag{4–3}$$

and the resulting magnetic flux densities are

$$\mathbf{B}_{aa'}(t) = B_M \sin \omega t \angle 0° \quad \text{T} \tag{4–7}$$
$$\mathbf{B}_{bb'}(t) = B_M \sin(\omega t - 120°) \angle 120° \quad \text{T} \tag{4–8}$$
$$\mathbf{B}_{cc'}(t) = B_M \sin(\omega t - 240°) \angle 240° \quad \text{T} \tag{4–9}$$

A simple MATLAB program that plots $\mathbf{B}_{aa'}$, $\mathbf{B}_{bb'}$, $\mathbf{B}_{cc'}$, and $\mathbf{B}_{net}$ as a function of time is shown below.

```
% M-file: mag_field.m
% M-file to calculate the net magetic field produced
% by a three-phase stator.

% Set up the basic conditions
bmax = 1;                % Normalize bmax to 1
freq = 60;               % 60 Hz
w = 2*pi*freq;           % angluar velocity (rad/s)

% First, generate the three component magnetic fields
t = 0:1/6000:1/60;
Baa = sin(w*t) .* (cos(0) + j*sin(0));
Bbb = sin(w*t-2*pi/3) .* (cos(2*pi/3) + j*sin(2*pi/3));
Bcc = sin(w*t+2*pi/3) .* (cos(-2*pi/3) + j*sin(-2*pi/3));

% Calculate Bnet
Bnet = Baa + Bbb + Bcc;

% Calculate a circle representing the expected maximum
% value of Bnet
circle = 1.5 * (cos(w*t) + j*sin(w*t));

% Plot the magnitude and direction of the resulting magnetic
% fields. Note that Baa is black, Bbb is blue, Bcc is
% magneta, and Bnet is red.
for ii = 1:length(t)

   % Plot the reference circle
   plot(circle,'k');
   hold on;

   % Plot the four magnetic fields
   plot([0 real(Baa(ii))],[0 imag(Baa(ii))],'k','LineWidth',2);
```

```
   plot([0 real(Bbb(ii))],[0 imag(Bbb(ii))],'b','LineWidth',2);
   plot([0 real(Bcc(ii))],[0 imag(Bcc(ii))],'m','LineWidth',2);
   plot([0 real(Bnet(ii))],[0 imag(Bnet(ii))],'r','LineWidth',3);
   axis square;
   axis([-2 2 -2 2]);
   drawnow;
   hold off;
end
```

When this program is executed, to draws lines corresponding to the three component magnetic fields as well as a line corresponding to the net magnetic field. Execute this program and observe the behavior of **B**$_{net}$.   ■

A more sophisticated version of this program that takes advantage of the MATLAB graphical user interface is available from the book's website.

4.2 | MAGNETOMOTIVE FORCE AND FLUX DISTRIBUTION ON AC MACHINES

In Section 4.1, the flux produced inside an AC machine was treated as if it were in free space. The direction of the flux density produced by a coil of wire was assumed to be perpendicular to the plane of the coil, with the direction of the flux given by the right-hand rule.

The *flux in a real machine does not behave in the simple manner assumed above*, since there is a ferromagnetic rotor in the center of the machine, with a small air gap between the rotor and the stator. The rotor can be cylindrical, like the one shown in Figure 4–5a, or it can have pole faces projecting out from its surface, as shown in Figure 4–5b. If the rotor is cylindrical, the machine is said to have *nonsalient pole*s; if the rotor has pole faces projecting out from it, the machine is said to have *salient poles*. Cylindrical rotor, or nonsalient-pole, machines are easier to understand and analyze than salient-pole machines, and this discussion will be restricted to machines with cylindrical rotors. Machines with salient poles are discussed in References 1, 3, and 4.

Refer to the cylindrical-motor machine in Figure 4–5a. The reluctance of the air gap in this machine is much higher than the reluctances of either the rotor or the stator, so *the flux density vector* **B** *takes the shortest possible path across the air gap* and jumps perpendicularly between the rotor and the stator.

To produce a sinusoidal voltage in a machine like this, *the magnitude of the flux density vector* **B** *must vary in a sinusoidal manner* along the surface of the air gap. The flux density will vary sinusoidally only if the magnetic field intensity **H** (and magnetomotive force $\mathcal{F}$) varies in a sinusoidal manner along the surface of the air gap (see Figure 4–6).

Figure 4–5 I (a) An AC machine with a cylindrical or nonsalient-pole rotor. (b) An AC machine with a salient-pole rotor.

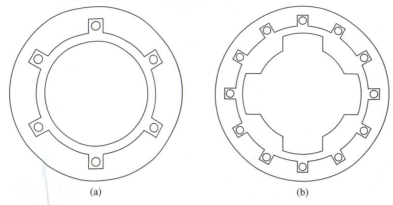

(a)   (b)

The most straightforward way to achieve a sinusoidal variation of magnetomotive force along the surface of the air gap is to distribute the turns of the winding that produces the magnetomotive force in closely spaced slots around the surface of the machine and to vary the number of conductors in each slot in a sinusoidal manner. Figure 4–7a shows such a winding, and Figure 4–7b shows the magnetomotive force resulting from the winding. The number of conductors in each slot is given by the equation

$$n_C = N_C \cos \alpha \tag{4–25}$$

where N_c is the number of conductors at an angle of 0°. As Figure 4–7b shows, this distribution of conductors produces a close approximation to a sinusoidal distribution of magnetomotive force. Furthermore, the more slots there are around the surface of the machine and the more closely spaced the slots are, the better this approximation becomes.

In practice, it is not possible to distribute windings exactly in accordance with Equation (4–25), since there are only a finite number of slots in a real machine and since only integral numbers of conductors can be included in each slot. The resulting magnetomotive force distribution is only approximately sinusoidal, and higher-order harmonic components will be present. Fractional-pitch windings are used to suppress these unwanted harmonic components. Fractional-pitch windings are beyond the scope of this book, but they are explained in a supplement available for download from the book's website.

Furthermore, it is often convenient for the machine designer to include equal numbers of conductors in each slot instead of varying the number in accordance with Equation (4–25). Distributed windings of this type are described in the supplement at the book's website; they have stronger high-order harmonic components than windings designed in accordance with Equation (4–25). Harmonic-suppression techniques are especially important for such windings.

Figure 4–6 | (a) A cylindrical rotor with sinusoidally varying air-gap flux density. (b) The magnetomotive force or magnetic field intensity as a function of angle α in the air gap. (c) The flux density as a function of angle α in the air gap.

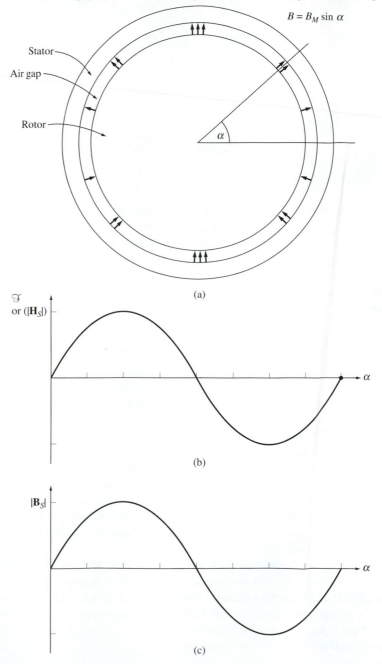

Figure 4–7 | (a) An AC machine with a distributed stator winding designed to produce a sinusoidally varying air-gap flux density. The number of conductors in each slot is indicated on the diagram. (b) The magnetomotive force distribution resulting from the winding, compared to an ideal distribution.

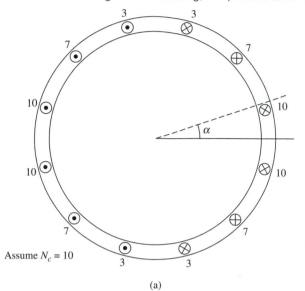

Assume $N_c = 10$

(a)

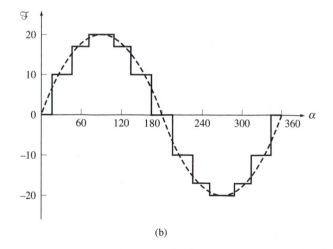

(b)

4.3 | INDUCED VOLTAGE IN AC MACHINES

Just as a three-phase set of currents in a stator can produce a rotating magnetic field, a rotating magnetic field can produce a three-phase set of voltages in the coils of a stator. The equations governing the induced voltage in a three-phase stator will be developed in this section. To make the development easier, we will begin by looking at just one single-turn coil and then expand the results to a more general three-phase stator.

The Induced Voltage in a Coil on a Two-Pole Stator

Figure 4–8 shows a *rotating* rotor with a sinusoidally distributed magnetic field in the center of a *stationary* coil.

We will assume that the magnitude of the flux density vector **B** in the air gap between the rotor and the stator varies sinusoidally with mechanical angle, while the direction of **B** is always radially outward. This sort of flux distribution is the ideal to which machine designers aspire. If α is the angle measured from the direction of the peak rotor flux density, then the magnitude of the flux density vector **B** at a point around the *rotor* is given by

$$B = B_M \cos \alpha \qquad (4\text{–}26)$$

Note that at some locations around the air gap the flux density vector will really point in toward the rotor; in those locations, the sign of Equation (4–26) is negative. Since the rotor is itself rotating within the stator at an angular velocity ω_m, the magnitude of the flux density vector **B** at any angle α around the *stator* is given by

$$\boxed{B = B_M \cos(\omega t - \alpha)} \qquad (4\text{–}27)$$

The equation for the induced voltage in a wire is

$$e_{\text{ind}} = (\mathbf{v} \times \mathbf{B}) \cdot \mathbf{l} \qquad (1\text{–}45)$$

where

$\mathbf{v}$ = velocity of the wire *relative to the magnetic field*

$\mathbf{B}$ = magnetic flux density vector

$\mathbf{l}$ = length of conductor in the magnetic field

However, this equation was derived for the case of a *moving wire* in a *stationary magnetic field*. In this case, the wire is stationary and the magnetic field is moving, so the equation does not directly apply. To use it, we must be in a frame of reference where the magnetic field appears to be stationary. If we "sit on the magnetic field" so that the field appears to be stationary, the sides of the coil will appear to go by at an apparent velocity $\mathbf{v}_{\text{rel}}$, and the equation can be applied. Figure 4–8b shows the vector magnetic field and velocities from the point of view of a stationary magnetic field and a moving wire.

The total voltage induced in the coil will be the sum of the voltages induced in each of its four sides. These voltages are determined on page 178.

Figure 4–8 | (a) A rotating rotor magnetic field inside a stationary stator coil. Detail of coil. (b) The vector magnetic flux densities and velocities on the sides of the coil. The velocities shown are from a frame of reference in which the magnetic field is stationary. (c) The flux density distribution in the air gap.

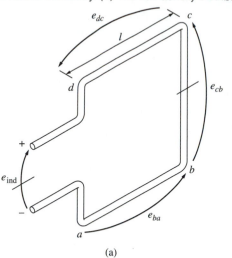

(a)

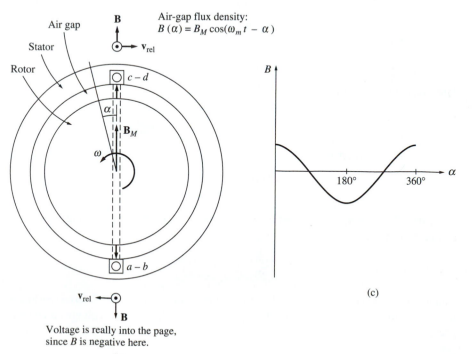

Air-gap flux density:
$B(\alpha) = B_M \cos(\omega_m t - \alpha)$

Voltage is really into the page, since B is negative here.

(b)

(c)

1. *Segment ab.* For segment *ab*, $\alpha = 180°$. Assuming that **B** is directed radially outward from the rotor, the angle between **v** and **B** in segment *ab* is 90°, while the quantity $\mathbf{v} \times \mathbf{B}$ is in the direction of **l**, so

$$\begin{aligned} e_{ba} &= (\mathbf{v} \times \mathbf{B}) \cdot \mathbf{l} \\ &= vBl \qquad \text{directed out of the page} \\ &= -v[B_M \cos(\omega_m t - 180°)]l \\ &= -vB_M l \cos(\omega_m t - 180°) \end{aligned} \qquad (4\text{–}28)$$

where the minus sign comes from the fact that the voltage is built up with a polarity opposite to the assumed polarity.

2. *Segment bc.* The voltage on segment *bc* is zero, since the vector quantity $\mathbf{v} \times \mathbf{B}$ is perpendicular to **l**, so

$$e_{cb} = (\mathbf{v} \times \mathbf{B}) \cdot \mathbf{l} = 0 \qquad (4\text{–}29)$$

3. *Segment cd.* For segment *cd*, the angle $\alpha = 0°$. Assuming that **B** is directed radially outward from the rotor, the angle between **v** and **B** in segment *cd* is 90°, while the quantity $\mathbf{v} \times \mathbf{B}$ is in the direction of **l**, so

$$\begin{aligned} e_{dc} &= (\mathbf{v} \times \mathbf{B}) \cdot \mathbf{l} \\ &= vBl \qquad \text{directed out of the page} \\ e_{dc} &= v(B_M \cos \omega_m t)l \\ &= vB_M l \cos \omega_m t \end{aligned} \qquad (4\text{–}30)$$

4. *Segment da.* The voltage on segment *da* is zero, since the vector quantity $\mathbf{v} \times \mathbf{B}$ is perpendicular to **l**, so

$$e_{ad} = (\mathbf{v} \times \mathbf{B}) \cdot \mathbf{l} = 0 \qquad (4\text{–}31)$$

Therefore, the total voltage on the coil will be

$$\begin{aligned} e_{\text{ind}} &= e_{ba} + e_{dc} \\ &= -vB_M l \cos(\omega_m t - 180°) + vB_M l \cos \omega_m t \end{aligned} \qquad (4\text{–}32)$$

Since $\cos \theta = -\cos(\theta - 180°)$,

$$\begin{aligned} e_{\text{ind}} &= vB_M l \cos \omega_m t + vB_M l \cos \omega_m t \\ &= 2vB_M l \cos \omega_m t \end{aligned} \qquad (4\text{–}33)$$

Since the velocity of the end conductors is given by $v = r\omega_m$, Equation (4–33) can be rewritten as

$$\begin{aligned} e_{\text{ind}} &= 2(r\omega_m)B_M l \cos \omega_m t \\ &= 2rlB_M \omega_m \cos \omega_m t \end{aligned}$$

Finally, the flux passing through the coil can be expressed as $\phi = 2rlB_M$ (see Problem 4–6), while $\omega_m = \omega_e = \omega$ for a two-pole stator, so the induced voltage can be expressed as

$$\boxed{e_{\text{ind}} = \phi\omega \cos \omega t} \qquad (4\text{–}34)$$

Equation (4–34) describes the voltage induced in a single-turn coil. If the coil in the stator has N_C turns of wire, then the total induced voltage of the coil will be

$$e_{\text{ind}} = N_C \phi \omega \cos \omega t \qquad (4\text{–}35)$$

Notice that the voltage produced in stator of this simple AC machine winding is sinusoidal with an amplitude that depends on the flux ϕ in the machine, the angular velocity ω of the rotor, and a constant depending on the construction of the machine (N_C in this simple case).

Note that Equation (4–35) contains the term $\cos \omega t$ instead of the $\sin \omega t$ found in some of the other equations in this chapter. The cosine term has no special significance compared to the sine—it resulted from our choice of reference direction for α in this derivation. If the reference direction for α had been rotated by 90° we would have had a $\sin \omega t$ term.

The Induced Voltage in a Three-Phase Set of Coils

If *three coils*, each of N_c turns, are placed around the rotor magnetic field as shown in Figure 4–9, then the voltages induced in each of them will be the same in magnitude but will differ in phase by 120°. The resulting voltages in each of the three coils are

$$
\begin{aligned}
e_{aa'}(t) &= N_C \phi \omega \sin \omega t & \text{V} \\
e_{bb'}(t) &= N_C \phi \omega \sin(\omega t - 120°) & \text{V} \\
e_{cc'}(t) &= N_C \phi \omega \sin(\omega t - 240°) & \text{V}
\end{aligned}
\qquad (4\text{–}36)
$$

Therefore, a three-phase set of currents can generate a uniform rotating magnetic field in a machine stator, and a uniform rotating magnetic field can generate a three-phase set of voltages in such a stator.

Figure 4–9 | The production of three-phase voltages from three coils spaced 120° apart.

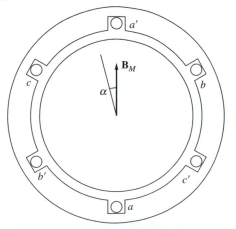

The RMS Voltage in a Three-Phase Stator

The peak voltage in any phase of a three-phase stator of this sort is

$$E_{max} = N_C \phi \omega \tag{4-37}$$

Since $\omega = 2\pi f$, this equation can also be written as

$$E_{max} = 2\pi N_C \phi f \tag{4-38}$$

Therefore, the rms voltage of any phase of this three-phase stator is

$$E_A = \frac{2\pi}{\sqrt{2}} N_C \phi f$$

$$\boxed{E_A = \sqrt{2}\pi N_C \phi f} \tag{4-39}$$

The rms voltage at the *terminals* of the machine will depend on whether the stator is Y or Δ connected. If the machine is Y connected, then the terminal voltage will be $\sqrt{3}$ times E_A; if the machine is Δ connected, then the terminal voltage will just be equal to E_A.

EXAMPLE 4–2

The following information is known about the simple two-pole generator in Figure 4–9. The peak flux density of the rotor magnetic field is 0.2 T; the mechanical rate of rotation of the shaft is 3600 r/min. The stator diameter of the machine is 0.5 m, its coil length is 0.3 m, and there are 15 turns per coil. The machine is Y connected.

a.　What are the three phase voltages of the generator as a function of time?
b.　What is the rms phase voltage of this generator?
c.　What is the rms terminal voltage of this generator?

■ **Solution**
The flux in this machine is given by

$$\phi = 2rlB = dlB$$

where d is the diameter and l is the length of the coil. Therefore, the flux in the machine is given by

$$\phi = (0.5 \text{ m})(0.3 \text{ m})(0.2 \text{ T}) = 0.03 \text{ Wb}$$

The speed of the rotor is given by

$$\omega = (3600 \text{ r/min})(2\pi \text{ rad})(1 \text{ min/60 s}) = 377 \text{ rad/s}$$

a.　The magnitudes of the peak phase voltages are thus

$$E_{max} = N_C \phi \omega$$
$$= (15 \text{ turns})(0.03 \text{ Wb})(377 \text{ rad/s}) = 169.7 \text{ V}$$

and the three phase voltages are

$$e_{aa'}(t) = 169.7 \sin 377t \quad \text{V}$$
$$e_{bb'}(t) = 169.7 \sin(377t - 120°) \quad \text{V}$$
$$e_{cc'}(t) = 169.7 \sin(377t - 240°) \quad \text{V}$$

b. The rms phase voltage of this generator is

$$E_A = \frac{E_{max}}{\sqrt{2}} = \frac{169.7 \text{ V}}{\sqrt{2}} = 120 \text{ V}$$

c. Since the generator is Y connected,

$$V_T = \sqrt{3}\, E_A = \sqrt{3}\,(120 \text{ V}) = 208 \text{ V} \quad \blacksquare$$

4.4 | INDUCED TORQUE IN AN AC MACHINE

In AC machines under normal operating conditions, there are two magnetic fields present—a magnetic field from the rotor circuit and another magnetic field from the stator circuit. The interaction of these two magnetic fields produces the torque in the machine, just as two permanent magnets near each other will experience a torque that causes them to line up.

Figure 4–10 shows a simplified AC machine with a sinusoidal stator flux distribution that peaks in the upward direction and a single coil of wire mounted on the rotor. The stator flux distribution in this machine is

$$B_S(\alpha) = B_S \sin \alpha \tag{4–40}$$

where B_S is the magnitude of the peak flux density; $B_S(\alpha)$ is positive when the flux density vector points radially outward from the rotor surface to the stator surface. How much torque is produced in the rotor of this simplified AC machine? To find out, we will analyze the force and torque on each of the two conductors separately.

The induced force on conductor 1 is

$$\mathbf{F} = i(\mathbf{l} \times \mathbf{B}) \tag{1–43}$$
$$= ilB_S \sin \alpha \quad \text{with direction as shown}$$

The torque on the conductor is

$$\tau_{ind,\,1} = (\mathbf{r} \times \mathbf{F})$$
$$= rilB_S \sin \alpha \quad \text{counterclockwise}$$

The induced force on conductor 2 is

$$\mathbf{F} = i(\mathbf{l} \times \mathbf{B}) \tag{1–43}$$
$$= ilB_S \sin \alpha \quad \text{with direction as shown}$$

The torque on the conductor is

$$\tau_{ind,\,2} = (\mathbf{r} \times \mathbf{F})$$
$$= rilB_S \sin \alpha \quad \text{counterclockwise}$$

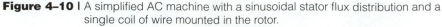

Figure 4–10 I A simplified AC machine with a sinusoidal stator flux distribution and a single coil of wire mounted in the rotor.

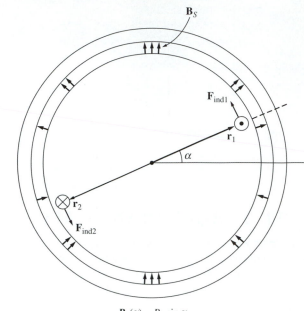

$$B_S(\alpha) = B_S \sin \alpha$$

Therefore, the torque on the rotor loop is

$$\boxed{\tau_{ind} = 2rilB_S \sin \alpha \qquad \text{counterclockwise}} \tag{4–41}$$

Equation (4–41) can be expressed in a more convenient form by examining Figure 4–11 and noting two facts:

1. The current i flowing in the rotor coil produces a magnetic field of its own. The direction of the peak of this magnetic field is given by the right-hand rule, and the magnitude of its magnetic field intensity $\mathbf{H}_R$ is directly proportional to the current flowing in the rotor:

$$H_R = Ci \tag{4–42}$$

   where C is a constant of proportionality.

2. The angle between the peak of the stator flux density $\mathbf{B}_S$ and the peak of the rotor magnetic field intensity $\mathbf{H}_R$ is γ. Furthermore,

$$\gamma = 180° - \alpha \tag{4–43}$$
$$\sin \gamma = \sin(180° - \alpha) = \sin \alpha \tag{4–44}$$

By combining these two observations, the torque on the loop can be expressed as

$$\tau_{ind} = KH_RB_S \sin \alpha \qquad \text{counterclockwise} \tag{4–45}$$

Figure 4–11

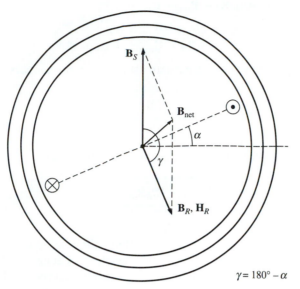

$\gamma = 180° - \alpha$

where K is a constant dependent on the construction of the machine. Note that both the magnitude and the direction of the torque can be expressed by the equation

$$\boxed{\tau_{\text{ind}} = K\mathbf{H}_R \times \mathbf{B}_S} \qquad (4\text{–}46)$$

Finally, since $B_R = \mu H_R$, this equation can be reexpressed as

$$\boxed{\tau_{\text{ind}} = k\mathbf{B}_R \times \mathbf{B}_S} \qquad (4\text{–}47)$$

where $k = K/\mu$. Note that in general k will not be constant, since the magnetic permeability μ varies with the amount of magnetic saturation in the machine.

Equation (4–47) can apply to any AC machine, not just to the simple one-loop rotor described above. Only the constant k will differ from machine to machine. This equation will be used only for a *qualitative* study of torque in AC machines, so the actual value of k is unimportant for our purposes.

The net magnetic field in this machine is the vector sum of the rotor and stator fields (assuming no saturation):

$$\mathbf{B}_{\text{net}} = \mathbf{B}_R + \mathbf{B}_S \qquad (4\text{–}48)$$

This fact can be used to produce an equivalent (and sometimes more useful) expression for the induced torque in the machine. From Equation (4–47)

$$\tau_{\text{ind}} = k\mathbf{B}_R \times \mathbf{B}_S \qquad (4\text{–}47)$$

But from Equation (4–48), $\mathbf{B}_S = \mathbf{B}_{net} - \mathbf{B}_R$, so

$$\begin{aligned}\tau_{ind} &= k\mathbf{B}_R \times (\mathbf{B}_{net} - \mathbf{B}_R) \\ &= k(\mathbf{B}_R \times \mathbf{B}_{net}) - k(\mathbf{B}_R \times \mathbf{B}_R)\end{aligned}$$

Since the cross product of any vector with itself is zero, this reduces to

$$\boxed{\tau_{ind} = k\mathbf{B}_R \times \mathbf{B}_{net}} \qquad (4\text{–}49)$$

so the induced torque can also be expressed as a cross product of $\mathbf{B}_R$ and $\mathbf{B}_{net}$ with the same constant k as before. The magnitude of this expression is

$$\boxed{\tau_{ind} = kB_R B_{net} \sin \delta} \qquad (4\text{–}50)$$

where δ is the angle between $\mathbf{B}_R$ and $\mathbf{B}_{net}$.

Equations (4–47) to (4–50) will be used to help develop a qualitative understanding of the torque in AC machines. For example, look at the simple synchronous machine in Figure 4–12. Its magnetic fields are rotating in a counterclockwise direction. What is the direction of the torque on the shaft of the machine's rotor? By applying the right-hand rule to Equation (4–47) or (4–49), the induced torque is found to be clockwise, or opposite the direction of rotation of the rotor. Therefore, this machine must be acting as a generator.

Figure 4–12 I A simplified synchronous machine showing its rotor and stator magnetic fields.

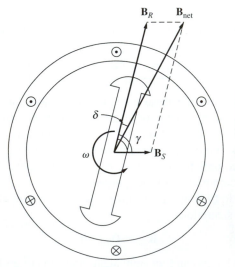

4.5 | WINDING INSULATION IN AN AC MACHINE

One of the most critical parts of an AC machine design is the insulation of its windings. If the insulation of a motor or generator breaks down, the machine shorts out. The repair of a machine with shorted insulation is quite expensive, if it is even possible. To prevent the winding insulation from breaking down as a result of overheating, it is necessary to limit the temperature of the windings. This can be partially done by providing a cooling air circulation over them, but ultimately the maximum winding temperature limits the maximum power that can be supplied continuously by the machine.

Insulation rarely fails from immediate breakdown at some critical temperature. Instead, the increase in temperature produces a gradual degradation of the insulation, making it subject to failure from another cause such as shock, vibration, or electrical stress. There was an old rule of thumb that said that the life expectancy of a motor with a given type of insulation is halved for each 10°C rise in temperature above the rated temperature of the winding. This rule still applies to some extent today.

To standardize the temperature limits of machine insulation, the National Electrical Manufacturers Association (NEMA) in the United States has defined a series of *insulation system classes*. Each insulation system class specifies the maximum temperature rise permissible for that class of insulation. There are three common NEMA insulation classes for integral-horsepower AC motors: B, F, and H. Each class represents a higher permissible winding temperature than the one before it. For example, the armature winding temperature rise above ambient temperature in one type of continuously operating AC induction motor must be limited to 80°C for class B, 105°C for class F, and 125°C for class H insulation.

The specific temperature specifications for each type of AC motor and generator are set out in great detail in NEMA Standard MG1-1993, *Motors and Generators*. Similar standards have been defined by the International Electrotechnical Commission (IEC) and by various national standards organizations in other countries.

4.6 | AC MACHINE POWER FLOWS AND LOSSES

AC generators take in mechanical power and produce electric power, while AC motors take in electric power and produce mechanical power. In either case, not all the power input to the machine appears in useful form at the other end—there is *always* some loss associated with the process.

The efficiency of an AC machine is defined by the equation

$$\eta = \frac{P_{out}}{P_{in}} \times 100\%$$

(4–51)

The difference between the input power and the output power of a machine is the losses that occur inside it. Therefore,

$$\eta = \frac{P_{in} - P_{loss}}{P_{in}} \times 100\%$$
(4–52)

The Losses in AC Machines

The losses that occur in AC machines can be divided into four basic categories:

1. Electrical or copper losses (I^2R losses)
2. Core losses
3. Mechanical losses
4. Stray load losses

Electrical or Copper Losses Copper losses are the resistive heating losses that occur in the stator (armature) and rotor (field) windings of the machine. The stator copper losses (SCL) in a three-phase AC machine are given by the equation

$$P_{SCL} = 3I_A^2 R_A$$
(4–53)

where I_A is the current flowing in each armature phase and R_A is the resistance of each armature phase.

The rotor copper losses (RCL) of a synchronous AC machine (induction machines will be considered separately in Chapter 7) are given by

$$P_{RCL} = 3I_F^2 R_F$$
(4–54)

where I_F is the current flowing in the field winding on the rotor and R_F is the resistance of the field winding. The resistance used in these calculations is usually the winding resistance at normal operating temperature.

Core Losses The core losses are the hysteresis losses and eddy current losses occurring in the metal of the motor. These losses were described in Chapter 1. These losses vary as the square of the flux density (B^2) and, for the stator, as the 1.5 power of the speed of rotation of the magnetic fields ($n^{1.5}$).

Mechanical Losses The mechanical losses in an AC machine are the losses associated with mechanical effects. There are two basic types of mechanical losses: *friction* and *windage*. Friction losses are losses caused by the friction of the bearings in the machine, while windage losses are caused by the friction between the moving parts of the machine and the air inside the motor's casing. These losses vary as the cube of the speed of rotation of the machine.

The mechanical and core losses of a machine are often lumped together and called the *no-load rotational loss* of the machine. At no load, all the input power must be used to overcome these losses. Therefore, measuring the input power to the stator of an AC machine acting as a motor at no load will give an approximate value for these losses.

Stray Losses (or Miscellaneous Losses) Stray losses are losses that cannot be placed in one of the previous categories, usually because of inaccuracies in modeling. No matter how carefully losses are accounted for, some always escape inclusion in one of the above categories. All such losses are lumped into stray losses. For many machines, stray losses are taken by convention to be 1 percent of full load.

The Power-Flow Diagram

One of the most convenient techniques for accounting for power losses in a machine is the *power-flow diagram*. A power-flow diagram for an AC generator is shown in Figure 4–13a. In this figure, mechanical power is input into the machine, and then the stray losses, mechanical losses, and core loses are subtracted. After they have been subtracted, the remaining power is ideally converted from mechanical to electrical form at the point labeled P_{conv}. The mechanical power that is converted is given by

$$P_{\text{conv}} = \tau_{\text{ind}}\omega_m \qquad (4\text{--}55)$$

and the same amount of electrical power is produced. However, this is not the power that appears at the machine's terminals. Before the terminals are reached, the electrical I^2R losses must be subtracted.

Figure 4–13 | (a) The power-flow diagram of a three-phase AC generator. (b) The power-flow diagram of a three-phase AC motor.

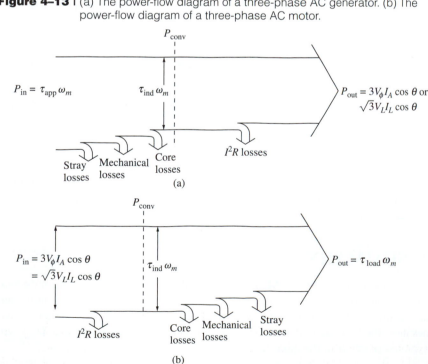

In the case of AC motors, this power-flow diagram is simply reversed. The power-flow diagram for a motor is shown in Figure 4–13b.

Example problems involving the calculation of AC motor and generator efficiencies will be given in the next two chapters.

4.7 | VOLTAGE REGULATION AND SPEED REGULATION

Generators are often compared to each other by a figure of merit called *voltage regulation*. Voltage regulation (VR) is a measure of the ability of a generator to keep a constant voltage at its terminals as load varies. It is defined by the equation

$$ \text{VR} = \frac{V_{\text{nl}} - V_{\text{fl}}}{V_{\text{fl}}} \times 100\% \qquad (4\text{–}56) $$

where V_{nl} is the no-load terminal voltage of the generator and V_{fl} is the full-load terminal voltage of the generator. It is a rough measure of the shape of the generator's voltage-current characteristic—a positive voltage regulation means a dropping characteristic, and a negative voltage regulation means a rising characteristic. A small VR is "better" in the sense that the voltage at the terminals of the generator is more constant with variations in load.

Similarly, motors are often compared to each other using a figure of merit called *speed regulation*. Speed regulation (SR) is a measure of the ability of a motor to keep a constant shaft speed with as load varies. It is defined by the equation

$$ \text{SR} = \frac{n_{\text{nl}} - n_{\text{fl}}}{n_{\text{fl}}} \times 100\% \qquad (4\text{–}57) $$

or

$$ \text{SR} = \frac{\omega_{\text{nl}} - \omega_{\text{fl}}}{\omega_{\text{fl}}} \times 100\% \qquad (4\text{–}58) $$

It is a rough measure of the shape of a motor's torque-speed characteristic—a positive speed regulation means that a motor's speed drops with increasing load, and a negative speed regulation means a motor's speed increases with increasing load. The magnitude of the speed regulation tells approximately how steep the slope of the torque-speed curve is.

4.8 | SUMMARY

There are two major types of AC machines: synchronous machines and induction machines. The principal difference between the two types is that synchronous machines require a DC field current to be supplied to their rotors, while induction machines have the field current induced in their rotors by transformer action. They will be explored in detail in the next three chapters.

A three-phase system of currents supplied to a system of three coils spaced 120 electrical degrees apart on a stator will produce a uniform rotating magnetic field within the stator. The *direction of rotation* of the magnetic field can be *reversed* by simply swapping the connections to any two of the three phases. Conversely, a rotating magnetic field will produce a three-phase set of voltages within such a set of coils.

In stators of more than two poles, one complete mechanical rotation of the magnetic fields produces more than one complete electrical cycle. For such a stator, one mechanical rotation produces $P/2$ electrical cycles. Therefore, the electrical angle of the voltages and currents in such a machine is related to the mechanical angle of the magnetic fields by

$$\theta_e = \frac{P}{2}\,\theta_m$$

The relationship between the electrical frequency of the stator and the mechanical rate of rotation of the magnetic fields is

$$f_e = \frac{n_m P}{120}$$

The types of losses that occur in AC machines are electrical or copper losses (I^2R losses), core losses, mechanical losses, and stray losses. Each of these losses was described in this chapter, along with the definition of overall machine efficiency. Finally, voltage regulation was defined for generators as

$$\boxed{\text{VR} = \frac{V_{nl} - V_{fl}}{V_{fl}} \times 100\%}$$

and speed regulation was defined for motors as

$$\boxed{\text{SR} = \frac{n_{nl} - n_{fl}}{n_{fl}} \times 100\%}$$

4.9 | QUESTIONS

4–1. What is the principal difference between a synchronous machine and an induction machine?

4–2. Why does switching the current flows in any two phases reverse the direction of rotation of a stator's magnetic field?

4–3. What is the relationship between electrical frequency and magnetic field speed for an AC machine?

4–4. What is the equation for the induced torque in an AC machine?

4.10 | PROBLEMS

4–1. Develop a table showing the speed of magnetic field rotation in AC machines of 2, 4, 6, 8, 10, 12, and 14 poles operating at frequencies of 50, 60, and 400 Hz.

4–2. A three-phase four-pole winding is installed in 12 slots on a stator. There are 40 turns of wire in each slot of the windings. All coils in each phase are connected in series, and the three phases are connected in Δ. The flux per pole in the machine is 0.060 Wb, and the speed of rotation of the magnetic field is 1800 r/min.

    a.   What is the frequency of the voltage produced in this winding?

    b.   What are the resulting phase and terminal voltages of this stator?

4–3. A three-phase Y-connected 50-Hz two-pole synchronous machine has a stator with 2000 turns of wire per phase. What rotor flux would be required to produce a terminal (line-to-line) voltage of 6 kV?

4–4. Modify the MATLAB in Example 4–1 by swapping the currents flowing in any two phases. What happens to the resulting net magnetic field?

4–5. If an AC machine has the rotor and stator magnetic fields shown in Figure P4–1, what is the direction of the induced torque in the machine? Is the machine acting as a motor or generator?

Figure P4–1 | The AC machine of Problem 4–5.

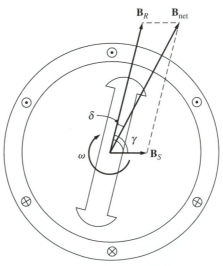

4–6. The flux density distribution over the surface of a two-pole stator of radius r and length l is given by

$$B = B_M \cos(\omega_m t - \alpha) \qquad (4\text{–}27)$$

Prove that the total flux under each pole face is

$$\phi = 2rlB_m$$

4.11 REFERENCES

1. Chapman, Stephen J.: *Electric Machinery Fundamentals*, 3rd ed., McGraw-Hill, Burr Ridge, Ill., 1999.

2. Del Toro, Vincent: *Electric Machines and Power Systems*, Prentice-Hall, Englewood Cliffs, N.J., 1985.

3. Fitzgerald, A. E., and Charles Kingsley: *Electric Machinery*, McGraw-Hill, New York, 1952.

4. Fitzgerald, A. E., Charles Kingsley, and S. D. Umans: *Electric Machinery*, 5th Ed., McGraw-Hill, New York, 1990.

5. International Electrotechnical Commission, *Rotating Electrical Machines Part 1: Rating and Performance*, IEC 34–1 (R1994), 1994.

6. Liwschitz-Garik, Michael, and Clyde Whipple: *Alternating-Current Machinery*, Van Nostrand, Princeton, N.J., 1961.

7. National Electrical Manufacturers Association: *Motors and Generators*, Publication MG1-1993, Washington, 1993.

8. Werninck, E. H. (ed.): *Electric Motor Handbook,* McGraw-Hill Book Company, London, 1978.

Synchronous Machines

Synchronous machines are AC machines that have a field circuit supplied by an external DC source. *Synchronous generators* or *alternators* are synchronous machines used to convert mechanical power to AC electrical power. *Synchronous motors* are synchronous machines used to convert AC electrical power to mechanical. This chapter explores the operation of synchronous generators and synchronous motors. The early sections of this chapter deal with synchronous machines acting as generators, while the later sections in the chapter deal with synchronous machines acting as motors.

5.1 | SYNCHRONOUS MACHINE CONSTRUCTION

In a synchronous generator, a DC current is applied to the rotor winding, which produces a rotor magnetic field. The rotor of the generator is then turned by a prime mover, producing a rotating magnetic field within the machine. This rotating magnetic field induces a three-phase set of voltages within the stator windings of the generator.

Synchronous motors reverse this process: a three-phase set of stator currents produces a rotating magnetic field, which causes the rotor magnetic field to align with it. Since the stator magnetic field is rotating, the rotor rotates as it tries to keep up with the moving stator magnetic fields, supplying mechanical power to a load.

Two terms commonly used to describe the windings on a machine are *field windings* and *armature windings*. In general, the term field windings applies to the windings that produce the main magnetic field in a machine, and the term armature windings applies to the windings where the main voltage is induced. For synchronous machines, the field windings are on the rotor, so the terms rotor windings and field windings are used interchangeably. Similarly, the terms stator windings and armature windings are used interchangeably.

The rotor of a synchronous machine is essentially a large electromagnet. The magnetic poles on the rotor can be of either salient or nonsalient construction. The term *salient* means protruding or sticking out, and a *salient pole* is a magnetic pole that sticks out from the surface of the rotor. On the other hand, a *nonsalient pole* is a magnetic pole constructed flush with the surface of the rotor. A nonsalient-pole rotor is shown in Figure 5–1, while a salient-pole rotor is shown in Figure 5–2. Nonsalient-pole rotors are normally used for two- and four-pole rotors, while salient-pole rotors are normally used for rotors with four or more poles. Because the rotor is subjected to changing magnetic fields, it is constructed of thin laminations to reduce eddy current losses.

Figure 5–1 | A nonsalient two-pole rotor for a synchronous machine.

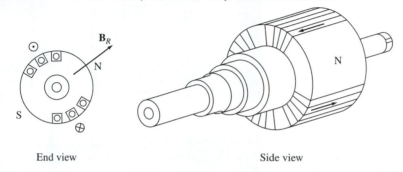

End view   Side view

A DC current must be supplied to the field circuit on the rotor. Since the rotor is rotating, a special arrangement is required to get the DC power to its field windings. There are two common approaches to supplying this DC power:

1. Supply the DC power from an external DC source to the rotor by means of *slip rings* and *brushes*.
2. Supply the DC power from a special DC power source mounted directly on the shaft of the synchronous machine.

Slip rings are metal rings completely encircling the shaft of a machine but insulated from it. One end of the DC rotor winding is tied to each of the two slip rings on the shaft of the synchronous machine, and a stationary brush rides on each slip ring. A "brush" is a block of graphite-like carbon compound that conducts electricity freely but has very low friction, so that it doesn't wear down the slip ring. If the positive end of a DC voltage source is connected to one brush and the negative end is connected to the other, then the same DC voltage will be applied to the field winding at all times regardless of the angular position or speed of the rotor.

Slip rings and brushes create a few problems when they are used to supply DC power to the field windings of a synchronous machine. They increase the amount of maintenance required on the machine, since the brushes must be checked for wear regularly. In addition, brush voltage drop can be the cause of significant power losses

Figure 5–2 | (a) A salient six-pole rotor for a synchronous machine. (b) Photograph of a salient eight-pole synchronous machine rotor showing the windings on the individual rotor poles. *(Courtesy of General Electric Company.)* (c) Photograph of a single salient pole from a rotor with the field windings not yet in place. *(Courtesy of General Electric Company.)* (d) A single salient pole shown after the field windings are installed but before it is mounted on the rotor. *(Courtesy of Westinghouse Electric Company.)*

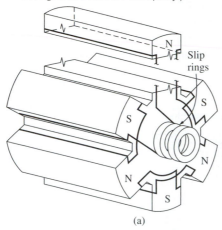

(a)

(b)

(c)

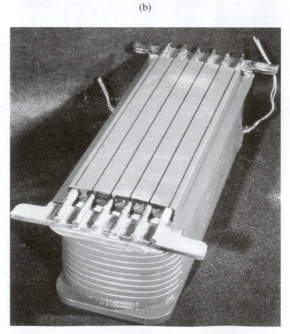

(d)

on machines with larger field currents. Despite these problems, slip rings and brushes are used on all smaller synchronous machines, because no other method of supplying the DC field current is cost-effective.

On larger generators and motors, *brushless exciters* are used to supply the DC field current to the machine. A brushless exciter is a small AC generator with its field circuit mounted on the stator and its armature circuit mounted on the rotor shaft. The three-phase output of the exciter generator is rectified to direct current by a three-phase rectifier circuit also mounted on the shaft of the generator and is then fed into the main DC field circuit. By controlling the small DC field current of the exciter generator (located on the stator), it is possible to adjust the field current on the main machine *without slip rings and brushes*. This arrangement is shown schematically in Figure 5–3, and a synchronous machine rotor with a brushless exciter mounted on the same shaft is shown in Figure 5–4. Since no mechanical contacts ever occur between the rotor and the stator, a brushless exciter requires much less maintenance than slip rings and brushes.

Figure 5–3 | A brushless exciter circuit. A small three-phase current is rectified and used to supply the field circuit of the exciter, which is located on the stator. The output of the armature circuit of the exciter (on the rotor) is then rectified and used to supply the field current of the main machine.

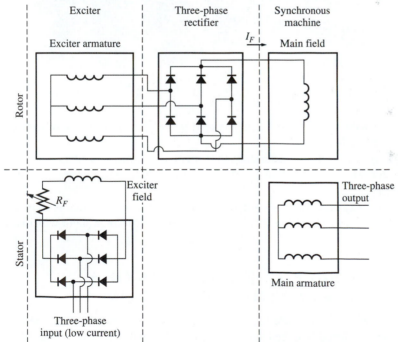

To make the excitation of a generator *completely* independent of any external power sources, a small pilot exciter is often included in the system. A *pilot exciter* is a small AC generator with *permanent magnets* mounted on the rotor shaft and a three-phase winding on the stator. It produces the power for the field circuit of the exciter, which in turn controls the field circuit of the main machine. If a pilot exciter is

Figure 5–4 | Photograph of a synchronous machine rotor with a brushless exciter mounted on the same shaft. Notice the rectifying electronics visible next to the armature of the exciter.

included on the generator shaft, then *no external electric power* is required to run the generator (see Figure 5–5).

Many synchronous generators that include brushless exciters also have slip rings and brushes, so that an auxiliary source of field DC current is available in emergencies.

A cutaway diagram of a complete large synchronous machine is shown in Figure 5–6. This drawing shows an eight-pole salient-pole rotor, a stator with distributed double-layer windings, and a brushless exciter.

5.2 | THE SPEED OF ROTATION OF A SYNCHRONOUS GENERATOR

Synchronous generators are by definition *synchronous*, meaning that the electrical frequency produced is locked in or synchronized with the mechanical rate of rotation of the generator. A synchronous generator's rotor consists of an electromagnet to which direct current is supplied. The rotor's magnetic field points in whatever direction the rotor is turned. Now, the rate of rotation of the magnetic fields in the machine is related to the stator electrical frequency by Equation (4–23):

$$f_e = \frac{n_m P}{120}$$
(4–23)

Figure 5–5 | A brushless excitation scheme that includes a pilot exciter. The permanent magnets of the pilot exciter produce the field current of the exciter, which in turn produces the field current of the main machine.

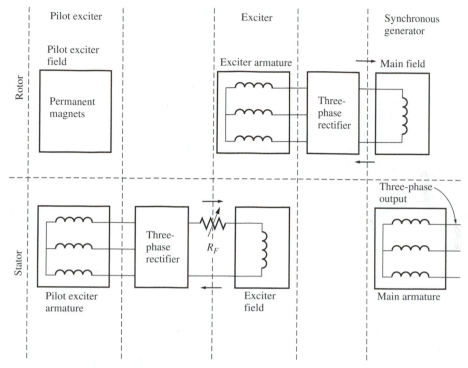

Figure 5–6 | A cutaway diagram of a large synchronous machine. Note the salient-pole construction and the on-shaft exciter. *(Courtesy of General Electric Company.)*

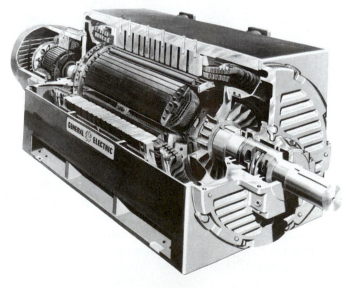

where

f_e = electrical frequency in Hz

n_m = mechanical speed of magnetic field, in r/min (= speed of rotor for synchronous machines)

P = number of poles

Since the rotor turns at the same speed as the magnetic field, *this equation relates the speed of rotor rotation to the resulting electrical frequency*. Electric power is generated at 50 or 60 Hz, so the generator must turn at a fixed speed depending on the number of poles on the machine. For example, to generate 60-Hz power in a two-pole machine, the rotor *must* turn at 3600 r/min. To generate 50-Hz power in a four-pole machine, the rotor *must* turn at 1500 r/min. The required rate of rotation for a given frequency can always be calculated from Equation (4–23).

Equation (4–23) shows that 60 Hz power can be produced by rotating the shaft at 3600 r/min for a 2-pole machine, at 1800 r/min for a 4-pole machine, and so forth. Therefore, the shaft of the generator can rotate at different speeds while still generating the same electrical frequency, if the number of poles in the machine is varied. This fact is important in the real world, because different sources of mechanical power rotate at different speeds. For example, steam turbines are most efficient when they rotate at high speeds, so steam turbines usually rotate at 3600 r/min and supply mechanical power to a 2-pole generator. On the other hand, water turbines are most efficient when they rotate at low speeds (200 to 300 r/min), so water turbines are usually connected to generators with many poles.

5.3 | THE INTERNAL GENERATED VOLTAGE OF A SYNCHRONOUS GENERATOR

In Chapter 4, the magnitude of the voltage induced in a given stator phase was found to be

$$E_A = \sqrt{2}\pi N_C \phi f \qquad (4\text{–}39)$$

This voltage depends on the flux ϕ in the machine, the frequency or speed of rotation, and the machine's construction. In solving problems with synchronous machines, this equation is sometimes rewritten in a simpler form that emphasizes the quantities that are variable during machine operation. This simpler form is

$$\boxed{E_A = K\phi\omega} \qquad (5\text{–}1)$$

where K is a constant representing the construction of the machine.

The internal generated voltage E_A is directly proportional to the flux and to the speed, but the flux itself depends on the current flowing in the rotor field circuit. The field circuit I_F is related to the flux ϕ in the manner shown in Figure 5–7a. Since E_A is directly proportional to the flux, the internal generated voltage E_A is related to the field current as shown in Figure 5–7b. This plot is called the *magnetization curve* or the *open-circuit characteristic* of the machine.

Figure 5–7 | (a) Plot of flux versus field current for a synchronous generator.
(b) The magnetization curve for the synchronous generator.

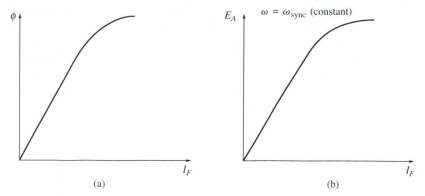

5.4 | THE EQUIVALENT CIRCUIT OF A SYNCHRONOUS GENERATOR

The voltage $\mathbf{E}_A$ is the internal generated voltage produced in one phase of a synchronous generator. However, this voltage $\mathbf{E}_A$ is *not* usually the voltage that appears at the terminals of the generator. In fact, the only time the internal voltage $\mathbf{E}_A$ is the same as the output voltage $\mathbf{V}_\phi$ of a phase is when there is no armature current flowing in the machine. Why is the output voltage $\mathbf{V}_\phi$ from a phase not equal to $\mathbf{E}_A$, and what is the relationship between the two voltages? The answer to these questions yields the model of a synchronous generator.

There are a number of factors that cause the difference between $\mathbf{E}_A$ and $\mathbf{V}_\phi$:

1. The distortion of the air-gap magnetic field by the current flowing in the stator, called *armature reaction*
2. The self-inductance of the armature coils
3. The resistance of the armature coils
4. The effect of salient-pole rotor shapes

We will explore the effects of the first three factors and derive a machine model from them. In this chapter, the effects of a salient-pole shape on the operation of a synchronous machine will be ignored; in other words, all the machines in this chapter are assumed to have nonsalient or cylindrical rotors. Making this assumption will cause the calculated answers to be slightly inaccurate if a machine does indeed have salient-pole rotors, but the errors are relatively minor. A brief discussion of the effects of rotor pole saliency can be found on the book's website.

The first effect mentioned, and normally the largest one, is armature reaction. When a synchronous generator's rotor is spun, a voltage $\mathbf{E}_A$ is induced in the generator's stator windings. If a load is attached to the terminals of the generator, a current flows. But a three-phase stator current flow will produce a magnetic field of its own in the machine. This *stator* magnetic field adds to the original rotor magnetic field, changing the resulting phase voltage. This effect is called *armature reaction* because the armature (stator) current affects the magnetic field that produced it in the first place.

To understand armature reaction, refer to Figure 5–8. Figure 5–8a shows a two-pole rotor spinning inside a three-phase stator. There is no load connected to the stator. The rotor magnetic field $\mathbf{B}_R$ produces an internal generated voltage $\mathbf{E}_A$ whose peak value coincides with the direction of $\mathbf{B}_R$. As was shown in the last chapter, the voltage will be positive out of the conductors at the top and negative into the conductors at the bottom of the figure. With no load on the generator, there is no armature current flow, and $\mathbf{E}_A$ will be equal to the phase voltage $\mathbf{V}_\phi$.

Figure 5–8 | The development of a model for armature reaction: (a) A rotating magnetic field produces the internal generated voltage $\mathbf{E}_A$. (b) The resulting voltage produces a lagging *current flow* when connected to a lagging load. (c) The stator current produces its own magnetic field $\mathbf{B}_S$, which produces its own voltage $\mathbf{E}_{\text{stat}}$ in the stator windings of the machine. (d) The field $\mathbf{B}_S$ adds to $\mathbf{B}_R$, distorting it into $\mathbf{B}_{\text{net}}$. The voltage $\mathbf{E}_{\text{stat}}$ adds to $\mathbf{E}_A$, producing $\mathbf{V}_\phi$ at the output of the phase.

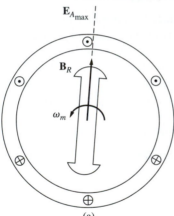

(a)

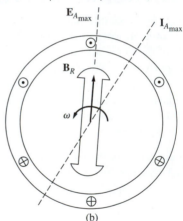

(b)

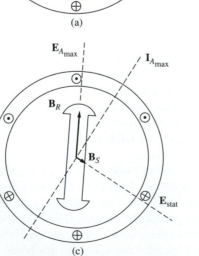

(c)

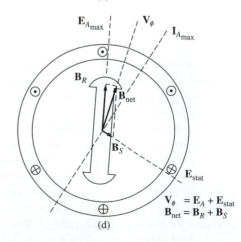

$$\mathbf{V}_\phi = \mathbf{E}_A + \mathbf{E}_{\text{stat}}$$
$$\mathbf{B}_{\text{net}} = \mathbf{B}_R + \mathbf{B}_S$$

(d)

Now suppose that the generator is connected to a lagging load. Because the load is lagging, the peak current will occur at an angle *behind* the peak voltage. This effect is shown in Figure 5–8b.

The current flowing in the stator windings produces a magnetic field of its own. This stator magnetic field is called $\mathbf{B}_S$ and its direction is given by the right-hand rule to be as shown in Figure 5–8c. The stator magnetic field $\mathbf{B}_S$ produces a voltage of its own in the stator, and this voltage is called $\mathbf{E}_{stat}$ on the figure.

With two voltages present in the stator windings, the total voltage in a phase is just the *sum* of the internal generated voltage $\mathbf{E}_A$ and the armature reaction voltage $\mathbf{E}_{stat}$:

$$\mathbf{V}_\phi = \mathbf{E}_A + \mathbf{E}_{stat} \tag{5-2}$$

The net magnetic field $\mathbf{B}_{net}$ is just the sum of the rotor and stator magnetic fields:

$$\mathbf{B}_{net} = \mathbf{B}_R + \mathbf{B}_S \tag{5-3}$$

Since the angles of $\mathbf{E}_A$ and $\mathbf{B}_R$ are the same, and the angles of $\mathbf{E}_{stat}$ and $\mathbf{B}_S$ are the same, the resulting magnetic field $\mathbf{B}_{net}$ will coincide with the net voltage $\mathbf{V}_\phi$. The resulting voltages and currents are shown in Figure 5–8d.

How can the effects of armature reaction on the phase voltage be modeled? First, note that the voltage $\mathbf{E}_{stat}$ lies at an angle of 90° behind the plane of maximum current $\mathbf{I}_A$. Second, the voltage $\mathbf{E}_{stat}$ is directly proportional to the current $\mathbf{I}_A$. If X is a constant of proportionality, then *the armature reaction voltage can be expressed as*

$$\mathbf{E}_{stat} = -jX\mathbf{I}_A \tag{5-4}$$

The voltage on a phase is thus

$$\boxed{\mathbf{V}_\phi = \mathbf{E}_A - jX\mathbf{I}_A} \tag{5-5}$$

Look at the circuit shown in Figure 5–9. The Kirchhoff's voltage law equation for this circuit is

$$\mathbf{V}_\phi = \mathbf{E}_A - jX\mathbf{I}_A \tag{5-6}$$

This is exactly the same equation as the one describing the armature reaction voltage. Therefore, the armature reaction voltage can be modeled as an inductor in series with the internal generated voltage.

Figure 5–9 I A simple circuit (see text).

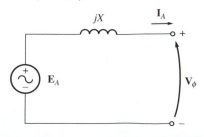

In addition to the effects of armature reaction, the stator coils have a self-inductance and a resistance. If the stator self-inductance is called L_A (and its corresponding reactance is called X_A) while the stator resistance is called R_A, then the total difference between $\mathbf{E}_A$ and $\mathbf{V}_\phi$ is given by

$$\mathbf{V}_\phi = \mathbf{E}_A - jX\mathbf{I}_A - jX_A\mathbf{I}_A - R_A\mathbf{I}_A \tag{5--7}$$

The armature reaction effects and the self-inductance in the machine are both represented by reactances, and it is customary to combine them into a single reactance, called the *synchronous reactance* of the machine:

$$X_S = X + X_A \tag{5--8}$$

Therefore, the final equation describing $\mathbf{V}_\phi$ is

$$\boxed{\mathbf{V}_\phi = \mathbf{E}_A - jX_S\mathbf{I}_A - R_A\mathbf{I}_A} \tag{5--9}$$

Figure 5–10 | The full equivalent circuit of a three-phase synchronous generator.

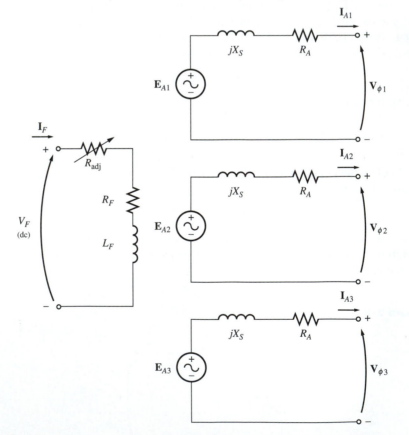

It is now possible to sketch the equivalent circuit of a three-phase synchronous generator. The full equivalent circuit of such a generator is shown in Figure 5–10. This figure shows a DC power source supplying the rotor field circuit, which is modeled by the coil's inductance and resistance in series. In series with R_F is an adjustable resistor R_{adj} which controls the flow of field current. The rest of the equivalent circuit consists of the models for each phase. Each phase has an internal generated voltage with a series inductance X_S (consisting of the sum of the armature

Figure 5–11 I The generator equivalent circuit connected in (a) Y and (b) Δ.

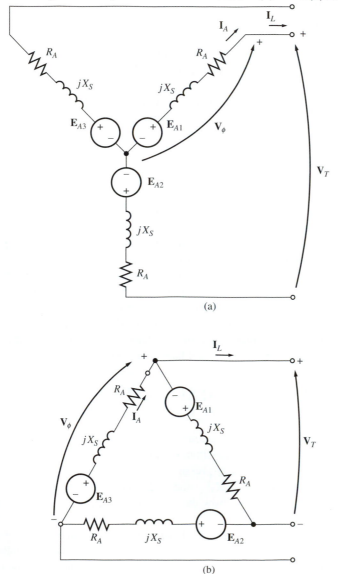

(a)

(b)

Figure 5–12 | The per-phase equivalent circuit of a synchronous generator. The internal field circuit resistance and the external variable resistance have been combined into a single resistor R_F.

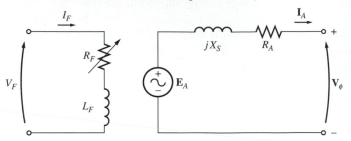

reactance and the coil's self-inductance) and a series resistance R_A. The voltages and currents of the three phases are 120° apart in angle, but otherwise the three phases are identical.

These three phases can be either Y or Δ connected as shown in Figure 5–11. If they are Y connected, then the terminal voltage V_T is related to the phase voltage by

$$V_T = \sqrt{3}V_\phi \tag{5–10}$$

If they are Δ connected, then

$$V_T = V_\phi \tag{5–11}$$

The fact that the three phases of a synchronous generator are identical in all respects except for phase angle normally leads to the use of a *per-phase equivalent circuit*. The per-phase equivalent circuit of this machine is shown in Figure 5–12. One important fact must be kept in mind when the per-phase equivalent circuit is used: The three phases have the same voltages and currents *only* when the loads attached to them are *balanced*. If the generator's loads are not balanced, more sophisticated techniques of analysis are required. One of these methods (the method of symmetrical components) is discussed later in this book.

5.5 | THE PHASOR DIAGRAM OF A SYNCHRONOUS GENERATOR

Because the voltages in a synchronous generator are AC voltages, they are usually expressed as phasors. Since phasors have both a magnitude and an angle, the relationship between them must be expressed by a two-dimensional plot. When the voltages within a phase ($\mathbf{E}_A$, $\mathbf{V}_\phi$, $jX_S\mathbf{I}_A$, and $R_A\mathbf{I}_A$) and the current $\mathbf{I}_A$ in the phase are plotted in such a fashion as to show the relationships among them, the resulting plot is called a *phasor diagram*.

For example, Figure 5–13 shows these relationships when the generator is supplying a load at unity power factor (a purely resistive load). From Equation (5–11), the total voltage $\mathbf{E}_A$ differs from the terminal voltage of the phase $\mathbf{V}_\phi$ by the resistive

Figure 5–13 | The phasor diagram of a synchronous generator at unity power factor.

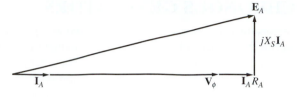

Figure 5–14 | The phasor diagram of a synchronous generator at (a) lagging and (b) leading power factor.

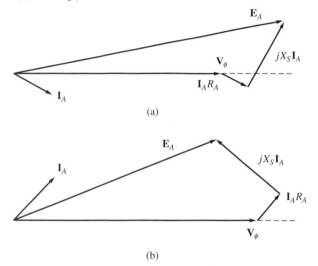

(a)

(b)

and inductive voltage drops. All voltages and currents are referenced to $\mathbf{V}_\phi$, which is arbitrarily assumed to be at an angle of 0°.

This phasor diagram can be compared to the phasor diagrams of generators operating at lagging and leading power factors. These phasor diagrams are shown in Figure 5–14. Notice that, *for a given phase voltage and armature current,* a larger internal generated voltage $\mathbf{E}_A$ is needed for lagging loads than for leading loads. Therefore, a larger field current is needed with lagging loads to get the same terminal voltage, because

$$E_A = K\phi\omega \tag{5-1}$$

and ω must be constant to keep a constant frequency.

Alternatively, *for a given field current and magnitude of load current, the terminal voltage is lower for lagging loads and higher for leading loads.*

In real synchronous machines, the synchronous reactance is normally *much* larger than the winding resistance R_A, so R_A is often neglected in the *qualitative* study of voltage variations. For accurate numerical results, R_A should of course be considered.

5.6 | POWER AND TORQUE IN SYNCHRONOUS GENERATORS

A synchronous generator is a synchronous machine used as a generator. It converts mechanical power to three-phase electric power. The source of mechanical power, the *prime mover*, may be a diesel engine, a steam turbine, a water turbine, or any similar device. Whatever the source, it must have the basic property that its speed is almost constant regardless of the power demand. If that were not so, then the resulting power system's frequency would wander.

Not all the mechanical power going into a synchronous generator becomes electric power out of the machine. The difference between output power and input power represents the losses of the machine. A power-flow diagram for a synchronous generator is shown in Figure 5–15. The input mechanical power is the shaft power in the generator $P_{in} = \tau_{app}\omega_m$, while the power converted from mechanical to electrical form internally is given by

$$P_{conv} = \tau_{ind}\omega_m \tag{5–12}$$

$$P_{conv} = 3E_A I_A \cos \gamma \tag{5–13}$$

where γ is the angle between $\mathbf{E}_A$ and $\mathbf{I}_A$. The difference between the input power to the generator and the power converted in the generator represents the mechanical, core, and stray losses of the machine.

Figure 5–15 | The power-flow diagram of a synchronous generator.

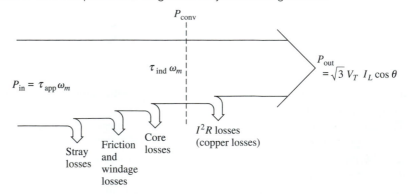

The real electric output power of the synchronous generator can be expressed in line quantities as

$$P_{out} = \sqrt{3}V_T I_L \cos \theta \tag{5–14}$$

and in phase quantities as

$$P_{out} = 3V_\phi I_A \cos \theta \tag{5–15}$$

The reactive power output can be expressed in line quantities as

$$Q_{out} = \sqrt{3}V_T I_L \sin \theta \tag{5–16}$$

or in phase quantities as

$$Q_{out} = 3V_\phi I_A \sin \theta \qquad (5\text{--}17)$$

Recall from Chapter 2 that the power factor angle θ in these four equations is the angle between $\mathbf{V}_\phi$ and $\mathbf{I}_A$, *not* the angle between $\mathbf{V}_T$ and $\mathbf{I}_L$.

In real synchronous machines of any size, the armature resistance R_A is more than 10 times smaller than the synchronous reactance X_S. If the armature resistance R_A is ignored (since $X_S \gg R_A$), then a very useful equation can be derived to approximate the output power of the generator. To derive this equation, examine the phasor diagram in Figure 5–16. Figure 5–16 shows a simplified phasor diagram of a generator with the stator resistance ignored. Notice that the vertical segment bc can be expressed as either $E_A \sin \delta$ or $X_S I_A \cos \theta$. Therefore,

$$I_A \cos \theta = \frac{E_A \sin \delta}{X_S}$$

and substituting into Equation (5–17),

$$\boxed{P = \frac{3V_\phi E_A \sin \delta}{X_S}} \qquad (5\text{--}18)$$

Since the resistances are assumed to be zero in Equation (5–18), there are no electrical losses in this generator, and this equation is both P_{conv} and P_{out}.

Figure 5–16 I Simplified phasor diagram with armature resistance ignored.

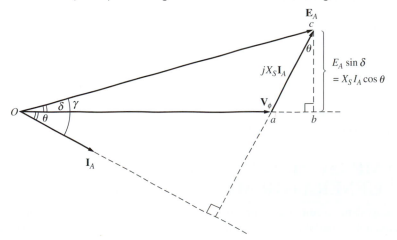

Equation (5–18) shows that the power produced by a synchronous generator depends on the angle δ between $\mathbf{V}_\phi$ and $\mathbf{E}_A$. The angle δ is known as the *torque angle*

of the machine. Notice also that the maximum power that the generator can supply occurs when $\delta = 90°$. At $\delta = 90°$, $\sin \delta = 1$, and

$$P_{max} = \frac{3V_\phi E_A}{X_S} \tag{5–19}$$

The maximum power indicated by this equation is called the *static stability limit* of the generator. Normally, real generators never even come close to that limit. Full-load torque angles of 15 to 20° are more typical of real machines.

Now take another look at Equations (5–15), (5–17), and (5–18). If $\mathbf{V}_\phi$ is assumed constant, then the *real power output is directly proportional* to the quantities $I_A \cos \theta$ and $E_A \sin \delta$, and the reactive power output is directly proportional to the quantity $I_A \sin \theta$. These facts are useful in plotting phasor diagrams of synchronous generators as loads change.

From Chapter 4, the induced torque in this generator can be expressed as

$$\tau_{ind} = k\mathbf{B}_R \times \mathbf{B}_S \tag{4–47}$$

or as

$$\tau_{ind} = k\mathbf{B}_R \times \mathbf{B}_{net} \tag{4–49}$$

The magnitude of Equation (4–49) can be expressed as

$$\tau_{ind} = kB_R B_{net} \sin \delta \tag{4–50}$$

where δ is the angle between the rotor and net magnetic fields (the so-called torque angle). Since $\mathbf{B}_R$ produces the voltage $\mathbf{E}_A$ and $\mathbf{B}_{net}$ produces the voltage $\mathbf{V}_\phi$ the angle δ between $\mathbf{E}_A$ *and* $\mathbf{V}_\phi$ is the same as the angle δ between $\mathbf{B}_R$ and $\mathbf{B}_{net}$.

An alternative expression for the induced torque in a synchronous generator can be derived from Equation (5–18). Because $P_{conv} = \tau_{ind}\omega_m$, the induced torque can be expressed as

$$\tau_{ind} = \frac{3V_\phi E_A \sin \delta}{\omega_m X_S} \tag{5–20}$$

This expression describes the induced torque in terms of electrical quantities, whereas Equation (4–49) gives the same information in terms of magnetic quantities.

5.7 | MEASURING SYNCHRONOUS GENERATOR MODEL PARAMETERS

The equivalent circuit of a synchronous generator that has been derived contains three quantities that must be determined in order to completely describe the behavior of a real synchronous generator:

1. The relationship between field current and flux (and therefore between the field current and E_A)
2. The synchronous reactance
3. The armature resistance

This section describes a simple technique for determining these quantities in a synchronous generator.

The first step in the process is to perform the *open-circuit test* on the generator. To perform this test, the generator is turned at the rated speed, the terminals are disconnected from all loads, and the field current is set to zero. Then the field current is gradually increased in steps, and the terminal voltage is measured at each step along the way. With the terminals open, $I_A = 0$, so E_A is equal to $\mathbf{V}_\phi$. It is thus possible to construct a plot of E_A or V_T versus I_F from this information. This plot is the so-called open-circuit characteristic (OCC) of a generator. With this characteristic, it is possible to find the internal generated voltage of the generator for any given field current. A typical open-circuit characteristic is shown in Figure 5–17a. Notice that at first the curve is almost perfectly linear, until some saturation is observed at high field currents. The unsaturated iron in the frame of the synchronous machine has a reluctance several thousand times lower than the air-gap reluctance, so at first almost *all* the magnetomotive force is across the air gap, and the resulting flux increase is linear. When the iron finally saturates, the reluctance of the iron increases

Figure 5–17 I (a) The open-circuit characteristic (OCC) of a synchronous generator. (b) The short-circuit characteristic (SCC) of a synchronous generator.

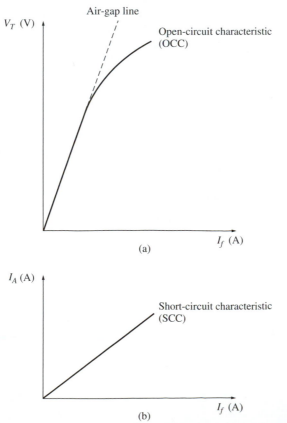

dramatically, and the flux increases much more slowly with an increase in magnetomotive force. The linear portion of an OCC is called the *air-gap line* of the characteristic.

The second step in the process is to conduct the *short-circuit test*. To perform the short-circuit test, adjust the field current to zero again and short-circuit the terminals of the generator through a set of ammeters. Then the armature current I_A or the line current I_L is measured as the field current is increased. Such a plot is called a *short-circuit characteristic* (SCC) and is shown in Figure 5–17b. It is essentially a straight line. To understand why this characteristic is a straight line, look at the equivalent circuit in Figure 5–12 when the terminals of the machine are short-circuited. Such a circuit is shown in Figure 5–18a. Notice that when the terminals are short-circuited, the armature current $\mathbf{I}_A$ is given by

$$\mathbf{I}_A = \frac{\mathbf{E}_A}{R_A + jX_S} \tag{5–21}$$

and its magnitude is given just by

$$I_A = \frac{E_A}{\sqrt{R_A^2 + X_S^2}} \tag{5–22}$$

The resulting phasor diagram is shown in Figure 5–18b, and the corresponding magnetic fields are shown in Figure 5–18c. Since $\mathbf{B}_S$ almost cancels $\mathbf{B}_R$, the net magnetic field $\mathbf{B}_{\text{net}}$ is *very* small (corresponding to internal resistive and inductive drops only). Since the net magnetic field in the machine is so small, the machine is unsaturated and the SCC is linear.

Figure 5–18 | (a) The equivalent circuit of a synchronous generator during the short-circuit test. (b) The resulting phasor diagram. (c) The magnetic fields during the short-circuit test.

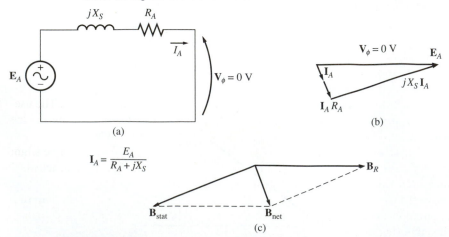

To understand what information these two characteristics yield, notice that, with V_ϕ equal to zero in Figure 5–18, the *internal machine impedance* is given by

$$Z_S = \sqrt{R_A^2 + X_S^2} = \frac{E_A}{I_A} \tag{5–23}$$

Since $X_S \gg R_A$, this equation reduces to

$$X_S \approx \frac{E_A}{I_A} = \frac{V_{\phi, OC}}{I_A} \tag{5–24}$$

If E_A and I_A are known for a given situation, then the synchronous reactance X_S can be found.

Therefore, an *approximate* method for determining the synchronous reactance X_S at a given field current is

1. Get the internal generated voltage E_A from the OCC at that field current.
2. Get the short-circuit current flow $I_{A, SC}$ at that field current from the SCC.
3. Find X_S by applying Equation (5–24).

There is a problem with this approach, however. The internal generated voltage E_A comes from the OCC, where the machine is partially *saturated* for large field currents, while I_A is taken from the SCC, where the machine is *unsaturated* at all field currents. Therefore, at higher field currents, the E_A taken from the OCC at a given field current is *not* the same as the E_A at the same field current under short-circuit conditions, and this difference makes the resulting value of X_S only approximate.

However, the answer given by this approach *is* accurate up to the point of saturation, so the *unsaturated synchronous reactance* $X_{S,u}$ of the machine can be found simply by applying Equation (5–24) at any field current in the linear portion (on the air-gap line) of the OCC curve.

The approximate value of synchronous reactance varies with the degree of saturation of the OCC, so the value of the synchronous reactance to be used in a given problem should be one calculated at the approximate load on the machine. A plot of approximate synchronous reactance as a function of field current is shown in Figure 5–19.

To get a more accurate estimation of the saturated synchronous reactance, refer to Section 5.3 of Reference 3.

If it is important to know a winding's resistance as well as its synchronous reactance, the resistance can be approximated by applying a DC voltage to the windings while the machine is stationary and measuring the resulting current flow. The use of DC voltage means that the reactance of the windings will be zero during the measurement process.

This technique is not perfectly accurate, since the AC resistance will be slightly larger than the DC resistance (as a result of the skin effect at higher frequencies). The measured value of the resistance can even be plugged into Equation (5–24) to improve the estimate of X_S, if desired. (Such an improvement is not much help in the approximate approach—saturation causes a much larger error in the X_S calculation than ignoring R_A does.)

Figure 5–19 | A sketch of the approximate synchronous reactance of a synchronous generator as a function of the field current in the machine. The constant value of reactance found at low values of field current is the *unsaturated* synchronous reactance of the machine.

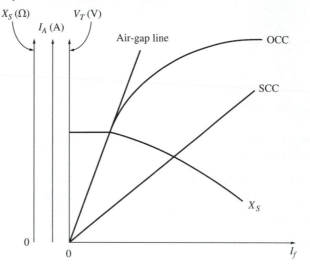

EXAMPLE 5–1

A 200-kVA, 480-V, 50-Hz, Y-connected synchronous generator with a rated field current of 5 A was tested, and the following data were taken:

1. $V_{T, OC}$ at the rated I_F was measured to be 540 V.
2. $I_{L, SC}$ at the rated I_F was found to be 300 A.
3. When a DC voltage of 10 V was applied to two of the terminals, a current of 25 A was measured.

Find the values of the armature resistance and the approximate synchronous reactance in ohms that would be used in the generator model at the rated conditions.

■ Solution
The generator described above is Y connected, so the direct current in the resistance test flows through *two phases* of the windings, as shown in Figure 5–20. Therefore, the resistance is given by

$$2R_A = \frac{V_{DC}}{I_{DC}}$$

$$R_A = \frac{V_{DC}}{2I_{DC}} = \frac{10 \text{ V}}{(2)(25 \text{ A})} = 0.2 \text{ } \Omega$$

Figure 5–20 I Configuration for measuring resistance in the synchronous machine.

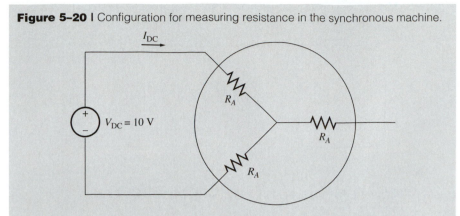

The internal generated voltage at the rated field current is equal to

$$E_A = V_{\phi, \, \text{OC}} = \frac{V_T}{\sqrt{3}}$$

$$= \frac{540 \text{ V}}{\sqrt{3}} = 311.8 \text{ V}$$

The short-circuit current I_A is just equal to the line current, since the generator is Y connected:

$$I_{A, \, \text{SC}} = I_{L, \, \text{SC}} = 300 \text{ A}$$

Therefore, the synchronous reactance at the rated field current can be calculated from Equation (5–23):

$$Z_S = \sqrt{R_A^2 + X_S^2} = \frac{E_A}{I_A} \qquad (5\text{–}23)$$

$$\sqrt{(0.2 \; \Omega)^2 + X_S^2} = \frac{311.8 \text{ V}}{300 \text{ A}}$$

$$= 1.039 \; \Omega$$

$$0.04 + X_S^2 = 1.08$$

$$X_S^2 = 1.04$$

$$= 1.02 \; \Omega$$

How much effect did the inclusion of R_A have on the estimate of X_S? Not much. If X_S is evaluated by Equation (5–24), the result is

$$X_S = \frac{E_A}{I_A} = \frac{311.8 \text{ V}}{300 \text{ A}} = 1.04 \; \Omega$$

Since the error in X_S due to ignoring R_A is much less than the error due to saturation effects, approximate calculations are normally done with Equation (5–24).

The resulting per-phase equivalent circuit is shown in Figure 5–21. ∎

Figure 5–21 | The per-phase equivalent circuit of the generator in Example 5–1.

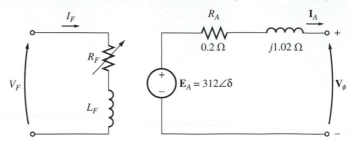

5.8 | THE SYNCHRONOUS GENERATOR OPERATING ALONE

The behavior of a synchronous generator under load varies greatly, depending on the power factor of the load and on whether the generator is operating alone or in parallel with other synchronous generators. In this section, we will study the behavior of synchronous generators operating alone. We will study the behavior of synchronous generators operating in parallel in Chapter 6.

Most synchronous generators in the world operate as parts of large power systems. The behavior of those generators will be discussed in Chapter 6. However, emergency generators and generators on isolated platforms such as ships may operate alone. In that case, they will respond to load changes as described in this section.

Throughout this section, concepts will be illustrated with simplified phasor diagrams ignoring the effect of R_A. In some of the numerical examples the resistance R_A will be included.

Unless otherwise stated in this section, the speed of the generators will be assumed constant, and all terminal characteristics are drawn assuming constant speed. Also, the rotor flux in the generators is assumed constant unless their field current is explicitly changed.

The Effect of Load Changes on a Synchronous Generator Operating Alone

To understand the operating characteristics of a synchronous generator operating alone, we will examine a generator supplying a load. A diagram of a single generator supplying a load is shown in Figure 5–22. What happens when we increase the load on this generator?

Figure 5–22 | A single generator supplying a load.

An increase in the load is an increase in the real and/or reactive power drawn from the generator. Such a load increase increases the load current drawn from the generator. Because the field resistor has not been changed, the field current is constant, and therefore the flux ϕ is constant. Since the prime mover also keeps a constant speed ω, the *magnitude of the internal generated voltage $E_A = K\phi\omega$ is constant.*

If E_A is constant, just what does vary with a changing load? The way to find out is to construct phasor diagrams showing an increase in the load, keeping the constraints on the generator in mind.

First, examine a generator operating at a lagging power factor. If more load is added at the *same power factor*, then $|\mathbf{I}_A|$ increases but remains at the same angle θ with respect to $\mathbf{V}_\phi$ as before. Therefore, the armature reaction voltage $jX_S\mathbf{I}_A$ is larger than before but at the same angle. Now since

$$\mathbf{E}_A = \mathbf{V}_\phi + jX_S\mathbf{I}_A$$

$jX_S\mathbf{I}_A$ must stretch between $\mathbf{V}_\phi$ at an angle of 0° and $\mathbf{E}_A$, which is constrained to be of the same magnitude as before the load increase. If these constraints are plotted on a phasor diagram, there is one and only one point at which the armature reaction voltage can be parallel to its original position while increasing in size. The resulting plot is shown in Figure 5–23a.

Figure 5–23 | The effect of an increase in generator loads at constant power factor on its terminal voltage: (a) Lagging power factor. (b) Unity power factor. (c) Leading power factor.

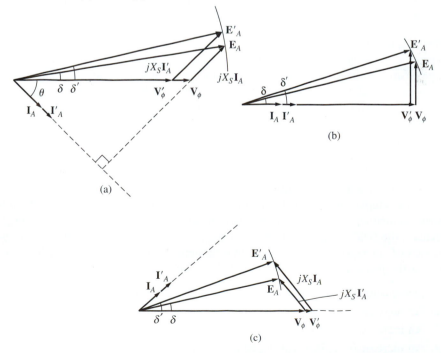

If the constraints are observed, then it is seen that as the load increases, the voltage $\mathbf{V}_\phi$ decreases rather sharply.

Now suppose the generator is loaded with unity-power-factor loads. What happens if new loads are added at the same power factor? With the same constraints as before, it can be seen that this time $\mathbf{V}_\phi$ decreases only slightly (see Figure 5–23b).

Finally, let the generator be loaded with leading-power-factor loads. If new loads are added at the same power factor this time, the armature reaction voltage lies outside its previous value, and $\mathbf{V}_\phi$ actually *rises* (see Figure 5–23c). In this last case, an increase in the load in the generator produced an increase in the terminal voltage. Such a result is not something one would expect on the basis of intuition alone.

General conclusions from this discussion of synchronous generator behavior are

1. If lagging loads ($+Q$ or inductive reactive power loads) are added to a generator, $\mathbf{V}_\phi$ and the terminal voltage V_T decrease significantly.

2. If unity-power-factor loads (no reactive power) are added to a generator, there is a slight decrease in $\mathbf{V}_\phi$ and the terminal voltage.

3. If leading loads ($-Q$ or capacitive reactive power loads) are added to a generator, $\mathbf{V}_\phi$ and the terminal voltage will rise.

A convenient way to compare the voltage behavior of two generators is by their *voltage regulation*. The voltage regulation (VR) of a generator is defined by the equation

$$\text{VR} = \frac{V_{\text{nl}} - V_{\text{fl}}}{V_{\text{fl}}} \times 100\% \tag{4–56}$$

where V_{nl} is the no-load voltage of the generator and V_{fl} is the full-load voltage of the generator. A synchronous generator operating at a lagging power factor has a fairly large positive voltage regulation, a synchronous generator operating at a unity power factor has a small positive voltage regulation, and a synchronous generator operating at a leading power factor often has a negative voltage regulation.

Normally, it is desirable to keep the voltage supplied to a load constant, even though the load itself varies. How can terminal voltage variations be corrected for? The obvious approach is to vary the magnitude of $\mathbf{E}_A$ to compensate for changes in the load. Recall that $E_A = K\phi\omega$. Since the frequency should not be changed in a normal system, E_A must be controlled by varying the flux in the machine.

For example, suppose that a lagging load is added to a generator. Then the terminal voltage will fall, as was previously shown. To restore it to its previous level, decrease the field resistor R_F. If R_F decreases, the field current will increase. An increase in I_F increases the flux, which in turn increases E_A, and an increase in E_A increases the phase and terminal voltage. This idea can be summarized as follows:

1. Decreasing the field resistance in the generator increases its field current.

2. An increase in the field current increases the flux in the machine.

3. An increase in the flux increases the internal generated voltage $E_A = K\phi\omega$.

4. An increase in E_A increases V_ϕ and the terminal voltage of the generator.

The process can be reversed to decrease the terminal voltage. It is possible to regulate the terminal voltage of a generator throughout a series of load changes simply by adjusting the field current.

Example Problems

The following three problems illustrate simple calculations involving voltages, currents, and power flows in synchronous generators. The first problem is an example that includes the armature resistance in its calculations, while the next two ignore R_A. Part of the first example problem addresses the question: *How must a generator's field current be adjusted to keep V_T constant as the load changes?* On the other hand, part of the second example problem asks the question: *If the load changes and the field is left alone, what happens to the terminal voltage?* You should compare the calculated behavior of the generators in these two problems to see if it agrees with the qualitative arguments of this section. Finally, the third example illustrates the use of a MATLAB program to derive the terminal characteristics of synchronous generator.

EXAMPLE 5–2

A 480-V, 60-Hz, Δ-connected, four-pole synchronous generator has the OCC shown in Figure 5–24a. This generator has a synchronous reactance of 0.1 Ω and an armature resistance of 0.015 Ω. At full load, the machine supplies 1200 A at 0.8 PF lagging. Under full-load conditions, the friction and windage losses are 40 kW, and the core losses are 30 kW. Ignore any field circuit losses.

a. What is the speed of rotation of this generator?

b. How much field current must be supplied to the generator to make the terminal voltage 480 V at no load?

c. If the generator is now connected to a load and the load draws 1200 A at 0.8 PF lagging, how much field current will be required to keep the terminal voltage equal to 480 V?

d. How much power is the generator now supplying? How much power is supplied to the generator by the prime mover? What is this machine's overall efficiency?

e. If the generator's load were suddenly disconnected from the line, what would happen to its terminal voltage?

f. Finally, suppose that the generator is connected to a load drawing 1200 A at 0.8 PF *leading*. How much field current would be required to keep V_T at 480 V?

■ **Solution**

This synchronous generator is Δ connected, so its phase voltage is equal to its line voltage $V_\phi = V_T$, while its phase current is related to its line current by the equation $I_L = \sqrt{3}I_\phi$.

Figure 5–24 | (a) Open-circuit characteristic of the generator in Example 5–2. (b) Phasor diagram of the generator in Example 5–2.

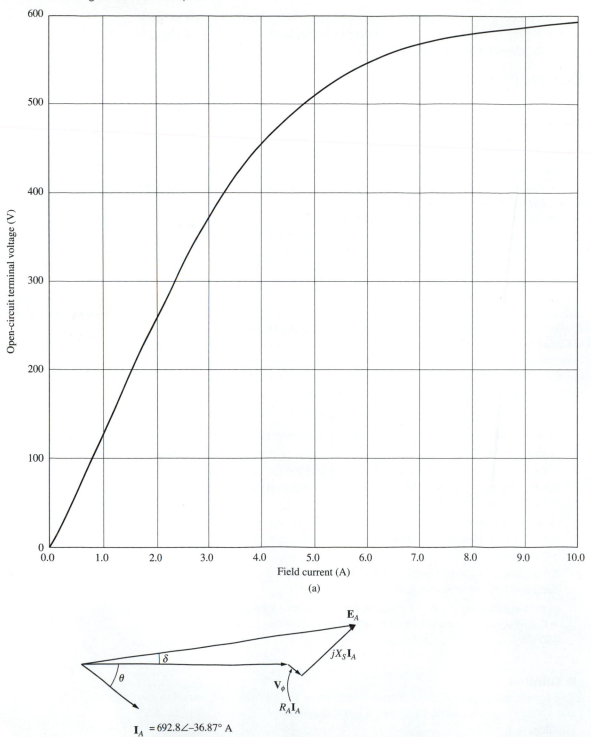

(a)

(b)

a. The relationship between the electrical frequency produced by a synchronous generator and the mechanical rate of shaft rotation is given by Equation (4–23):

$$f_e = \frac{n_m P}{120} \qquad (4\text{--}23)$$

Therefore,

$$n_m = \frac{120 \, f_e}{P}$$
$$= \frac{120(60 \text{ Hz})}{4 \text{ poles}} = 1800 \text{ r/min}$$

b. In this machine, $V_T = V_\phi$. Since the generator is at no load, $I_A = 0$ and $\mathbf{E}_A = \mathbf{V}_\phi$. Therefore, $V_T = V_\phi = E_A = 480$ V, and from the open-circuit characteristic, $I_F = 4.5$ A.

c. If the generator is supplying 1200 A, then the armature current in the machine is

$$I_A = \frac{1200 \text{ A}}{\sqrt{3}} = 692.8 \text{ A}$$

The phasor diagram for this generator is shown in Figure 5–24b. If the terminal voltage is adjusted to be 480 V, the size of the internal generated voltage $\mathbf{E}_A$ is given by

$$\begin{aligned}
\mathbf{E}_A &= \mathbf{V}_\phi + R_A \mathbf{I}_A + jX_S \mathbf{I}_A \\
&= 480\angle 0° \text{ V} + (0.015 \text{ }\Omega)(692.8\angle -36.87° \text{ A}) + (j0.1 \text{ }\Omega)(692.8\angle -36.87° \text{ A}) \\
&= 480\angle 0° \text{ V} + 10.39\angle -36.87° \text{ V} + 69.28\angle -53.13° \text{ V} \\
&= 529.9 + j49.2 \text{ V} = 532\angle 5.3° \text{ V}
\end{aligned}$$

To keep the terminal voltage at 480 V, $\mathbf{E}_A$ must be adjusted to 532 V. From Figure 5–24, the required field current is 5.7 A.

d. The power that the generator is now supplying can be found from Equation (5–16):

$$\begin{aligned}
P_{out} &= \sqrt{3} V_T I_L \cos \theta \qquad (5\text{--}14) \\
&= \sqrt{3}(480 \text{ V})(1200 \text{ A}) \cos 36.87° \\
&= 798 \text{ kW}
\end{aligned}$$

To determine the power input to the generator, use the power-flow diagram (Figure 5–15). From the power-flow diagram, the mechanical input power is given by

$$P_{in} = P_{out} + P_{elec \, loss} + P_{core \, loss} + P_{mech \, loss} + P_{stray \, loss}$$

The stray losses were not specified here, so they will be ignored. In this generator, the electrical losses are

$$\begin{aligned}
P_{elec \, loss} &= 3I_A^2 R_A \\
&= 3(692.8 \text{ A})^2(0.015 \text{ }\Omega) = 21.6 \text{ kW}
\end{aligned}$$

The core losses are 30 kW, and the friction and windage losses are 40 kW, so the total input power to the generator is

$$P_{in} = 798 \text{ kW} + 21.6 \text{ kW} + 30 \text{ kW} + 40 \text{ kW} = 889.6 \text{ kW}$$

Therefore, the machine's overall efficiency is

$$\eta = \frac{P_{out}}{P_{in}} \times 100\% = \frac{798 \text{ kW}}{889.6 \text{ kW}} \times 100\% = 89.7\%$$

e. If the generator's load were suddenly disconnected from the line, the current I_A would drop to zero, making $\mathbf{E}_A = \mathbf{V}_\phi$. Since the field current has not changed, $|\mathbf{E}_A|$ has not changed and $\mathbf{V}_\phi$ and V_T must rise to equal $\mathbf{E}_A$. Therefore, if the load were suddenly dropped, the terminal voltage of the generator would rise to 532 V.

f. If the generator were loaded down with 1200 A at 0.8 PF leading while the terminal voltage was 480 V, then the internal generated voltage would have to be

$$\begin{aligned}
\mathbf{E}_A &= \mathbf{V}_\phi + R_A \mathbf{I}_A + j X_S \mathbf{I}_A \\
&= 480\angle 0° \text{ V} + (0.015 \ \Omega)(692.8\angle 36.87° \text{ A}) + (j0.1 \ \Omega)(692.8\angle 36.87° \text{ A}) \\
&= 480\angle 0° \text{ V} + 10.39\angle 36.87° \text{ V} + 69.28\angle 126.87° \text{ V} \\
&= 446.7 + j61.7 \text{ V} = 451\angle 7.1° \text{ V}
\end{aligned}$$

Therefore, the internal generated voltage E_A must be adjusted to provide 451 V if V_T is to remain 480 V. According to the open-circuit characteristic, the field current would have to be adjusted to 4.1 A. ∎

Which type of load (leading or lagging) needed a larger field current to maintain the rated voltage? Which type of load (leading or lagging) placed more thermal stress on the generator? Why?

EXAMPLE 5–3

A 480-V, 50-Hz, Y-connected, six-pole synchronous generator has a per-phase synchronous reactance of 1.0 Ω. Its full-load armature current is 60 A at 0.8 PF lagging. This generator has friction and windage losses of 1.5 kW and core losses of 1.0 kW at 60 Hz at full load. Since the armature resistance is being ignored, assume that the $I^2 R$ losses are negligible. The field current has been adjusted so that the terminal voltage is 480 V at no load.

a. What is the speed of rotation of this generator?
b. What is the terminal voltage of this generator if the following are true?
    1. It is loaded with the rated current at 0.8 PF lagging.
    2. It is loaded with the rated current at 1.0 PF.
    3. It is loaded with the rated current at 0.8 PF leading.

c. What is the efficiency of this generator (ignoring the unknown electrical losses) when it is operating at the rated current and 0.8 PF lagging?

d. How much shaft torque must be applied by the prime mover at full load? How large is the induced countertorque?

e. What is the voltage regulation of this generator at 0.8 PF lagging? At 1.0 PF? At 0.8 PF leading?

■ **Solution**

This generator is Y connected, so its phase voltage is given by $V_\phi = V_T/\sqrt{3}$. That means that when V_T is adjusted to 480 V, $V_\phi = 277$ V. The field current has been adjusted so that $V_{Tnl} = 480$ V, so $V_\phi = 277$ V. At *no load*, the armature current is zero, so the armature reaction voltage and the $I_A R_A$ drops are zero. Since $\mathbf{I}_A = 0$, the internal generated voltage $E_A = V_\phi = 277$ V. The internal generated voltage E_A ($= K\phi\omega$) varies only when the field current changes. Since the problem states that the field current is adjusted initially and then left alone, the magnitude of the internal generated voltage is $E_A = 277$ V and will not change in this example.

a. The speed of rotation of a synchronous generator in revolutions per minute is given by Equation (4–23):

$$f_e = \frac{n_m P}{120} \qquad\qquad (4\text{–}23)$$

Therefore,

$$n_m = \frac{120\, f_3}{P}$$

$$= \frac{120(50\ \text{Hz})}{6\ \text{poles}} = 1000\ \text{r/min}$$

Alternatively, the speed expressed in radians per second is

$$\omega_m = (1000\ \text{r/min})\left(\frac{1\ \text{min}}{60\ \text{s}}\right)\left(\frac{2\pi\ \text{rad}}{1\ \text{r}}\right)$$

$$= 104.7\ \text{rad/s}$$

b.

1. If the generator is loaded down with rated current at 0.8 PF lagging, the resulting phasor diagram looks like the one shown in Figure 5–25a. In this phasor diagram, we know that $\mathbf{V}_\phi$ is at an angle of $0°$, that the magnitude of $\mathbf{E}_A$ is 277 V, and that the quantity $jX_S\mathbf{I}_A$ is

$$jX_S\mathbf{I}_A = j(1.0\ \Omega)(60\angle{-36.87°}\ \text{A}) = 60\angle{53.13°}\ \text{V}$$

The two quantities not known on the voltage diagram are the magnitude of $\mathbf{V}_\phi$ and the angle δ of $\mathbf{E}_A$. To find these values, the easiest approach is to construct a right triangle on the phasor diagram, as shown in the figure. From Figure 5–25a, the right triangle gives

$$E_A^2 = (V_\phi + X_S I_A \sin\theta)^2 + (X_S I_A \cos\theta)^2$$

Therefore, the phase voltage at the rated load and 0.8 PF lagging is

$$
\begin{aligned}
(277 \text{ V})^2 &= [V_\phi + (1.0 \ \Omega)(60 \text{ A}) \sin 36.87°]^2 + [(1.0 \ \Omega)(60 \text{ A}) \cos 36.87°]^2 \\
76{,}729 &= (V_\phi + 36)^2 + 2304 \\
74{,}425 &= (V_\phi + 36)^2 \\
272.8 &= V_\phi + 36 \\
V_\phi &= 236.8 \text{ V}
\end{aligned}
$$

Since the generator is Y connected, $V_T = \sqrt{3}V_\phi = 410$ V.

Figure 5–25 | Generator phasor diagrams for Example 5–3: (a) Lagging power factor. (b) Unity power factor. (c) Leading power factor.

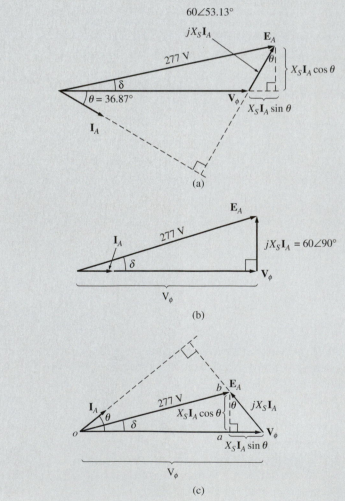

2. If the generator is loaded with the rated current at unity power factor, then the phasor diagram will look like Figure 5–25b. To find $\mathbf{V}_\phi$ here the right triangle is

$$E_A^2 = (V_\phi + X_S I_A)^2 + (X_S I_A)^2$$
$$(277 \text{ V})^2 = V_\phi^2 + [(1.0\ \Omega)(60 \text{ A})]^2$$
$$76{,}729 = V_\phi^2 + 3600$$
$$V_\phi^2 = 73{,}129$$
$$= 270.4 \text{ V}$$

Therefore, $V_T = \sqrt{3}V_\phi = 468.4$ V.

3. When the generator is loaded with the rated current at 0.8 PF leading, the resulting phasor diagram is the one shown in Figure 5–25c. To find $\mathbf{V}_\phi$ in this situation, we construct the triangle OAB shown in the figure. The resulting equation is

$$E_A^2 = (V_\phi - X_S I_A \sin \theta)^2 + (X_S I_A \cos \theta)^2$$

Therefore, the phase voltage at the rated load and 0.8 PF leading is

$$(277 \text{ V})^2 = [V_\phi - (1.0\ \Omega)(60 \text{ A}) \sin 36.87°]^2 + [(1.0\ \Omega)(60 \text{ A}) \cos 36.87°]^2$$
$$76{,}729 = (V_\phi - 36)^2 + 2304$$
$$74{,}425 = (V_\phi - 36)^2$$
$$272.8 = V_\phi - 36$$
$$V_\phi = 308.8 \text{ V}$$

Since the generator is Y connected, $V_T = \sqrt{3}V_\phi = 535$ V.

c. The output power of this generator at 60 A and 0.8 PF lagging is

$$P_{out} = 3V_\phi I_A \cos \theta$$
$$= 3(236.8 \text{ V})(60 \text{ A})(0.8) = 34.1 \text{ kW}$$

The mechanical input power is given by

$$P_{in} = P_{out} + P_{elec\ loss} + P_{core\ loss} + P_{mech\ loss}$$
$$= 34.1 \text{ kW} + 0 + 1.0 \text{ kW} + 1.5 \text{ kW} = 36.6 \text{ kW}$$

The efficiency of the generator is thus

$$\eta = \frac{P_{out}}{P_{in}} \times 100\% = \frac{34.1 \text{ kW}}{36.6 \text{ kW}} \times 100\% = 93.2\%$$

d. The input torque to this generator is given by the equation

$$P_{in} = \tau_{app}\omega_m$$

so

$$\tau_{app} = \frac{P_{in}}{\omega_m} = \frac{36.6 \text{ kW}}{125.7 \text{ rad/s}} = 291.2 \text{ N-m}$$

The induced countertorque is given by

$$P_{conv} = \tau_{ind}\omega_m$$

so
$$\tau_{ind} = \frac{P_{conv}}{\omega_m} = \frac{34.1 \text{ kW}}{125.7 \text{ rad/s}} = 271.3 \text{ N-m}$$

e.   The voltage regulation of a generator is defined as

$$P_{conv} = \tau_{ind}\,\omega_m \qquad\qquad (4\text{--}56)$$

By this definition, the voltage regulation for the lagging, unity, and leading power-factor cases is

1.   Lagging case: $VR = \dfrac{480 \text{ V} - 410 \text{ V}}{410 \text{ V}} \times 100\% = 17.1\%$

2.   Unity case: $VR = \dfrac{480 \text{ V} - 468 \text{ V}}{468 \text{ V}} \times 100\% = 2.6\%$

3.   Leading case: $VR = \dfrac{480 \text{ V} - 535 \text{ V}}{535 \text{ V}} \times 100\% = -10.3\%$ ∎

In Example 5–3, lagging loads resulted in a drop in terminal voltage, unity-power-factor loads caused little effect on V_T, and leading loads resulted in an increase in terminal voltage.

EXAMPLE 5–4

Assume that the generator of Example 5–3 is operating at no load with a terminal voltage of 480 V. Plot the terminal characteristic (terminal voltage versus line current) of this generator as its armature current varies from no-load to full load at a power factor of (a) 0.8 lagging and (b) 0.8 leading. Assume that the field current remains constant at all times.

■ Solution

The terminal characteristic of a generator is a plot of its terminal voltage versus line current. Since this generator is Y connected, its phase voltage is given by $V_\phi = V_T/\sqrt{3}$. If V_T is adjusted to 480 V at no load conditions, then $V_\phi = E_A = 277$ V. Because the field current remains constant, E_A will remain 277 V at all times. The output current I_L from this generator will be the same as its armature current I_A because it is Y connected.

a.   If the generator is loaded with a 0.8 PF lagging current, the resulting phasor diagram looks like the one shown in Figure 5–25a. In this phasor diagram, we know that $\mathbf{V}_\phi$ is at an angle of 0°, that the magnitude of $\mathbf{E}_A$ is 277 V, and that the quantity $jX_S\mathbf{I}_A$ stretches between $\mathbf{V}_\phi$ and $\mathbf{E}_A$ as shown. The two quantities not known on the phasor diagram are the magnitude of $\mathbf{V}_\phi$ and the angle δ of $\mathbf{E}_A$. To find V_ϕ, the easiest approach is to construct a right triangle on the phasor diagram, as shown in the figure. From Figure 5–25a, the right triangle gives

$$E_A^2 = (V_\phi + X_S I_A \sin\theta)^2 + (X_S I_A \cos\theta)^2$$

This equation can be used to solve for V_ϕ as a function of the current I_A:

$$V_\phi = \sqrt{E_A^2 - (X_S I_A \cos \theta)^2} - X_S I_A \sin \theta$$

A simple MATLAB m-file can be used to calculate V_ϕ (and hence V_T) as a function of current. Such an m-file is shown below.

```
% M-file: term_char_a.m
% M-file to plot the terminal characteristics of the
%   generator of Example 5-4 with an 0.8 PF lagging load.

% First, initialize the current amplitudes (21 values
% in the range 0-60 A)
i_a = (0:1:20) * 3;

% Now initialize all other values
v_phase = zeros(1,21);
e_a = 277.0;
x_s = 1.0;
theta = 36.87 * (pi/180); % Converted to radians

% Now calculate v_phase for each current level
for ii = 1:21
   v_phase(ii) = sqrt(e_a^2 - (x_s * i_a(ii) * cos(theta))^2) ...
                      - (x_s * i_a(ii) * sin(theta));
end

% Calculate terminal voltage from the phase voltage
v_t = v_phase * sqrt(3);

% Plot the terminal characteristic, remembering the
% the line current is the same as i_a
plot(i_a,v_t,'Color','k','Linewidth',2.0);
xlabel('Line Current (A)','Fontweight','Bold');
ylabel('Terminal Voltage (V)','Fontweight','Bold');
title ('Terminal Characteristic for 0.8 PF lagging load', ...
       'Fontweight','Bold');
grid on;
axis([0 60 400 500]);
```

The plot resulting when this m-file is executed is shown in Figure 5-26a.

b. If the generator is loaded with a 0.8 PF leading current, the resulting phasor diagram looks like the one shown in Figure 5-25c. To find V_ϕ, the easiest approach is to construct a right triangle on the phasor diagram, as shown in the figure. From Figure 5-25c, the right triangle gives

$$E_A^2 = (V_\phi - X_S I_A \sin \theta)^2 + (X_S I_A \cos \theta)^2$$

This equation can be used to solve for V_ϕ as a function of the current I_A:

$$V_\phi = \sqrt{E_A^2 - (X_S I_A \cos \theta)^2} + X_S I_A \sin \theta$$

Figure 5–26 | (a) Terminal characteristic for the generator of Example 5–4 when loaded with an 0.8 PF lagging load. (b) Terminal characteristic for the generator when loaded with an 0.8 PF leading load.

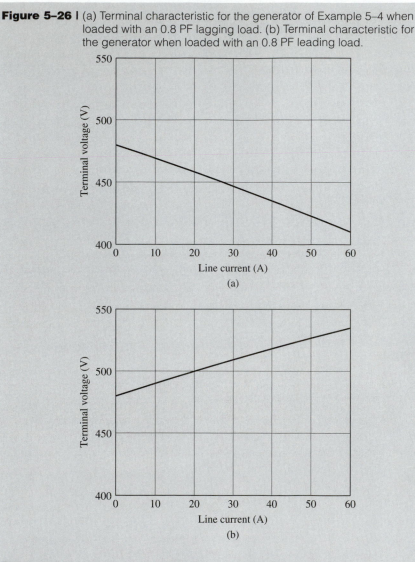

This equation can be used to calculate and plot the terminal characteristic in a manner similar to that in part (a) above. The resulting terminal characteristic is shown in Figure 5–26b. ■

5.9 | SYNCHRONOUS MOTORS

Synchronous motors are synchronous machines used to convert electric power to mechanical power. To understand the basic concept of a synchronous motor, look at Figure 5–27, which shows a two-pole synchronous motor. The field current I_F of the motor produces a steady-state magnetic field $\mathbf{B}_R$. A three-phase set of voltages is applied to the stator of the machine, which produces a three-phase current flow in the windings.

Figure 5–27 | A two-pole synchronous motor.

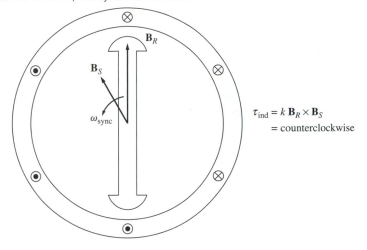

$$\tau_{ind} = k\, \mathbf{B}_R \times \mathbf{B}_S$$
$$= \text{counterclockwise}$$

As we saw in Chapter 4, a three-phase set of currents in an armature winding produces a uniform rotating magnetic field $\mathbf{B}_S$. Therefore, there are two magnetic fields present in the machine, and the *rotor field will tend to line up with the stator field,* just as two bar magnets will tend to line up if placed near each other. Since the stator magnetic field is rotating, the rotor magnetic field (and the rotor itself) will constantly try to catch up. The larger the angle between the two magnetic fields (up to a certain maximum), the greater the torque on the rotor of the machine. The basic principle of synchronous motor operation is that the rotor "chases" the rotating stator magnetic field around in a circle, never quite catching up with it.

Since a synchronous motor is the same physical machine as a synchronous generator, all of the basic speed, power, and torque equations of Chapters 4 and earlier in this chapter apply to synchronous motors also.

The Equivalent Circuit of a Synchronous Motor

A synchronous motor is the same in all respects as a synchronous generator, except that the direction of power flow is reversed. Since the direction of power flow in the machine is reversed, the direction of current flow in the stator of the motor may be expected to reverse also. Therefore, the equivalent circuit of a synchronous motor is exactly the same as the equivalent circuit of a synchronous generator, *except* that the reference direction of $\mathbf{I}_A$ is *reversed.* The per-phase equivalent circuit is shown in Figure 5–28. As before, the three phases of the equivalent circuit may be either Y or Δ connected.

Figure 5–28 | The per-phase equivalent circuit of a synchronous motor.

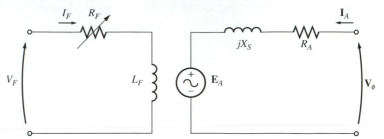

Because of the change in direction of $\mathbf{I}_A$, the Kirchhoff's voltage law equation for the equivalent circuit changes too. Writing a Kirchhoff's voltage law equation for the new equivalent circuit yields

$$\mathbf{V}_\phi = \mathbf{E}_A + jX_S\mathbf{I}_A + R_A\mathbf{I}_A \tag{5–25}$$

or

$$\mathbf{E}_A = \mathbf{V}_\phi - jX_S\mathbf{I}_A - R_A\mathbf{I}_A \tag{5–26}$$

This is exactly the same as the equation for a generator, except that the sign on the current term has been reversed.

The Synchronous Motor from a Magnetic Field Perspective

The phasor diagram of a generator connected to a large power system and operating with a large field current is shown in Figure 5–29a, and the corresponding magnetic field diagram is shown in Figure 5–29b. As described before, $\mathbf{B}_R$ corresponds to (produces) $\mathbf{E}_A$, $\mathbf{B}_{net}$ corresponds to (produces) $\mathbf{V}_\phi$, and $\mathbf{B}_S$ corresponds to $\mathbf{E}_{stat}$ ($= -jX_S\mathbf{I}_A$). The rotation of both the phasor diagram and magnetic field diagram is counterclockwise in the figure, following the standard mathematical convention of increasing angle.

The induced torque in the generator can be found from the magnetic field diagram. From Equations (4–49) and (4–50) the induced torque is given by

$$\tau_{ind} = k\mathbf{B}_R \times \mathbf{B}_{net} \tag{4–49}$$

$$\tau_{ind} = kB_RB_{net} \sin \delta \tag{4–50}$$

Notice that from the magnetic field diagram *the induced torque in this machine is clockwise*, opposing the direction of rotation. In other words, the induced torque in the generator is a countertorque, opposing the rotation caused by the external applied torque τ_{app}.

Suppose that, instead of turning the shaft in the direction of motion, the prime mover suddenly loses power and starts to drag on the machine's shaft. What happens to the machine now? The rotor slows down because of the drag on its shaft and falls behind the net magnetic field in the machine (see Figure 5–30a). As the rotor, and therefore $\mathbf{B}_R$, slows down and falls behind $\mathbf{B}_{net}$ the operation of the machine suddenly

Figure 5–29 | (a) Phasor diagram of a synchronous generator operating at a lagging power factor. (b) The corresponding magnetic field diagram.

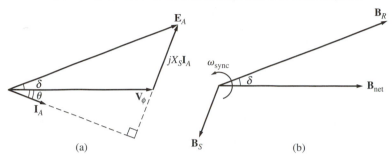

(a)   (b)

changes. By Equation (4–49), when $\mathbf{B}_R$ is behind $\mathbf{B}_{net}$, the induced torque's direction reverses and becomes counterclockwise. In other words, the machine's torque is now in the direction of motion, and the machine is acting as a motor. The increasing torque angle δ results in a larger and larger torque in the direction of rotation, until eventually the motor's induced torque equals the load torque on its shaft. At that point, the machine will be operating at steady state and synchronous speed again, but now as a motor.

The phasor diagram corresponding to generator operation is shown in Figure 5–29a, and the phasor diagram corresponding to motor operation is shown in Figure 5–30a. The reason that the quantity $jX_S\mathbf{I}_A$ points from $\mathbf{V}_\phi$, to $\mathbf{E}_A$ in the generator and from $\mathbf{E}_A$ to $\mathbf{V}_\phi$ in the motor is that the reference direction of $\mathbf{I}_A$ was reversed in the definition of the motor equivalent circuit. The basic difference between motor and generator operation in synchronous machines can be seen either in the magnetic field diagram or in the phasor diagram. *In a generator,* $\mathbf{E}_A$ lies ahead of $\mathbf{V}_\phi$, and $\mathbf{B}_R$ lies ahead of $\mathbf{B}_{net}$. *In a motor*, $\mathbf{E}_A$ lies behind $\mathbf{V}_\phi$, and $\mathbf{B}_R$ lies behind $\mathbf{B}_{net}$. In a motor the induced torque is in the direction of motion, and in a generator the induced torque is a countertorque opposing the direction of motion.

5.10 | STEADY-STATE SYNCHRONOUS MOTOR OPERATION

This section explores the behavior of synchronous motors under varying conditions of load and field current as well as the question of power-factor correction with synchronous motors. The following discussions will generally ignore the armature resistance of the motors for simplicity. However, R_A will be considered in some of the worked numerical calculations.

The Synchronous Motor Torque–Speed Characteristic Curve

Synchronous motors supply power to loads that are basically constant-speed devices. They are usually connected to power systems *very* much larger than the individual motors, so the power systems appear as constant voltage sources to the motors. This means that the terminal voltage and the system frequency will be constant regardless of the amount of power drawn by the motor. The speed of rotation

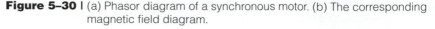

Figure 5–30 | (a) Phasor diagram of a synchronous motor. (b) The corresponding magnetic field diagram.

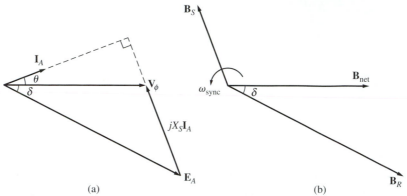

(a)   (b)

of the motor is locked to the applied electrical frequency, so the speed of the motor will be constant regardless of the load. The resulting torque-speed characteristic curve is shown in Figure 5–31. The steady-state speed of the motor is constant from no load all the way up to the maximum torque that the motor can supply (called the *pullout torque*), so the speed regulation of this motor [Equation (4–68)] is 0 percent. The torque equation is

$$\tau_{\text{ind}} = kB_R B_{\text{net}} \sin \delta \tag{4–50}$$

or

$$\tau_{\text{ind}} = \frac{3V_\phi E_A \sin \delta}{\omega_m X_S} \tag{5–20}$$

The maximum or pullout torque occurs when $\delta = 90°$. Normal full-load torques are much less than that, however. In fact, the pullout torque may typically be 3 times the full-load torque of the machine.

When the torque on the shaft of a synchronous motor exceeds the pullout torque, the rotor can no longer remain locked to the stator and net magnetic fields. Instead, the rotor starts to slip behind them. As the rotor slows down, the stator magnetic field "laps" it repeatedly, and the direction of the induced torque in the rotor reverses with each pass. The resulting huge torque surges, first one way and then the other way, cause the whole motor to vibrate severely. The loss of synchronization after the pullout torque is exceeded is known as *slipping poles*.

The maximum or pullout torque of the motor is given by

$$\tau_{\text{max}} = kB_R B_{\text{net}} \tag{5–27}$$

or

$$\tau_{\text{max}} = \frac{3V_\phi E_A}{\omega_m X_S} \tag{5–28}$$

These equations indicate that the larger the field current (and hence E_A), the greater the maximum torque of the motor. There is therefore a stability advantage in operating the motor with a large field current or a large E_A.

Figure 5–31 I The torque–speed characteristic of a synchronous motor. Since the speed of the motor is constant, its speed regulation is zero.

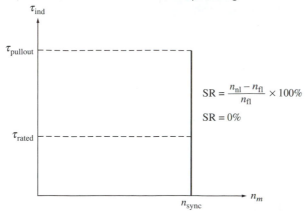

$$SR = \frac{n_{nl} - n_{fl}}{n_{fl}} \times 100\%$$

$$SR = 0\%$$

The Effect of Load Changes on a Synchronous Motor

If a load is attached to the shaft of a synchronous motor, the motor will develop enough torque to keep the motor and its load turning at a synchronous speed. What happens when the load is changed on a synchronous motor?

To find out, examine a synchronous motor operating initially with a leading power factor, as shown in Figure 5–32. If the load on the shaft of the motor is increased, the rotor will initially slow down. As it does, the torque angle δ becomes larger, and the induced torque increases. The increase in induced torque eventually speeds the rotor back up, and the motor again turns at synchronous speed but with a larger torque angle δ.

What does the phasor diagram look like during this process? To find out, examine the constraints on the machine during a load change. Figure 5–32a shows the motor's phasor diagram before the loads are increased.

Recall that the terminal voltage and frequency supplied to the motor are maintained constant by the motor's power supply. The internal generated voltage E_A is equal to $K\phi\omega$ and so depends on *only* the field current in the machine and the speed of the machine. Since the speed is constrained to be constant by the input power supply, and since no one has touched the field circuit, the field current is constant as well. Therefore, $|\mathbf{E}_A|$ *must be constant as the load changes.*

Since we have assumed armature resistance is negligible, the power converted from electrical to mechanical form in the motor will be the same as the input power of the motor. This power is given by either of the following two equations

$$P = 3V_\phi I_A \cos \theta \tag{5–15}$$

$$P = \frac{3V_\phi E_A \sin \delta}{X_S} \tag{5–18}$$

Since the phase voltage of the motor is held constant by the motor's power supply, the quantities $I_A \cos \theta$ and $E_A \sin \delta$ on the phasor diagram must be *directly proportional* to the power supplied by the motor.

Figure 5–32 | (a) Phasor diagram of a motor operating at a leading power factor. (b) The effect of an increase in load on the operation of a synchronous motor.

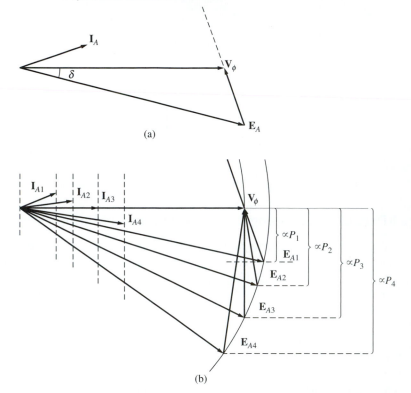

(a)

(b)

When the power supplied by the motor increases, the distances proportional to power ($E_A \sin \delta$ and $I_A \cos \theta$) will increase, but the magnitude of $\mathbf{E}_A$ must remain constant. As the load increases, $\mathbf{E}_A$ swings down in the manner shown in Figure 5–32b. As $\mathbf{E}_A$ swings down further and further, the quantity $jX_S\mathbf{I}_A$ has to increase to reach from the tip of $\mathbf{E}_A$ to $\mathbf{V}_\phi$, and therefore the armature current $\mathbf{I}_A$ also increases. Notice that the power-factor angle θ changes too, becoming less and less leading and then more and more lagging.

EXAMPLE 5–5

A 208-V, 45-kVA, 0.8-PF-leading, Δ-connected, 60-Hz synchronous machine has a synchronous reactance of 2.5 Ω and a negligible armature resistance. Its friction and windage losses are 1.5 kW, and its core losses are 1.0 kW. Initially, the shaft is supplying a 15-hp load, and the motor's power factor is 0.80 leading.

a.   Sketch the phasor diagram of this motor, and find the values of $\mathbf{I}_A$, I_L, and $\mathbf{E}_A$.

b.   Assume that the shaft load is now increased to 30 hp. Sketch the behavior of the phasor diagram in response to this change.

c.   Find $\mathbf{I}_A$, I_L, and $\mathbf{E}_A$ after the load change. What is the new motor power factor?

■ **Solution**

a.   Initially, the motor's output power is 15 hp. This corresponds to an output of

$$P_{out} = (15 \text{ hp})(0.746 \text{ KW/hp}) = 11.19 \text{ kW}$$

Therefore, the electric power supplied to the machine is

$$P_{in} = P_{out} + P_{mech\,loss} + P_{core\,loss} + P_{elec\,loss}$$
$$= 11.19 \text{ kW} + 1.5 \text{ kW} + 1.0 \text{ kW} + 0 \text{ kW} = 13.69 \text{ kW}$$

Note that we have assumed $P_{elec\,loss} = 0$ because the problem states that the armature resistance is negligible. Since the motor's power factor is 0.80 leading, the resulting line current flow is

$$I_L = \frac{P_{in}}{\sqrt{3} V_T \cos \theta}$$
$$= \frac{13.69 \text{ kW}}{\sqrt{3}(208 \text{ V})(0.80)} = 47.5 \text{ A}$$

and the armature current is $I_L/\sqrt{3}$, with 0.8 leading power factor, which gives the result

$$\mathbf{I}_A = 27.4 \angle 36.87° \text{ A}$$

To find $\mathbf{E}_A$, apply Kirchhoff's voltage law [Equation (5–26)]:

$$\mathbf{E}_A = \mathbf{V}_\phi - jX_S\mathbf{I}_A$$
$$= 208 \angle 0° \text{ V} - (j2.5 \text{ }\Omega)(27.4 \angle 36.87° \text{ A})$$
$$= 208 \angle 0° \text{ V} - 68.5 \angle 126.87° \text{ V}$$
$$= 249.1 - j54.8 \text{ V} = 255 \angle -12.4° \text{ V}$$

The resulting phasor diagram is shown in Figure 5–33a.

b.   As the power on the shaft is increased to 30 hp, the shaft slows momentarily, and the internal generated voltage $\mathbf{E}_A$ swings out to a larger angle δ while maintaining a constant magnitude. The resulting phasor diagram is shown in Figure 5–33b.

c.   After the load changes, the electric input power of the machine becomes

$$P_{in} = P_{out} + P_{mech\,loss} + P_{core\,loss} + P_{elec\,loss}$$
$$= (30 \text{ hp})(0.746 \text{ kW/hp}) + 1.5 \text{ kW} + 1.0 \text{ kW} + 0 \text{ kW}$$
$$= 24.88 \text{ kW}$$

From the equation for power in terms of torque angle [Equation (5–18)], it is possible to find the magnitude of the angle δ (remember that the magnitude of $\mathbf{E}_A$ is constant):

Figure 5–33 | (a) The motor phasor diagram for Example 5–5a. (b) The motor phasor diagram for Example 5–5b.

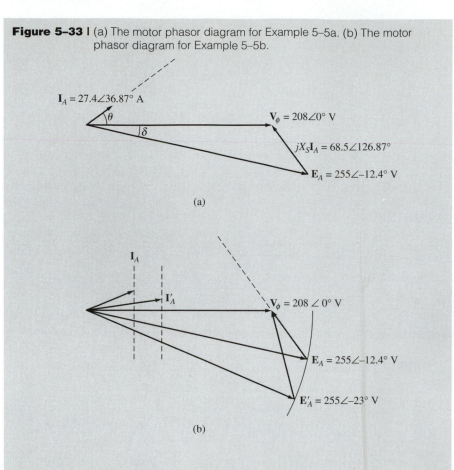

(a)

(b)

$$P = \frac{3V_\phi E_A \sin\delta}{X_S} \qquad (5\text{–}18)$$

so

$$\delta = \sin^{-1}\frac{X_S P}{3V_\phi E_A}$$

$$= \sin^{-1}\frac{(2.5\ \Omega)(24.88\ \text{kW})}{3(208\ \text{V})(255\ \text{V})}$$

$$= \sin^{-1} 0.391 = 23°$$

The internal generated voltage thus becomes $E_A = 355\angle -23°$ V. Therefore, I_A will be given by

$$I_A = \frac{V_\phi - E_A}{jX_S}$$

$$= \frac{208\angle 0°\ \text{V} - 255\angle -23°\ \text{V}}{j2.5\ \Omega}$$

$$= \frac{103.1\angle 105°\ \text{V}}{j2.5\ \Omega} = 41.2\angle 15°\ \text{A}$$

and I_L will become

$$I_L = \sqrt{3}I_A = 71.4 \text{ A}$$

The final power factor will be cos $(-15°)$ or 0.966 leading. ∎

The Effect of Field Current Changes on a Synchronous Motor

We have seen how a change in shaft load on a synchronous motor affects the motor. There is one other quantity on a synchronous motor that can be readily adjusted—its field current. What effect does a change in field current have on a synchronous motor?

To find out, look at Figure 5–34. Figure 5–34a shows a synchronous motor initially operating at a lagging power factor. Now, increase its field current and see what happens to the motor.

Figure 5–34 I (a) A synchronous motor operating at a lagging power factor. (b) The effect of an increase in field current on the operation of this motor.

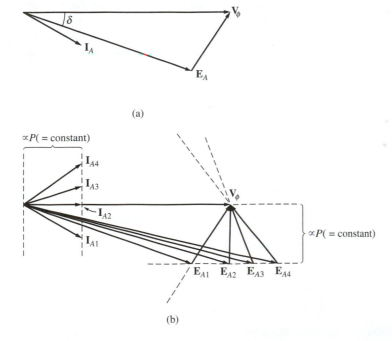

Note that *an increase in field current increases the magnitude of* $\mathbf{E}_A$ *but does not affect the real power supplied by the motor.* The power supplied by the motor changes only when the shaft load torque changes. Since a change in I_A does not affect the shaft speed n_m, and since the load attached to the shaft is unchanged, the real power supplied is unchanged. Of course, V_T is also constant, since it is kept constant by the power source supplying the motor. The distances proportional to power on the phasor diagram ($E_A \sin \delta$ and $I_A \cos \theta$) must therefore be constant. When the field current is increased, $\mathbf{E}_A$ must increase, but it can only do so by sliding out along the line of constant power. This effect is shown in Figure 5–34b.

Notice that as the value of $\mathbf{E}_A$ increases, the magnitude of the armature current $\mathbf{I}_A$ first decreases and then increases again. At low $\mathbf{E}_A$, the armature current is lagging, and the motor is an inductive load. It is acting like an inductor-resistor combination, consuming reactive power Q. As the field current is increased, the armature current eventually lines up with $\mathbf{V}_\phi$, and the motor looks purely resistive. As the field current is increased further, the armature current becomes leading, and the motor becomes a capacitive load. It is now acting like a capacitor-resistor combination, consuming negative reactive power $-Q$ or, alternatively, supplying reactive power Q to the system.

A plot of I_A versus I_F for a synchronous motor is shown in Figure 5–35. Such a plot is called a *synchronous motor V curve*, for the obvious reason that it is shaped like the letter V. There are several V curves drawn, corresponding to different real power levels. For each curve, the minimum armature current occurs at unity power factor, when only real power is being supplied to the motor. At any other point on the curve, some reactive power is being supplied to or by the motor as well. For field currents *less* than the value giving minimum I_A, the armature current is lagging, consuming Q. For field currents *greater* than the value giving the minimum I_A, the armature current is leading, supplying Q to the power system as a capacitor would.

Figure 5–35 I Synchronous motor V curves.

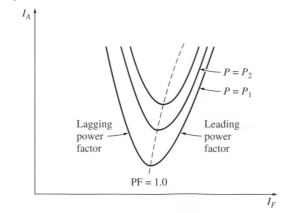

Therefore, by controlling the field current of a synchronous motor, the *reactive power* supplied to or consumed by the power system can be controlled.

When the projection of the phasor $\mathbf{E}_A$ onto $\mathbf{V}_\phi$ ($E_A \cos \delta$) is *shorter* than $\mathbf{V}_\phi$ itself, a synchronous motor has a lagging current and consumes Q. Since the field current is small in this situation, the motor is said to be *underexcited*. On the other hand, when the projection of $\mathbf{E}_A$ onto $\mathbf{V}_\phi$ is *longer* than $\mathbf{V}_\phi$ itself, a synchronous motor has a leading current and supplies Q to the power system. Since the field current is large in this situation, the motor is said to be *overexcited*. Phasor diagrams illustrating these concepts are shown in Figure 5–36.

Figure 5–36 | (a) The phasor diagram of an *underexcited* synchronous motor. (b) The phasor diagram of an *overexcited* synchronous motor.

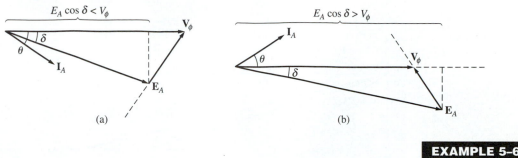

(a)   (b)

EXAMPLE 5–6

The 208-V, 45-kVA, 0.8-PF-leading, Δ-connected, 60-Hz synchronous motor of the previous example is supplying a 15-hp load with an initial power factor of 0.85 PF lagging. The field current I_F at these conditions is 4.0 A.

a. Sketch the initial phasor diagram of this motor, and find the values I_A and E_A.

b. If the motor's flux is increased by 25 percent, sketch the new phasor diagram of the motor. What are E_A, I_A, and the power factor of the motor now?

c. Assume that the flux in the motor varies linearly with the field current I_F. Make a plot of I_A versus I_F for the synchronous motor with a 15-hp load.

■ **Solution**

a. From the previous example, the electric input power with all the losses included is P_{in} = 13.69 kW. Since the motor's power factor is 0.85 lagging, the resulting armature current flow is

$$I_A = \frac{P_{in}}{3V_\phi \cos \theta}$$

$$= \frac{13.69 \text{ kW}}{3(208 \text{ V})(0.85)} = 25.8 \text{ A}$$

The angle θ is $\cos^{-1} 0.85 = 31.8°$, so the phasor current I_A is equal to

$$I_A = 25.8\angle-31.8° \text{ A}$$

To find E_A apply Kirchhoff's voltage law [Equation (5–26)]:

$$E_A = V_\phi - jX_S I_A$$
$$= 208\angle0° \text{ V} - (j2.5 \text{ }\Omega)(25.8\angle-31.8° \text{ A})$$
$$= 208\angle0° \text{ V} - 64.5\angle58.2° \text{ V}$$
$$= 182\angle-17.5° \text{ V}$$

The resulting phasor diagram is shown in Figure 5–37, together with the results for part (b).

b. If the flux ϕ is increased by 25 percent, then $E_A = K\phi\omega$ will increase by 25 percent too:

$$E_{A2} = 1.25 E_{A1} = 1.25(182 \text{ V}) = 227.5 \text{ V}$$

Figure 5–37 | The phasor diagram of the motor in Example 5–6.

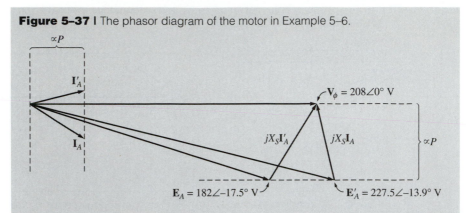

However, the power supplied to the load must remain constant. Since the distance $E_A = \sin \delta$ is proportional to the power, that distance on the phasor diagram must be constant from the original flux level to the new flux level. Therefore,

$$E_{A1} \sin \delta_1 = E_{A2} \sin \delta_2$$

$$\delta_2 = \sin^{-1}\left(\frac{E_{A1}}{E_{A2}} \sin \delta_1\right)$$

$$\delta_2 = \sin^{-1}\left[\frac{182\text{ V}}{227.5\text{ V}} \sin(-17.5°)\right] = -13.9°$$

The armature current can now be found from Kirchhoff's voltage law:

$$\mathbf{I}_{A2} = \frac{\mathbf{V}_\phi - \mathbf{E}_{A2}}{jX_S}$$

$$= \frac{208\angle 0°\text{ V} - 227.5\angle -13.9°\text{ V}}{j2.5\ \Omega}$$

$$= \frac{56.2\angle 103.2°\text{ V}}{j2.5\ \Omega} = 22.5\angle 13.2°\text{ A}$$

Finally, the motor's power factor is now

$$PF = \cos(13.2°) = 0.974 \quad \text{leading}$$

The resulting phasor diagram is also shown in Figure 5–37.

c.   Because the flux is assumed to vary linearly with field current, E_A will also vary linearly with field current. We know that E_A is 182 V for a field current of 4.0 A, so E_A for any given field current can be found from the ratio

$$\frac{E_{A2}}{182\text{ V}} = \frac{I_{F2}}{4.0\text{ A}}$$

or
$$E_{A2} = 45.5 I_{F2} \tag{5–29}$$

The torque angle δ for any given field current can be found from the fact that the power supplied to the load must remain constant:

$$E_{A1} \sin \delta_1 = E_{A2} \sin \delta_2$$

so
$$\delta_2 = \sin^{-1}\left(\frac{E_{A1}}{E_{A2}} \sin \delta_1\right) \tag{5–30}$$

These two pieces of information give us the phasor voltage $\mathbf{E}_A$. Once $\mathbf{E}_A$ is available, the new armature current can be calculated from Kirchhoff's voltage law:

$$\mathbf{I}_{A2} = \frac{\mathbf{V}_\phi - \mathbf{E}_{A2}}{jX_S} \tag{5–31}$$

A MATLAB m-file to calculate and plot I_A versus I_F using Equations (5–29) through (5–31) is shown below:

```
% M-file: v_curve.m
% M-file create a plot of armature current versus field
%   current for the synchronous motor of Example 5-6.

% First, initialize the field current values (21 values
% in the range 3.8-5.8 A)
i_f = (38:1:58) / 10;

% Now initialize all other values
i_a = zeros(1,21);              % Pre-allocate i_a array
x_s = 2.5;                      % Synchronous reactance
v_phase = 208;                  % Phase voltage at 0 degrees
delta1 = -17.5 * pi/180;        % delta 1 in radians
e_a1 = 182 * (cos(delta1) + j * sin(delta1));

% Calculate the armature current for each value
for ii = 1:21
   % Calculate magnitude of e_a2
   e_a2 = 45.5 * i_f(ii);

   % Calculate delta2
   delta2 = asin ( abs(e_a1) / abs(e_a2) * sin(delta1) );

   % Calculate the phasor e_a2
   e_a2 = e_a2 * (cos(delta2) + j * sin(delta2));

   % Calculate i_a
   i_a(ii) = ( v_phase - e_a2 ) / ( j * x_s);
end

% Plot the v-curve
plot(i_f,abs(i_a),'Color','k','Linewidth',2.0);
xlabel('Field Current (A)','Fontweight','Bold');
ylabel('Armature Current (A)','Fontweight','Bold');
title ('Synchronous Motor V-Curve','Fontweight','Bold');
grid on;
```

The plot produced by this m-file is shown in Figure 5–38. Note that for a field current of 4.0 A, the armature current is 25.8 A. This result agrees with part (a) of this example. ■

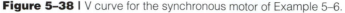

Figure 5–38 | V curve for the synchronous motor of Example 5–6.

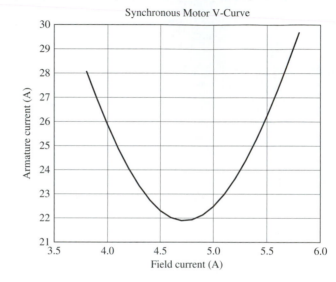

The Synchronous Motor and Power-Factor Correction

Figure 5–39 shows a large power system whose output is connected through a transmission line to an industrial plant at a distant point. The industrial plant shown consists of three loads. Two of the loads are induction motors with lagging power factors, and the third load is a synchronous motor with a variable power factor.

What does the ability to set the power factor of one of the loads do for the power system? To find out, examine the following example problem.

EXAMPLE 5–7

The power system in Figure 5–39 operates at 480 V. Load 1 is an induction motor consuming 100 kW at 0.78 PF lagging, and load 2 is an induction motor consuming 200 kW at 0.8 PF lagging. Load 3 is a synchronous motor whose real power consumption is 150 kW.

a.   If the synchronous motor is adjusted to operate at 0.85 PF lagging, what is the transmission line current in this system?

b.   If the synchronous motor is adjusted to operate at 0.85 PF leading, what is the transmission line current in this system?

Figure 5–39 | A simple power system consisting of an infinite bus supplying an industrial plant through a transmission line.

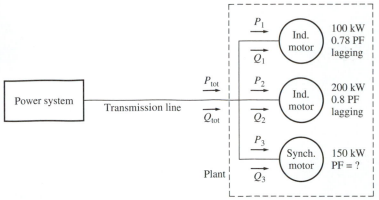

c. Assume that the transmission line losses are given by

$$P_{LL} = 3I_L^2 R_L \quad \text{line loss}$$

where LL stands for line losses. How do the transmission losses compare in the two cases?

■ **Solution**

a. In the first case, the real power of load 1 is 100 kW, and the reactive power of load 1 is

$$Q_1 = P_1 \tan \theta$$
$$= (100 \text{ kW}) \tan (\cos^{-1} 0.78) = (100 \text{ kW}) \tan 38.7°$$
$$= 80.2 \text{ kVAR}$$

The real power of load 2 is 200 kW, and the reactive power of load 2 is

$$Q_2 = P_2 \tan \theta$$
$$= (200 \text{ kW}) \tan (\cos^{-1} 0.80) = (200 \text{ kW}) \tan 36.87°$$
$$= 150 \text{ kVAR}$$

The real power load 3 is 150 kW, and the reactive power of load 3 is

$$Q_3 = P_3 \tan \theta$$
$$= (150 \text{ kW}) \tan (\cos^{-1} 0.85) = (150 \text{ kW}) \tan 31.8°$$
$$= 93 \text{ kVAR}$$

Thus, the total real load is

$$P_{tot} = P_1 + P_2 + P_3$$
$$= 100 \text{ kW} + 200 \text{ kW} + 150 \text{ kW} = 450 \text{ kW}$$

and the total reactive load is

$$Q_{tot} = Q_1 + Q_2 + Q_3$$
$$= 80.2 \text{ kVAR} + 150 \text{ kVAR} + 93 \text{ kVAR} = 323.2 \text{ kVAR}$$

The equivalent system power factor is thus

$$PF = \cos \theta = \cos\left(\tan^{-1}\frac{Q}{P}\right) = \cos\left(\tan^{-1}\frac{323.2 \text{ kVAR}}{450 \text{ kW}}\right)$$
$$= \cos 35.7° = 0.812 \quad \text{lagging}$$

Finally, the line current is given by

$$I_L = \frac{P_{tot}}{\sqrt{3}V_L \cos \theta} = \frac{450 \text{ kW}}{\sqrt{3}(480 \text{ V})(0.812)} = 667 \text{ A}$$

b. The real and reactive powers of loads 1 and 2 are unchanged, as is the real power of load 3. The reactive power of load 3 is

$$Q_3 = P_3 \tan \theta$$
$$= (150 \text{ kW}) \tan(-\cos^{-1} 0.85) = (150 \text{ kW}) \tan(-31.8°)$$
$$= -93 \text{ kVAR}$$

Thus, the total real load is

$$P_{tot} = P_1 + P_2 + P_3$$
$$= 100 \text{ kW} + 200 \text{ kW} + 150 \text{ kW} = 450 \text{ kW}$$

and the total reactive load is

$$Q_{tot} = Q_1 + Q_2 + Q_3$$
$$= 80.2 \text{ kVAR} + 150 \text{ kVAR} - 93 \text{ kVAR} = 137.2 \text{ kVAR}$$

The equivalent system power factor is thus

$$PF = \cos \theta = \cos\left(\tan^{-1}\frac{Q}{P}\right) = \cos\left(\tan^{-1}\frac{137.2 \text{ kVAR}}{450 \text{ kW}}\right)$$
$$= \cos 16.96° = 0.957 \quad \text{lagging}$$

Finally, the line current is given by

$$I_L = \frac{P_{tot}}{\sqrt{3}V_L \cos \theta} = \frac{450 \text{ kW}}{\sqrt{3}(480 \text{ V})(0.957)} = 566 \text{ A}$$

c. The transmission losses in the first case are

$$P_{LL} = 3I_L^2 R_L = 3(667 \text{ A})^2 R_L = 1,344,700 R_L$$

The transmission losses in the second case are

$$P_{LL} = 3I_L^2 R_L = 3(566 \text{ A})^2 R_L = 961,070 R_L$$

Notice that in the second case the transmission power losses are 28 percent less than in the first case, while the power supplied to the loads is the same. ∎

As seen in the preceding example, the ability to adjust the power factor of one or more loads in a power system can significantly affect the operating efficiency of the power system. The lower the power factor of a system, the greater the losses in the power lines feeding it. Most loads on a typical power system are induction motors, so power systems are almost invariably lagging in power factor. Having one or more leading loads (overexcited synchronous motors) on the system can be useful for the following reasons:

1. A leading load can supply some reactive power Q for nearby lagging loads, instead of it coming from the generator. Since the reactive power does not have to travel over the long and fairly high resistance transmission lines, the transmission line current is reduced and the power system losses are much lower. (This was shown by the previous example.)

2. Since the transmission lines carry less current, they can be smaller for a given rated power flow. A lower equipment current rating reduces the cost of a power system significantly.

3. In addition, requiring a synchronous motor to operate with a leading power factor means that the motor must be run *overexcited*. This mode of operation increases the motor's maximum torque and reduces the chance of accidentally exceeding the pullout torque.

The use of synchronous motors or other equipment to increase the overall power factor of a power system is called *power-factor correction*. Since a synchronous motor can provide power-factor correction and lower power system costs, many loads that can accept a constant-speed motor (even though they do not necessarily *need* one) are driven by synchronous motors. Even though a synchronous motor may cost more than an induction motor on an individual basis, the ability to operate a synchronous motor at leading power factors for power-factor correction saves money for industrial plants. This results in the purchase and use of synchronous motors.

Any synchronous motor that exists in a plant is run overexcited as a matter of course to achieve power-factor correction and to increase its pullout torque. However, running a synchronous motor overexcited requires a high field current and flux, which causes significant rotor heating. An operator must be careful not to overheat the field windings by exceeding the rated field current.

5.11 | STARTING SYNCHRONOUS MOTORS

Section 5.10 explained the behavior of a synchronous motor under steady-state conditions. In that section, the motor was always assumed to be initially turning at *synchronous speed*. What has not yet been considered is the question: How did the motor get to synchronous speed in the first place?

To understand the nature of the starting problem, refer to Figure 5–40. This figure shows a 60-Hz synchronous motor at the moment power is applied to its stator windings. The rotor of the motor is stationary, and therefore the magnetic field $\mathbf{B}_R$ is stationary. The stator magnetic field $\mathbf{B}_S$ is starting to sweep around the motor at synchronous speed.

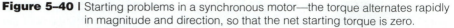

Figure 5–40 | Starting problems in a synchronous motor—the torque alternates rapidly in magnitude and direction, so that the net starting torque is zero.

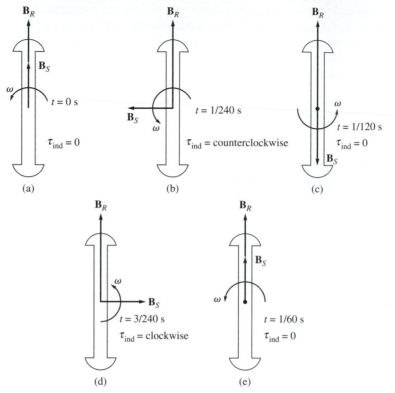

Figure 5–40a shows the machine at time $t = 0$ s, when $\mathbf{B}_R$ and $\mathbf{B}_S$ are exactly lined up. By the induced-torque equation

$$\tau_{ind} = k\mathbf{B}_R \times \mathbf{B}_S \tag{4–47}$$

the induced torque on the shaft of the rotor is zero. Figure 5–40b shows the situation at time $t = 1/240$ s. In such a short time, the rotor has barely moved, but the stator magnetic field now points to the left. By the induced-torque equation, the torque on the shaft of the rotor is now *counterclockwise*. Figure 5–40c shows the situation at time $t = 2/240$ s. At that point $\mathbf{B}_R$ and $\mathbf{B}_S$ point in opposite directions, and τ_{ind} again equals zero. At $t = 3/240$ s, the stator magnetic field now points to the right, and the resulting torque is *clockwise*.

Finally, at $t = 4/240$ s, the stator magnetic field is again lined up with the rotor magnetic field, and $\tau_{ind} = 0$. During one electrical cycle, the torque was first counterclockwise and then clockwise, and the average torque over the complete cycle was zero. What happens to the motor is that it vibrates heavily with each electrical cycle and finally overheats.

Such an approach to synchronous motor starting is hardly satisfactory—managers tend to frown on employees who burn up their expensive equipment. So just how *can* a synchronous motor be started?

Three basic approaches can be used to safely start a synchronous motor:

1. *Reduce the speed of the stator magnetic field* to a low enough value that the rotor can accelerate and lock in with it during one half-cycle of the magnetic field's rotation. This can be done by reducing the frequency of the applied electric power. Such frequency control used to be difficult, but it can now be easily accomplished with modern solid-state motor controllers.

2. *Use an external prime mover* to accelerate the synchronous motor up to synchronous speed, go through the paralleling procedure, and bring the machine on the line as a generator. Then, turning off or disconnecting the prime mover will make the synchronous machine a motor.

3. *Use damper windings or amortisseur windings*. The function of damper windings and their use in motor starting will be explained below.

Motor Starting by Using Amortisseur or Damper Windings

By far the most popular way to start a synchronous motor is to employ *amortisseur* or *damper* windings. Amortisseur windings are special bars laid into notches carved in the face of a synchronous motor's rotor and then shorted out on each end by a large *shorting ring*. A pole face with a set of amortisseur windings is shown in Figure 5–41, and amortisseur windings are visible in Figures 5–2 and 5–4.

Figure 5–41 | A rotor field pole for a synchronous machine showing amortisseur windings in the pole face. *(Courtesy of General Electric Company.)*

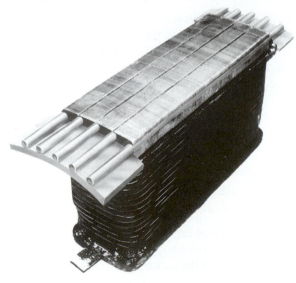

To understand what a set of amortisseur windings does in a synchronous motor, examine the stylized salient two-pole rotor shown in Figure 5–42. This rotor shows an amortisseur winding, with the shorting bars on the ends of the two rotor pole faces connected by wires. (This is not quite the way normal machines are constructed, but it will serve beautifully to illustrate the point of the windings.)

Figure 5–42 | A simplified diagram of a salient two-pole machine showing amortisseur windings.

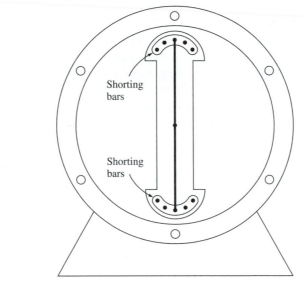

Assume initially that *the main rotor field winding is disconnected* and that a three-phase set of voltages is applied to the stator of this machine. When the power is first applied at time $t = 0$ s, assume that the magnetic field $\mathbf{B}_S$ is vertical, as shown in Figure 5–43a. As the magnetic field $\mathbf{B}_S$ sweeps along in a counterclockwise direction, it induces a voltage in the bars of the amortisseur winding given by Equation (1–45):

$$e_{\text{ind}} = (\mathbf{v} \times \mathbf{B}) \cdot \mathbf{l} \qquad (1\text{–}45)$$

where

$\quad\mathbf{v}$ = velocity of the bar *relative to the magnetic field*

$\quad\mathbf{B}$ = magnetic flux density vector

$\quad\mathbf{l}$ = length of conductor in the magnetic field

The bars at the top of the rotor are moving to the right *relative to the magnetic field*, so the resulting direction of the induced voltage is out of the page. Similarly, the induced voltage is into the page in the bottom bars. These voltages produce a current

Figure 5–43 | The development of a unidirectional torque with synchronous motor amortisseur windings.

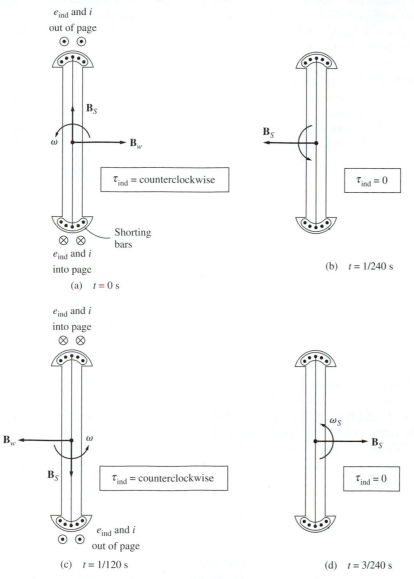

(a) $t = 0$ s

(b) $t = 1/240$ s

(c) $t = 1/120$ s

(d) $t = 3/240$ s

flow out of the top bars and into the bottom bars, resulting in a winding magnetic field $\mathbf{B}_W$ pointing to the right. By the induced-torque equation

$$\tau_{ind} = k\mathbf{B}_W \times \mathbf{B}_S$$

the resulting torque on the bars (and the rotor) is *counterclockwise*.

Figure 5–43b shows the situation at $t = 1/240$ s. Here, the stator magnetic field has rotated 90° while the rotor has barely moved (it simply cannot speed up in so short a time). At this point, the voltage induced in the amortisseur windings is zero, because **v** is parallel to **B**. With no induced voltage, there is no current in the windings, and the induced torque is zero.

Figure 5–43c shows the situation at $t = 2/240$ s. Now the stator magnetic field has rotated another 90°, and the rotor still has not moved yet. The induced voltage [given by Equation (1–45)] in the amortisseur windings is out of the page in the bottom bars and into the page in the top bars. The resulting current flow is out of the page in the bottom bars and into the page in the top bars, causing a magnetic field $\mathbf{B}_W$ to point to the left. The resulting induced torque, given by

$$\tau_{\text{ind}} = k\mathbf{B}_W \times \mathbf{B}_S$$

is counterclockwise.

Finally, Figure 5–43d shows the situation at time $t = 3/240$ s. Here, as at $t = 1/240$ s, the induced torque is zero.

Notice that sometimes the torque is counterclockwise and sometimes it is essentially zero, but it is *always unidirectional*. Since there is a net torque in a single direction, the motor's rotor speeds up. (This is entirely different from starting a synchronous motor with its normal field current, since in that case torque is first clockwise and then counterclockwise, averaging out to zero. In this case, torque is *always* in the same direction, so there is a nonzero average torque.)

Although the motor's rotor will speed up, it can never quite reach synchronous speed. This is easy to understand. Suppose that a rotor is turning at synchronous speed. Then the speed of the stator magnetic field $\mathbf{B}_S$ is the same as the rotor's speed, and there is *no relative motion* between $\mathbf{B}_S$ and the rotor. If there is no relative motion, the induced voltage in the windings will be zero, the resulting current flow will be zero, and the winding magnetic field will be zero. Therefore, there will be no torque on the rotor to keep it turning. Even though a rotor cannot speed up all the way to synchronous speed, it can get close. It gets close enough to n_{sync} that the regular field current can be turned on, and the rotor will pull into step with the stator magnetic fields.

In a real machine, the field windings are not open circuited during the starting procedure. If the field windings were open circuited, then very high voltages would be produced in them during starting. If the field winding is short circuited during starting, no dangerous voltages are produced, and the induced field current actually contributes extra starting torque to the motor.

To summarize, if a machine has amortisseur windings, it can be started by the following procedure:

1. Disconnect the field windings from their DC power source and short them out.

2. Apply a three-phase voltage to the stator of the motor, and let the rotor accelerate up to near-synchronous speed. The motor should have no load on its shaft, so that its speed can approach n_{sync} as closely as possible.

3. Connect the DC field circuit to its power source. After this is done, the motor will lock into step at synchronous speed, and loads may then be added to its shaft.

5.12 | THE RELATIONSHIP BETWEEN SYNCHRONOUS GENERATORS AND SYNCHRONOUS MOTORS

A synchronous generator is a synchronous machine that converts mechanical power to electrical power, while a synchronous motor is a synchronous machine that converts electrical power to mechanical power. In fact, they are both the same physical machine.

A synchronous machine can supply real power to or consume real power from a power system and can supply reactive power to or consume reactive power from a power system. All four combinations of real and reactive power flows are possible, and Figure 5–44 shows the phasor diagrams for these conditions.

Figure 5–44 | Phasor diagrams showing the generation and consumption of real power P and reactive power Q by synchronous generators and motors.

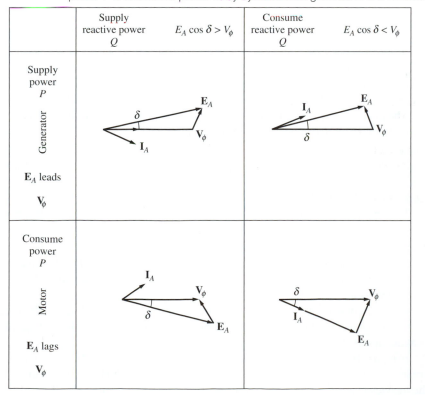

Notice from the figure that

1. The distinguishing characteristic of a synchronous generator (supplying *P*) is that $\mathbf{E}_A$ *lies ahead of* $\mathbf{V}_\phi$ while for a motor $\mathbf{E}_A$ *lies behind* $\mathbf{V}_\phi$.
2. The distinguishing characteristic of a machine supplying reactive power *Q* is that $E_A \cos \delta > V_\phi$ regardless of whether the machine is acting as a generator or as a motor. A machine that is consuming reactive power *Q* has $E_A \cos \delta < V_\phi$.

5.13 | SYNCHRONOUS MACHINE RATINGS

There are certain basic limits to the speed and power that may be obtained from a synchronous motor or generator. These limits are expressed as *ratings* of the machine. The purpose of the ratings is to protect the machine from damage by improper operation. To this end, each machine has a number of ratings listed on a nameplate attached to it.

Typical ratings on a synchronous machine are *voltage, frequency, speed, apparent power (kilovolt-amperes), power factor, field current,* and *service factor.* These ratings, and the interrelationships among them, will be discussed in the following sections.

The Voltage, Speed, and Frequency Ratings

The rated frequency of a synchronous machine depends on the power system to which it is connected. The commonly used power system frequencies today are 50 Hz (in Europe, Asia, etc.), 60 Hz (in the Americas), and 400 Hz (in special-purpose and control applications). Once the operating frequency is known, there is only one possible rotational speed for a given number of poles. The fixed relationship between frequency and speed is given by Equation (4–23)

$$f_e = \frac{n_m P}{120} \tag{4–23}$$

as previously described.

Perhaps the most obvious rating is the voltage at which a synchronous machine is designed to operate. A generator's voltage depends on the flux, the speed of rotation, and the mechanical construction of the machine. For a given mechanical frame size and speed, the higher the desired voltage, the higher the machine's required flux. However, flux cannot be increased forever, since there is always a maximum allowable field current.

Another consideration in setting the maximum allowable voltage is the breakdown value of the winding insulation—normal operating voltages must not approach breakdown too closely.

Is it possible to operate a synchronous machine rated for one frequency at a different frequency? For example, is it possible to operate a 60-Hz generator at 50 Hz? The answer is a *qualified* yes, as long as certain conditions are met. Basically, the problem is that there is a maximum flux achievable in any given machine, and since

$E_A = K\phi\omega$, the maximum allowable E_A changes when the speed is changed. Specifically, if a 60-Hz generator is to be operated at 50 Hz, then the operating voltage must be *derated* to 50/60, or 83.3 percent, of its original value. Just the opposite effect happens when a 50-Hz generator is operated at 60 Hz.

Apparent Power and Power-Factor Ratings

There are two factors that determine the power limits of electric machines. One is the mechanical torque on the shaft of the machine, and the other is the heating of the machine's windings. In all practical synchronous motors and generators, the shaft is strong enough mechanically to handle a much larger steady-state power than the machine is rated for, so the practical steady-state limits are set by heating in the machine's windings.

There are two windings in a synchronous machine, and each one must be protected from overheating. These two windings are the armature winding and the field winding. Since heating is caused by I^2R losses, this fact means that there is a maximum acceptable armature current and a maximum acceptable field current. We will now discuss how these maximum currents affect the ratings of synchronous machines. (Note that the following discussion is from the point of view of synchronous generators, but the results are equally valid for both motors and generators.)

The maximum acceptable armature current sets the apparent power rating for a generator, since the apparent power S is given by

$$S = 3V_\phi I_A \tag{5-32}$$

If the rated voltage is known, then the maximum acceptable armature current determines the rated kilovoltamperes of the generator:

$$S_{\text{rated}} = 3V_{\phi, \text{rated}}I_{A, \text{max}} \tag{5-33}$$

or
$$S_{\text{rated}} = \sqrt{3}V_{L, \text{rated}}I_{L, \text{max}} \tag{5-34}$$

It is important to realize that, for heating the armature windings, *the power factor of the armature current is irrelevant*. The heating effect of the stator copper losses is given by

$$P_{\text{SCL}} = 3I_A^2 R_A \tag{5-35}$$

and is independent of the angle of the current with respect to $\mathbf{V}_\phi$. Because the current angle is irrelevant to the armature heating, synchronous generators are rated in kilovolt-amperes instead of kilowatts.

The other winding of concern is the field winding. The field copper losses are given by

$$P_{\text{RCL}} = I_F^2 R_F \tag{5-36}$$

so the maximum allowable heating sets a maximum field current for the machine. Since $E_A = K\phi\omega$, this sets the maximum acceptable size for E_A.

The effect of having a maximum I_F and a maximum E_A translates directly into a restriction on the lowest acceptable power factor of the generator when it is operating at the rated kilovolt-amperes. Figure 5–45 shows the phasor diagram of a synchronous generator with the rated voltage and armature current. The current can assume many different angles, as shown. The internal generated voltage $\mathbf{E}_A$ is the sum of $\mathbf{V}_\phi$ and $jX_S\mathbf{I}_A$. Notice that for some possible current angles the required E_A exceeds $E_{A,\text{max}}$. If the generator were operated at the rated armature current and these power factors, the field winding would burn up.

Figure 5–45 | How the rotor field current limit sets the rated power factor of a generator.

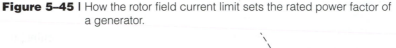
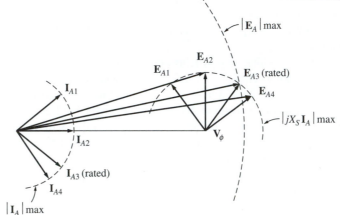

The angle of $\mathbf{I}_A$ that requires the maximum possible $\mathbf{E}_A$ while $\mathbf{V}_\phi$ remains at the rated value gives the rated power factor of the generator. It is possible to operate the generator at a lower (more lagging) power factor than the rated value, but only by cutting back on the kilovolt-amperes supplied by the generator.

Synchronous motors are usually rated in terms of real output power and the lowest power factor at full-load conditions. These two ratings together specify the maximum apparent power consumed by the motor, thus protecting the armature windings.

Short-Time Operation and Service Factor

The most important limit in the steady-state operation of a synchronous machine is the heating of its armature and field windings. However, the heating limit usually occurs at a point much less than the maximum power that the machine is magnetically and mechanically able to supply. In fact, a typical synchronous machine is often able to supply up to 300 percent of its rated power for a while (until its windings burn up). This ability to supply power above the rated amount is used to supply momentary power surges during motor starting and similar load transients.

It is also possible to use a synchronous machine at powers exceeding the rated values for longer periods of time, as long as the windings do not have time to heat up too much before the excess load is removed. For example, a generator that could supply 1 MW indefinitely would be able to supply 1.5 MW for 1 minute without serious harm and for progressively longer periods at lower power levels. However, the load must finally be removed, or the windings will overheat. The higher the power over the rated value, the shorter the time a machine can tolerate it.

The maximum temperature rise that a machine can stand depends on the *insulation class* of its windings. There are four standard insulation classes: A, B, F, and H. While there is some variation in acceptable temperature, depending on a machine's particular construction and the method of temperature measurement, these classes generally correspond to temperature rises of 60, 80, 105, and 125°C, respectively, above ambient temperature. The higher the insulation class of a given machine, the greater the power that can be drawn out of it without overheating its windings.

Overheating of windings is a *very serious* problem in a motor or generator. It was an old rule of thumb that for each 10°C temperature rise above the rated winding temperature, the average lifetime of a machine is cut in half. Modern insulating materials are less susceptible to breakdown than that, but temperature rises still drastically shorten their lives. For this reason, a synchronous machine should not be overloaded unless absolutely necessary.

A question related to the overheating problem is: Just how well is the power requirement of a machine known? Before installation, there are often only approximate estimates of load. Because of this, general-purpose machines usually have a *service factor*. The service factor is defined as the ratio of the actual maximum power of the machine to its nameplate rating. A machine with a service factor of 1.15 can actually be operated at 115 percent of the rated load indefinitely without harm. The service factor on a machine provides a margin of error in case the loads were improperly estimated.

5.14 | SUMMARY

A synchronous generator is a device for converting mechanical power from a prime mover to AC electric power at a specific voltage and frequency. The term *synchronous* refers to the fact that this machine's electrical frequency is locked in or synchronized with its mechanical rate of shaft rotation. The synchronous generator is used to produce the vast majority of electric power used throughout the world.

The internal generated voltage of this machine depends on the rate of shaft rotation and on the magnitude of the field flux. The phase voltage of the machine differs from the internal generated voltage by the effects of armature reaction in the generator and also by the internal resistance and reactance of the armature windings. The terminal voltage of the generator will either equal the phase voltage or be related to it by $\sqrt{3}$, depending on whether the machine is Δ or Y connected.

When a generator operates alone, the real and reactive powers that must be supplied are determined by the load attached to it, and the governor set points and field current control the frequency and terminal voltage, respectively.

A synchronous motor is the same physical machine as a synchronous generator, except that the direction of real power flow is reversed. Since synchronous motors are usually connected to power systems containing generators much larger than the motors, the frequency and terminal voltage of a synchronous motor are basically fixed.

The speed of a synchronous motor is constant from no load to the maximum possible load on the motor. The speed of rotation is

$$n_m = n_{\text{sync}} = \frac{120 f_e}{P}$$

The maximum possible power and torque that a synchronous motor can produce are

$$P_{\text{max}} = \frac{3V_\phi E_A}{X_S} \tag{5–19}$$

$$\tau_{\text{max}} = \frac{3V_\phi E_A}{\omega_m X_S} \tag{5–28}$$

If this value is exceeded, the rotor will not be able to stay locked in with the stator magnetic fields, and the motor will *slip poles*.

If the field current of a synchronous motor is varied while its shaft load remains constant, then the reactive power supplied or consumed by the motor will vary. If $E_A \cos \delta > V_\phi$, the motor will supply reactive power, while if $E_A \cos \delta < V_\phi$, the motor will consume reactive power.

A synchronous motor has no net starting torque and so cannot start by itself. There are three main ways to start a synchronous motor:

1. Reduce the stator frequency to a safe starting level.
2. Use an external prime mover.
3. Put amortisseur or damper windings on the motor to accelerate it to near-synchronous speed before a direct current is applied to the field windings.

If damper windings are present on a motor, they will also increase the stability of the motor during load transients.

A synchronous generator's ability to produce electric power is primarily limited by heating within the machine. When the generator's windings overheat, the life of the machine can be severely shortened. Since here are two different windings (armature and field), there are two separate constraints on the generator. The maximum allowable heating in the armature windings sets the maximum kilovolt-amperes allowable from the machine, and the maximum allowable heating in the field windings sets the maximum size of E_A. The maximum size of E_A and the maximum size of I_A together set the rated power factor of the generator.

The ratings of a synchronous motor are similar, except that output power and input power factor are specified. Together with the rated voltage of the machine, these quantities are sufficient to limit the maximum currents in both the armature and field windings.

5.15 | QUESTIONS

5–1. Why is the frequency of a synchronous generator locked into its rate of shaft rotation?

5–2. Why does an alternator's voltage drop sharply when it is loaded down with a lagging load?

5–3. Why does an alternator's voltage rise when it is loaded down with a leading load?

5–4. Sketch the phasor diagrams and magnetic field relationships for a synchronous generator operating at (a) unity power factor, (b) lagging power factor, (c) leading power factor.

5–5. Explain how the synchronous impedance and armature resistance can be determined in a synchronous generator.

5–6. What is the difference between a synchronous motor and a synchronous generator?

5–7. What is the speed regulation of a synchronous motor?

5–8. When would a synchronous motor be used even though its constant-speed characteristic is not needed?

5–9. Why can't a synchronous motor start by itself?

5–10. What techniques are available to start a synchronous motor?

5–11. What are amortisseur windings? Why is the torque produced by them unidirectional at starting, while the torque produced by the main field winding alternates direction?

5–12. Explain, using phasor diagrams, what happens to a synchronous motor as its field current is varied. Derive a synchronous motor V curve from the phasor diagram.

5–13. Is a synchronous motor's field circuit in more danger of overheating when it is operating at a leading or at a lagging power factor? Explain, using phasor diagrams.

5–14. A synchronous motor is operating at a fixed real load, and its field current is increased. If the armature current falls, was the motor initially operating at a lagging or a leading power factor?

5–15. Why must a 60-Hz generator be derated if it is to be operated at 50 Hz? How much derating must be done?

5–16. Would you expect a 400-Hz generator to be larger or smaller than a 60-Hz generator of the same power and voltage rating? Why?

5–17. Why is overheating such a serious matter for a synchronous machine?

5–18. What are short-time ratings? Why are they important in regular generator operation?

5.16 | PROBLEMS

5–1. At a location in Europe, it is necessary to supply 300 kW of 60-Hz power. The only power sources available operate at 50 Hz. It is decided to generate the power by means of a motor-generator set consisting of a synchronous motor driving a synchronous generator. How many poles should each of the two machines have in order to convert 50-Hz power to 60-Hz power?

5–2. A 480-V, 200-kVA, 0.8 power factor lagging, 60-Hz, two-pole Y-connected synchronous generator has a synchronous reactance of 0.4 Ω and an armature resistance of 0.04 Ω. At 60 Hz, its friction and windage losses are 6 kW, and its core losses are 4 kW. The field circuit has a DC voltage of 200 V, and the maximum I_F is 10 A. The resistance of the field circuit is adjustable over the range from 20 to 200 Ω. The OCC of this generator is shown in Figure P5–1.

    a.   How much field current is required to make V_T equal to 2300 V when the generator is running at no load?

    b.   What is the internal generated voltage of this machine at rated conditions?

    c.   How much field current is required to make V_T equal to 2300 V when the generator is running at rated conditions?

    d.   How much power and torque must the generator's prime mover be capable of supplying?

Figure P5–1 | The open-circuit characteristic for the generator in Problem 5–2.

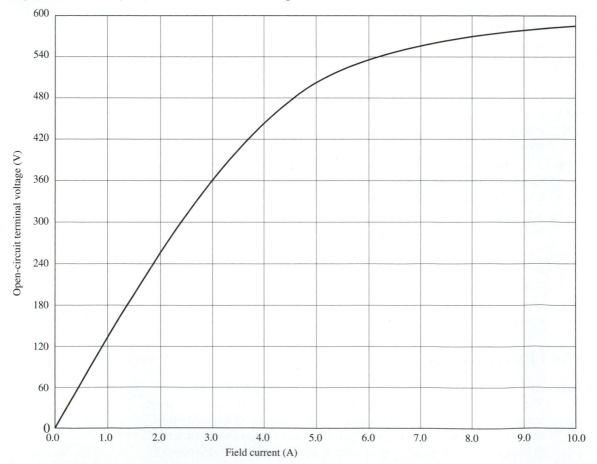

5–3. Assume that the field current of the generator in Problem 5–2 is adjusted to achieve rated voltage (480 V) at full load conditions in each of the questions below.

    a.   What is the efficiency of the generator at rated load?

    b.   What is the voltage regulation of the generator if it is loaded to rated kilovolt-amperes with 0.8 PF lagging loads?

    c.   What is the voltage regulation of the generator if it is loaded to rated kilovolt-amperes with 0.8 PF leading loads?

    d.   What is the voltage regulation of the generator if it is loaded to rated kilovolt-amperes with unity-power-factor loads?

    e.   Use MATLAB to plot the terminal voltage of the generator as a function of load for all three power factors.

5–4. Assume that the field current of the generator in Problem 5–2 has been adjusted so that it supplies rated voltage when loaded with rated current at unity power factor.

    a.   What is the torque angle δ of the generator when supplying rated current at unity power factor?

    b.   What is the maximum power that this generator can deliver to a unity power factor load when the field current is adjusted to the current value?

    c.   When this generator is running at full load with unity power factor, how close is it to the static stability limit of the machine?

    d.   Plot the torque angle δ as a function of the power supplied by the generator for these conditions.

Problems 5–5 to 5–14 refer to a two-pole Y-connected synchronous generator rated at 300 kVA, 480 V, 60 Hz, and 0.85 PF lagging. Its armature resistance R_A is 0.04 Ω. The core losses of this generator at rated conditions are 10 kW, and the friction and windage losses are 13 kW. The open-circuit and short-circuit characteristics are shown in Figure P5–2.

5–5. (a) What is the saturated synchronous reactance of this generator at the rated conditions? (b) What is the unsaturated synchronous reactance of this generator? (c) Plot the saturated synchronous reactance of this generator as a function of load.

5–6. (a) What are the rated current and internal generated voltage of this generator? (b) What field current does this generator require to operate at the rated voltage, current, and power factor?

5–7. What is the voltage regulation of this generator at the rated current and power factor?

5–8. If this generator is operating at the rated conditions and the load is suddenly removed, what will the terminal voltage be?

5–9. What are the electrical losses in this generator at rated conditions?

Figure P5–2 | The OCC and the SCC for the generator in Problems 5–5 to 5–14.

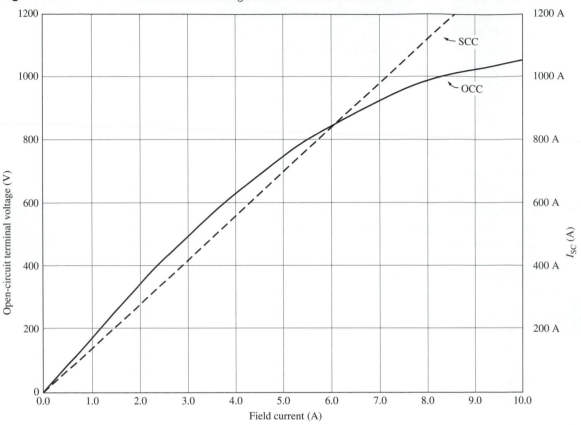

5–10. If this machine is operating at rated conditions, what input torque must be applied to the shaft of this generator? Express your answer both in newton-meters and in pound-feet.

5–11. What is the torque angle δ of this generator at rated conditions?

5–12. Assume that the generator field current is adjusted to supply 480 V under rated conditions. What is the static stability limit of this generator? (*Note:* You may ignore R_A to make this calculation easier.) How close is the full-load condition of this generator to the static stability limit?

5–13. Assume that the generator field current is adjusted to supply 480 V under rated conditions. Plot the power supplied by the generator as a function of the torque angle δ.

5–14. Assume that the generator's field current is adjusted so that the generator supplies rated voltage at the rated load current and power factor. If the field current and the magnitude of the load current are held constant, how will the terminal voltage change as the load power factor varies from 0.85 PF lagging to 0.85 PF leading? Make a plot of the terminal voltage versus the load power factor.

5–15. A 100-MVA, 11.5-kV, 0.85 PF lagging, 50-Hz, two-pole, Y-connected synchronous generator has a per-unit synchronous reactance of 0.8 and a per-unit armature resistance of 0.012.

    a.   What are its synchronous reactance and armature resistance in ohms?

    b.   What is the magnitude of the internal generated voltage E_A at the rated conditions? What is its torque angle δ at these conditions?

    c.   Ignoring losses in this generator, what torque must be applied to its shaft by the prime mover at full load?

5–16. A three-phase, Y-connected synchronous generator is rated 120 MVA, 13.2 kV, 0.8 PF lagging, and 60 Hz. Its synchronous reactance is 0.7 Ω, and its resistance may be ignored.

    a.   What is its voltage regulation?

    b.   What would the voltage and apparent power rating of this generator be if it were operated at 50 Hz with the same armature and field losses as it had at 60 Hz?

    c.   What would the voltage regulation of the generator be at 50 Hz?

5–17. A 25-MVA, three-phase, 13.5-kV, two-pole, 60-Hz synchronous generator was tested by the open-circuit test, and its air-gap voltage was extrapolated with the following results:

Open-circuit test					
Field current, A	320	365	380	475	570
Line voltage, kV	13.0	13.8	14.1	15.2	16.0
Extrapolated air-gap voltage, kV	15.4	17.5	18.3	22.8	27.4

The short-circuit test was then performed with the following results:

Short-circuit test					
Field current, A	320	365	380	475	570
Armature current, A	1040	1190	1240	1550	1885

The armature resistance is 0.24 Ω per phase.

    a.   Find the unsaturated synchronous reactance of this generator in ohms per phase and in per-unit.

    b.   Find the approximate saturated synchronous reactance X_S at a field current of 380 A. Express the answer both in ohms per phase and in per-unit.

    c.   Find the approximate saturated synchronous reactance at a field current of 475 A. Express the answer both in ohms per phase and in per-unit.

    d.   Find the short-circuit ratio for this generator.

5–18. A 480-V, six-pole synchronous motor draws 80 A from the line at unity power factor and full load. Assuming that the motor is lossless, answer the following questions:

    a.   What is the output torque of this motor? Express the answer both in newton-meters and in pound-feet.

    b.   What must be done to change the power factor to 0.8 leading? Explain your answer, using phasor diagrams.

    c.   What will the magnitude of the line current be if the power factor is adjusted to 0.8 leading?

5–19. A 480-V, 400-hp, 0.8 PF leading, eight-pole, Δ-connected synchronous motor has a synchronous reactance of 1.0 Ω and negligible armature resistance. Ignore its friction, windage, and core losses for the purposes of this problem.

    a.   If this motor is initially supplying 400 hp at 0.8 PF lagging, what are the magnitudes and angles of $\mathbf{E}_A$ and $\mathbf{I}_A$?

    b.   How much torque is this motor producing? What is the torque angle δ? How near is this value to the maximum possible induced torque of the motor for this field current setting?

    c.   If $|\mathbf{E}_A|$ is increased by 15 percent, what is the new magnitude of the armature current? What is the motor's new power factor?

    d.   Calculate and plot the motor's V curve for this load condition.

5–20. A 2300-V, 2000-hp, 0.8 PF leading, 60-Hz, two-pole, Y-connected synchronous motor has a synchronous reactance of 1.5 Ω and an armature resistance of 0.3 Ω. At 60 Hz, its friction and windage losses are 50 kW, and its core losses are 40 kW. The field circuit has a DC voltage of 200 V, and the maximum I_F is 10 A. The open-circuit characteristic of this motor is shown in Figure P5–3. Answer the following questions about the motor, assuming that it is being supplied by an infinite bus.

    a.   How much field current would be required to make this machine operate at unity power factor when supplying full load?

    b.   What is the motor's efficiency at full load and unity power factor?

    c.   If the field current were increased by 5 percent, what would the new value of the armature current be? What would the new power factor be? How much reactive power is being consumed or supplied by the motor?

    d.   What is the maximum torque this machine is theoretically capable of supplying at unity power factor? At 0.8 PF leading?

5–21. Plot the V curves (I_A versus I_F) for the synchronous motor of Problem 5–20 at no-load, half-load, and full-load conditions. (Note that an electronic version of the open-circuit characteristics in Figure P5–3 is available at the book's website. It may simplify the calculations required by this problem.)

Figure P5–3 | The open-circuit characteristic for the motor in Problems 5–20 and 5–21.

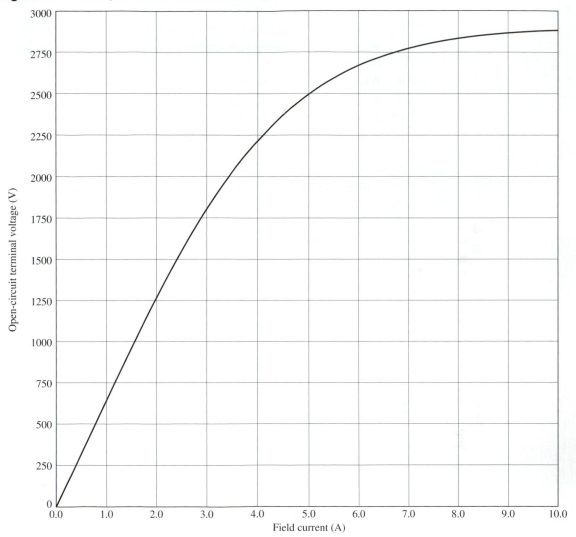

5–22. If a 60-Hz synchronous motor is to be operated at 50 Hz, will its synchronous reactance be the same as at 60 Hz, or will it change? *(Hint:* Think about the derivation of X_S.)

5–23. A 480-V, 100-kW, 0.85 PF leading, 50-Hz, six-pole Y-connected synchronous motor has a synchronous reactance of 1.5 Ω and a negligible armature resistance. The rotational losses are also to be ignored. This motor is to be operated over a continuous range of speeds from 300 to 1000 r/min, where the speed changes are to be accomplished by controlling the system frequency with a solid-state drive.

    a.   Over what range must the input frequency be varied to provide this speed control range?

    b.   How large is E_A at the motor's rated conditions?

    c.   What is the maximum power the motor can produce at the rated conditions?

    d.   What is the largest E_A could be at 300 r/min?

    e.   Assuming that the applied voltage V_ϕ is derated by the same amount as E_A, what is the maximum power the motor could supply at 300 r/min?

    f.   How does the power capability of a synchronous motor relate to its speed?

5–24. A 208-V, Y-connected synchronous motor is drawing 50 A at unity power factor from a 208-V power system. The field current flowing under these conditions is 2.7 A. Its synchronous reactance is 0.8 Ω. Assume a linear open-circuit characteristic.

    a.   Find the torque angle δ.

    b.   How much field current would be required to make the motor operate at 0.78 PF leading?

    c.   What is the new torque angle in part (b)?

5–25. A synchronous machine has a synchronous reactance of 2.0 Ω per phase and an armature resistance of 0.4 Ω per phase. If $\mathbf{E}_A = 460\angle -8°$ V and $\mathbf{V}_\phi = 480\angle 0°$ V, is this machine a motor or a generator? How much power P is this machine consuming from or supplying to the electrical system? How much reactive power Q is this machine consuming from or supplying to the electrical system?

5–26. Figure P5–4 shows a synchronous motor phasor diagram for a motor operating at a leading power factor with no R_A. For this motor, the torque angle is given by

$$\tan \delta = \frac{X_S I_A \cos \theta}{V_\phi + X_S I_A \sin \theta}$$

$$\delta = \tan^{-1}\left(\frac{X_S I_A \cos \theta}{V_\phi + X_S I_A \sin \theta}\right)$$

Derive an equation for the torque angle of the synchronous motor *if the armature resistance is included.*

5–27. A 2300-V, 400-hp, 60-Hz, eight-pole, Y-connected synchronous motor has a rated power factor of 0.85 leading. At full load, the efficiency is 85 percent. The armature resistance is 0.4 Ω, and the synchronous reactance is 4.4 Ω. Find the following quantities for this machine when it is operating at full load:

    a.   Output torque

    b.   Input power

Figure P5–4 | Phasor diagram of a motor at a leading power factor.

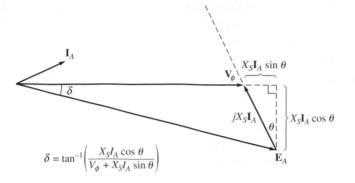

$$\delta = \tan^{-1}\left(\frac{X_S I_A \cos\theta}{V_\phi + X_S I_A \sin\theta}\right)$$

   c. n_m

   d. $\mathbf{E}_A$

   e. $|\mathbf{I}_A|$

   f. P_{conv}

   g. $P_{mech} + P_{core} + P_{stray}$

5–28. A 440-V, three-phase, Y-connected synchronous motor has a synchronous reactance of 1.5 Ω per phase. The field current has been adjusted so that the torque angle δ is 20° when the power supplied by the generator is 90 kW.

   a.   What is the magnitude of the internal generated voltage $\mathbf{E}_A$ in this machine?

   b.   What are the magnitude and angle of the armature current in the machine? What is the motor's power factor?

   c.   If the field current remains constant, what is the absolute maximum power this motor could supply?

5–29. A 100-hp, 440-V, 0.8 PF leading, Δ-connected synchronous motor has an armature resistance of 0.6 Ω and a synchronous reactance of 3.0 Ω. Its efficiency at full load is 89 percent.

   a.   What is the input power to the motor at rated conditions?

   b.   What is the line current of the motor at rated conditions? What is the phase current of the motor at rated conditions?

   c.   What is the reactive power consumed by or supplied by the motor at rated conditions?

   d.   What is the internal generated voltage $\mathbf{E}_A$ of this motor at rated conditions?

   e.   What are the stator copper losses in the motor at rated conditions?

   f.   What is P_{conv} at rated conditions?

   g.   If E_A is decreased by 10 percent, how much reactive power will be consumed by or supplied by the motor?

5–30. Answer the following questions about the machine of Problem 5–28.

    a.   If $\mathbf{E}_A = 430\angle 13.5°$ V and $\mathbf{V}_\phi = 440\angle 0°$ V, is this machine consuming real power from or supplying real power to the power system? Is it consuming reactive power from or supplying reactive power to the power system?

    b.   Calculate the real power P and reactive power Q supplied or consumed by the machine under the conditions in part (a). Is the machine operating within its ratings under these circumstances?

    c.   If $\mathbf{E}_A = 470\angle -12°$ V and $\mathbf{V}_\phi = 440\angle 0°$ V, is this machine consuming real power from or supplying real power to the power system? Is it consuming reactive power from or supplying reactive power to the power system?

    d.   Calculate the real power P and reactive power Q supplied or consumed by the machine under the conditions in part (c). Is the machine operating within its ratings under these circumstances?

5.17 | REFERENCES

1. Chapman, Stephen J.: *Electric Machinery Fundamentals,* 3rd ed., McGraw-Hill, Burr Ridge, Ill., 1999.

2. Chaston, A. N.: *Electric Machinery,* Reston Publishing, Reston, Va., 1986.

3. Del Toro, V.: *Electric Machines and Power Systems,* Prentice-Hall, Englewood Cliffs, N.J., 1985.

4. Fitzgerald, A. E., and C. Kingsley, Jr.: *Electric Machinery,* McGraw-Hill, New York, 1952.

5. Fitzgerald, A. E., C. Kingsley, Jr., and S. D. Umans: *Electric Machinery,* 5th ed., McGraw-Hill, New York, 1990.

6. Kosow, Irving L.: *Electric Machinery and Transformers,* Prentice-Hall, Englewood Cliffs, N.J., 1972.

7. Liwschitz-Garik, Michael, and Clyde Whipple: *Alternating-Current Machinery,* Van Nostrand, Princeton, N.J., 1961.

8. McPherson, George: *An Introduction to Electrical Machines and Transformers,* Wiley, New York, 1981.

9. Werninck, E. H. (ed.): *Electric Motor Handbook,* McGraw-Hill, London, 1978.

Parallel Operation of Synchronous Generators

In today's world, an isolated synchronous generator supplying its own load independently of other generators is very rare. Such a situation is found in only a few out-of-the-way applications such as emergency generators. For all usual generator applications, there is more than one generator operating in parallel to supply the power demanded by the loads. An extreme example of this situation is the U.S. power grid, in which literally thousands of generators share the load on the system.

Why are synchronous generators operated in parallel? There are several major advantages to such operation:

1. Several generators can supply a bigger load than one machine by itself.
2. Having many generators increases the reliability of the power system, since the failure of any one of them does not cause a total power loss to the load.
3. Having many generators operating in parallel allows one or more of them to be removed for shutdown and preventive maintenance.
4. If only one generator is used and it is not operating at near full load, then it will be relatively inefficient. But with several smaller machines it is possible to operate only a fraction of them. The ones that do operate are operating near full load and thus more efficiently.

6.1 | PARALLELING GENERATORS

This section explores the requirements for paralleling AC generators and then looks at the behavior of synchronous generators operated in parallel.

The Conditions Required for Paralleling

Figure 6–1 shows a synchronous generator G_1 supplying power to a load, with another generator G_2 about to be paralleled with G_1 by closing of the switch S_1. What conditions must be met before the switch can be closed and the two generators connected?

Figure 6–1 I A generator being paralleled with a running power system.

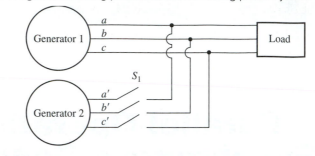

If the switch is closed arbitrarily at some moment, the generators are liable to be severely damaged, and the load may lose power. If the voltages are not exactly the same in each conductor being tied together, there will be a *very* large current flow when the switch is closed. To avoid this problem, each of the three phases must have *exactly the same voltage magnitude and phase angle* as the conductor to which it is connected. In other words, the voltage in phase *a* must be *exactly* the same as the voltage in phase *a'*, and so forth for phases *b–b'* and *c–c'*. To achieve this match, the following *paralleling conditions* must be met:

1. The rms *line voltages* of the two generators must be equal.
2. The two generators must have the same *phase sequence.*
3. The phase angles of the two *a* phases must be equal.
4. The frequency of the new generator, called the *oncoming generator,* must be slightly higher than the frequency of the running system.

These paralleling conditions require some explanation. Condition 1 is obvious— in order for two sets of voltages to be identical, they must of course have the same rms magnitude of voltage. The voltage in phases *a* and *a'* will be completely identical at all times if both their magnitudes and their angles are the same, which explains condition 3.

Condition 2 ensures that the sequence in which the phase voltages peak in the two generators is the same. If the phase sequence is different (as shown in Figure 6–2a), then even though one pair of voltages (the *a* phases) are in phase, the other two pairs of voltages are 120° out of phase. If the generators were connected in this manner, there would be no problem with phase *a*, but huge currents would flow in phases *b* and *c*, damaging both machines. To correct a phase sequence problem, simply swap the connections on any two of the three phases on one of the machines.

If the frequencies of the generators are not very nearly equal when they are connected together, large power transients will occur until the generators stabilize at a common frequency. The frequencies of the two machines must be very nearly equal, but they cannot be exactly equal. They must differ by a small amount so that the phase angles of the oncoming machine will change slowly with respect to the phase angles of the running system. In that way, the angles between the voltages can be observed and switch S_1 can be closed when the systems are exactly in phase.

The General Procedure for Paralleling Generators

Suppose that generator G_2 is to be connected to the running system shown in Figure 6–2. The following steps should be taken to accomplish the paralleling.

Figure 6–2 | (a) The two possible phase sequences of a three-phase system. (b) The three-light-bulb method for checking phase sequence.

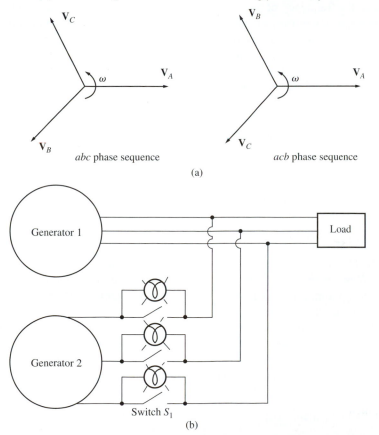

abc phase sequence

acb phase sequence

(a)

Generator 1

Load

Generator 2

Switch S_1

(b)

First, using voltmeters, the field current of the oncoming generator should be adjusted until its terminal voltage is equal to the line voltage of the running system.

Second, the phase sequence of the oncoming generator must be compared to the phase sequence of the running system. The phase sequence can be checked in a number of different ways. One way is to alternately connect a small induction motor to the terminals of each of the two generators. If the motor rotates in the same direction each time, then the phase sequence is the same for both generators. If the motor rotates in opposite directions, then the phase sequences differ, and two of the conductors on the incoming generator must be reversed.

Another way to check the phase sequence is the *three-light-bulb method.* In this approach, three light bulbs are stretched across the open terminals of the switch connecting the generator to the system as shown in Figure 6–2b. As the phase changes between the two systems, the light bulbs first get bright (large phase difference) and then get dim (small phase difference). *If all three bulbs get bright and dark together, then the systems have the same phase sequence.* If the bulbs brighten in succession, then the systems have the opposite phase sequence, and one of the sequences must be reversed.

Next, the frequency of the oncoming generator is adjusted to be slightly higher than the frequency of the running system. This is done first by watching a frequency meter until the frequencies are close and then by observing changes in phase between the systems. The oncoming generator is adjusted to a slightly higher frequency so that when it is connected, it will come on the line supplying power as a generator, instead of consuming it as a motor would (this point will be explained later).

Once the frequencies are very nearly equal, the voltages in the two systems will change phase with respect to each other very slowly. The phase changes are observed, and when the phases angles are equal, the switch connecting the two systems together is shut.

How can one tell when the two systems are finally in phase? A simple way is to watch the three light bulbs described above in connection with the discussion of phase sequence. When the three light bulbs all go out, the voltage difference across them is zero and the systems are in phase. This simple scheme works, but it is not very accurate. A better approach is to employ a synchroscope. A *synchroscope* is a meter that measures the difference in phase angle between the a phases of the two systems. The face of a synchroscope is shown in Figure 6–3. The dial shows the phase difference between the two *a* phases, with 0 (meaning in phase) at the top and 180° at the bottom. Since the frequencies of the two systems are slightly different, the phase angle on the meter changes slowly. If the oncoming generator or system is faster than the running system (the desired situation), then the phase angle advances and the synchroscope needle rotates clockwise. If the oncoming machine is slower, the needle rotates counterclockwise. When the synchroscope needle is in the vertical position, the voltages are in phase, and the switch can be shut to connect the systems.

Figure 6–3 | A synchroscope.

Notice, though, that a *synchroscope checks the relationships on only one phase*. It gives no information about phase sequence.

In large generators belonging to power systems, this whole process of paralleling a new generator to the line is automated, and a computer does this job. For smaller generators, though, the operator manually goes through the paralleling steps just described.

6.2 | FREQUENCY–POWER AND VOLTAGE–REACTIVE POWER CHARACTERISTICS OF A SYNCHRONOUS GENERATOR

All generators are driven by a *prime mover,* which is the generator's source of mechanical power. The most common type of prime mover is a steam turbine, but other types include diesel engines, gas turbines, water turbines, and even wind turbines.

Regardless of the original power source, all prime movers tend to behave in a similar fashion—as the power drawn from them increases, the speed at which they turn decreases. The decrease in speed is in general nonlinear, but some form of governor mechanism is usually included to make the decrease in speed linear with an increase in power demand.

Whatever governor mechanism is present on a prime mover, it will always be adjusted to provide a slight drooping characteristic with increasing load. The speed droop (SD) of a prime mover is defined by the equation

$$\text{SD} = \frac{n_{\text{nl}} - n_{\text{fl}}}{n_{\text{fl}}} \times 100\%$$

(6–1)

where n_{nl} is the no-load prime-mover speed and n_{fl} is the full-load prime-mover speed. Most generator prime movers have a speed droop of 2 to 4 percent, as defined in Equation (6–1). In addition, most governors have some type of set point adjustment to allow the no-load speed of the turbine to be varied. A typical speed versus power plot is shown in Figure 6–4.

Since the shaft speed is related to the resulting electrical frequency by Equation (4-23),

$$f_e = \frac{n_m P}{120}$$

(4–23)

the power output of a synchronous generator is related to its frequency. An example plot of frequency versus power is shown in Figure 6–4b. Frequency-power characteristics of this sort play an essential role in the parallel operation of synchronous generators.

The relationship between frequency and power can be described quantitatively by the equation

$$P = s_P(f_{\text{nl}} - f_{\text{sys}})$$

(6–2)

Figure 6–4 | (a) The speed versus power curve for a typical prime mover. (b) The resulting frequency versus power curve for the generator.

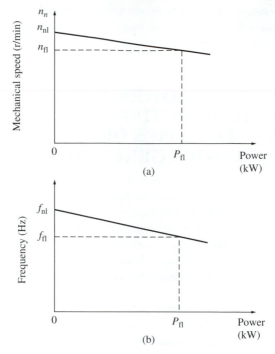

(a)

(b)

where

$\qquad P$ = power output of the generator

$\qquad f_{\text{nl}}$ = no-load frequency of the generator

$\qquad f_{\text{sys}}$ = operating frequency of system

$\qquad s_P$ = slope of curve, kW/Hz or MW/Hz

A similar relationship can be derived for the reactive power Q and terminal voltage V_T. As previously seen, when a lagging load is added to a synchronous generator, its terminal voltage drops. Likewise, when a leading load is added to a synchronous generator, its terminal voltage increases. It is possible to make a plot of terminal voltage versus reactive power, and such a plot has a drooping characteristic like the one shown in Figure 6–5. This characteristic is not intrinsically linear, but many generator voltage regulators include a feature to make it so. The characteristic curve can be moved up and down by changing the no-load terminal voltage set point on the voltage regulator. As with the frequency–power characteristic, this curve plays an important role in the parallel operation of synchronous generators.

The relationship between the terminal voltage and reactive power can be expressed by an equation similar to the frequency-power relationship [Equation (5–28)] if the voltage regulator produces an output that is linear with changes in reactive power.

Figure 6–5 | The curve of terminal voltage V_T versus reactive power Q for a synchronous generator.

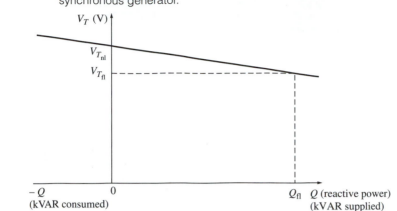

It is important to realize that when a single generator is operating alone, the real power P and reactive power Q supplied by the generator will be the amount demanded by the load attached to the generator—the P and Q supplied cannot be controlled by the generator's controls. Therefore, for any given real power, the governor set points control the generator's operating frequency f_e and for any given reactive power, the field current controls the generator's terminal voltage V_T.

Figure 6–6 shows a generator supplying a load. A second load is to be connected in parallel with the first one. The generator has a no-load frequency of 61.0 Hz and a slope s_P of 1 MW/Hz. Load 1 consumes a real power of 1000 kW at 0.8 PF lagging, while load 2 consumes a real power of 800 kW at 0.707 PF lagging.

a. Before the switch is closed, what is the operating frequency of the system?

b. After load 2 is connected, what is the operating frequency of the system?

c. After load 2 is connected, what action could an operator take to restore the system frequency to 60 Hz?

Figure 6–6 | The power system in Example 6–1.

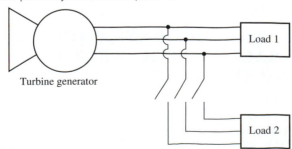

■ Solution

This problem states that the slope of the generator's characteristic is 1 MW/Hz and that its no-load frequency is 61 Hz. Therefore, the power produced by the generator is given by

$$P = s_P(f_{nl} - f_{sys}) \qquad (6\text{--}2)$$

so

$$f_{sys} = f_{nl} - \frac{P}{s_P}$$

a.   The initial system frequency is given by

$$f_{sys} = f_{nl} - \frac{P}{s_P}$$

$$= 61 \text{ Hz} - \frac{1000 \text{ kW}}{1 \text{ MW/Hz}} = 61 \text{ Hz} - 1 \text{ Hz} = 60 \text{ Hz}$$

b.   After load 2 is connected,

$$f_{sys} = f_{nl} - \frac{P}{s_P}$$

$$= 61 \text{ Hz} - \frac{1800 \text{ kW}}{1 \text{ MW/Hz}} = 61 \text{ Hz} - 1.8 \text{ Hz} = 59.2 \text{ Hz}$$

c.   After the load is connected, the system frequency falls to 59.2 Hz. To restore the system to its proper operating frequency, the operator should increase the governor no-load set points by 0.8 Hz, to 61.8 Hz. This action will restore the system frequency to 60 Hz.   ■

To summarize, when a generator is operating by itself supplying the system loads, then

1. The real and reactive power supplied by the generator will be the amount demanded by the attached load.
2. The governor set points of the generator will control the operating frequency of the power system.
3. The field current (or the field regulator set points) control the terminal voltage of the power system.

This is the situation found in isolated generators in remote field environments.

6.3 | OPERATION OF GENERATORS IN PARALLEL WITH LARGE POWER SYSTEMS

When a synchronous generator is connected to a power system, the power system is often so large that *nothing* the operator of the generator does will have much of an effect on the power system. An example of this situation is the connection of a single generator to the U.S. power grid. The U.S. power grid is so large that no reasonable action on the part of one generator can cause an observable change in overall grid frequency.

This idea is idealized in the concept of an infinite bus. An *infinite bus* is a power system so large that its voltage and frequency do not vary regardless of how much real and reactive power is drawn from or supplied to it. The power–frequency characteristic of such a system is shown in Figure 6–7a, and the reactive power–voltage characteristic is shown in Figure 6–7b.

Figure 6–7 | The curves of frequency versus power and terminal voltage versus reactive power for an infinite bus.

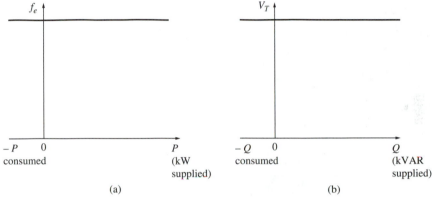

To understand the behavior of a generator connected to such a large system, examine a system consisting of a generator and an infinite bus in parallel supplying a load. Assume that the generator's prime mover has a governor mechanism, but that the field is controlled manually by a resistor. It is easier to explain generator operation without considering an automatic field current regulator, so this discussion will ignore the slight differences caused by the field regulator when one is present. Such a system is shown in Figure 6–8a.

When a generator is connected in parallel with another generator or a large system, *the frequency and terminal voltage of all the machines must be the same*, since their output conductors are tied together. Therefore, their real power–frequency and reactive power–voltage characteristics can be plotted back to back, with a common vertical axis. Such a sketch, sometimes informally called a *house diagram*, is shown in Figure 6–8b.

Assume that the generator has just been paralleled with the infinite bus according to the procedure described previously. Then the generator will be essentially "floating" on the line, supplying a small amount of real power and little or no reactive power. This situation is shown in Figure 6–9.

Suppose the generator had been paralleled to the line but, instead of being at a slightly higher frequency than the running system, it was at a slightly lower frequency. In this case, when paralleling is completed, the resulting situation is shown in Figure 6–10. Notice that here the no-load frequency of the generator is less than

Figure 6–8 I (a) A synchronous generator operating in parallel with an infinite bus. (b) The frequency versus power diagram (or *house diagram*) for a synchronous generator in parallel with an infinite bus.

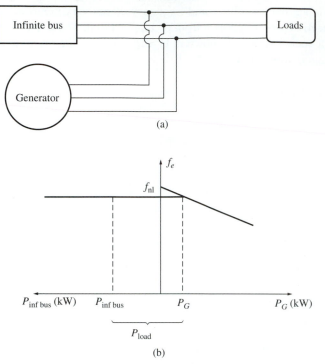

(a)

(b)

Figure 6–9 I The frequency versus power diagram at the moment just after paralleling.

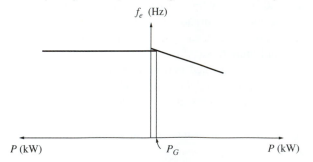

the system's operating frequency. At this frequency, the power supplied by the generator is actually negative. In other words, when the generator's no-load frequency is less than the system's operating frequency, the generator actually consumes electric power and runs as a motor. It is to ensure that a generator comes on line supplying power instead of consuming it that the oncoming machine's frequency is adjusted higher than the running system's frequency. *Many real generators have a reverse-power trip connected to them, so it is imperative that they be paralleled with their*

Figure 6–10 | The frequency versus power diagram if the no-load frequency of the generator were slightly *less* than system frequency before paralleling.

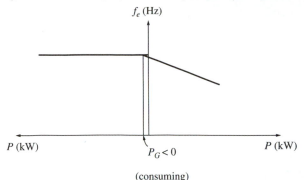

frequency higher than that of the running system. If such a generator ever starts to consume power, it will be automatically disconnected from the line.

Once the generator has been connected, what happens when its governor set points are increased? The effect of this increase is to shift the no-load frequency of the generator upward. Since the frequency of the system is unchanged (the frequency of an infinite bus cannot change), the power supplied by the generator increases. This is shown by the house diagram in Figure 6–11a and by the phasor diagram in Figure 6–11b. Notice in the phasor diagram that $E_A \sin \delta$ (which is proportional to the power supplied as long as V_T is constant) has increased, while the magnitude of $E_A (= K\phi\omega)$ remains constant, since both the field current I_F and the speed of rotation ω are unchanged. As the governor set points are further increased, the no-load frequency increases and the power supplied by the generator increases. As the power output increases, E_A remains at constant magnitude while $E_A \sin \delta$ is further increased.

What happens in this system if the power output of the generator is increased until it exceeds the power consumed by the load? If this occurs, the extra power generated flows back into the infinite bus. The infinite bus, by definition, can supply or consume any amount of power without a change in frequency, so the extra power is consumed.

After the real power of the generator has been adjusted to the desired value, the phasor diagram of the generator looks like Figure 6–11b. Notice that at this time the generator is actually operating at a slightly leading power factor, so it is acting as a capacitor, supplying negative reactive power. Alternatively, the generator can be said to be consuming reactive power. How can the generator be adjusted so that it will supply some reactive power Q to the system? This can be done by adjusting the field current of the machine. To understand why this is true, it is necessary to consider the constraints on the generator's operation under these circumstances.

The first constraint on the generator is that *the power must remain constant* when I_F is changed. The power into a generator is given by the Equation $P_{in} = \tau_{ind}\omega_m$. Now, the prime mover of a synchronous generator has a fixed torque–speed characteristic for any given governor setting. This curve changes only when the governor set points are changed. Since the generator is tied to an infinite bus, its speed *cannot*

Figure 6–11 | The effect of increasing the governor's set points on (a) the house diagram; (b) the phasor diagram.

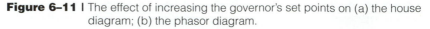

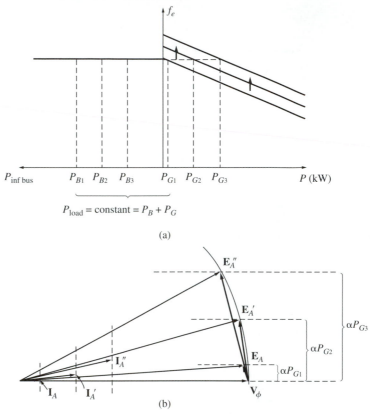

(b)

change. If the generator's speed does not change and the governor set points have not been changed, the power supplied by the generator must remain constant.

If the power supplied is constant as the field current is changed, then the distances proportional to the power in the phasor diagram ($I_A \cos \theta$ and $E_A \sin \delta$) cannot change. When the field current is increased, the flux ϕ increases, and therefore E_A ($= K\phi\!\uparrow\!\omega$) increases. If E_A increases, but $E_A \sin \delta$ must remain constant, then the phasor $\mathbf{E}_A$ must "slide" along the line of constant power, as shown in Figure 6–12. Since $\mathbf{V}_\phi$ is constant, the angle of $jX_s\mathbf{I}_A$ changes as shown, and therefore the angle and magnitude of $\mathbf{I}_A$ change. Notice that as a result the distance proportional to Q ($I_A \sin \theta$) increases. In other words, *increasing the field current in a synchronous generator operating in parallel with an infinite bus increases the reactive power output of the generator.*

To summarize, when a generator is operating in parallel with an infinite bus:

1. The frequency and terminal voltage of the generator are controlled by the system to which it is connected.

Figure 6–12 | The effect of increasing the generator's field current on the phasor diagram of the machine.

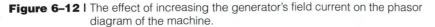
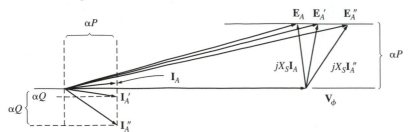

2. The governor set points of the generator control the real power supplied by the generator to the system.

3. The field current in the generator controls the reactive power supplied by the generator to the system.

This situation is much the way real generators operate when connected to a very large power system.

6.4 | OPERATION OF GENERATORS IN PARALLEL WITH OTHER GENERATORS OF THE SAME SIZE

When a single generator operated alone, the real and reactive powers (P and Q) supplied by the generator were fixed, constrained to be equal to the power demanded by the load, and the frequency and terminal voltage were varied by the governor set points and the field current. When a generator operated in parallel with an infinite bus, the frequency and terminal voltage were constrained to be constant by the infinite bus, and the real and reactive powers were varied by the governor set points and the field current. What happens when a synchronous generator is connected in parallel not with an infinite bus, but rather with another generator of the same size? What will be the effect of changing governor set points and field currents?

If a generator is connected in parallel with another one of the same size, the resulting system is as shown in Figure 6–13a. In this system, the basic constraint is that *the sum of the real and reactive powers supplied by the two generators must equal the P and Q demanded by the load.* The system frequency is not constrained to be constant, and neither is the power of a given generator constrained to be constant. The power–frequency diagram for such a system immediately after G_2 has been paralleled to the line is shown in Figure 6–13b. Here, the total power P_{tot} (which is equal to P_{load}) is given by

$$P_{\text{tot}} = P_{\text{load}} = P_{G1} + P_{G2} \qquad (6\text{–}3)$$

and the total reactive power is given by

$$Q_{\text{tot}} = Q_{\text{load}} = Q_{G1} + Q_{G2} \qquad (6\text{–}4)$$

Figure 6–13 | (a) A generator connected in parallel with another machine of the same size. (b) The corresponding house diagram at the moment generator 2 is paralleled with the system. (c) The effect of increasing generator 2's governor set points on the operation of the system. (d) The effect of increasing generator 2's field current on the operation of the system.

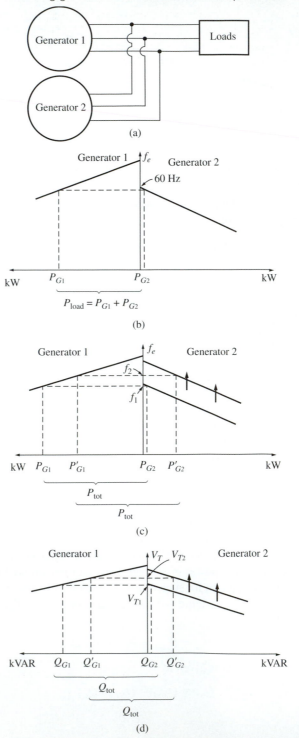

What happens if the governor set points of G_2 are increased? When the governor set points of G_2 are increased, the power–frequency curve of G_2 shifts upward, as shown in Figure 6–13c. Remember, the total power supplied to the load must not change. At the original frequency f_1, the power supplied by G_1 and G_2 will now be larger than the load demand, so the system cannot continue to operate at the same frequency as before. In fact, there is only one frequency at which the sum of the powers out of the two generators is equal to P_{load}. That frequency f_2 is higher than the original system operating frequency. At that frequency, G_2 supplies more power than before, and G_1 supplies less power than before.

Therefore, when two generators are operating together, an increase in governor set points on one of them

1. *Increases the system frequency.*
2. *Increases the power supplied by that generator, while reducing the power supplied by the other one.*

What happens if the field current of G_2 is increased? The resulting behavior is analogous to the real-power situation and is shown in Figure 6–13d. When two generators are operating together and the field current of G_2 is increased,

1. *The system terminal voltage is increased.*
2. *The reactive power Q supplied by that generator is increased, while the reactive power supplied by the other generator is decreased.*

If the slopes and no-load frequencies of the generator's speed droop (frequency–power) curves are known, then the powers supplied by each generator and the resulting system frequency can be determined quantitatively. Example 6–2 shows how this can be done.

EXAMPLE 6–2

Figure 6–13a shows two generators supplying a load. Generator 1 has a no-load frequency of 61.5 Hz and a slope s_{P1} of 1 MW/Hz. Generator 2 has a no-load frequency of 61.0 Hz and a slope s_{P2} of 1 MW/Hz. The two generators are supplying a real load totaling 2.5 MW at 0.8 PF lagging. The resulting system power-frequency or house diagrams are shown in Figure 6–14.

a. At what frequency is this system operating, and how much power is supplied by each of the two generators?
b. Suppose an additional 1-MW load were attached to this power system. What would the new system frequency be, and how much power would G_1 and G_2 supply now?
c. With the system in the configuration described in part (b), what will the system frequency and generator powers be if the governor set points on G_2 are increased by 0.5 Hz?

Figure 6–14 | The house diagram for the system in Example 6–2.

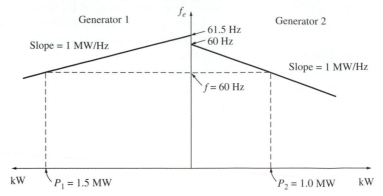

■ Solution

The power produced by a synchronous generator with a given slope and no-load frequency is given by Equation (6–2):

$$P_1 = s_{P1}(f_{nl,\,1} - f_{sys})$$
$$P_2 = s_{P2}(f_{nl,\,2} - f_{sys})$$

Since the total power supplied by the generators must equal the power consumed by the loads,

$$P_{load} = P_1 + P_2$$

These equations can be used to answer all the questions asked.

a. In the first case, both generators have a slope of 1 MW/Hz, and G_1 has a no-load frequency of 61.5 Hz, while G_2 has a no-load frequency of 61.0 Hz. The total load is 2.5 MW. Therefore, the system frequency can be found as follows:

$$
\begin{aligned}
P_{load} &= P_1 + P_2 \\
&= s_{P1}(f_{nl,\,1} - f_{sys}) + s_{P2}(f_{nl,\,2} - f_{sys}) \\
2.5 \text{ MW} &= (1 \text{ MW/Hz})(61.5 \text{ Hz} - f_{sys}) + (1 \text{ MW/Hz})(61 \text{ Hz} - f_{sys}) \\
&= 61.5 \text{ MW} - (1 \text{ MW/Hz})f_{sys} + 61 \text{ MW} - (1 \text{ MW/Hz})f_{sys} \\
&= 122.5 \text{ MW} - (2 \text{ MW/Hz})f_{sys} \\
f_{sys} &= \frac{122.5 \text{ MW} - 2.5 \text{ MW}}{2 \text{ MW/Hz}} = 60.0 \text{ Hz}
\end{aligned}
$$

The resulting powers supplied by the two generators are

$$
\begin{aligned}
P_1 &= s_{P1}(f_{nl,\,1} - f_{sys}) \\
&= (1 \text{ MW/Hz})(61.5 \text{ Hz} - 60.0 \text{ Hz}) = 1.5 \text{ MW} \\
P_2 &= s_{P2}(f_{nl,\,2} - f_{sys}) \\
f_2 &= (1 \text{ MW/Hz})(61.0 \text{ Hz} - 60.0 \text{ Hz}) = 1 \text{ MW}
\end{aligned}
$$

b. When the load is increased by 1 MW, the total load becomes 3.5 MW. The new system frequency is now given by

$$P_{load} = s_{P1}(f_{nl,1} - f_{sys}) + s_{P2}(f_{nl,2} - f_{sys})$$
$$3.5 \text{ MW} = (1 \text{ MW/Hz})(61.5 \text{ Hz} - f_{sys}) + (1 \text{ MW/Hz})(61 \text{ Hz} - f_{sys})$$
$$= 61.5 \text{ MW} - (1 \text{ MW/Hz})f_{sys} + 61 \text{ MW} - (1 \text{ MW/Hz})f_{sys}$$
$$= 122.5 \text{ MW} - (2 \text{ MW/Hz})f_{sys}$$
$$f_{sys} = \frac{122.5 \text{ MW} - 3.5 \text{ MW}}{(2 \text{ MW/Hz})} = 59.5 \text{ Hz}$$

The resulting powers are

$$P_1 = s_{P1}(f_{nl,1} - f_{sys})$$
$$= (1 \text{ MW/Hz})(61.5 \text{ Hz} - 59.5 \text{ Hz}) = 2.0 \text{ MW}$$
$$P_2 = s_{P2}(f_{nl,2} - f_{sys})$$
$$= (1 \text{ MW/Hz})(61.0 \text{ Hz} - 59.5 \text{ Hz}) = 1.5 \text{ MW}$$

c. If the no-load governor set points of G_2 are increased by 0.5 Hz, the new system frequency becomes

$$P_{load} = s_{P1}(f_{nl,1} - f_{sys}) + s_{P2}(f_{nl,2} - f_{sys})$$
$$3.5 \text{ MW} = (1 \text{ MW/Hz})(61.5 \text{ Hz} - f_{sys}) + (1 \text{ MW/Hz})(61.5 \text{ Hz} - f_{sys})$$
$$= 123 \text{ MW} - (2 \text{ MW/Hz})f_{sys}$$
$$f_{sys} = \frac{123 \text{ MW} - 3.5 \text{ MW}}{(2 \text{ MW/Hz})} = 59.75 \text{ Hz}$$

The resulting powers are

$$P_1 = P_2 = s_{P1}(f_{nl,1} - f_{sys})$$
$$= (1 \text{ MW/Hz})(61.5 \text{ Hz} - 59.75 \text{ Hz}) = 1.75 \text{ MW}$$

Notice that the system frequency rose, the power of G_2 rose, and the power of G_1 fell. ■

When two generators of similar size are operating in parallel, a change in the governor set points of one of them changes both the system frequency and the power sharing between them. It would normally be desired to adjust only one of these quantities at a time. How can the power sharing of the power system be adjusted independently of the system frequency, and vice versa?

The answer is very simple. An increase in governor set points on one generator increases that machine's power and increases system frequency. A decrease in governor set points on the other generator decreases that machine's power and decreases the system frequency. Therefore, to adjust power sharing without changing the system frequency, *increase the governor set points of one generator and simultaneously decrease the governor set points of the other generator* (see Figure 6–15a). Similarly, *to adjust the system frequency without changing the power sharing, simultaneously increase or decrease both governor set points* (see Figure 6–15b).

Reactive power and terminal voltage adjustments work in an analogous fashion. To shift the reactive power sharing without changing V_T, *simultaneously increase the*

Figure 6–15 | (a) Shifting power sharing without affecting system frequency.
(b) Shifting system frequency without affecting power sharing.
(c) Shifting reactive power sharing without affecting terminal voltage.
(d) Shifting terminal voltage without affecting reactive power sharing.

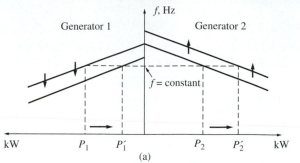

(a)

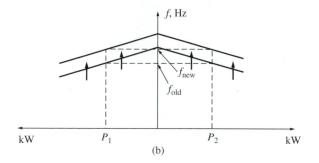

(b)

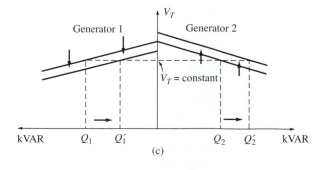

(c)

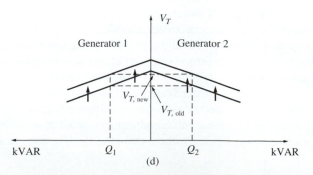

(d)

field current on one generator and decrease the field current on the other (see Figure 6–15c). To change the terminal voltage without affecting the reactive power sharing, *simultaneously increase or decrease both field currents* (see Figure 6–15d).

To summarize, in the case of two generators operating together:

1. The system is constrained in that the total power supplied by the two generators together must equal the amount consumed by the load. Neither f_{sys} nor V_T is constrained to be constant.

2. To adjust the real power sharing between generators without changing f_{sys}, simultaneously increase the governor set points on one generator while decreasing the governor set points on the other. The machine whose governor set point was increased will assume more of the load.

3. To adjust f_{sys} without changing the real power sharing, simultaneously increase or decrease both generators' governor set points.

4. To adjust the reactive power sharing between generators without changing V_T, simultaneously increase the field current on one generator while decreasing the field current on the other. The machine whose field current was increased will assume more of the reactive load.

5. To adjust V_T without changing the reactive power sharing, simultaneously increase or decrease both generators' field currents.

It is very important that any synchronous generator intended to operate in parallel with other machines have a *drooping* frequency–power characteristic. If two generators have flat or nearly flat characteristics, then the power sharing between them can vary widely with only the tiniest changes in no-load speed. This problem is illustrated by Figure 6–16. Notice that even very tiny changes in f_{nl} in one of the generators would cause wild shifts in power sharing. To ensure good control of power sharing between generators, they should have speed droops in the range of 2 to 5 percent.

Figure 6–16 | Two synchronous generators with flat frequency–power characteristics. A very tiny change in the no-load frequency of either of these machines could cause huge shifts in the power sharing.

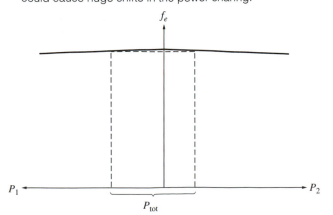

6.5 | SUMMARY

The way in which a synchronous generator operates in a real power system depends on the constraints on it. When a generator operates alone, the real and reactive powers that must be supplied are determined by the load attached to it, and the governor set points and field current control the frequency and terminal voltage, respectively. When the generator is connected to an infinite bus, its frequency and voltage are fixed, so the governor set points and field current control the real and reactive power flow from the generator. In real systems containing generators of approximately equal size, the governor set points affect both frequency and power flow, and the field current affects both terminal voltage and reactive power flow.

6.6 | QUESTIONS

6–1. What conditions are necessary for paralleling two synchronous generators?

6–2. Why must the oncoming generator on a power system be paralleled at a higher frequency than that of the running system?

6–3. What is an infinite bus? What constraints does it impose on a generator paralleled with it?

6–4. How can the real power sharing between two generators be controlled without affecting the system's frequency? How can the reactive power sharing between two generators be controlled without affecting the system's terminal voltage?

6–5. How can the system frequency of a large power system be adjusted without affecting the power sharing among the system's generators?

6–6. How can the concepts of Section 6.4 be expanded to calculate the system frequency and power sharing among three or more generators operating in parallel?

6.7 | PROBLEMS

6–1. A 480-V, 400-kVA, 0.85 PF lagging, 50-Hz, four-pole Δ-connected generator is driven by a 500-hp diesel engine and is used as a standby or emergency generator. This machine can also be paralleled with the normal power supply (a very large power system) if desired.

    a. What are the conditions required for paralleling the emergency generator with the existing power system? What is the generator's rate of shaft rotation after paralleling occurs?

    b. If the generator is connected to the power system and is initially floating on the line, sketch the resulting magnetic fields and phasor diagram.

    c. The governor setting on the diesel is now increased. Show by both house diagrams and phasor diagrams what happens to the generator. How much reactive power does the generator supply now?

   d.   With the diesel generator now supplying real power to the power system, what happens to the generator as its field current is increased and decreased? Show this behavior with both phasor diagrams and house diagrams.

6–2. A 13.5-kV, 20-MVA, 0.8 PF lagging, 60-Hz, two-pole Y-connected steam-turbine generator has a synchronous reactance of 5.0 Ω per phase and an armature resistance of 0.5 Ω per phase. This generator is operating in parallel with a large power system (infinite bus).

   a.   What is the magnitude of E_A at rated conditions?

   b.   What is the torque angle of the generator at rated conditions?

   c.   If the field current is constant, what is the maximum power possible out of this generator? How much reserve power or torque does this generator have at full load?

   d.   At the absolute maximum power possible, how much reactive power will this generator be supplying or consuming? Sketch the corresponding phasor diagram. (Assume I_F is still unchanged.)

6–3. A 480-V, 200-kW, two-pole, three-phase, 50-Hz synchronous generator's prime mover has a no-load speed of 3040 r/min and a full-load speed of 2975 r/min. It is operating in parallel with a 480-V, 150-kW, four-pole, 50-Hz synchronous generator whose prime mover has a no-load speed of 1500 r/min and a full-load speed of 1485 r/min. The loads supplied by the two generators consist of 200 kW at 0.85 PF lagging.

   a.   Calculate the speed droops of generator 1 and generator 2.

   b.   Find the operating frequency of the power system.

   c.   Find the power being supplied by each of the generators in this system.

   d.   If V_T is 460 V, what must the generators' operators do to correct for the low terminal voltage?

6–4. Three physically identical synchronous generators are operating in parallel. They are all rated for a full load of 3 MW at 0.8 PF lagging. The no-load frequency of generator A is 61 Hz, and its speed droop is 3.4 percent. The no-load frequency of generator B is 61.5 Hz, and its speed droop is 3 percent. The no-load frequency of generator C is 60.5 Hz, and its speed droop is 2.6 percent.

   a.   If a total load consisting of 8 MW is being supplied by this power system, what will the system frequency be and how will the power be shared among the three generators?

   b.   Create a plot showing the power supplied by each generator as a function of the total power supplied to all loads (you may use MATLAB to create this plot). At what load does one of the generators exceed its ratings? Which generator exceeds its ratings first?

   c.   Is this power sharing in (a) acceptable? Why or why not?

   d.   What actions could an operator take to improve the real power sharing among these generators?

6–5. A paper mill has installed three steam generators (boilers) to provide process steam and also to use some of its waste products as an energy source. Since there is extra capacity, the mill has installed three 5-MW turbine–generators to take advantage of the situation. Each generator is a 4160-V, 6250-kVA, 0.85 PF lagging, two-pole, Y-connected synchronous generator with a synchronous reactance of 0.75 Ω and an armature resistance of 0.04 Ω. Generators 1 and 2 have a characteristic power–frequency slope s_P of 2.5 MW/Hz, and generators 2 and 3 have a slope of 3 MW/Hz.

    a.   If the no-load frequency of each of the three generators is adjusted to 61 Hz, how much power will the three machines be supplying when actual system frequency is 60 Hz?

    b.   What is the maximum power the three generators can supply in this condition without the ratings of one of them being exceeded? At what frequency does this limit occur? How much power does each generator supply at that point?

    c.   What would have to be done to get all three generators to supply their rated real and reactive powers at an overall operating frequency of 60 Hz?

    d.   What would the internal generated voltages of the three generators be under this condition?

6–6. Two identical 300-kVA, 480-V synchronous generators are connected in parallel to supply a load. The prime movers of the two generators happen to have different speed droop characteristics. When the field currents of the two generators are equal, one delivers 200 A at 0.9 PF lagging, while the other delivers 180 A at 0.75 PF lagging.

    a.   What are the real power and the reactive power supplied by each generator to the load?

    b.   What is the overall power factor of the load?

    c.   In what direction must the field current on each generator be adjusted in order for them to operate at the same power factor?

6–7. A generating station for a power system consists of four 120-MVA, 15-kV, 0.85 PF lagging synchronous generators with identical speed droop characteristics operating in parallel. The governors on the generators' prime movers are adjusted to produce a 3-Hz drop from no load to full load. Three of these generators are each supplying a steady 75 MW at a frequency of 60 Hz, while the fourth generator (called the *slack generator*) handles all incremental load changes on the system while maintaining the system's frequency at 60 Hz.

    a.   At a given instant, the total system loads are 260 MW at a frequency of 60 Hz. What are the no-load frequencies of each of the system's generators?

    b.   If the system load rises to 290 MW and the generator's governor set points do not change, what will the new system frequency be?

    c.   To what frequency must the no-load frequency of the swing generator be adjusted in order to restore the system frequency to 60 Hz?

    d.   If the system is operating at the conditions described in part (c), what would happen if the swing generator were tripped off the line (disconnected from the power line)?

6–8.   Suppose that you were an engineer planning a new electric cogeneration facility for a plant with excess process steam. You have a choice of either two 10-MW turbine–generators or a single 20-MW turbine generator. What would be the advantages and disadvantages of each choice?

6.8 | REFERENCES

1. Chapman, Stephen J.: *Electric Machinery Fundamentals*, 3rd ed., McGraw-Hill, Burr Ridge, Ill., 1999.

2. Del Toro, V.: *Electric Machines and Power Systems*, Prentice-Hall, Englewood Cliffs, N.J., 1985.

3. Fitzgerald, A. E., and C. Kingsley, Jr.: *Electric Machinery*, McGraw-Hill, New York, 1952.

Induction Motors

In Chapter 5, we saw how amortisseur windings on a synchronous motor could develop a starting torque without the necessity of supplying an external field current to them. In fact, amortisseur windings work so well that a motor could be built without the synchronous motor's main DC field circuit at all. A machine with only amortisseur windings is called an *induction machine*. Such machines are called induction machines because the rotor voltage (which produces the rotor current and the rotor magnetic field) is *induced* in the rotor windings instead of being physically connected by wires. The distinguishing feature of an induction motor is that *no DC field current is required* to run the machine.

Although it is possible to use an induction machine as either a motor or a generator, it has many disadvantages as a generator and so is rarely used in that manner. For this reason, induction machines are usually referred to as induction motors.

7.1 | INDUCTION MOTOR CONSTRUCTION

An induction motor has the same physical stator as a synchronous machine, with a different rotor construction. A typical two-pole stator is shown in Figure 7–1. It looks (and is) the same as a synchronous machine stator. There are two different types of induction motor rotors that can be placed inside the stator. One is called a *squirrel-cage* rotor or simply a cage rotor while the other is called a *wound rotor*.

Figures 7–2 and 7–3 show *squirrel-cage* induction motor rotors. A squirrel-cage induction motor rotor consists of a series of conducting bars laid into slots carved in the face of the rotor and shorted at either end by large *shorting rings*. This design is referred to as a squirrel-cage rotor because the conductors, if examined by themselves, would look like one of the exercise wheels that squirrels or hamsters run on.

The other type of rotor is a wound rotor. A *wound rotor* has a complete set of three-phase windings that are mirror images of the windings on the stator. The three phases of the rotor windings are usually Y connected, and the ends of the three rotor wires are tied to slip rings on the rotor's shaft. The rotor windings are shorted

Figure 7–1 I The stator of a typical induction motor, showing the stator windings. *(Courtesy of MagneTek Inc.)*

Figure 7–2 I (a) Sketch of squirrel-cage rotor. (b) A typical squirrel-cage rotor. *(Courtesy of General Electric Company.)*

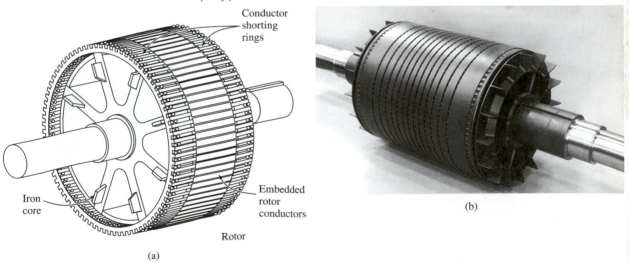

through brushes riding on the slip rings. Wound-rotor induction motors therefore have their rotor currents accessible at the stator brushes, where they can be examined and where extra resistance can be inserted into the rotor circuit. It is possible to take advantage of this feature to modify the torque–speed characteristic of the motor. Two wound rotors are shown in Figure 7–4, and a complete wound-rotor induction motor is shown in Figure 7–5.

Figure 7–3 | (a) Cutaway diagram of a typical small squirrel-cage induction motor. *(Courtesy of MagneTek, Inc.)* (b) Cutaway diagram of a typical large squirrel-cage induction motor. *(Courtesy of General Electric Company.)*

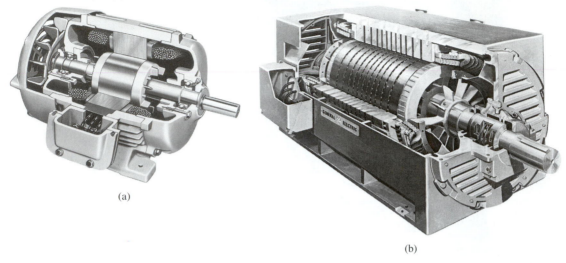

(a)

(b)

Figure 7–4 | Typical wound rotors for induction motors. Notice the slip rings and the bars connecting the rotor windings to the slip rings. *(Courtesy of General Electric Company.)*

Wound rotor induction motors are more expensive than squirrel cage induction motors, and they require much more maintenance because of the wear associated with their brushes and slip rings. As a result, wound rotor induction motors are rarely used.

7.2 | BASIC INDUCTION MOTOR CONCEPTS

Induction motor operation is basically the same as that of amortisseur windings on synchronous motors. That basic operation will now be reviewed, and some important induction motor terms will be defined.

Figure 7–5 | Cutaway diagram of a wound-rotor induction motor. Notice the brushes and slip rings. Also notice that the rotor windings are skewed to eliminate slot harmonics. *(Courtesy of MagneTek, Inc.)*

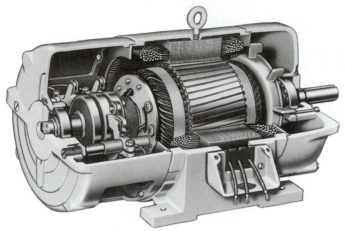

The Development of Induced Torque in an Induction Motor

Figure 7–6 shows a squirrel-cage induction motor. A three-phase set of voltages has been applied to the stator, and a three-phase set of stator currents is flowing. These currents produce a magnetic field $\mathbf{B}_S$, which is rotating in a counterclockwise direction. The speed of the magnetic field's rotation is given by

$$n_{\text{sync}} = \frac{120 f_e}{P} \tag{7–1}$$

where f_e is the system frequency in hertz and P is the number of poles in the machine. This rotating magnetic field $\mathbf{B}_S$ passes over the rotor bars and induces a voltage in them.

The voltage induced in a given rotor bar is given by the equation

$$e_{\text{ind}} = (\mathbf{v} \times \mathbf{B}) \cdot \mathbf{l} \tag{1–45}$$

where

$\mathbf{v}$ = velocity of the bar *relative to the magnetic field*

$\mathbf{B}$ = magnetic flux density vector

$\mathbf{l}$ = length of conductor in the magnetic field

It is the *relative* motion of the rotor compared to the stator magnetic field that produces induced voltage in a rotor bar. The velocity of the upper rotor bars relative to the magnetic field is to the right, so the induced voltage in the upper bars is out of the page, while the induced voltage in the lower bars is into the page. This results in a current flow out of the upper bars and into the lower bars. However, since the rotor assembly is inductive, the peak rotor current lags behind the peak rotor voltage (see Figure 7–6b). The rotor current flow produces a rotor magnetic field $\mathbf{B}_R$.

Figure 7–6 I The development of induced torque in an induction motor. (a) The rotating stator field $\mathbf{B}_S$ induces a voltage in the rotor bars. (b) The rotor voltage produces a rotor current flow, which lags behind the voltage because of the inductance of the rotor. (c) The rotor current produces a rotor magnetic field $\mathbf{B}_R$ lagging 90° behind itself, and $\mathbf{B}_R$ interacts with $\mathbf{B}_S$ to produce a counterclockwise torque in the machine.

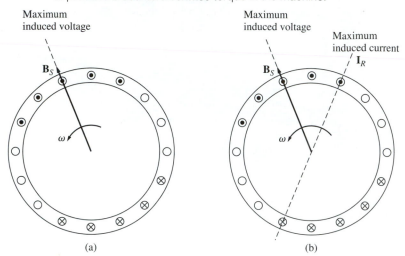

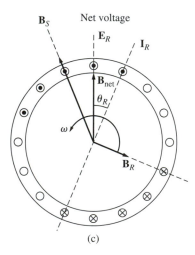

Finally, since the induced torque in the machine is given by

$$\tau_{\text{ind}} = k\mathbf{B}_R \times \mathbf{B}_S \tag{4–47}$$

the resulting torque is counterclockwise. Since the rotor induced torque is counterclockwise, the rotor accelerates in that direction.

There is a finite upper limit to the motor's speed, however. If the induction motor's rotor were turning at *synchronous speed*, then the rotor bars would be stationary *relative to the magnetic field* and there would be no induced voltage. If e_{ind} were equal to 0, then there would be no rotor current and no rotor magnetic field. With no rotor magnetic field, the induced torque would be zero, and the rotor would slow down as a result of friction losses. An induction motor can thus speed up to near-synchronous speed, but it can never exactly reach synchronous speed.

Note that in normal operation *both the rotor and stator magnetic fields* $\mathbf{B}_R$ *and* $\mathbf{B}_S$ *rotate together at speed synchronous speed* n_{sync}, *while the rotor itself turns at a slower speed*.

The Concept of Rotor Slip

The voltage induced in a rotor bar of an induction motor depends on the speed of the rotor *relative to the magnetic fields*. Since the behavior of an induction motor depends on the rotor's voltage and current, it is often more logical to talk about this relative speed. Two terms are commonly used to define the relative motion of the rotor and the magnetic fields. One is *slip speed*, defined as the difference between synchronous speed and rotor speed:

$$n_{slip} = n_{sync} - n_m \qquad (7\text{--}2)$$

where

n_{slip} = slip speed of the machine

n_{sync} = speed of the magnetic fields

n_m = mechanical shaft speed of motor

The other term used to describe the relative motion is *slip*, which is the relative speed expressed on a per-unit (or sometimes a percentage) basis. That is, slip is defined as

$$s = \frac{n_{slip}}{n_{sync}} \qquad (7\text{--}3)$$

$$s = \frac{n_{sync} - n_m}{n_{sync}} \qquad (7\text{--}4)$$

This equation can also be expressed in terms of angular velocity ω (radians per second) as

$$s = \frac{\omega_{sync} - \omega_m}{\omega_{sync}} \qquad (7\text{--}5)$$

Notice that if the rotor turns at synchronous speed, $s = 0$, while if the rotor is stationary, $s = 1$. All normal motor speeds fall somewhere between those two limits.

It is possible to express the mechanical speed of the rotor shaft in terms of synchronous speed and slip. Solving Equations (7–4) and (7–5) for mechanical speed yields

$$n_m = (1 - s)n_{\text{sync}} \qquad (7\text{–}6)$$

or

$$\omega_m = (1 - s)\omega_{\text{sync}} \qquad (7\text{–}7)$$

These equations are useful in the derivation of induction motor torque and power relationships.

The Electrical Frequency on the Rotor

An induction motor works by inducing voltages and currents in the rotor of the machine, and for that reason it has sometimes been called a *rotating transformer*. Like a transformer, the primary (stator) induces a voltage in the secondary (rotor), but *unlike* a transformer, the secondary frequency is not necessarily the same as the primary frequency.

If the rotor of a motor is locked so that it cannot move, then the rotor will have the same frequency as the stator. On the other hand, if the rotor turns at synchronous speed, the frequency on the rotor will be zero. What will the rotor frequency be for any arbitrary rate of rotor rotation?

At $n_m = 0$ r/min, the rotor frequency $f_r = f_e$, and the slip $s = 1$. At $n_m = n_{\text{sync}}$, the rotor frequency $f_r = 0$ Hz, and the slip $s = 0$. For any speed in between, the rotor frequency is directly proportional to the *difference* between the speed of the magnetic field n_{sync} and the speed of the rotor n_m. Since the slip of the rotor is defined as

$$s = \frac{n_{\text{sync}} - n_m}{n_{\text{sync}}} \qquad (7\text{–}4)$$

the rotor frequency can be expressed as

$$f_r = sf_e \qquad (7\text{–}8)$$

Several alternative forms of this expression exist that are sometimes useful. One of the more common expressions is derived by substituting Equation (7–4) for the slip into Equation (7–8) and then substituting Equation (7–1) for n_{sync} in the denominator of the expression:

$$f_r = \frac{n_{\text{sync}} - n_m}{n_{\text{sync}}} f_e$$

But $n_{\text{sync}} = 120 f_e / P$ [from Equation (7–1)], so

$$f_r = (n_{\text{sync}} - n_m) \frac{P}{120 f_e} f_e$$

Therefore,

$$\boxed{f_r = \frac{P}{120}(n_{\text{sync}} - n_m)} \qquad (7\text{–}9)$$

EXAMPLE 7–1

A 208-V, 10-hp, four-pole, 60-Hz Y-connected induction motor has a full-load slip of 5 percent.

a. What is the synchronous speed of this motor?

b. What is the rotor speed of this motor at the rated load?

c. What is the rotor frequency of this motor at the rated load?

d. What is the shaft torque of this motor at the rated load?

■ **Solution**

a. The synchronous speed of this motor is

$$n_{\text{sync}} = \frac{120 \, f_e}{P} \qquad (7\text{–}1)$$

$$= \frac{120(60 \text{ Hz})}{4 \text{ poles}} = 1800 \text{ r/min}$$

b. The rotor speed of the motor is given by

$$n_m = (1 - s)n_{\text{sync}} \qquad (7\text{–}6)$$

$$= (1 - 0.95)(1800 \text{ r/min}) = 1710 \text{ r/min}$$

c. The rotor frequency of this motor is given by

$$f_r = sf_e = (0.05)(60 \text{ Hz}) = 3 \text{ Hz} \qquad (7\text{–}8)$$

Alternatively, the frequency can be found from Equation (7–9):

$$f_r = \frac{P}{120}(n_{\text{sync}} - n_m) \qquad (7\text{–}9)$$

$$= \frac{4}{120}(1800 \text{ r/min} - 1710 \text{ r/min}) = 3 \text{ Hz}$$

d. The shaft load torque is given by

$$\tau_{\text{load}} = \frac{P_{\text{out}}}{\omega_m}$$

$$= \frac{(10 \text{ hp})(746 \text{ W/hp})}{(1710 \text{ r/min})(2\pi \text{ rad/r})(1 \text{ min/60 s})} = 41.7 \text{ N-m}$$

The shaft load torque in English units is given by Equation (1–17):

$$\tau_{\text{load}} = \frac{5252P}{n}$$

where τ is in pound-feet, P is in horsepower, and n_m is in revolutions per minute. Therefore,

$$\tau_{\text{load}} = \frac{5252(10 \text{ hp})}{1710 \text{ r/min}} = 30.7 \text{ lb-ft}$$

7.3 | THE EQUIVALENT CIRCUIT OF AN INDUCTION MOTOR

An induction motor relies for its operation on the induction of voltages and currents in its rotor circuit from the stator circuit (transformer action). Because the induction of voltages and currents in the rotor circuit of an induction motor is essentially a transformer operation, the equivalent circuit of an induction motor will turn out to be very similar to the equivalent circuit of a transformer. An induction motor is called a *singly excited* machine (as opposed to a *doubly excited* synchronous machine), since power is supplied to only the stator circuit. Because an induction motor does not have an independent field circuit, its model will not contain an internal voltage source such as the internal generated voltage $\mathbf{E}_A$ in a synchronous machine.

It is possible to derive the equivalent circuit of an induction motor from a knowledge of transformers and from what we already know about the variation of rotor frequency with speed in induction motors. The induction motor model will be developed by starting with the transformer model in Chapter 2 and then deciding how to take the variable rotor frequency and other similar induction motor effects into account.

The Transformer Model of an Induction Motor

A transformer per-phase equivalent circuit, representing the operation of an induction motor, is shown in Figure 7–7.

Figure 7–7 | The transformer model or an induction motor, with rotor and stator connected by an ideal transformer of turns ratio a_{eff}.

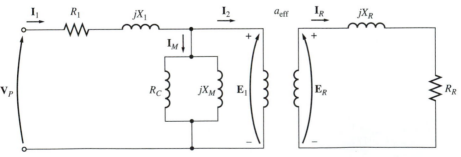

Like any transformer, there is a certain resistance and self-inductance in the primary (stator) windings, which must be represented in the equivalent circuit of the machine. The stator resistance will be called R_1, and the stator leakage reactance will be called X_1. These two components appear right at the input to the machine model.

Also, like any transformer with an iron core, the flux in the machine is related to the integral of the applied voltage $\mathbf{E}_1$. The curve of magnetomotive force versus flux (magnetization curve) for this machine is compared to a similar curve for a power transformer in Figure 7–8. Notice that the slope of the induction motor's magneto-motive force–flux curve is much shallower than the curve of a good transformer. This is because there must be an air gap in an induction motor, which greatly increases the reluctance of the flux path and therefore reduces the coupling between primary and secondary windings. The higher reluctance caused by the air gap means that a higher magnetizing current is required to obtain a given flux level. Therefore, the magnetiz-ing reactance X_M in the equivalent circuit will have a much smaller value (or the sus-ceptance B_M will have a much larger value) than it would in an ordinary transformer.

Figure 7–8 | The magnetization curve of an induction motor compared to that of a transformer.

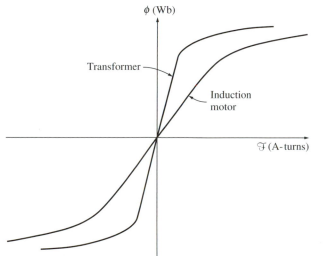

ϕ (Wb)

Transformer

Induction motor

$\mathcal{F}$ (A-turns)

The primary internal stator voltage $\mathbf{E}_1$ is coupled to the secondary $\mathbf{E}_R$ by an ideal transformer with an effective turns ratio a_{eff}. The effective turns ratio a_{eff} is fairly easy to determine for a wound-rotor motor—it is basically the ratio of the conductors per phase on the stator to the conductors per phase on the rotor, modified by any pitch and distribution factor differences. It is rather difficult to see a_{eff} clearly in the case of a squirrel-case rotor motor because there are no distinct windings on the squirrel-cage rotor. In either case, there *is* an effective turns ratio for the motor.

The voltage $\mathbf{E}_R$ produced in the rotor in turn produces a current flow in the shorted rotor (or secondary) circuit of the machine.

The primary impedances and the magnetization current of the induction motor are very similar to the corresponding components in a transformer equivalent circuit. An induction motor equivalent circuit differs from a transformer equivalent circuit primarily in the effects of varying rotor frequency on the rotor voltage $\mathbf{E}_R$ and the rotor impedances R_R and jX_R.

The Rotor Circuit Model

In an induction motor, when the voltage is applied to the stator windings, a voltage is induced in the rotor windings of the machine. In general, *the greater the relative motion between the rotor and the stator magnetic fields, the greater the resulting rotor voltage and rotor frequency.* The largest relative motion occurs when the rotor is stationary, called the *locked-rotor* or *blocked-rotor* condition, so the largest voltage and rotor frequency are induced in the rotor at that condition. The smallest voltage (0 V) and frequency (0 Hz) occur when the rotor moves at the same speed as the stator magnetic field, resulting in no relative motion. The magnitude and frequency of the voltage induced in the rotor at any speed between these extremes is *directly proportional to the slip of the rotor.* Therefore, if the magnitude of the induced rotor voltage at locked-rotor conditions is called E_{LR}, the magnitude of the induced voltage at any slip will be given by the equation

$$E_R = sE_{\mathrm{LR}} \tag{7–10}$$

and the frequency of the induced voltage at any slip will be given by the equation

$$f_r = sf_e \tag{7–8}$$

This voltage is induced in a rotor containing both resistance and reactance. The rotor resistance R_R is a constant (except for the skin effect), independent of slip, while the rotor reactance is affected in a more complicated way by slip.

The reactance of an induction motor rotor depends on the inductance of the rotor and the frequency of the voltage and current in the rotor. With a rotor inductance of L_R, the rotor reactance is given by

$$X_R = \omega_r L_R = 2\pi f_r L_R$$

By Equation (7–8), $f_r = sf_e$, so

$$\begin{aligned} X_R &= 2\pi sf_e L_R \\ &= s(2\pi f_e L_R) \\ &= sX_{\mathrm{LR}} \end{aligned} \tag{7–11}$$

where X_{LR} is the locked-rotor rotor reactance.

The resulting rotor equivalent circuit is shown in Figure 7–9. The rotor current flow can be found as

$$\mathbf{I}_R = \frac{\mathbf{E}_R}{R_R + jX_R}$$

$$\boxed{\mathbf{I}_R = \frac{s\mathbf{E}_{\mathrm{LR}}}{R_R + jsX_{\mathrm{LR}}}} \tag{7–12}$$

or

$$\boxed{\mathbf{I}_R = \frac{\mathbf{E}_{\mathrm{LR}}}{R_R/s + jX_{\mathrm{LR}}}} \tag{7–13}$$

Notice from Equation (7–13) that it is possible to treat all of the rotor effects due to varying rotor speed as being caused by a *varying impedance* supplied with power from a constant-voltage source $\mathbf{E}_{LR}$. The equivalent rotor impedance from this point of view is

$$\mathbf{Z}_{R,\,eq} = R_R/s + jX_{LR} \tag{7–14}$$

and the rotor equivalent circuit using this convention is shown in Figure 7–10. In the equivalent circuit in Figure 7–10, the rotor voltage is a constant $\mathbf{E}_{LR}$ volts and the rotor impedance $\mathbf{Z}_{R,eq}$ contains all the effects of varying rotor slip. A plot of the current flow in the rotor as developed in Equations (7–12) and (7–13) is shown in Figure 7–11.

Figure 7–9 | The rotor circuit model of an induction motor.

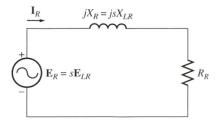

Figure 7–10 | The rotor circuit model with all the frequency (slip) effects concentrated in resistor R_R.

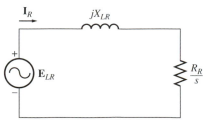

Notice that at very low slips the resistive term $R_R/s \gg X_{LR}$, so the rotor resistance predominates and the rotor current varies *linearly* with slip. At high slips, X_{LR} is much larger than R_R/s, and the rotor current *approaches a steady-state value* as the slip becomes very large.

The Final Equivalent Circuit

To produce the final per-phase equivalent circuit for an induction motor, it is necessary to refer the rotor part of the model over to the stator side. The rotor circuit model that will be referred to the stator side is the model shown in Figure 7–10, which has all the speed variation effects concentrated in the impedance term.

Figure 7–11 I Rotor current as a function of rotor speed.

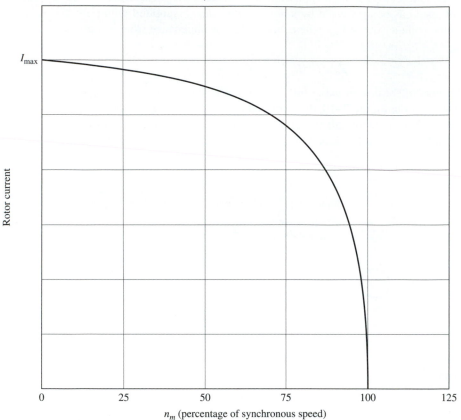

In an ordinary transformer, the voltages, currents, and impedances on the secondary side of the device can be referred to the primary side by means of the turns ratio of the transformer:

$$\mathbf{V}_P = \mathbf{V}'_S = a\mathbf{V}_S \qquad (7\text{–}15)$$

$$\mathbf{I}_P = \mathbf{I}'_S = \frac{\mathbf{I}_S}{a} \qquad (7\text{–}16)$$

and $$\mathbf{Z}'_S = a^2\mathbf{Z}_S \qquad (7\text{–}17)$$

where the prime refers to the referred values of voltage, current, and impedance.

Exactly the same sort of transformation can be done for the induction motor's rotor circuit. If the effective turns ratio of an induction motor is a_{eff}, then the transformed rotor voltage becomes

$$\mathbf{E}_1 = \mathbf{E}'_R = a_{\text{eff}}\mathbf{E}_{\text{LR}} \qquad (7\text{–}18)$$

the rotor current becomes

$$\mathbf{I}_2 = \frac{\mathbf{I}_R}{a_{\text{eff}}} \qquad (7\text{–}19)$$

and the rotor impedance becomes

$$\mathbf{Z}_2 = a_{\text{eff}}^2 \left(\frac{R_R}{s} + j X_{\text{LR}} \right)$$ (7–20)

If we now make the following definitions:

$$R_2 = a_{\text{eff}}^2 R_R$$ (7–21)
$$X_2 = a_{\text{eff}}^2 X_{\text{LR}}$$ (7–22)

then the final per-phase equivalent circuit of the induction motor is as shown in Figure 7–12.

Figure 7–12 | The per-phase equivalent circuit of an induction motor.

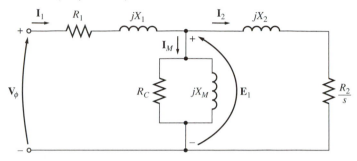

The rotor resistance R_R and the locked-rotor rotor reactance X_{LR} are very difficult or impossible to determine directly on squirrel-cage rotors, and the effective turns ratio a_{eff} is also difficult to obtain for squirrel-cage rotors. Fortunately, though, it is possible to make measurements that will directly give the *referred resistance and reactance R_2 and X_2*, even though R_R, X_{LR}, and a_{eff} are not known separately. The measurement of induction motor parameters will be taken up in Section 7.9.

7.4 | POWER AND TORQUE IN AN INDUCTION MOTOR

Because induction motors are singly excited machines, their power and torque relationships are considerably different from the relationships in the synchronous machines previously studied. This section reviews the power and torque relationships in induction motors.

Losses and the Power-Flow Diagram

An induction motor can be basically described as a rotating transformer. Its input is a three-phase set of voltages and currents. For an ordinary transformer, the output is electric power from the secondary windings. The secondary windings in an induction motor (the rotor) are shorted out, so no electrical output exists from normal induction motors. Instead, the output is mechanical. The relationship between the input electric power and the output mechanical power of this motor is shown in the power-flow diagram in Figure 7–13.

Figure 7–13 | The power-flow diagram of an induction motor.

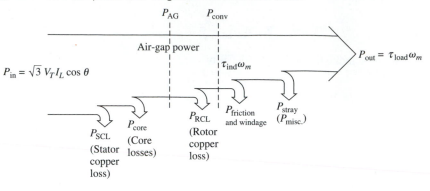

The input power to an induction motor P_{in} is in the form of three-phase electric voltages and currents. The first losses encountered in the machine are I^2R losses in the stator windings (the *stator copper loss* P_{SCL}). Then some amount of power is lost as hysteresis and eddy currents in the stator (P_{core}). The power remaining at this point is transferred to the rotor of the machine across the air gap between the stator and rotor. This power is called the *air-gap power* P_{AG} of the machine. After the power is transferred to the rotor, some of it is lost as I^2R losses (the *rotor copper loss* P_{RCL}), and the rest is converted from electrical to mechanical form (P_{conv}). Finally, friction and windage losses $P_{F\&W}$ and stray losses P_{misc} are subtracted. The remaining power is the output of the motor P_{out}.

The *core losses* do not always appear in the power-flow diagram at the point shown in Figure 7–13. Because of the nature of core losses, where they are accounted for in the machine is somewhat arbitrary. The core losses of an induction motor come partially from the stator circuit and partially from the rotor circuit. Since an induction motor normally operates at a speed near synchronous speed, the relative motion of the magnetic fields over the rotor surface is quite slow, and the rotor core losses are very tiny compared to the stator core losses. Since the largest fraction of the core losses comes from the stator circuit, all the core losses are lumped together at that point on the diagram. These losses are represented in the induction motor equivalent circuit by the resistor R_C (or the conductance G_C). If core losses are just given by a number (X watts) instead of as a circuit element, they are often lumped together with the mechanical losses and subtracted at the point on the diagram where the mechanical losses are located.

The *higher* the speed of an induction motor, the *higher* its friction, windage, and stray losses will be. On the other hand, the *higher* the speed of the motor (up to n_{sync}), the *lower* its core losses. Therefore, these three categories of losses are sometimes lumped together and called *rotational losses*. The total rotational losses of a motor are often considered to be constant with changing speed, since the component losses change in opposite directions with a change in speed.

EXAMPLE 7–2

A 480-V, 60 Hz, 50-hp, three-phase induction motor is drawing 60 A at 0.85 PF lagging. The stator copper losses are 2 kW, and the rotor copper losses are 700 W. The friction and windage losses are 600 W, the core losses are 1800 W, and the stray losses are negligible. Find the following quantities:

a. The air-gap power P_{AG}
b. The power converted P_{conv}
c. The output power P_{out}
d. The efficiency of the motor

■ **Solution**

To answer these questions, refer to the power-flow diagram for an induction motor (Figure 7–13).

a. The air-gap power is just the input power minus the stator I^2R losses. The input power is given by

$$P_{in} = \sqrt{3}V_T I_L \cos \theta$$
$$= \sqrt{3}(480 \text{ V})(60 \text{ A})(0.85) = 42.4 \text{ kW}$$

From the power-flow diagram, the air-gap power is given by

$$P_{AG} = P_{in} - P_{SCL} - P_{core}$$
$$= 42.4 \text{ kW} - 2 \text{ kW} - 1.8 \text{ kW} = 38.6 \text{ kW}$$

b. From the power-flow diagram, the power converted from electrical to mechanical form is

$$P_{conv} = P_{AG} - P_{RCL}$$
$$= 38.6 \text{ kW} - 700 \text{ W} = 37.9 \text{ kW}$$

c. From the power-flow diagram, the output power is given by

$$P_{out} = P_{conv} - P_{F\&W} - P_{misc}$$
$$= 37.9 \text{ kW} - 600 \text{ W} - 0 \text{ W} = 37.3 \text{ kW}$$

or, in horsepower,

$$P_{out} = (37.3 \text{ kW}) \frac{1 \text{ hp}}{0.746 \text{ kW}} = 50 \text{ hp}$$

d. Therefore, the induction motor's efficiency is

$$\eta = \frac{P_{out}}{P_{in}} \times 100\%$$
$$= \frac{37.3 \text{ kW}}{42.4 \text{ kW}} \times 100\% = 88\%$$

■

Power and Torque in an Induction Motor

Figure 7–12 shows the per-phase equivalent circuit of an induction motor. If the equivalent circuit is examined closely, it can be used to derive the power and torque equations governing the operation of the motor.

The input current to a phase of the motor can be found by dividing the input voltage by the total equivalent impedance:

$$\mathbf{I}_1 = \frac{V_\phi}{\mathbf{Z}_{eq}} \tag{7–23}$$

where

$$\mathbf{Z}_{eq} = R_1 + jX_1 + \cfrac{1}{G_C - jB_M + \cfrac{1}{R_2/s + jX_2}} \tag{7–24}$$

Therefore, the stator copper losses, the core losses, and the rotor copper losses can be found. The stator copper losses in the three phases are given by

$$\boxed{P_{\text{SCL}} = 3I_1^2 R_1} \tag{7–25}$$

The core losses are given by

$$\boxed{P_{\text{core}} = 3E_1^2 G_C} \tag{7–26}$$

so the air-gap power can be found as

$$\boxed{P_{\text{AG}} = P_{\text{in}} - P_{\text{SCL}} - P_{\text{core}}} \tag{7–27}$$

Look closely at the equivalent circuit of the rotor. The *only* element in the equivalent circuit where the air-gap power can be consumed is in the resistor R_2/s. Therefore, the *air-gap power* can also be given by

$$\boxed{P_{\text{AG}} = 3I_2^2 \frac{R_2}{s}} \tag{7–28}$$

The actual resistive losses in the rotor circuit are given by the equation

$$P_{\text{RCL}} = 3I_R^2 R_R \tag{7–29}$$

Since power is unchanged when referred across an ideal transformer, the rotor copper losses can also be expressed as

$$\boxed{P_{\text{RCL}} = 3I_2^2 R_2} \tag{7–30}$$

After stator copper losses, core losses, and rotor copper losses are subtracted from the input power to the motor, the remaining power is converted from electrical to mechanical form. This power converted, which is sometimes called *developed mechanical power*, is given by

$$P_{conv} = P_{AG} - P_{RCL}$$

$$= 3I_2^2 \frac{R_2}{s} - 3I_2^2 R_2$$

$$= 3I_2^2 R_2 \left(\frac{1}{s} - 1\right)$$

$$\boxed{P_{conv} = 3I_2^2 R_2 \left(\frac{1-s}{s}\right)} \tag{7-31}$$

Notice from Equations (7–28) and (7–30) that the rotor copper losses are equal to the air-gap power times the slip:

$$P_{RCL} = sP_{AG} \tag{7-32}$$

Therefore, the lower the slip of the motor, the lower the rotor losses in the machine. Note also that if the rotor is not turning, the slip $s = 1$ and the *air-gap power is entirely consumed in the rotor.* This is logical, since if the rotor is not turning, the output power P_{out} ($= \tau_{load}\omega_m$) must be zero. Since $P_{conv} = P_{AG} - P_{RCL}$, this also gives another relationship between the air-gap power and the power converted from electrical to mechanical form:

$$P_{conv} = P_{AG} - P_{RCL}$$

$$= P_{AG} - sP_{AG}$$

$$\boxed{P_{conv} = (1 - s)P_{AG}} \tag{7-33}$$

Finally, if the friction and windage losses and the stray losses are known, the output power can be found as

$$\boxed{P_{out} = P_{conv} - P_{F\&W} - P_{misc}} \tag{7-34}$$

The *induced torque* τ_{ind} in a machine was defined as the torque generated by the internal electric-to-mechanical power conversion. This torque differs from the torque actually available at the terminals of the motor by an amount equal to the friction and windage torques in the machine. The induced torque is given by the equation

$$\tau_{ind} = \frac{P_{conv}}{\omega_m} \tag{7-35}$$

This torque is also called the *developed torque* of the machine.

The induced torque of an induction motor can be expressed in a different form as well. Equation (7–7) expresses actual speed in terms of synchronous speed and slip, while Equation (7–33) expresses P_{conv} in terms of P_{AG} and slip. Substituting these two equations into Equation (7–35) yields

$$\tau_{\text{ind}} = \frac{(1-s)P_{\text{AG}}}{(1-s)\omega_{\text{sync}}}$$

$$\boxed{\tau_{\text{ind}} = \frac{P_{\text{AG}}}{\omega_{\text{sync}}}} \tag{7–36}$$

The last equation is especially useful because it expresses induced torque directly in terms of air-gap power and *synchronous speed*, which does not vary. A knowledge of P_{AG} thus directly yields τ_{ind}.

Separating the Rotor Copper Losses and the Power Converted in an Induction Motor's Equivalent Circuit

Part of the power coming across the air gap in an induction motor is consumed in the rotor copper losses, and part of it is converted to mechanical power to drive the motor's shaft. It is possible to separate the two uses of the air-gap power and to indicate them separately on the motor equivalent circuit.

Equation (7–28) gives an expression for the total air-gap power in an induction motor, while Equation (7–30) gives the actual rotor losses in the motor. The air-gap power is the power that would be consumed in a resistor of value R_2/s, while the rotor copper losses are the power that would be consumed in a resistor of value R_2. The difference between them is P_{conv}, which must therefore be the power consumed in a resistor of value

$$R_{\text{conv}} = \frac{R_2}{s} - R_2 = R_2\left(\frac{1}{s} - 1\right)$$

$$\boxed{R_{\text{conv}} = R_2\left(\frac{1-s}{s}\right)} \tag{7–37}$$

Per-phase equivalent circuit with the rotor copper losses and the power converted to mechanical form separated into distinct elements is shown in Figure 7–14.

Figure 7–14 | The per-phase equivalent circuit with rotor losses and P_{conv} separated.

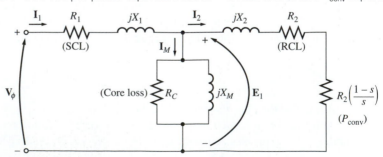

EXAMPLE 7–3

A 460-V, 25-hp, 60-Hz, four-pole, Y-connected induction motor has the following impedances in ohms per phase referred to the stator circuit:

$$R_1 = 0.641 \ \Omega \qquad R_2 = 0.332 \ \Omega$$
$$X_1 = 1.106 \ \Omega \qquad X_2 = 0.464 \ \Omega \qquad R_M = 26.3 \ \Omega$$

The total rotational losses are 1100 W and are assumed to be constant. The core loss is lumped in with the rotational losses. For a rotor slip of 2.2 percent at the rated voltage and rated frequency, find the motor's

a. Speed

b. Stator current

c. Power factor

d. P_{conv} and P_{out}

e. τ_{ind} and τ_{load}

f. Efficiency

■ Solution

The per-phase equivalent circuit of this motor is shown in Figure 7–12, and the power-flow diagram is shown in Figure 7–13. Since the core losses are lumped together with the friction and windage losses and the stray losses, they will be treated like the mechanical losses and be subtracted after P_{conv} in the power-flow diagram.

a. The synchronous speed is

$$n_{sync} = \frac{120 \ f_e}{P} = \frac{120(60 \ \text{Hz})}{4 \ \text{poles}} = 1800 \ \text{r/min}$$

or

$$\omega_{sync} = (1800 \ \text{r/min})\left(\frac{2\pi \ \text{rad}}{1 \ \text{r}}\right)\left(\frac{1 \ \text{min}}{60 \ \text{s}}\right) = 188.5 \ \text{rad/s}$$

The rotor's mechanical shaft speed is

$$n_m = (1 - s)n_{sync}$$
$$= (1 - 0.022)(1800 \ \text{r/min}) = 1760 \ \text{r/min}$$

or

$$\omega_m = (1 - s)\omega_{sync}$$
$$= (1 - 0.022)(188.5 \ \text{rad/s}) = 184.4 \ \text{rad/s}$$

b. To find the stator current, get the equivalent impedance of the circuit. The first step is to combine the referred rotor impedance in parallel with the magnetization branch, and then to add the stator impedance to that combination in series. The referred rotor impedance is

$$\mathbf{Z}_2 = \frac{R_2}{s} + jX_2$$

$$= \frac{0.332}{0.022} + j0.464$$

$$= 15.09 + j0.464 \ \Omega = 15.10\angle 1.76° \ \Omega$$

The combined magnetization plus rotor impedance is given by

$$\mathbf{Z}_f = \frac{1}{1/jX_M + 1/\mathbf{Z}_2}$$

$$= \frac{1}{-j0.038 + 0.0662\angle -1.76°}$$

$$= \frac{1}{0.0773\angle -31.1°} = 12.94\angle 31.1° \ \Omega$$

Therefore, the total impedance is

$$\mathbf{Z}_{tot} = \mathbf{Z}_{stat} + \mathbf{Z}_f$$

$$= 0.641 + j1.106 + 12.94\angle 31.1° \ \Omega$$

$$= 11.72 + j7.79 = 14.07\angle 33.6° \ \Omega$$

The resulting stator current is

$$\mathbf{I}_1 = \frac{V_\phi}{\mathbf{Z}_{tot}}$$

$$= \frac{266\angle 0° \ V}{14.07\angle 33.6° \ \Omega} = 18.88\angle -33.6° \ A$$

c. The motor power factor is

$$PF = \cos 33.6° = 0.833 \qquad \text{lagging}$$

d. The input power to this motor is

$$P_{in} = \sqrt{3} \ V_T I_L \cos \theta$$

$$= \sqrt{3}(460 \ V)(18.88 \ A)(0.833) = 12{,}530 \ W$$

The stator copper losses in this machine are

$$P_{SCL} = 3I_1^2 R_1 \qquad\qquad (7\text{–}25)$$

$$= 3(18.88 \ A)^2(0.641 \ \Omega) = 685 \ W$$

The air-gap power is given by

$$P_{AG} = P_{in} - P_{SCL} = 12{,}530 \ W - 685 \ W = 11{,}845 \ W$$

Therefore, the power converted is

$$P_{conv} = (1 - s)P_{AG} = (1 - 0.022)(11{,}845 \ W) = 11{,}585 \ W$$

The power P_{out} is given by

$$P_{out} = P_{conv} - P_{rot} = 11{,}585 \ W - 1100 \ W = 10{,}485 \ W$$

$$= (10{,}485 \ W)\left(\frac{1 \ hp}{746 \ W}\right) = 14.1 \ hp$$

e. The induced torque is given by

$$\tau_{ind} = \frac{P_{AG}}{\omega_{sync}}$$

$$= \frac{11,845 \text{ W}}{188.5 \text{ rad/s}} = 62.8 \text{ N} \cdot \text{m}$$

and the output torque is given by

$$\tau_{load} = \frac{P_{out}}{\omega_m}$$

$$= \frac{10,485 \text{ W}}{184.4 \text{ rad/s}} = 56.9 \text{ N-m}$$

(In English units, these torques are 46.3 and 41.9 lb-ft, respectively.)

f. The motor's efficiency at this operating condition is

$$\eta = \frac{P_{out}}{P_{in}} \times 100\%$$

$$= \frac{10,485 \text{ W}}{12,530 \text{ W}} \times 100\% = 83.7\%$$ ∎

7.5 | INDUCTION MOTOR TORQUE–SPEED CHARACTERISTICS

How does the torque of an induction motor change as the load changes? How much torque can an induction motor supply at starting conditions? How much does the speed of an induction motor drop as its shaft load increases? To find out the answers to these and similar questions, it is necessary to clearly understand the relationships among the motor's torque, speed, and power.

It is possible to use the equivalent circuit of an induction motor and the power-flow diagram for the motor to derive a general expression for induced torque as a function of speed. The induced torque in an induction motor is given by Equation (7–35) or (7–36):

$$\tau_{ind} = \frac{P_{conv}}{\omega_m} \qquad (7\text{–}35)$$

$$\tau_{ind} = \frac{P_{AG}}{\omega_{sync}} \qquad (7\text{–}36)$$

The latter equation is especially useful, since the synchronous speed is a constant for a given frequency and number of poles. Since ω_{sync} is constant, a knowledge of the air-gap power gives the induced torque of the motor.

The air-gap power is the power crossing the gap from the stator circuit to the rotor circuit. It is equal to the power absorbed in the resistance R_2/s. How can this power be found?

Refer to the equivalent circuit given in Figure 7–15. In this figure, the air-gap power supplied to one phase of the motor can be seen to be

$$P_{AG,\,1\phi} = I_2^2 \frac{R_2}{s}$$

Therefore, the total air-gap power is

$$P_{AG} = 3I_2^2 \frac{R_2}{s}$$

If I_2 can be determined, then the air-gap power and the induced torque will be known.

Figure 7–15 | Per-phase equivalent circuit of an induction motor.

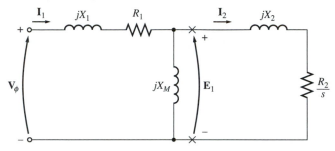

Although there are several ways to solve the circuit in Figure 7–15 for the current I_2, perhaps the easiest one is to determine the Thevenin equivalent of the portion of the circuit to the left of the X's in the figure. Thevenin's theorem states that any linear circuit that can be separated by two terminals from the rest of the system can be replaced by a single voltage source in series with an equivalent impedance. If this were done to the induction motor equivalent circuit, the resulting circuit would be a simple series combination of elements as shown in Figure 7–16c.

To calculate the Thevenin equivalent of the input side of the induction motor equivalent circuit, first open-circuit the terminals at the X's and find the resulting open-circuit voltage present there. Then, to find the Thevenin impedance, kill (short-circuit) the phase voltage and find the $\mathbf{Z}_{eq}$ seen "looking" into the terminals.

Figure 7–16a shows the open terminals used to find the Thevenin voltage. By the voltage divider rule,

$$\mathbf{V}_{TH} = \mathbf{V}_\phi \frac{\mathbf{Z}_M}{\mathbf{Z}_M + \mathbf{Z}_1}$$

$$= \mathbf{V}_\phi \frac{jX_M}{R_1 + jX_1 + jX_M}$$

The magnitude of the Thevenin voltage $\mathbf{V}_{TH}$ is

$$\boxed{V_{TH} = V_\phi \frac{X_M}{\sqrt{R_1^2 + (X_1 + X_M)^2}}} \tag{7–38}$$

Figure 7–16 I (a) The Thevenin equivalent voltage of an induction motor input circuit. (b) The Thevenin equivalent impedance of the input circuit. (c) The resulting simplified equivalent circuit of an induction motor.

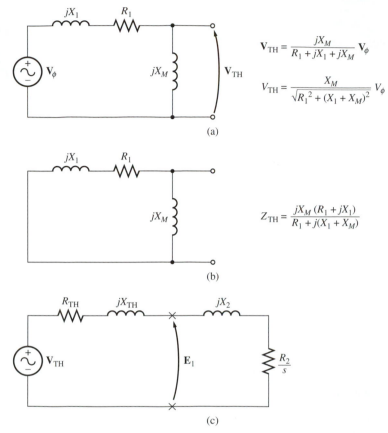

$$\mathbf{V}_{TH} = \frac{jX_M}{R_1 + jX_1 + jX_M} \mathbf{V}_\phi$$

$$V_{TH} = \frac{X_M}{\sqrt{R_1^2 + (X_1 + X_M)^2}} V_\phi$$

(a)

$$\mathbf{Z}_{TH} = \frac{jX_M (R_1 + jX_1)}{R_1 + j(X_1 + X_M)}$$

(b)

(c)

Since the magnetization reactance $X_M \gg X_1$ and $X_M \gg R_1$, the magnitude of the Thevenin voltage is approximately

$$\boxed{V_{TH} \approx V_\phi \frac{X_M}{X_1 + X_M}} \tag{7–39}$$

to quite good accuracy.

Figure 7–16b shows the input circuit with the input voltage source killed. The two impedances are in parallel, and the Thevenin impedance is given by

$$\mathbf{Z}_{TH} = \frac{\mathbf{Z}_1 \mathbf{Z}_M}{\mathbf{Z}_1 + \mathbf{Z}_M} \tag{7–40}$$

This impedance reduces to

$$\mathbf{Z}_{TH} = R_{TH} + jX_{TH} = \frac{jX_M (R_1 + jX_1)}{R_1 + j(X_1 + X_M)} \tag{7–41}$$

Because $X_M \gg X_1$ and $X_M + X_1 \gg R_1$, the Thevenin resistance and reactance are approximately given by

$$R_{TH} \approx R_1 \left(\frac{X_M}{X_1 + X_M}\right)^2 \tag{7–42}$$

$$X_{TH} \approx X_1 \tag{7–43}$$

The resulting equivalent circuit is shown in Figure 7–16c. From this circuit, the current $\mathbf{I}_2$ is given by

$$\mathbf{I}_2 = \frac{\mathbf{V}_{TH}}{\mathbf{Z}_{TH} + \mathbf{Z}_2} \tag{7–44}$$

$$= \frac{\mathbf{V}_{TH}}{R_{TH} + R_2/s + jX_{TH} + jX_2} \tag{7–45}$$

The magnitude of this current is

$$I_2 = \frac{V_{TH}}{\sqrt{(R_{TH} + R_2/s)^2 + (X_{TH} + X_2)^2}} \tag{7–46}$$

The air-gap power is therefore given by

$$P_{AG} = 3I_2^2 \frac{R_2}{s}$$

$$= \frac{3V_{TH}^2 R_2/s}{(R_{TH} + R_2/s)^2 + (X_{TH} + X_2)^2} \tag{7–47}$$

and the rotor-induced torque is given by

$$\tau_{ind} = \frac{P_{AG}}{\omega_{sync}}$$

$$\tau_{ind} = \frac{3V_{TH}^2 R_2/s}{\omega_{sync}[(R_{TH} + R_2/s)^2 + (X_{TH} + X_2)^2]} \tag{7–48}$$

A plot of induction motor torque as a function of speed (and slip) is shown in Figure 7–17, and a plot showing speeds both above and below the normal motor range is shown in Figure 7–18.

Comments on the Induction Motor Torque–Speed Curve

The induction motor torque–speed characteristic curve plotted in Figures 7–17 and 7–18 provides several important pieces of information about the operation of induction motors. This information is summarized below:

1. The induced torque of the motor is zero at synchronous speed. This fact has been discussed previously.

Figure 7–17 I A typical induction motor torque–speed characteristic curve.

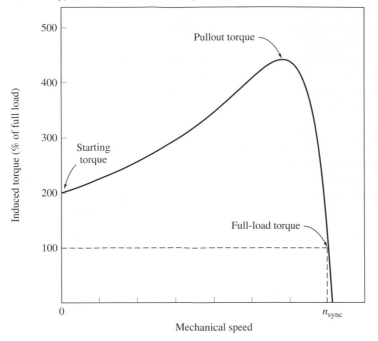

Figure 7–18 I Induction motor torque–speed characteristic curve, showing the extended operating ranges (braking region and generator region).

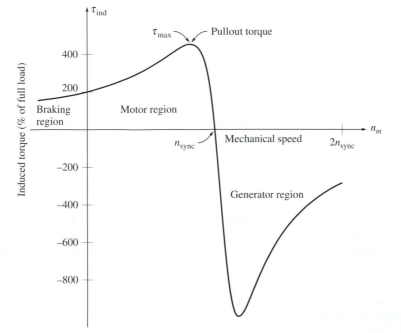

2. The torque–speed curve is nearly linear between no load and full load. In this range, the rotor resistance is much larger than the rotor reactance, so the rotor current, the rotor magnetic field, and the induced torque increase linearly with increasing slip.

3. There is a maximum possible torque that cannot be exceeded. This torque, called the *pullout torque* or *breakdown torque*, is 2 to 3 times the rated full-load torque of the motor. The next section of this chapter contains a method for calculating pullout torque.

4. The starting torque on the motor is slightly larger than its full-load torque, so this motor will start carrying any load that it can supply at full power.

5. Notice that the torque on the motor for a given slip varies as the square of the applied voltage. This fact is useful in one form of induction motor speed control that will be described later.

6. If the rotor of the induction motor is driven faster than synchronous speed, then the direction of the induced torque in the machine reverses and the machine becomes a *generator*, converting mechanical power to electric power.

7. If the motor is turning backward relative to the direction of the magnetic fields, the induced torque in the machine will stop the machine very rapidly and will try to rotate it in the other direction. Since reversing the direction of magnetic field rotation is simply a matter of switching any two stator phases, this fact can be used as a way to very rapidly stop an induction motor. The act of switching two phases in order to stop the motor very rapidly is called *plugging*.

The power converted to mechanical form in an induction motor is equal to

$$P_{conv} = \tau_{ind}\omega_m$$

and is shown plotted in Figure 7–19. Notice that the peak power supplied by the induction motor occurs at a different speed than the maximum torque, and, of course, no power is converted to mechanical form when the rotor is at zero speed.

Maximum (Pullout) Torque in an Induction Motor

Since the induced torque is equal to P_{AG}/ω_{sync}, the maximum possible torque occurs when the air-gap power is maximum. Since the air-gap power is equal to the power consumed in the resistor R_2/s, the *maximum induced torque will occur when the power consumed by that resistor is maximum*.

When is the power supplied to R_2/s at its maximum? Refer to the simplified equivalent circuit in Figure 7–16c. In a situation where the angle of the load impedance is fixed, the maximum power transfer theorem states that maximum power transfer to the load resistor R_2/s will occur when the *magnitude* of that impedance is equal to the *magnitude* of the source impedance. The equivalent source impedance in the circuit is

$$\mathbf{Z}_{source} = R_{TH} + jX_{TH} + jX_2 \qquad (7\text{–}49)$$

so the maximum power transfer occurs when

Figure 7–19 | Induced torque and power converted versus motor speed in revolutions per minute for an example four-pole induction motor.

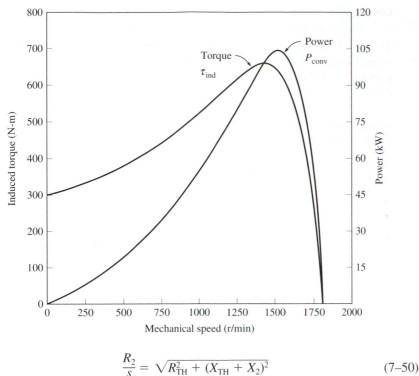

$$\frac{R_2}{s} = \sqrt{R_{TH}^2 + (X_{TH} + X_2)^2} \tag{7-50}$$

Solving Equation (7–50) for slip, we see that *the slip at pullout torque is given by*

$$s_{max} = \frac{R_2}{\sqrt{R_{TH}^2 + (X_{TH} + X_2)^2}} \tag{7-51}$$

Notice that the referred rotor resistance R_2 appears only in the numerator, so the slip of the rotor at maximum torque is directly proportional to the rotor resistance.

The value of the maximum torque can be found by inserting the expression for the slip at maximum torque into the torque equation [Equation (7–48)]. The resulting equation for the maximum or pullout torque is

$$\tau_{max} = \frac{3V_{TH}^2}{2\omega_{sync}[R_{TH} + \sqrt{R_{TH}^2 + (X_{TH} + X_2)^2}]} \tag{7-52}$$

This torque is proportional to the square of the supply voltage and is also inversely related to the size of the stator impedances and the rotor reactance. The smaller a machine's reactances, the larger the maximum torque it is capable of achieving. Note that *slip* at which the maximum torque occurs is directly proportional to rotor resistance [Equation (7–51)], but the *value* of the maximum torque is independent of the value of rotor resistance [Equation (7–52)].

The torque–speed characteristic for a wound-rotor induction motor is shown in Figure 7–20. Recall that it is possible to insert resistance into the rotor circuit of a wound rotor because the rotor circuit is brought out to the stator through slip rings. Notice on the figure that as the rotor resistance is increased, the pullout speed of the motor decreases, but the maximum torque remains constant.

It is possible to take advantage of this characteristic of wound-rotor induction motors to start very heavy loads. If a resistance is inserted into the rotor circuit, the maximum torque can be adjusted to occur at starting conditions. Therefore, the maximum possible torque would be available to start heavy loads. On the other hand, once the load is turning, the extra resistance can be removed from the circuit, and the maximum torque will move up to near-synchronous speed for regular operation.

Figure 7–20 I The effect of varying rotor resistance on the torque–speed characteristic of a wound-rotor induction motor.

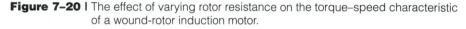

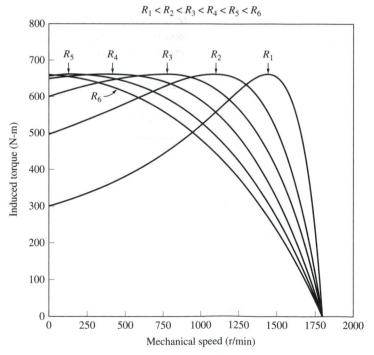

EXAMPLE 7–4

A two-pole, 50-Hz induction motor supplies 15 kW to a load at a speed of 2950 r/min.

a.   What is the motor's slip?

b.   What is the induced torque in the motor in N · m under these conditions?

c.   What will the operating speed of the motor be if its torque is doubled?

d.   How much power will be supplied by the motor when the torque is doubled?

■ **Solution**

a. The synchronous speed of this motor is

$$n_{sync} = \frac{120 \, f_e}{P} = \frac{120(50 \text{ Hz})}{2 \text{ poles}} = 3000 \text{ r/min}$$

Therefore, the motor's slip is

$$s = \frac{n_{sync} - n_m}{n_{sync}} (\times 100\%) \qquad (7\text{-}4)$$

$$= \frac{3000 \text{ r/min} - 2950 \text{ r/min}}{3000 \text{ r/min}} (\times 100\%)$$

$$= 0.0167 \text{ or } 1.67\%$$

b. The induced torque in the motor must be assumed equal to the load torque, and P_{conv} must be assumed equal to P_{load}, since no value was given for mechanical losses. The torque is thus

$$\tau_{ind} = \frac{P_{conv}}{\omega_m}$$

$$= \frac{15 \text{ kW}}{(2950 \text{ r/min})(2\pi \text{ rad/r})(1 \text{ min/60 s})} = 48.6 \text{ N-m}$$

c. In the low-slip region, the torque–speed curve is linear, and the induced torque is directly proportional to slip. Therefore, if the torque doubles, then the new slip will be 3.33 percent. The operating speed of the motor is thus

$$n_m = (1 - s)n_{sync} = (1 - 0.0333)(3000 \text{ r/min}) = 2900 \text{ r/min}$$

d. The power supplied by the motor is given by

$$P_{conv} = \tau_{ind} \, \omega_m$$

$$= (97.2 \text{ N-m})(2900 \text{ r/min})(2\pi \text{ rad/r})(1 \text{ min/60 s})$$

$$= 29.5 \text{ kW}$$

■

EXAMPLE 7–5

A 460-V, 25-hp, 60-Hz, four-pole, Y-connected wound-rotor induction motor has the following impedances in ohms per phase referred to the stator circuit:

$$R_1 = 0.641 \, \Omega \qquad R_2 = 0.332 \, \Omega$$
$$X_1 = 1.106 \, \Omega \qquad X_2 = 0.464 \, \Omega \qquad X_M = 26.3 \, \Omega$$

a. What is the maximum torque of this motor? At what speed and slip does it occur?

b. What is the starting torque of this motor?

c. When the rotor resistance is doubled, what is the speed at which the maximum torque now occurs? What is the new starting torque of the motor?

d. Calculate and plot the torque–speed characteristics of this motor both with the original rotor resistance and with the rotor resistance doubled.

■ Solution

The Thevenin voltage of this motor is

$$V_{TH} = \frac{X_M}{\sqrt{R_1^2 + (X_1 + X_M)^2}} V_\phi \tag{7-38}$$

$$= \frac{(266 \text{ V})(26.3 \text{ }\Omega)}{\sqrt{(0.641 \text{ }\Omega)^2 + (1.106 \text{ }\Omega + 26.3 \text{ }\Omega)^2}} = 255.2 \text{ V}$$

The Thevenin resistance is

$$R_{TH} \approx R_1 \left(\frac{X_M}{X_1 + X_M}\right)^2 \tag{7-42}$$

$$\approx (0.641 \text{ }\Omega)\left(\frac{263 \text{ }\Omega}{1.106 \text{ }\Omega + 263 \text{ }\Omega}\right)^2 = 0.590 \text{ }\Omega$$

The Thevenin reactance is

$$X_{TH} \approx X_1 = 1.106 \text{ }\Omega \tag{7-43}$$

a. The slip at which maximum torque occurs is given by Equation (7–51):

$$s_{max} = \frac{R_2}{\sqrt{R_{TH}^2 + (X_{TH} + X_2)^2}} \tag{7-51}$$

$$= \frac{0.332 \text{ }\Omega}{\sqrt{(0.590 \text{ }\Omega)^2 + (1.106 \text{ }\Omega + 0.464 \text{ }\Omega)^2}} = 0.198$$

This corresponds to a mechanical speed of

$$n_m = (1 - s)n_{sync} = (1 - 0.198)(1800 \text{ r/min}) = 1444 \text{ r/min}$$

The torque at this speed is

$$\tau_{max} = \frac{3V_{TH}^2}{2\omega_{sync}[R_{TH} + \sqrt{R_{TH}^2 + (X_{TH} + X_2)^2}]} \tag{7-52}$$

$$= \frac{3(255.2 \text{ V})^2}{2(188.5 \text{ rad/s})[0.590 \text{ V} + \sqrt{0.590 \text{ }\Omega)^2 + (1.106 \text{ }\Omega + 0.464 \text{ }\Omega)^2}]}$$

$$= 229 \text{ N-m}$$

b. The starting torque of this motor is the torque when the slip $s = 1$:

$$\tau_{start} = \frac{3V_{TH}^2 R_2}{\omega_{sync}[(R_{TH} + R_2)^2 + (X_{TH} + X_2)^2]} \tag{7-53}$$

$$= \frac{3(255.2 \text{ }\Omega)^2(0.332 \text{ }\Omega)}{(188.5 \text{ rad/s})[(0.590 \text{ }\Omega + 0.332 \text{ }\Omega)^2 + (1.106 \text{ }\Omega + 0.464 \text{ }\Omega)^2]}$$

$$= 104 \text{ N-m}$$

c. If the rotor resistance is doubled, then the slip at maximum torque doubles, too. Therefore,

$$s_{max} = 0.396$$

and the speed at maximum torque is

$$n_m = (1 - s)n_{sync} = (1 - 0.396)(1800 \text{ r/min}) = 1087 \text{ r/min}$$

The maximum torque is still

$$\tau_{max} = 229 \text{ N-m}$$

The starting torque is now

$$\tau_{start} = \frac{3(255.2 \ \Omega)^2(0.664 \ \Omega)}{(188.5 \text{ rad/s})[(0.590 \ \Omega + 0.664 \ \Omega)^2 + (1.106 \ \Omega + 0.464 \ \Omega)^2]}$$
$$= 170 \text{ N-m}$$

d. We will create a MATLAB m-file to calculate and plot the torque–speed characteristic of the motor both with the original rotor resistance and with the doubled rotor resistance. The m-file will calculate the Thevenin impedance using the exact equations for V_{TH} and Z_{TH} [Equations (7–38) and (7–41)] instead of the approximate equations, because the computer can easily perform the exact calculations. It will then calculate the induced torque using Equation (7–48) and plot the results. The resulting m-file is shown below.

```
% M-file: torque_speed_curve.m
% M-file create a plot of the torque-speed curve of the
% induction motor of Example 7-5.

% First, initialize the values needed in this program.
r1 = 0.641;                  % Stator resistance
x1 = 1.106;                  % Stator reactance
r2 = 0.332;                  % Rotor resistance
x2 = 0.464;                  % Rotor reactance
xm = 26.3;                   % Magnetization branch reactance
v_phase = 460 / sqrt(3); % Phase voltage
n_sync = 1800;               % Synchronous speed (r/min)
w_sync = 188.5;              % Synchronous speed (rad/s)

% Calculate the Thevenin voltage and impedance from Equations
% 7-41a and 7-43.
v_th = v_phase * ( xm / sqrt(r1^2 + (x1 + xm)^2) );
z_th = ((j*xm) * (r1 + j*x1)) / (r1 + j*(x1 + xm));
r_th = real(z_th);
x_th = imag(z_th);

% Now calculate the torque-speed characteristic for many
% slips between 0 and 1. Note that the first slip value
% is set to 0.001 instead of exactly 0 to avoid divide-
% by-zero problems.
s = (0:1:50) / 50;           % Slip
s(1) = 0.001;
nm = (1 - s) * n_sync;       % Mechanical speed
```

```
% Calculate torque for original rotor resistance
for ii = 1:51
    t_ind1(ii) = (3 * v_th^2 * r2 / s(ii)) / ...
           (w_sync * ((r_th + r2/s(ii))^2 + (x_th + x2)^2) );
end

% Calculate torque for doubled rotor resistance
for ii = 1:51
    t_ind2(ii) = (3 * v_th^2 * (2*r2) / s(ii)) / ...
           (w_sync * ((r_th + (2*r2)/s(ii))^2 + (x_th + x2)^2) );
end

% Plot the torque-speed curve
plot(nm,t_ind1,'Color','k','LineWidth',2.0);
hold on;
plot(nm,t_ind2,'Color','k','LineWidth',2.0,'LineStyle','-.');
xlabel('\itn_{m}','Fontweight','Bold');
ylabel('\tau_{ind}','Fontweight','Bold');
title ('Induction Motor Torque-Speed
Characteristic','Fontweight','Bold');
legend ('Original R_{2}','Doubled R_{2}');
grid on;
hold off;
```

The resulting torque–speed characteristics are shown in Figure 7–21. Note that the peak torque and starting torque values on the curves match the calculations of parts (a) through (c). Also, note that the starting torque of the motor rose as R_2 increased. ∎

Figure 7–21 | Torque–speed characteristics for the motor of Example 7–5.

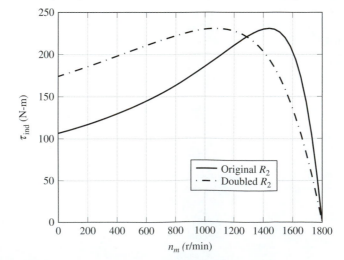

7.6 | VARIATIONS IN INDUCTION MOTOR TORQUE–SPEED CHARACTERISTICS

Section 7.5 contained the derivation of the torque–speed characteristic for an induction motor. In fact, several characteristic curves were shown, depending on the rotor resistance. Example 7–5 illustrated an induction motor designer's dilemma—if a rotor is designed with high resistance, then the motor's starting torque is quite high, but the slip is also quite high at normal operating conditions. Recall that $P_{conv} = (1 - s)P_{AG}$, so *the higher the slip, the smaller the fraction of air-gap power actually converted to mechanical form*, and thus the lower the motor's efficiency. A motor with high rotor resistance has a good starting torque but poor efficiency at normal operating conditions. On the other hand, a motor with low rotor resistance has a low starting torque and high starting current, but its efficiency at normal operating conditions is quite high. An induction motor designer is forced to compromise between the conflicting requirements of high starting torque and good efficiency.

One possible solution to this difficulty was suggested in passing in Section 7.5: Use a wound-rotor induction motor and insert extra resistance into the rotor during starting. The extra resistance could be completely removed for better efficiency during normal operation. Unfortunately, wound-rotor motors are more expensive, need more maintenance, and require a more complex automatic control circuit than squirrel-cage rotor motors. Also, it is sometimes important to completely seal a motor when it is placed in a hazardous or explosive environment, and this is easier to do with a completely self-contained rotor. It would be nice to figure out some way to add extra rotor resistance at starting and to remove it during normal running without slip rings and *without operator or control circuit intervention.*

Figure 7–22 illustrates the desired motor characteristic. This figure shows two wound-rotor motor characteristics, one with high resistance and one with low resistance. At high slips, the desired motor should behave like the high-resistance

Figure 7–22 | A torque–speed characteristic curve combining high-resistance effects at low speeds (high slip) with low-resistance effects at high speed (low slip).

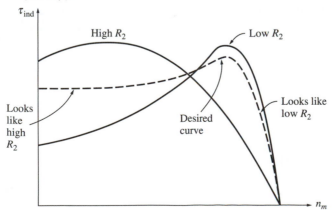

wound-rotor motor curve; at low slips, it should behave like the low-resistance wound-rotor motor curve.

Fortunately, it is possible to accomplish just this effect by properly taking advantage of *leakage reactance* in induction motor rotor design.

Control of Motor Characteristics by Squirrel-Cage Rotor Design

The reactance X_2 in an induction motor equivalent circuit represents the referred form of the rotor's leakage reactance. Recall that leakage reactance is the reactance due to the rotor flux lines that do not also couple with the stator windings. In general, the farther away from the stator a rotor bar or part of a bar is, the greater its leakage reactance, since a smaller percentage of the bar's flux will reach the stator. Therefore, if the bars of a squirrel-cage rotor are placed near the surface of the rotor, they will have only a small leakage flux and the reactance X_2 will be small in the equivalent circuit. On the other hand, if the rotor bars are placed deeper into the rotor surface, there will be more leakage and the rotor reactance X_2 will be larger.

For example, Figure 7–23a is a photograph of a rotor lamination showing the cross section of the bars in the rotor. The rotor bars in the figure are quite large and are placed near the surface of the rotor. Such a design will have a low resistance (because of its large cross section) and a low leakage reactance and X_2 (due to the bar's location near the stator). Because of the low rotor resistance, the pullout torque will be quite near synchronous speed [see Equation (7–51)], and the motor will be quite efficient. Remember that

$$P_{\text{conv}} = (1 - s)P_{\text{AG}} \qquad (7\text{–}33)$$

so very little of the air-gap power is lost in the rotor resistance. However, since R_2 is small, the motor's starting torque will be small, and its starting current will be high. This type of design is called the National Electrical Manufacturers Association (NEMA) design class A. It is more or less a typical induction motor, and its characteristics are basically the same as those of a wound-rotor motor with no extra resistance inserted. Its torque–speed characteristic is shown in Figure 7–24.

Figure 7–23d, however, shows the cross section of an induction motor rotor with *small* bars placed near the surface of the rotor. Since the cross-sectional area of the bars is small, the rotor resistance is relatively high. Since the bars are located near the stator, the rotor leakage reactance is still small. This motor is very much like a wound-rotor induction motor with extra resistance inserted into the rotor. Because of the large rotor resistance, this motor has a pullout torque occurring at a high slip, and its starting torque is quite high. A squirrel-cage motor with this type of rotor construction is called NEMA design class D. Its torque–speed characteristic is also shown in Figure 7–24.

Deep-Bar and Double-Cage Rotor Designs

Both of the previous rotor designs are essentially similar to a wound-rotor motor with a set rotor resistance. How can a *variable* rotor resistance be produced to combine the

Figure 7–23 | Laminations from typical squirrel-cage induction motor rotors, showing the cross section of the rotor bars: (a) NEMA design class A—large bars near the surface. (b) NEMA design class B—large, deep rotor bars. (c) NEMA design class C—double-cage rotor design. (d) NEMA design class D—small bars near the surface. *(Courtesy of MagneTek, Inc.)*

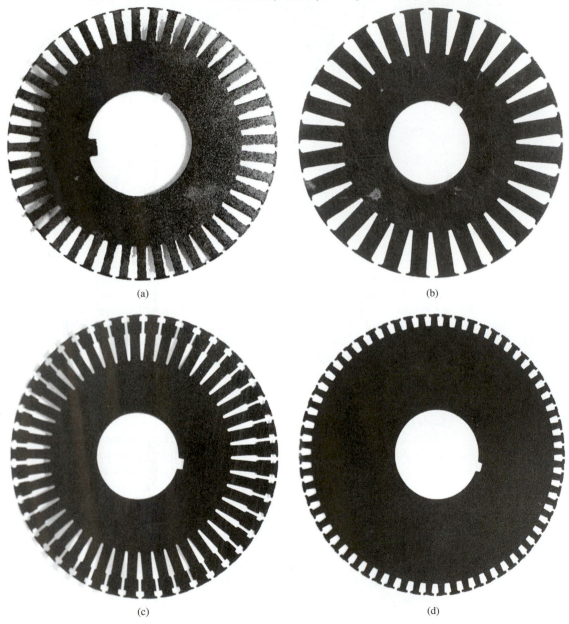

(a)   (b)

(c)   (d)

high starting torque and low starting current of a class D design with the low normal operating slip and high efficiency of a class A design?

Figure 7–24 | Typical torque–speed curves for different rotor designs.

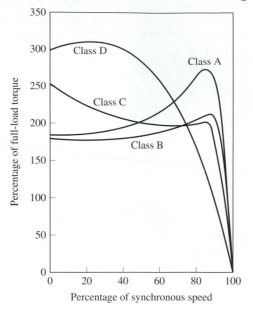

It is possible to produce a variable rotor resistance by the use of deep rotor bars or double-cage rotors. The basic concept is illustrated with a deep-bar rotor in Figure 7–25. Figure 7–25a shows a current flowing through the upper part of a deep rotor bar. Since current flowing in that area is tightly coupled to the stator, the leakage inductance is small for this region. Figure 7–25b shows current flowing deeper in the bar. Here, the leakage inductance is higher. Since all parts of the rotor bar are in parallel electrically, the bar essentially represents a series of parallel electric circuits, the upper ones having a smaller inductance and the lower ones having a larger inductance (Figure 7–25c).

At low slip, the rotor's frequency is very small, and the reactances of all the parallel paths through the bar are small compared to their resistances. The impedances of all parts of the bar are approximately equal, so current flows through all parts of the bar equally. The resulting large cross-sectional area makes the rotor resistance quite small, resulting in good efficiency at low slips. At high slip (starting conditions), the reactances are large compared to the resistances in the rotor bars, so all the current is forced to flow in the low-reactance part of the bar near the stator. Since the *effective* cross section is lower, the rotor resistance is higher than before. With a high rotor resistance at starting conditions, the starting torque is higher and the starting current is lower than in a class A design. A typical torque–speed characteristic for this construction is the design class B curve in Figure 7–24.

A cross-sectional view of a double-cage rotor is shown in Figure 7–23c. It consists of a large, low-resistance set of bars buried deeply in the rotor and a small, high-resistance set of bars set at the rotor surface. It is similar to the deep-bar rotor, except that the difference between low-slip and high-slip operation is even more

Figure 7–25 | Flux linkage in a deep-bar rotor. (a) For a current flowing in the top of the bar, the flux is tightly linked to the stator, and leakage inductance is small. (b) For a current flowing in the bottom of the bar, the flux is loosely linked to the stator, and leakage inductance is large. (c) Resulting equivalent circuit of the rotor bar as a function of depth in the rotor.

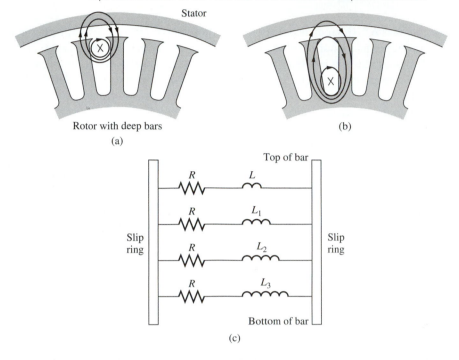

exaggerated. At starting conditions, only the small bar is effective, and the rotor resistance is *quite* high. This high resistance results in a large starting torque. However, at normal operating speeds, both bars are effective, and the resistance is almost as low as in a deep-bar rotor. Double-cage rotors of this sort are used to produce NEMA class B and class C characteristics. Possible torque–speed characteristics for a rotor of this design are designated design class B and design class C in Figure 7–24.

Double-cage rotors have the disadvantage that they are more expensive than the other types of squirrel-cage rotors, but they are cheaper than wound-rotor designs. They allow some of the best features possible with wound-rotor motors (high starting torque with a low starting current and good efficiency at normal operating conditions) at a lower cost and without the need of maintaining slip rings and brushes.

Induction Motor Design Classes

It is possible to produce a large variety of torque–speed curves by varying the rotor characteristics of induction motors. To help industry select appropriate motors for varying applications in the integral-horsepower range, NEMA in the United States and the International Electrotechnical Commission (IEC) in Europe have defined a series of standard designs with different torque–speed curves. These standard designs are referred to as *design classes*, and an individual motor may be referred to as a design class *X* motor. It is these NEMA and IEC design classes that were referred to earlier. Figure 7–24 shows typical torque–speed curves for the four standard NEMA design classes. The characteristic features of each standard design class are given below.

Design Class A　Design class A motors are the standard motor design, with a normal starting torque, a normal starting current, and low slip. The full-load slip of design A motors must be less than 5 percent and must be less than that of a design B motor of equivalent rating. The pullout torque is 200 to 300 percent of the full-load torque and occurs at a low slip (less than 20 percent). The starting torque of this design is at least the rated torque for larger motors and is 200 percent or more of the rated torque for smaller motors. The principal problem with this design class is its extremely high inrush current on starting. Current flows at starting are typically 500 to 800 percent of the rated current. In sizes above about 7.5 hp, some form of reduced-voltage starting must be used with these motors to prevent voltage dip problems on starting in the power system they are connected to. In the past, design class A motors were the standard design for most applications below 7.5 hp and above about 200 hp, but they have largely been replaced by design class B motors in recent years. Typical applications for these motors are driving fans, blowers, pumps, lathes, and other machine tools.

Design Class B　Design class B motors have a normal starting torque, a lower starting current, and low slip. This motor produces about the same starting torque as the class A motor with about 25 percent less current. The pullout torque is greater than or equal to 200 percent of the rated load torque, but less than that of the class A design because of the increased rotor reactance. Rotor slip is still relatively low (less than 5 percent) at full load. Applications are similar to those for design A, but design B is preferred because of its lower starting-current requirements. Design class B motors have largely replaced design class A motors in new installations.

Design Class C　Design class C motors have a high starting torque with low starting currents and low slip (less than 5 percent) at full load. The pullout torque is slightly lower than that for class A motors, while the starting torque is up to 250 percent of the full-load torque. These motors are built from double-cage rotors, so they are more expensive than motors in the previous classes. They are used for high-starting-torque loads, such as loaded pumps, compressors, and conveyors.

Design Class D　Design class D motors have a very high starting torque (275 percent or more of the rated torque) and a low starting current, but they also have a high slip at full load. They are essentially ordinary class A induction motors, but with the rotor bars made smaller and with a higher-resistance material. The high rotor resistance shifts the peak torque to a very low speed. It is even possible for the highest torque to occur at zero speed (100 percent slip). Full-load slip for these motors is quite high because of the high rotor resistance. It is typically 7 to 11 percent, but may go as high as 17 percent or more. These motors are used in applications requiring the acceleration of extremely high inertia loads, especially large flywheels used in punch presses or shears. In such applications, these motors gradually accelerate a large flywheel up to full speed, which then drives the punch. After a punching operation, the motor then reaccelerates the flywheel over a fairly long time for the next operation.

EXAMPLE 7–6

A 460-V, 30-hp, 60-Hz, four-pole, Y-connected induction motor has two possible rotor designs: a single-cage rotor and a double-cage rotor. (The stator is identical for either rotor design.) The motor with the single-cage rotor may be modeled by the following impedances in ohms per phase referred to the stator circuit:

$$R_1 = 0.641 \ \Omega \qquad R_2 = 0.300 \ \Omega$$
$$X_1 = 0.750 \ \Omega \qquad X_2 = 0.500 \ \Omega \qquad X_M = 26.3 \ \Omega$$

The motor with the double-cage rotor may be modeled is a tightly coupled, high-resistance outer cage in parallel with a loosely coupled, low-resistance inner cage (similar to the structure of Figure 7–25c). The stator and magnetization resistance and reactances will be identical with those in the single-cage design.

The resistance and reactance of the rotor outer cage are:

$$R_{2o} = 3.200 \ \Omega \qquad X_{2o} = 0.500 \ \Omega$$

Note that the resistance is high because the outer bar has a small cross section, while the reactance is the same as the reactance of the single-cage rotor, since the outer cage is very close to the stator, and the leakage reactance is small.

The resistance and reactance of the inner cage are:

$$R_{2i} = 0.400 \ \Omega \qquad X_{2i} = 3.300 \ \Omega$$

Here the resistance is low because the bars have a large cross-sectional area, but the leakage reactance is quite high.

Calculate the torque–speed characteristics associated with the two rotor designs. How do they compare?

■ Solution

The torque–speed characteristic of the motor with the single-cage rotor can be calculated in exactly the same manner as Example 7–5. The torque–speed characteristic of the motor with the double-cage rotor can also be calculated in the same fashion, *except* that at each slip the rotor resistance and reactance will be the parallel combination of the impedances of the inner and outer cages. At low slips, the rotor reactance will be relatively unimportant, and the large inner cage will pay a major part in the machine's operation. At high slips, the high reactance of the inner cage almost removes it from the circuit.

A MATLAB m-file to calculate and plot the two torque–speed characteristics is shown below.

```
% M-file: torque_speed_2.m
% M-file create and plot of the torque-speed curve of an
%    induction motor with a double-cage rotor design.
```

```
% First, initialize the values needed in this program.
r1 = 0.641;              % Stator resistance
x1 = 0.750;              % Stator reactance
r2 = 0.300;              % Rotor resistance for single-
                         %   cage motor
r2i = 0.400;             % Rotor resistance for inner
                         %   cage of double-cage motor
r2o = 3.200;             % Rotor resistance for outer
                         %   cage of double-cage motor
x2 = 0.500;              % Rotor reactance for single-
                         %   cage motor
x2i = 3.300;             % Rotor reactance for inner
                         %   cage of double-cage motor
x2o = 0.500;             % Rotor reactance for outer
                         %   cage of double-cage motor
xm = 26.3;               % Magnetization branch reactance
v_phase = 460 / sqrt(3); % Phase voltage
n_sync = 1800;           % Synchronous speed (r/min)
w_sync = 188.5;          % Synchronous speed (rad/s)

% Calculate the Thevenin voltage and impedance from Equations
% 7-41a and 7-43.
v_th = v_phase * ( xm / sqrt(r1^2 + (x1 + xm)^2) );
z_th = ((j*xm) * (r1 + j*x1)) / (r1 + j*(x1 + xm));
r_th = real(z_th);
x_th = imag(z_th);

% Now calculate the motor speed for many slips between
% 0 and 1. Note that the first slip value is set to
% 0.001 instead of exactly 0 to avoid divide-by-zero
% problems.
s = (0:1:50) / 50;       % Slip
s(1) = 0.001;            % Avoid division-by-zero
nm = (1 - s) * n_sync;   % Mechanical speed

% Calculate torque for the single-cage rotor.
for ii = 1:51
   t_ind1(ii) = (3 * v_th^2 * r2 / s(ii)) / ...
           (w_sync * ((r_th + r2/s(ii))^2 + (x_th + x2)^2) );
end

% Calculate resistance and reactance of the double-cage
% rotor at this slip, and then use those values to
% calculate the induced torque.
for ii = 1:51
   y_r = 1/(r2i + j*s(ii)*x2i) + 1/(r2o + j*s(ii)*x2o);
   z_r = 1/y_r;              % Effective rotor impedance
   r2eff = real(z_r);        % Effective rotor resistance
   x2eff = imag(z_r);        % Effective rotor reactance
   % Calculate induced torque for double-cage rotor.
   t_ind2(ii) = (3 * v_th^2 * r2eff / s(ii)) / ...
   (w_sync * ((r_th + r2eff/s(ii))^2 + (x_th + x2eff)^2) );
end
```

```
% Plot the torque-speed curves
plot(nm,t_ind1,'Color','k','LineWidth',2.0);
hold on;
plot(nm,t_ind2,'Color','k','LineWidth',2.0,'LineStyle','-.');
xlabel('\itn_{m}','Fontweight','Bold');
ylabel('\tau_{ind}','Fontweight','Bold');
title ('Induction Motor Torque-Speed
Characteristics','Fontweight','Bold');
legend ('Single-Cage Design','Double-Cage Design');
grid on;
hold off;
```

The resulting torque–speed characteristics are shown in Figure 7–26. Note that the double-cage design has a slightly higher slip in the normal operating range, a smaller maximum torque and a higher starting torque compared to the corresponding single-cage rotor design. This behavior matches our theoretical discussions in this section. ■

Figure 7–26 I Comparison of torque–speed characteristics for the single- and double-cage rotors of Example 7–6.

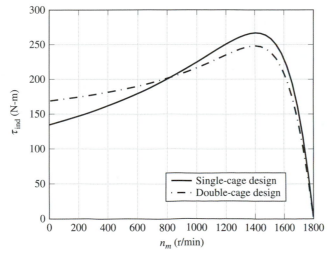

7.7 I STARTING INDUCTION MOTORS

Induction motors do not present the types of starting problems that synchronous motors do. In many cases, induction motors can be started by simply connecting them to the power line. However, there are sometimes good reasons for not doing this. For example, the starting current required may cause such a dip in the power system voltage that *across-the-line starting* is not acceptable.

For wound-rotor induction motors, starting can be achieved at relatively low currents by inserting extra resistance in the rotor circuit during starting. This extra resistance not only increases the starting torque but also reduces the starting current.

For squirrel-cage induction motors, the starting current can vary widely depending primarily on the motor's rated power and on the effective rotor resistance at starting conditions. To estimate the rotor current at starting conditions, all squirrel-cage motors now have a starting *code letter* (not to be confused with their *design class* letter) on their nameplates. The code letter sets limits on the amount of current the motor can draw at starting conditions.

These limits are expressed in terms of the starting apparent power of the motor as a function of its horsepower rating. Figure 7–27 is a table containing the starting kilovolt-amperes per horsepower for each code letter.

To determine the starting current for an induction motor, read the rated voltage, horsepower, and code letter from its nameplate. Then the starting apparent power for the motor will be

$$S_{\text{start}} = (\text{rated horsepower})(\text{code letter factor}) \qquad (7\text{–}54)$$

and the starting current can be found from the equation

$$I_L = \frac{S_{\text{start}}}{\sqrt{3}V_T} \qquad (7\text{–}55)$$

Figure 7–27 | Table of NEMA code letters, indicating the starting kilovoltamperes per horsepower of rating for a motor. Each code letter extends up to, but does not include, the lower bound of the next higher class. *(Reproduced by permission from Motors and Generators, NEMA Publication MG-1, copyright 1987 by NEMA.)*

Nominal code letter	Locked rotor, kVA/hp	Nominal code letter	Locked rotor, kVA/hp
A	0–3.15	L	9.00–10.00
B	3.15–3.55	M	10.00–11.00
C	3.55–4.00	N	11.20–12.50
D	4.00–4.50	P	12.50–14.00
E	4.50–5.00	R	14.00–16.00
F	5.00–5.60	S	16.00–18.00
G	5.60–6.30	T	18.00–20.00
H	6.30–7.10	U	20.00–22.40
J	7.10–8.00	V	22.40 and up
K	8.00–9.00		

EXAMPLE 7–7

What is the starting current of a 15-hp, 208-V, code letter F, three-phase induction motor?

■ **Solution**

According to Figure 7–27, the maximum kilovoltamperes per horsepower is 5.6. Therefore, the maximum starting kilovoltamperes of this motor is

$$S_{start} = (15 \text{ hp})(5.6) = 84 \text{ kVA}$$

The starting current is thus

$$I_L = \frac{S_{start}}{\sqrt{3} V_T} \tag{7-55}$$

$$= \frac{84 \text{ kVA}}{\sqrt{3}(208 \text{ V})} = 233 \text{ A} \qquad ■$$

If necessary, the starting current of an induction motor may be reduced by a starting circuit. However, if this is done, it will also reduce the starting torque of the motor.

One way to reduce the starting current is to insert extra inductors or resistors into the power line during starting. While formerly common, this approach is rare today. An alternative approach is to reduce the motor's terminal voltage during starting by using autotransformers to step it down. Figure 7–28 shows a typical reduced-voltage starting circuit using autotransformers. During starting, contacts 1 and 3 are shut, supplying a lower voltage to the motor. Once the motor is nearly up to speed, those contacts are opened and contacts 2 are shut. These contacts put full line voltage across the motor.

Figure 7–28 | An autotransformer starter for an induction motor.

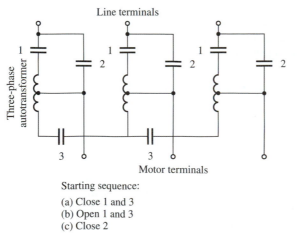

Starting sequence:

(a) Close 1 and 3
(b) Open 1 and 3
(c) Close 2

It is important to realize that, while the starting current is reduced in direct proportion to the decrease in terminal voltage, the starting torque decreases as the *square* of the applied voltage. Therefore, only a certain amount of current reduction can be done if the motor is to start with a shaft load attached.

7.8 | SPEED CONTROL OF INDUCTION MOTORS

Until the advent of modern solid-state drives, induction motors in general were not good machines for applications requiring considerable speed control. The normal operating range of a typical induction motor (design classes A, B, and C) is confined to less than 5 percent slip, and the speed variation over that range is more or less directly proportional to the load on the shaft of the motor. Even if the slip could be made larger, the efficiency of the motor would become very poor, since the rotor copper losses are directly proportional to the slip on the motor (remember that $P_{\text{RCL}} = sP_{\text{AG}}$).

There are really only two techniques by which the speed of an induction motor can be controlled. One is to vary the synchronous speed, which is the speed of the stator and rotor magnetic fields, since the rotor speed always remains near n_{sync}. The other technique is to vary the slip of the motor for a given load. Each of these approaches will be taken up in more detail below.

The synchronous speed of an induction motor is given by

$$n_{\text{sync}} = \frac{120 f_e}{P} \tag{7–1}$$

so the only ways in which the synchronous speed of the machine can be varied are (1) by changing the electrical frequency and (2) by changing the number of poles on the machine. Slip control may be accomplished by varying either the rotor resistance or the terminal voltage of the motor.

Induction Motor Speed Control by Pole Changing

In the days before modern solid-state control circuits were common, the stator windings of induction motors were often constructed so that the number of poles in the stator windings could be changed. This technique is largely obsolete now, so it will not be discussed further here. See Reference 2 for details on this topic.

Speed Control by Changing the Line Frequency

If the electrical frequency applied to the stator of an induction motor is changed, the rate of rotation of its magnetic fields n_{sync} will change in direct proportion to the change in electrical frequency, and the no-load point on the torque–speed characteristic curve will change with it (see Figure 7–29). The synchronous speed of the motor at rated conditions is known as the *base speed*. By using variable frequency control, it is possible to adjust the speed of the motor either above or below base speed. A properly designed variable-frequency induction motor drive can be *very* flexible. It can control the speed of an induction motor over a range from as little as 5 percent of base speed up to about twice base speed. However, it is important to maintain certain voltage and torque limits on the motor as the frequency is varied, to ensure safe operation.

When a motor is running at speeds below its base speed, it is necessary to reduce the terminal voltage applied to the stator for proper operation. The terminal voltage

Figure 7–29 | Variable-frequency speed control in an induction motor: (a) The family of torque–speed characteristic curves for speeds below base speed, assuming that the line voltage is derated linearly with frequency. (b) The family of torque–speed characteristic curves for speeds above base speed, assuming that the line voltage is held constant. (c) The torque–speed characteristic curves for all frequencies.

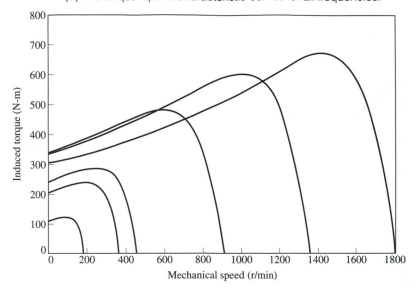

(a)

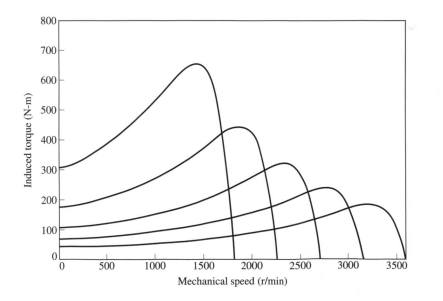

(b)

(continued)

Figure 7-29 I *(continued)*

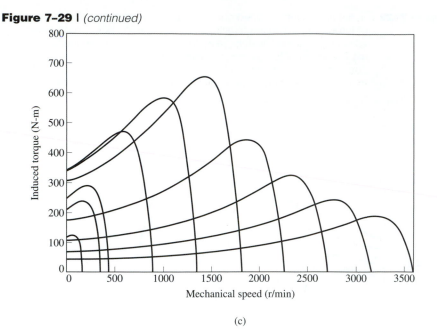

(c)

applied to the stator should be decreased linearly with decreasing stator frequency. This process is called *derating*. If it is not done, the steel in the core of the induction motor will saturate and excessive magnetization currents will flow in the machine.

To understand the necessity for derating, recall that an induction motor is basically a rotating transformer. As with any transformer, the flux in the core of an induction motor can be found from Faraday's law:

$$v(t) = N \frac{d\phi}{dt} \qquad (1\text{--}36)$$

If a voltage is applied to the core, the resulting flux ϕ is

$$\phi(t) = \frac{1}{N_P} \int v(t) \, dt$$

$$= \frac{1}{N_P} \int V_M \sin \omega t \, dt$$

$$\boxed{\phi(t) = -\frac{V_M}{\omega N_P} \cos \omega t} \qquad (7\text{--}56)$$

Note that the electrical frequency appears in the *denominator* of this expression. Therefore, if the electrical frequency applied to the stator *decreases* by 10 percent while the magnitude of the voltage applied to the stator remains constant, the flux in the core of the motor will *increase* by about 10 percent and the magnetization current of the motor will increase. In the unsaturated region of the motor's magnetization curve, the increase in magnetization current will also be about 10 percent. However, in the saturated region of the motor's magnetization curve, a 10 percent increase in flux requires a much larger increase in magnetization current. Induction motors are normally designed to operate near the saturation point on their magnetization curves, so the increase in flux due to a decrease in frequency will cause excessive magnetization currents to flow in the motor. (This same problem was observed in transformers; see Section 3.12.)

To avoid excessive magnetization currents, it is customary to decrease the applied stator voltage in direct proportion to the decrease in frequency whenever the frequency falls below the rated frequency of the motor. Since the applied voltage v appears in the numerator of Equation (7–56) and the frequency ω appears in the denominator of Equation (7–56), the two effects counteract each other, and the magnetization current is unaffected.

When the voltage applied to an induction motor is varied linearly with frequency below the base speed, the flux in the motor will remain approximately constant. Therefore, the maximum torque that the motor can supply remains fairly high. However, the maximum power rating of the motor must be decreased linearly with decreases in frequency to protect the stator circuit from overheating. The power supplied to a three-phase induction motor is given by

$$P = \sqrt{3}V_L I_L \cos \theta$$

If the voltage V_L is decreased, then the maximum power P must also be decreased, or else the current flowing in the motor will become excessive, and the motor will overheat.

Figure 7–29a shows a family of induction motor torque–speed characteristic curves for speeds below base speed, assuming that the magnitude of the stator voltage varies linearly with frequency.

When the electrical frequency applied to the motor exceeds the rated frequency of the motor, the stator voltage is held constant at the rated value. Although saturation considerations would permit the voltage to be raised above the rated value under these circumstances, it is limited to the rated voltage to protect the winding insulation of the motor. The higher the electrical frequency above base speed, the larger the denominator of Equation (7–57) becomes. Since the numerator term is held constant above rated frequency, the resulting flux in the machine decreases and the maximum torque decreases with it. Figure 7–29b shows a family of induction motor torque–speed characteristic curves for speeds above base speed, assuming that the stator voltage is held constant.

If the stator voltage is varied linearly with frequency below base speed and is held constant at rated value above base speed, then the resulting family of torque–speed characteristics is as shown in Figure 7–29c. The rated speed for the motor shown in Figure 7–29 is 1800 r/min.

In the past, the principal disadvantage of electrical frequency control as a method of speed changing was that a dedicated generator or mechanical frequency changer was required to make it operate. This problem has disappeared with the development of modern solid-state variable-frequency motor drives. In fact, changing the line frequency with solid-state motor drives has become the method of choice for induction motor speed control. Note that this method can be used with *any* induction motor, unlike the pole-changing technique, which requires a motor with special stator windings.

Speed Control by Changing the Line Voltage

The torque developed by an induction motor is proportional to the square of the applied voltage. If a load has a torque–speed characteristic such as the one shown in Figure 7–30, then the speed of the motor may be controlled over a limited range by varying the line voltage. This method of speed control is sometimes used on small motors driving fans.

Figure 7–30 | Variable-line-voltage speed control in an induction motor.

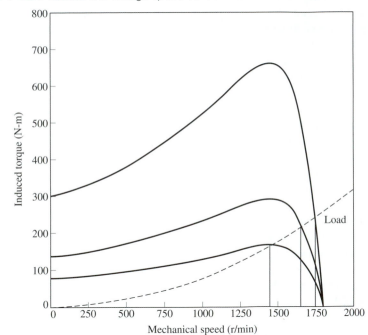

Speed Control by Changing the Rotor Resistance

In wound-rotor induction motors, it is possible to change the shape of the torque–speed curve by inserting extra resistances into the rotor circuit of the machine. The resulting torque–speed characteristic curves are shown in Figure 7–31. If the torque–speed curve of the load is as shown in the figure, then changing the rotor resistance will change the operating speed of the motor. However, inserting extra resistances into the rotor circuit of an induction motor seriously reduces the efficiency of the machine. Such a method of speed control is normally used only for short periods because of this efficiency problem.

Figure 7–31 I Speed control by varying the rotor resistance of a wound-rotor induction motor.

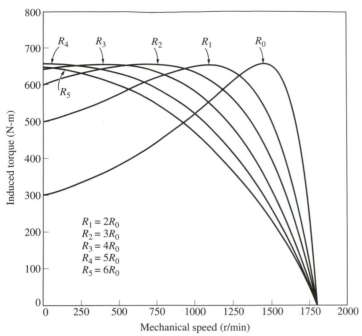

7.9 I DETERMINING CIRCUIT MODEL PARAMETERS

The equivalent circuit of an induction motor is a very useful tool for determining the motor's response to changes in load. However, if a model is to be used for a real machine, it is necessary to determine what the element values are that go into the model. How can R_1, R_2, X_1, X_2, and X_M be determined for a real motor?

These pieces of information may be found by performing a series of tests on the induction motor that are analogous to the short-circuit and open-circuit tests in a transformer. The tests must be performed under precisely controlled conditions, since the resistances vary with temperature and the rotor resistance also varies with rotor frequency. The exact details of how each induction motor test must be performed in order to achieve accurate results are described in IEEE Standard 112. Although the details of the tests are very complicated, the concepts behind them are relatively straightforward and will be explained here.

The DC Test for Stator Resistance

The rotor resistance R_2 plays an extremely critical role in the operation of an induction motor. Among other things, R_2 determines the shape of the torque–speed curve, determining the speed at which the pullout torque occurs. A standard motor test called the *locked-rotor test* can be used to determine the total motor circuit resistance (this test is taken up in the next section). However, this test finds only the *total* resistance. To find the rotor resistance R_2 accurately, it is necessary to know R_1 so that it can be subtracted from the total.

There is a test for R_1 independent of R_2, X_1, and X_2. This test is called the *DC test*. Basically, a DC voltage is applied to the stator windings of an induction motor. Because the current is DC, there is no induced voltage in the rotor circuit and no resulting rotor current flow. Also, the reactance of the motor is zero at direct current. Therefore, the only quantity limiting current flow in the motor is the stator resistance, and that resistance can be determined.

The basic circuit for the DC test is shown in Figure 7–32. This figure shows a DC power supply connected to two of the three terminals of a Y-connected induction motor. To perform the test, the current in the stator windings is adjusted to the rated value, and the voltage between the terminals is measured. The current in the stator windings is adjusted to the rated value in an attempt to heat the windings to the same temperature they would have during normal operation (remember, winding resistance is a function of temperature).

Figure 7–32 | Test circuit for a DC resistance test.

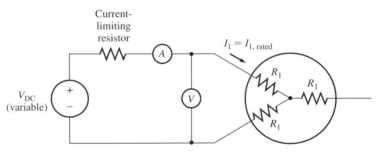

The current in Figure 7–32 flows through two of the windings, so the total resistance in the current path is $2R_1$. Therefore,

$$2R_1 = \frac{V_{DC}}{I_{DC}}$$

or

$$R_1 = \frac{V_{DC}}{2I_{DC}} \tag{7–57}$$

With this value of R_1 the stator copper losses at no load may be determined, and the rotational losses may be found as the difference between the input power at no load and the stator copper losses.

The value of R_1 calculated in this fashion is not completely accurate, since it neglects the skin effect that occurs when an AC voltage is applied to the windings. More details concerning corrections for temperature and skin effect can be found in IEEE Standard 112.

The No-Load Test

The no-load test of an induction motor measures the rotational losses of the motor and provides information about its magnetization current. The test circuit for this test is shown in Figure 7–33a. Wattmeters, a voltmeter, and three ammeters are connected to an induction motor, which is allowed to spin freely. The only load on the motor is the friction and windage losses, so all P_{conv} in this motor is consumed by mechanical losses, and the slip of the motor is very small (possibly as small as 0.001 or less). The equivalent circuit of this motor is shown in Figure 7–33b. With its very small slip, the resistance corresponding to its power converted, $R_2(1 - s)/s$, is much much larger than the resistance corresponding to the rotor copper losses R_2 and much larger than the rotor reactance X_2. In this case, the equivalent circuit reduces approximately to the second circuit in Figure 7–33b. There, the output resistor is in parallel with the magnetization reactance X_M and the core losses R_C.

In this motor at no-load conditions, the input power measured by the meters must equal the losses in the motor. The rotor copper losses are negligible because the current I_2 is *extremely* small [because of the large load resistance $R_2(1 - s)/s$], so they may be neglected. The stator copper losses are given by

$$\boxed{P_{SCL} = 3I_1^2 R_1} \tag{7–25}$$

so the input power must equal

$$\begin{aligned} P_{in} &= P_{SCL} + P_{core} + P_{F\&W} + P_{misc} \\ &= 3I_1^2 R_1 + P_{rot} \end{aligned} \tag{7–58}$$

where is the rotational losses of the motor:

$$P_{rot} = P_{core} + P_{F\&W} + P_{misc} \tag{7–59}$$

Thus, given the input power to the motor, the rotational losses of the machine may be determined.

The equivalent circuit that describes the motor operating in this condition contains resistors R_C and $R_2(1 - s)/s$ in parallel with the magnetizing reactance X_M. The current needed to establish a magnetic field is quite large in an induction motor, because of the high reluctance of its air gap, so the reactance X_M will be much smaller than the resistances in parallel with it and the overall input power factor will be very small. With the large lagging current, most of the voltage drop will be across the inductive components in the circuit. The equivalent input impedance is thus approximately

$$\boxed{|\mathbf{Z}_{eq}| = \frac{V_\phi}{I_{1,\,nl}} \approx X_1 + X_M} \tag{7–60}$$

and if X_1 can be found in some other fashion, the magnetizing impedance X_M will be known for the motor.

Figure 7–33 | The no-load test of an induction motor: (a) Test circuit. (b) The resulting motor equivalent circuit. Note that at no load the motor's impedance is essentially the series combination of R_1, jX_1, and jX_M.

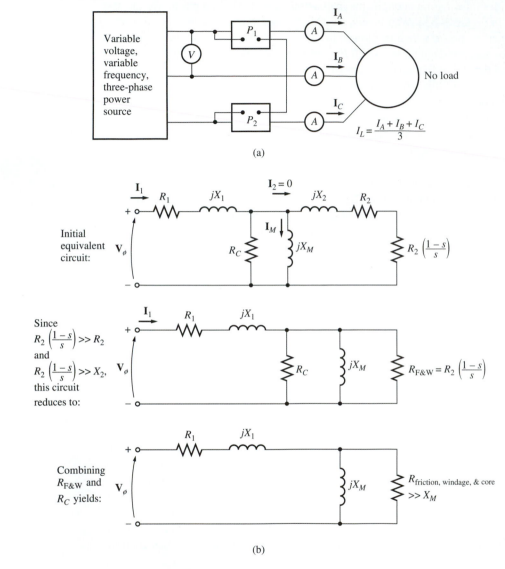

(a)

(b)

The Locked-Rotor Test

The third test that can be performed on an induction motor to determine its circuit parameters is called the *locked-rotor test*, or sometimes the *blocked-rotor test*. This test corresponds to the short-circuit test on a transformer. In this test, the rotor is locked or blocked so that it *cannot* move, a voltage is applied to the motor, and the resulting voltage, current, and power are measured.

Figure 7–34a shows the connections for the locked-rotor test. To perform the blocked-rotor test, an AC voltage is applied to the stator, and the current flow is adjusted to be approximately full-load value. When the current is full-load value, the voltage, current, and power flowing into the motor are measured. The equivalent circuit for this test is shown in Figure 7–34b. Notice that since the rotor is not moving, the slip $s = 1$, and so the rotor resistance R_2/s is just equal to R_2 (quite a small value). Since R_2 and X_2 are so small, almost all the input current will flow through them, instead of through the much larger magnetizing reactance X_M. Therefore, the circuit under these conditions looks like a series combination of X_1, R_1, X_2, and R_2.

Figure 7–34 | The locked-rotor test for an induction motor: (a) Test circuit.
(b) Motor equivalent circuit.

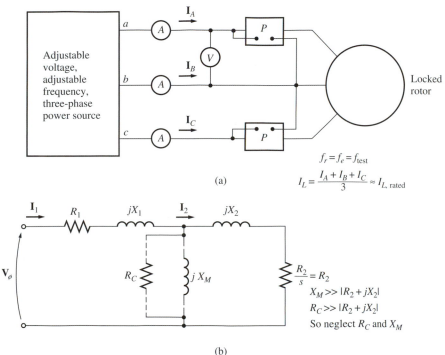

There is one problem with this test, however. In normal operation, the stator frequency is the line frequency of the power system (50 or 60 Hz). At starting conditions, the rotor is also at line frequency. However, at normal operating conditions, the slip of most motors is only 2 to 4 percent, and the resulting rotor frequency is in the range of 1 to 3 Hz. This creates a problem in that *the line frequency does not represent the normal operating conditions of the rotor.* Since effective rotor resistance is a strong function of frequency for design class B and C motors, the incorrect rotor frequency can lead to misleading results in this test. A typical compromise is to use a frequency 25 percent or less of the rated frequency. While this approach is acceptable for essentially constant resistance rotors (design classes A and D), it leaves a lot to be desired when one is trying to find the normal rotor resistance of a variable-resistance rotor. Because of these and similar problems, a great deal of care must be exercised in taking measurements for these tests.

After a test voltage and frequency have been set up, the current flow in the motor is quickly adjusted to about the rated value, and the input power, voltage, and current are measured before the rotor can heat up too much. The input power to the motor is given by

$$P = \sqrt{3} V_T I_L \cos \theta$$

so the locked-rotor power factor can be found as

$$\text{PF} = \cos \theta = \frac{P_{\text{in}}}{\sqrt{3} V_T I_L} \qquad (7\text{–}61)$$

and the impedance angle θ is just equal to $\cos^{-1} \text{PF}$.

The magnitude of the total impedance in the motor circuit at this time is

$$|\mathbf{Z}_{\text{LR}}| = \frac{V_\phi}{I_1} = \frac{V_T}{\sqrt{3} I_L} \qquad (7\text{–}62)$$

and the angle of the total impedance is θ. Therefore,

$$\begin{aligned}
\mathbf{Z}_{\text{LR}} &= R_{\text{LR}} + j X'_{\text{LR}} \\
&= |\mathbf{Z}_{\text{LR}}| \cos \theta + j |\mathbf{Z}_{\text{LR}}| \sin \theta \qquad (7\text{–}63)
\end{aligned}$$

The locked-rotor resistance R_{LR} is equal to

$$R_{\text{LR}} = R_1 + R_2 \qquad (7\text{–}64)$$

while the locked-rotor reactance X'_{LR} is equal to

$$X'_{\text{LR}} = X'_1 + X'_2 \qquad (7\text{–}65)$$

where X'_1 and X'_2 are the stator and rotor reactances *at the test frequency*, respectively.

The rotor resistance R_2 can now be found as

$$R_2 = R_{\text{LR}} - R_1 \qquad (7\text{–}66)$$

where R_1 was determined in the DC test. The total rotor reactance referred to the stator can also be found. Since the reactance is directly proportional to the frequency, the total equivalent reactance at the normal operating frequency can be found as

$$X'_{\text{LR}} = \frac{f_{\text{rated}}}{f_{\text{test}}} X'_{\text{LR}} = X_1 + X_2 \qquad (7\text{–}67)$$

Unfortunately, there is no simple way to separate the contributions of the stator and rotor reactances from each other. Over the years, experience has shown that motors of certain design types have certain proportions between the rotor and stator reactances. Figure 7–35 summarizes this experience. In normal practice, it really does not matter just how X_{LR} is broken down, since the reactance appears as the sum $X_1 + X_2$ in all the torque equations.

Figure 7–35 | Rules of thumb for dividing rotor and stator circuit reactance.

Rotor design	X_1 and X_2 as functions of X_{LR}	
	X_1	X_2
Wound rotor	$0.5X_{LR}$	$0.5X_{LR}$
Design A	$0.5X_{LR}$	$0.5X_{LR}$
Design B	$0.4X_{LR}$	$0.6X_{LR}$
Design C	$0.3X_{LR}$	$0.7X_{LR}$
Design D	$0.5X_{LR}$	$0.5X_{LR}$

EXAMPLE 7–8

The following test data were taken on a 7.5-hp, four-pole, 208-V, 60-Hz, design A, Y-connected induction motor having a rated current of 28 A.
DC test:

$$V_{DC} = 13.6 \text{ V} \qquad I_{DC} = 28.0 \text{ A}$$

No-load test:

$$V_T = 208 \text{ V} \qquad f = 60 \text{ Hz}$$
$$I_A = 8.12 \text{ A} \qquad P_{in} = 420 \text{ W}$$
$$I_B = 8.20 \text{ A}$$
$$I_C = 8.18 \text{ A}$$

Locked-rotor test:

$$V_T = 25 \text{ V} \qquad f = 15 \text{ Hz}$$
$$I_A = 28.1 \text{ A} \qquad P_{in} = 920 \text{ W}$$
$$I_B = 28.0 \text{ A}$$
$$I_C = 27.6 \text{ A}$$

a.   Sketch the per-phase equivalent circuit for this motor.
b.   Find the slip at the pullout torque, and find the value of the pullout torque itself.

■ Solution

a.   From the DC test,

$$R_1 = \frac{V_{DC}}{2I_{DC}} = \frac{13.6 \text{ V}}{2(28.0 \text{ A})} = 0.243 \text{ }\Omega$$

From the no-load test,

$$I_{L,\,av} = \frac{8.12\ A + 8.20\ A + 8.18\ A}{3} = 8.17\ A$$

$$V_{\phi,\,nl} = \frac{208\ V}{\sqrt{3}} = 120\ V$$

Therefore,

$$|\mathbf{Z}_{nl}| = \frac{120\ V}{8.17\ A} = 14.7\ \Omega = X_1 + X_M$$

When X_1 is known, X_M can be found. The stator copper losses are

$$P_{SCL} = 3I_1^2 R_1 = 3(8.17\ A)^2(0.243\ \Omega) = 48.7\ W$$

Therefore, the no-load rotational losses are

$$P_{rot} = P_{in,\,nl} - P_{SCL,\,nl}$$
$$P_{rot} = 420\ W - 48.7\ W = 371.3\ W$$

From the locked-rotor test,

$$I_{L,\,av} = \frac{28.1\ A + 28.0\ A + 27.6\ A}{3} = 27.9\ A$$

The locked-rotor impedance is

$$|\mathbf{Z}_{LR}| = \frac{V_\phi}{I_A} = \frac{V_T}{\sqrt{3}I_A} = \frac{25\ V}{\sqrt{3}(27.9\ A)} = 0.517\ \Omega$$

and the impedance angle θ is

$$\theta = \cos^{-1}\frac{P_{in}}{\sqrt{3}V_T I_L}$$
$$= \cos^{-1}\frac{920\ W}{\sqrt{3}(25\ V)(27.9\ A)}$$
$$= \cos^{-1} 0.792 = 40.4°$$

Therefore, $R_{LR} = 0.517 \cos 40.4° = 0.394\ \Omega = R_1 + R_2$. Since $R_1 = 0.243\ \Omega$, R_2 must be $0.151\ \Omega$. The reactance at 15 Hz is

$$X'_{LR} = 0.517 \sin 40.4° = 0.335\ \Omega$$

The equivalent reactance at 60 Hz is

$$X_{LR} = \frac{f_{rated}}{f_{test}} X'_{LR} = \left(\frac{60\ Hz}{15\ Hz}\right)0.335\ \Omega = 1.34\ \Omega$$

For design class A induction motors, this reactance is assumed to be divided equally between the rotor and stator, so

$$X_1 = X_2 = 0.67\ \Omega$$
$$X_M = |\mathbf{Z}_{nl}| - X_1 = 14.7\ \Omega - 0.67\ \Omega = 14.03\ \Omega$$

The final per-phase equivalent circuit is shown in Figure 7–36.

Figure 7–36 | Motor per-phase equivalent circuit for Example 7–8.

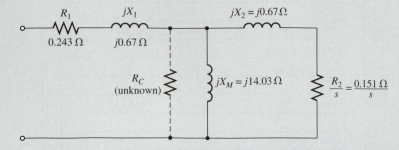

b.   For this equivalent circuit, the Thevenin equivalents are found from Equations (7–38), (7–42), and (7–43) to be

$$V_{TH} = 114.6 \text{ V} \quad R_{TH} = 0.221 \ \Omega \quad X_{TH} = 0.67 \ \Omega$$

Therefore, the slip at the pullout torque is given by

$$s_{max} = \frac{R_2}{\sqrt{R_{TH}^2 + (X_{TH} + X_2)^2}} \tag{7-51}$$

$$= \frac{0.151 \ \Omega}{\sqrt{(0.243 \ \Omega)^2 + (0.67 \ \Omega + 0.67 \ \Omega)^2}} = 0.111 = 11.1\%$$

The maximum torque of this motor is given by

$$\tau_{max} = \frac{3V_{TH}^2}{2\omega_{sync} \ [R_{TH} + \sqrt{R_{TH}^2 + (X_{TH} + X_2)^2}]} \tag{7-52}$$

$$= \frac{3(114.6 \text{ V})^2}{2(188.5 \text{ rad/s})[0.221 \ \Omega + \sqrt{(0.221 \ \Omega)^2 + (0.67 \ \Omega + 0.67 \ \Omega)^2}]}$$

$$= 66.2 \text{ N-m} \qquad\blacksquare$$

7.10 | INDUCTION MOTOR RATINGS

A nameplate for a typical high-efficiency integral-horsepower induction motor is shown in Figure 7–37. The most important ratings present on the nameplate are

1. Output power
2. Voltage
3. Current
4. Power factor
5. Speed
6. Nominal efficiency
7. NEMA design class
8. Starting code

Figure 7–37 | The nameplate of a typical high-efficiency induction motor. *(Courtesy of MagneTek, Inc.)*

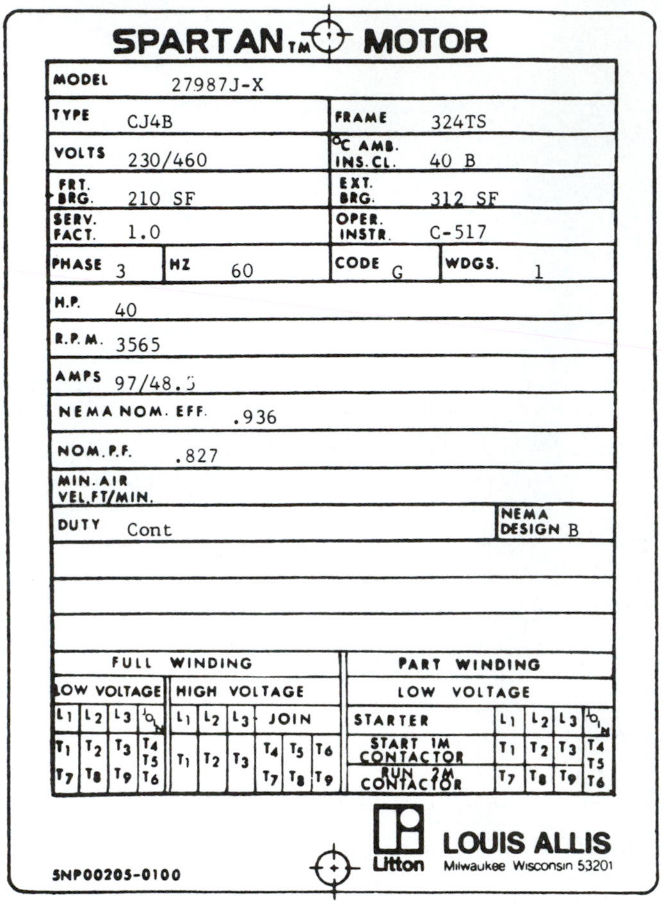

SPARTAN™ MOTOR

MODEL	27987J-X		
TYPE	CJ4B	FRAME	324TS
VOLTS	230/460	°C AMB. INS.CL.	40 B
FRT. BRG.	210 SF	EXT. BRG.	312 SF
SERV. FACT.	1.0	OPER. INSTR.	C-517
PHASE 3	HZ 60	CODE G	WDGS. 1

H.P.	40
R.P.M.	3565
AMPS	97/48.5
NEMA NOM. EFF.	.936
NOM. P.F.	.827
MIN. AIR VEL. FT/MIN.	
DUTY	Cont

NEMA DESIGN B

FULL WINDING		PART WINDING	
LOW VOLTAGE	HIGH VOLTAGE	LOW VOLTAGE	

LOW VOLTAGE				HIGH VOLTAGE							PART — LOW VOLTAGE					

L1	L2	L3	L0M	L1	L2	L3	JOIN	STARTER	L1	L2	L3	L0M
T1	T2	T3	T4 T5	T1	T2	T3	T4 T5 T6	START 1M CONTACTOR	T1	T2	T3	T4 T5
T7	T8	T9	T6				T7 T8 T9	RUN 2M CONTACTOR	T7	T8	T9	T6

LOUIS ALLIS
Litton   Milwaukee Wisconsin 53201

5NP00205-0100

A nameplate for a typical standard-efficiency induction motor would be similar, except that it might not show a nominal efficiency.

The voltage limit on the motor is based on the maximum acceptable magnetization current flow, since the higher the voltage gets, the more saturated the motor's iron becomes and the higher its magnetization current becomes. Just as in the case of transformers and synchronous machines, a 60-Hz induction motor may be used on a 50-Hz power system, but only if the voltage rating is decreased by an amount proportional to the decrease in frequency. This derating is necessary because the flux in the core of the motor is proportional to the integral of the applied voltage. To keep the maximum flux in the core constant while the period of integration is increasing, the average voltage level must decrease.

The current limit on an induction motor is based on the maximum acceptable heating in the motor's windings, and the power limit is set by the combination of the voltage and current ratings with the machine's power factor and efficiency.

NEMA design classes and starting code letters were discussed in previous sections of this chapter.

7.11 | SUMMARY

The induction motor is the most popular type of AC motor because of its simplicity and ease of operation. An induction motor does not have a separate field circuit; instead, it depends on transformer action to induce voltages and currents in its field circuit. In fact, an induction motor is basically a rotating transformer. Its equivalent circuit is similar to that of a transformer, except for the effects of varying speed.

An induction motor normally operates at a speed near synchronous speed, but it can never operate at exactly n_{sync}. There must always be some relative motion in order to induce a voltage in the induction motor's field circuit. The rotor voltage induced by the relative motion between the rotor and the stator magnetic field produces a rotor current, and that rotor current interacts with the stator magnetic field to produce the induced torque in the motor.

In an induction motor, the slip or speed at which the maximum torque occurs can be controlled by varying the rotor resistance. The *value* of that maximum torque is independent of the rotor resistance. A high rotor resistance lowers the speed at which maximum torque occurs and thus increases the starting torque of the motor. However, it pays for this starting torque by having very poor speed regulation in its normal operating range. A low rotor resistance, on the other hand, reduces the motor's starting torque while improving its speed regulation. Any normal induction motor design must be a compromise between these two conflicting requirements.

One way to achieve such a compromise is to employ deep-bar or double-cage rotors. These rotors have a high effective resistance at starting and a low effective resistance under normal running conditions, thus yielding both a high starting torque and good speed regulation in the same motor. The same effect can be achieved with a wound-rotor induction motor if the rotor field resistance is varied.

Speed control of induction motors can be accomplished by changing the number of poles on the machine, by changing the applied electrical frequency, by changing the applied terminal voltage, or by changing the rotor resistance in the case of a wound-rotor induction motor.

7.12 | QUESTIONS

7–1. What are slip and slip speed in an induction motor?

7–2. How does an induction motor develop torque?

7–3. Why is it impossible for an induction motor to operate at synchronous speed?

7–4. Sketch and explain the shape of a typical induction motor torque–speed characteristic curve.

7–5. What equivalent circuit element has the most direct control over the speed at which the pullout torque occurs?

7–6. What is a deep-bar squirrel-cage rotor? Why is it used? What NEMA design class(es) can be built with it?

7–7. What is a double-cage squirrel-cage rotor? Why is it used? What NEMA design class(es) can be built with it?

7–8. Describe the characteristics and uses of wound-rotor induction motors and of each NEMA design class of squirrel-cage motors.

7–9. Why is the efficiency of an induction motor (wound rotor or squirrel cage) so poor at high slips?

7–10. Name and describe four means of controlling the speed of induction motors.

7–11. Why is it necessary to reduce the voltage applied to an induction motor as electrical frequency is reduced?

7–12. Why is terminal voltage speed control limited in operating range?

7–13. What are starting code letters? What do they say about the starting current of an induction motor?

7–14. What information is learned in a locked-rotor test?

7–15. What information is learned in a no-load test?

7–16. Two 480-V, 100-hp induction motors are manufactured. One is designed for 50-Hz operation, and one is designed for 60-Hz operation, but they are otherwise similar. Which of these machines is larger?

7–17. An induction motor is running at the rated conditions. If the shaft load is now increased, how do the following quantities change?
  a. Mechanical speed
  b. Slip
  c. Rotor induced voltage
  d. Rotor current
  e. Rotor frequency
  f. P_{RCL}
  g. Synchronous speed

7.13 | PROBLEMS

7–1. A DC test is performed on a 460-V, Δ-connected, 100-hp induction motor. If $V_{DC} = 21$ V and $I_{DC} = 72$ A, what is the stator resistance R_1? *Why is this so?*

7–2. A 220-V, three-phase, six-pole, 50-Hz induction motor is running at a slip of 3.5 percent. Find:
  a. The speed of the magnetic fields in revolutions per minute
  b. The speed of the rotor in revolutions per minute
  c. The slip speed of the rotor
  d. The rotor frequency in hertz

7–3. Answer the questions in Problem 7–2 for a 480-V, three-phase, four-pole, 60-Hz induction motor running at a slip of 0.025.

7–4. A three-phase, 60-Hz induction motor runs at 715 r/min at no load and at 670 r/min at full load.

 a. How many poles does this motor have?

 b. What is the slip at rated load?

 c. What is the speed at one-quarter of the rated load?

 d. What is the rotor's electrical frequency at one-quarter the rated load?

7–5. A 50-kW, 440-V, 50-Hz, two-pole induction motor has a slip of 6 percent when operating a full-load conditions. At full-load conditions, the friction and windage losses are 520 W, and the core losses are 500 W. Find the following values for full-load conditions:

 a. The shaft speed n_m

 b. The output power in watts

 c. The load torque τ_{load} in newton-meters

 d. The induced torque τ_{ind} in newton-meters

 e. The rotor frequency in hertz

7–6. A three-phase, 60-Hz, two-pole induction motor runs at a no-load speed of 3580 r/min and a full-load speed of 3440 r/min. Calculate the slip and the electrical frequency of the rotor at no-load and full-load conditions. What is the speed regulation of this motor [Equation (4–57)]?

7–7. A 208-V, four-pole, 60-Hz, Y-connected, wound-rotor induction motor is rated at 15 hp. Its equivalent circuit components are

$$R_1 = 0.220 \ \Omega \qquad R_2 = 0.127 \ \Omega \qquad X_M = 15.0 \ \Omega$$
$$X_1 = 0.430 \ \Omega \qquad X_2 = 0.430 \ \Omega$$
$$P_{mech} = 300 \ W \qquad P_{misc} \approx 0 \qquad P_{core} = 200 \ W$$

For a slip of 0.05, find

 a. The line current I_L

 b. The stator copper losses P_{SCL}

 c. The air-gap power P_{AG}

 d. The power converted from electrical to mechanical form P_{conv}

 e. The induced torque τ_{ind}

 f. The load torque τ_{load}

 g. The overall machine efficiency

 h. The motor speed in revolutions per minute and radians per second

7–8. For the motor in Problem 7–7, what is the slip at the pullout torque? What is the pullout torque of this motor?

7–9. (a) Calculate and plot the torque–speed characteristic of the motor in Problem 7–7. (b) Calculate and plot the output power versus speed curve of the motor in Problem 7–7.

7–10. For the motor of Problem 7–7, how much additional resistance (referred to the stator circuit) would it be necessary to add to the rotor circuit to make the maximum torque occur at starting conditions (when the shaft is not moving)? Plot the torque–speed characteristic of this motor with the additional resistance inserted.

7–11. If the motor in Problem 7–7 is to be operated on a 50-Hz power system, what must be done to its supply voltage? Why? What will the equivalent circuit component values be at 50 Hz? Answer the questions in Problem 7–7 for operation at 50 Hz with a slip of 0.05 and the proper voltage for this machine.

7–12. Figure 7–16a shows a simple circuit consisting of a voltage source, a resistor, and two reactances. Find the Thevenin equivalent voltage and impedance of this circuit at the terminals. Then derive the expressions for the magnitude of $\mathbf{V}_{TH}$ and for R_{TH} given in Equations (7–39) and (7–42).

7–13. Figure P7–1 shows a simple circuit consisting of a voltage source, two resistors, and two reactances in parallel with each other. If the resistor R_L is allowed to vary but all the other components are constant, at what value of R_L will the maximum possible power be supplied to it? *Prove* your answer. (*Hint:* Derive an expression for load power in terms of V, R_S, X_S, R_L and X_L and take the partial derivative of that expression with respect to R_L.) Use this result to derive the expression for the pullout torque [Equation (7–52)].

Figure P7–1 | Circuit for Problem 7–13.

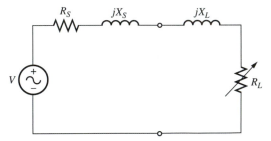

7–14. A 440-V, 50-Hz, six-pole, Y-connected induction motor is rated at 75 kW. The equivalent circuit parameters are

$$R_1 = 0.082 \ \Omega \qquad R_2 = 0.070 \ \Omega \qquad X_M = 7.2 \ \Omega$$
$$X_1 = 0.19 \ \Omega \qquad X_2 = 0.18 \ \Omega$$
$$P_{F\&W} = 1.3 \text{ kW} \qquad P_{misc} = 150 \text{ W} \qquad P_{core} = 1.4 \text{ kW}$$

For a slip of 0.04, find

a. The line current I_L

b. The stator power factor

c. The rotor power factor

d. The stator copper losses P_{SCL}

e. The air-gap power P_{AG}

 f. The power converted from electrical to mechanical form P_{conv}

 g. The induced torque τ_{ind}

 h. The load torque τ_{load}

 i. The overall machine efficiency η

 j. The motor speed in revolutions per minute and radians per second

7–15. For the motor in Problem 7–14, what is the pullout torque? What is the slip at the pullout torque? What is the rotor speed at the pullout torque?

7–16. If the motor in Problem 7–14 is to be driven from a 440-V, 60-Hz power supply, what will the pullout torque be? What will the slip be at pullout?

7–17. Plot the following quantities for the motor in Problem 7–14 as slip varies from 0% to 10%: (a) τ_{ind}; (b) P_{conv}; (c) P_{out}; (d) efficiency η. At what slip does P_{out} equal the rated power of the machine?

7–18. A 208-V, six-pole, Y-connected, 25-hp, design class B induction motor is tested in the laboratory, with the following results:

No load: 208 V, 22.0 A, 1200 W, 60 Hz

Locked rotor: 24.6 V, 64.5 A, 2200 W, 15 Hz

DC test: 13.5 V, 64 A

Find the equivalent circuit of this motor, and plot its torque–speed characteristic curve.

7–19. A 208-V, four-pole, 10-hp, 60-Hz, Y-connected, three-phase induction motor develops its full-load induced torque at 3.8 percent slip when operating at 60 Hz and 208 V. The per-phase circuit model impedances of the motor are

$$R_1 = 0.33\ \Omega \qquad X_M = 16\ \Omega$$
$$X_1 = 0.42\ \Omega \qquad X_2 = 0.42\ \Omega$$

Mechanical, core, and stray losses may be neglected in this problem.

 a. Find the value of the rotor resistance R_2.

 b. Find τ_{max}, s_{max}, and the rotor speed at maximum torque for this motor.

 c. Find the starting torque of this motor.

 d. What code letter factor should be assigned to this motor?

7–20. Answer the following questions about the motor in Problem 7–19.

 a. If this motor is started from a 208-V infinite bus, how much current will flow in the motor at starting?

 b. If transmission line with an impedance of $0.50 + j0.35\ \Omega$ per phase is used to connect the induction motor to the infinite bus, what will the starting current of the motor be? What will the motor's terminal voltage be on starting?

 c. If an ideal 1.2:1 step-down autotransformer is connected between the transmission line and the motor, what will the current be in the transmission line during starting? What will the voltage be at the motor end of the transmission line during starting?

7–21. In this chapter, we learned that a step-down autotransformer could be used to reduce the starting current drawn by an induction motor. While this technique works, an autotransformer is relatively expensive. A much less expensive way to reduce the starting current is to use a device called *Y–Δ starter*. If an induction motor is normally Δ connected, it is possible to reduce its phase voltage V_ϕ (and hence its starting current) by simply reconnecting the stator windings in Y during starting, and then restoring the connections to Δ when the motor comes up to speed. Answer the following questions about this type of starter.

    a.   How would the phase voltage at starting compare with the phase voltage under normal running conditions?

    b.   How would the starting current of the Y-connected motor compare to the starting current if the motor remained in a Δ-connection during starting?

7–22. A 460-V, 50-hp, six-pole, Δ-connected, 60-Hz, three-phase induction motor has a full-load slip of 4 percent, an efficiency of 91 percent, and a power factor of 0.87 lagging. At start-up, the motor develops 1.75 times the full-load torque but draws 7 times the rated current at the rated voltage. This motor is to be started with an autotransformer reduced-voltage starter.

    a.   What should the output voltage of the starter circuit be to reduce the starting torque until it equals the rated torque of the motor?

    b.   What will the motor starting current and the current drawn from the supply be at this voltage?

7–23. A wound-rotor induction motor is operating at rated voltage and frequency with its slip rings shorted and with a load of about 25 percent of the rated value for the machine. If the rotor resistance of this machine is doubled by inserting external resistors into the rotor circuit, explain what happens to the following:

    a.   Slip s

    b.   Motor speed n_m

    c.   The induced voltage in the rotor

    d.   The rotor current

    e. τ_{ind}

    f. P_{out}

    g. R_{RCL}

    h.   Overall efficiency η

7–24. Answer the following questions about a 460-V, Δ-connected, two-pole, 100-hp, 60-Hz, starting code letter F induction motor:

a. What is the maximum current starting current that this machine's controller must be designed to handle?

b. If the controller is designed to switch the stator windings from a Δ connection to a Y connection during starting, what is the maximum starting current that the controller must be designed to handle?

c. If a 1.25:1 step-down autotransformer starter is used during starting, what is the maximum starting current that it must be designed to handle?

7–25. When it is necessary to stop an induction motor very rapidly, many induction motor controllers reverse the direction of rotation of the magnetic fields by switching any two stator leads. When the direction of rotation of the magnetic fields is reversed, the motor develops an induced torque opposite to the current direction of rotation, so it quickly stops and tries to start turning in the opposite direction. If power is removed from the stator circuit at the moment when the rotor speed goes through zero, then the motor has been stopped very rapidly. This technique for rapidly stopping an induction motor is called *plugging*. The motor of Problem 7–19 is running at rated conditions and is to be stopped by plugging.

a. What is the slip s before plugging?

b. What is the frequency of the rotor before plugging?

c. What is the induced torque τ_{ind} before plugging?

d. What is the slip s immediately after switching the stator leads?

e. What is the frequency of the rotor immediately after switching the stator leads?

f. What is the induced torque τ_{ind} immediately after switching the stator leads?

7.14 | REFERENCES

1. Alger, Phillip: *Induction Machines*, 2nd ed., Gordon and Breach, New York, 1970.

2. Chapman, Stephen J.: *Electric Machinery Fundamentals*, 3rd ed., McGraw-Hill, Burr Ridge, Ill., 1999.

3. Del Toro, V.: *Electric Machines and Power Systems*, Prentice-Hall, Englewood Cliffs, N.J., 1985.

4. Fitzgerald, A. E., and C. Kingsley, Jr.: *Electric Machinery*, McGraw-Hill, New York, 1952.

5. Fitzgerald, A. E., C. Kingsley, Jr., and S. D. Umans: *Electric Machinery*, 5th ed., McGraw-Hill, New York. 1990.

6. Institute of Electrical and Electronics Engineers: *Standard Test Procedure for Polyphase Induction Motors and Generators*, IEEE Standard 112-1996, IEEE, New York, 1996.

7. Kosow, Irving L.: *Control of Electric Motors*, Prentice-Hall, Englewood Cliffs, N.J., 1972.

8. National Electrical Manufacturers Association: *Motors and Generators*, Publication No. MG1-1993, NEMA, Washington, 1993.

9. Slemon, G. R., and A. Straughen: *Electric Machines*, Addison-Wesley, Reading. Mass., 1980.

10. Vithayathil, Joseph: *Power Electronics: Principles and Applications*, McGraw-Hill, New York, 1995.

11. Werninck, E. H. (ed.): *Electric Motor Handbook*, McGraw-Hill, London, 1978.

DC Motors

The earliest power systems in the United States were DC systems, but by the 1890s AC power systems were clearly winning out over DC systems. Despite this fact, DC motors continued to be a significant fraction of the machinery purchased each year through the 1960s (that fraction has declined in the last 30 years). Why were DC motors so common, when DC power systems themselves were fairly rare?

There were several reasons for the continued popularity of DC motors. One was that DC power systems are still common in cars, trucks, and aircraft. When a vehicle has a DC power system, it makes sense to consider using DC motors. Another application for DC motors was a situation in which wide variations in speed are needed. Before the widespread use of power electronic rectifier-inverters, DC motors were unexcelled in speed control applications. Even if no DC power source were available, solid-state rectifier and chopper circuits were used to create the necessary DC power, and DC motors were used to provide the desired speed control. (Today, induction motors with solid-state drive packages are the preferred choice over DC motors for most speed control applications. However, there are still some applications where DC motors are preferred.)

Today, DC motors will usually be restricted to specialized uses on platforms such as automobiles where there is a DC power source. However, there are a *lot* of automobiles in the world, so huge quantities of DC motors are manufactured every year.

While DC motors are still quite common in certain applications, DC generators are extremely rare. They will not be covered in this book. For information about DC generators, see Reference 1.

Most DC machines are like AC machines in that they have AC voltages and currents within them—DC machines have a DC output only because a mechanism exists that converts the internal AC voltages to DC voltages at their terminals. Since this mechanism is called a commutator, DC machinery is also known as *commutating machinery*.

The fundamental principles involved in the operation of DC machines are very simple. Unfortunately, they are usually somewhat obscured by the complicated construction of real machines. This chapter will introduce some of the basic principles of DC motors, but will not get bogged down in the low-level details. See Reference 1 for a more detailed discussion of DC machinery.

8.1 | THE SIMPLEST POSSIBLE DC MACHINE: A SIMPLE ROTATING LOOP BETWEEN CURVED POLE FACES

The simplest possible rotating DC machine is shown in Figure 8–1. It consists of a single loop of wire rotating about a fixed axis. The rotating part of this machine is called the *rotor*, and the stationary part is called the *stator*. The magnetic field for the machine is supplied by the magnetic north and south poles shown on the stator in Figure 8–1.

Notice that the loop of rotor wire lies in a slot carved in a ferromagnetic core. The iron rotor, together with the curved shape of the pole faces, provides a constant-width air gap between the rotor and stator. Remember from Chapter 1 that the reluctance of air is very much higher than the reluctance of the iron in the machine. To minimize the reluctance of the flux path through the machine, the magnetic flux must take the shortest possible path through the air between the pole face and the rotor surface.

Since the magnetic flux must take the shortest path through the air, it is *perpendicular* to the rotor surface everywhere under the pole faces. Also, since the air gap is of uniform width, the reluctance is the same everywhere under the pole faces. The uniform reluctance means that the magnetic flux density is constant everywhere under the pole faces.

The Voltage Induced in a Rotating Loop

If the rotor of this machine is rotated, a voltage will be induced in the wire loop. To determine the magnitude and shape of the voltage, examine Figure 8–2. The loop of wire shown is rectangular, with sides *ab* and *cd* perpendicular to the plane of the page and with sides *bc* and *da* parallel to the plane of the page. The magnetic field is constant and perpendicular to the surface of the rotor everywhere under the pole faces and rapidly falls to zero beyond the edges of the poles.

To determine the total voltage e_{tot} on the loop, examine each segment of the loop separately and sum all the resulting voltages. The voltage on each segment is given by Equation (1–45):

$$e_{ind} = (\mathbf{v} \times \mathbf{B}) \cdot \mathbf{l} \tag{1–45}$$

1. *Segment ab.* In this segment, the velocity of the wire is tangential to the path of rotation. The magnetic field **B** points out perpendicular to the rotor surface everywhere under the pole face and is zero beyond the edges of the pole face.

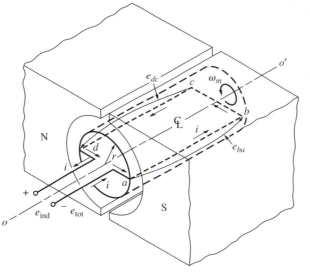

(a)

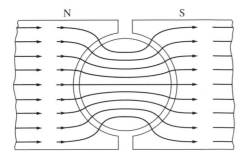

(b)

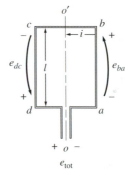

(c)

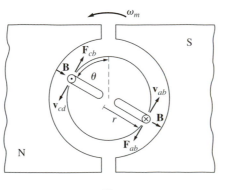

(d)

Figure 8–2 | Derivation of an equation for the voltages induced in the loop.

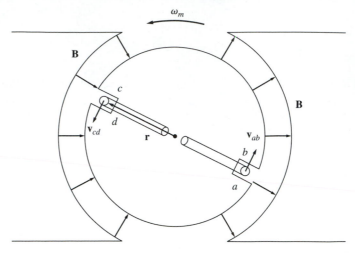

Under the pole face, velocity $\mathbf{v}$ is perpendicular to $\mathbf{B}$, and the quantity $\mathbf{v} \times \mathbf{B}$ points into the page. Therefore, the induced voltage on the segment is

$$e_{ba} = (\mathbf{v} \times \mathbf{B}) \cdot \mathbf{l}$$
$$= \begin{cases} vBl & \text{positive into page} & \text{under the pole face} \\ 0 & & \text{beyond the pole edges} \end{cases} \quad (8\text{–}1)$$

2. *Segment bc.* In this segment, the quantity $\mathbf{v} \times \mathbf{B}$ is either into or out of the page, while length $\mathbf{l}$ is in the plane of the page, so $\mathbf{v} \times \mathbf{B}$ is perpendicular to $\mathbf{l}$. Therefore the voltage in segment *bc* will be zero:

$$e_{cb} = 0 \quad (8\text{–}2)$$

3. *Segment cd.* In this segment, the velocity of the wire is tangential to the path of rotation. The magnetic field $\mathbf{B}$ points *in* perpendicular to the rotor surface everywhere under the pole face and is zero beyond the edges of the pole face. Under the pole face, velocity $\mathbf{v}$ is perpendicular to $\mathbf{B}$, and the quantity $\mathbf{v} \times \mathbf{B}$ points out of the page. Therefore, the induced voltage on the segment is

$$e_{dc} = (\mathbf{v} \times \mathbf{B}) \cdot \mathbf{l}$$
$$= \begin{cases} vBl & \text{positive out of page} & \text{under the pole face} \\ 0 & & \text{beyond the pole edges} \end{cases} \quad (8\text{–}3)$$

4. *Segment da.* Just as in segment be, $\mathbf{v} \times \mathbf{B}$ is perpendicular to $\mathbf{l}$. Therefore the voltage in this segment will be zero too:

$$e_{ad} = 0 \quad (8\text{–}4)$$

The total induced voltage on the loop e_{ind} is given by

$$e_{ind} = e_{ba} + e_{cb} + e_{dc} + e_{ad}$$

$$e_{ind} = \begin{cases} 2vBl & \text{under the pole faces} \\ 0 & \text{beyond the pole edges} \end{cases} \tag{8–5}$$

When the loop rotates through 180°, segment ab is under the north pole face instead of the south pole face. At that time, the direction of the voltage on the segment reverses, but its magnitude remains constant. The resulting voltage e_{tot} is shown as a function of time in Figure 8–3.

Figure 8–3 | The output voltage of the loop.

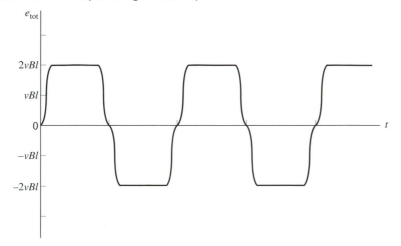

There is an alternative way to express Equation (8–5), which clearly relates the behavior of the single loop to the behavior of larger, real DC machines. To derive this alternative expression, examine Figure 8–4. Notice that the tangential velocity v of the edges of the loop can be expressed as

$$v = r\omega$$

where r is the radius from axis of rotation out to the edge of the loop and ω is the angular velocity of the loop. Substituting this expression into Equation (8–5) gives

$$e_{ind} = \begin{cases} 2r\omega Bl & \text{under the pole faces} \\ 0 & \text{beyond the pole edges} \end{cases}$$

$$e_{ind} = \begin{cases} 2rlB\omega & \text{under the pole faces} \\ 0 & \text{beyond the pole edges} \end{cases}$$

Figure 8–4 | Derivation of an alternative form of the induced voltage equation.

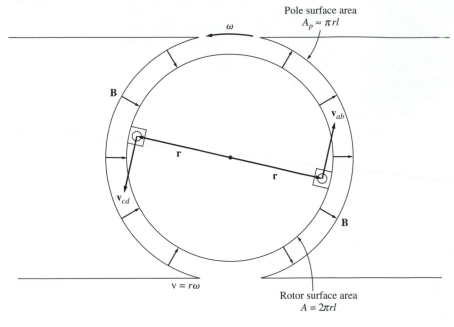

Notice also from Figure 8–4 that the rotor surface is a cylinder, so the area of the rotor surface A is just equal to $2\pi rl$. Since there are two poles, the area of the rotor *under each pole* (ignoring the small gaps between poles) is $A_P = \pi rl$. Therefore,

$$e_{\text{ind}} = \begin{cases} \dfrac{2}{\pi} A_P B \omega & \text{under the pole faces} \\ 0 & \text{beyond the pole edges} \end{cases}$$

Since the flux density B is constant everywhere in the air gap under the pole faces, the total flux under each pole is just the area of the pole times its flux density:

$$\phi = A_P B$$

Therefore, the final form of the voltage equation is

$$e_{\text{ind}} = \begin{cases} \dfrac{2}{\pi} \phi \omega & \text{under the pole faces} \\ 0 & \text{beyond the pole edges} \end{cases} \tag{8–6}$$

Thus, *the voltage generated in the machine is equal to the product of the flux inside the machine and the speed of rotation of the machine,* multiplied by a constant representing the mechanical construction of the machine. In general, the voltage in any real machine will depend on the same three factors:

1. The flux in the machine
2. The speed of rotation
3. A constant representing the construction of the machine

Getting DC Voltage out of the Rotating Loop

Figure 8–3 is a plot of the voltage e_{tot} generated by the rotating loop. As shown, the voltage out of the loop is alternately a constant positive value and a constant negative value. How can this machine be made to produce a DC voltage instead of the AC voltage it now has?

One way to do this is shown in Figure 8–5a. Here two semicircular conducting segments are added to the end of the loop, and two fixed contacts are set up at an angle such that, at the instant when the voltage in the loop is zero, the contacts short-circuit the two segments. In this fashion, *every time the voltage of the loop switches direction, the contacts also switch connections, and the output of the contacts is always built up in the same way* (Figure 8–5b). This connection-switching process is known as *commutation*. The rotating semicircular segments are called *commutator segments*, and the fixed contacts are called *brushes*.

The Induced Torque in the Rotating Loop

Suppose a battery is now connected to the machine in Figure 8–5. The resulting configuration is shown in Figure 8–6. How much torque will be produced in the loop when the switch is closed and a current is allowed to flow into it? To determine the torque, look at the close-up of the loop shown in Figure 8–6b.

The approach to take in determining the torque on the loop is to look at one segment of the loop at a time and then sum the effects of all the individual segments. The force on a segment of the loop is given by Equation (1–43):

$$\mathbf{F} = i(\mathbf{l} \times \mathbf{B}) \tag{1–43}$$

and the torque on the segment is given by

$$\tau = rF \sin \theta \tag{1–6}$$

where θ is the angle between $\mathbf{r}$ and $\mathbf{F}$. The torque is essentially zero whenever the loop is beyond the pole edges.

While the loop is under the pole faces, the torque is

1. *Segment ab.* In segment *ab*, the current from the battery is directed out of the page. The magnetic field under the pole face is pointing radially out of the rotor, so the force on the wire is given by

$$
\begin{aligned}
\mathbf{F}_{ab} &= i(\mathbf{l} \times \mathbf{B}) \\
&= ilB \qquad \text{tangent to direction of motion}
\end{aligned} \tag{8–7}
$$

The torque on the rotor caused by this force is

$$
\begin{aligned}
\tau_{ab} &= rF \sin \theta \\
&= r(ilB) \sin 90° \\
&= rilB \qquad \text{CCW}
\end{aligned} \tag{8–8}
$$

Figure 8–5 I Producing a DC output from the machine with a commutator and brushes. (a) Perspective view. (b) The resulting output voltage.

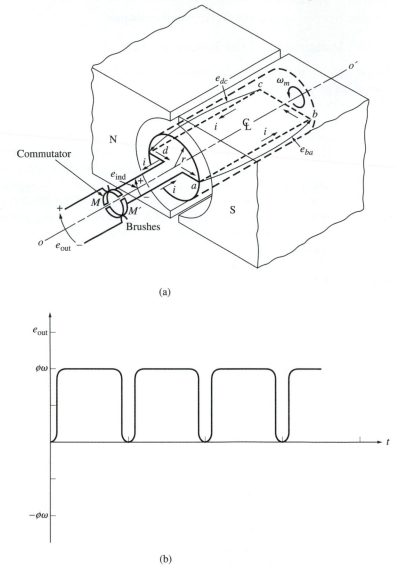

(a)

(b)

2. *Segment bc.* In segment *bc*, the current from the battery is flowing from the upper left to the lower right in the picture. The force induced on the wire is given by

$$\mathbf{F}_{bc} = i(\mathbf{l} \times \mathbf{B})$$
$$= 0 \qquad \text{since } \mathbf{l} \text{ is parallel to } \mathbf{B} \qquad (8\text{–}9)$$

Figure 8–6 | Derivation of an equation for the induced torque in the loop. Note that the iron core is not shown in part (b) for clarity.

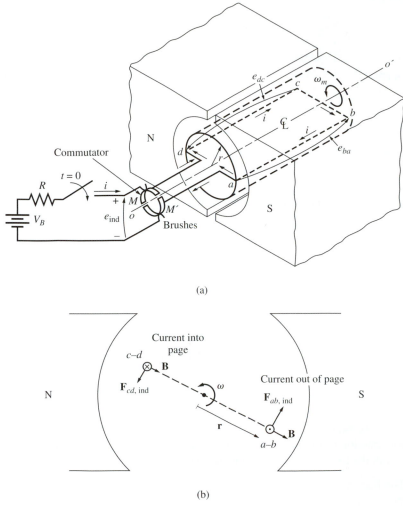

(a)

(b)

Therefore,

$$\tau_{bc} = 0 \qquad (8\text{--}10)$$

3. *Segment cd.* In segment *cd*, the current from the battery is directed into the page. The magnetic field under the pole face is pointing radially into the rotor, so the force on the wire is given by

$$\mathbf{F}_{cd} = i(\mathbf{l} \times \mathbf{B})$$
$$= ilB \qquad \text{tangent to direction of motion} \qquad (8\text{--}11)$$

The torque on the rotor caused by this force is

$$\tau_{cd} = rF \sin \theta$$
$$= r(ilB) \sin 90°$$
$$= rilB \qquad \text{CCW} \tag{8–12}$$

4. *Segment da.* In segment *da,* the current from the battery is flowing from the upper left to the lower right in the picture. The force induced on the wire is given by

$$\mathbf{F}_{da} = i(\mathbf{l} \times \mathbf{B})$$
$$= 0 \qquad \text{since } \mathbf{l} \text{ is parallel to } \mathbf{B} \tag{8–13}$$

Therefore,

$$\tau_{da} = 0 \tag{8–14}$$

The resulting total induced torque on the loop is given by

$$\tau_{\text{ind}} = \tau_{ab} + \tau_{bc} + \tau_{cd} + \tau_{da}$$

$$\tau_{\text{ind}} = \begin{cases} 2rilB & \text{under the pole faces} \\ 0 & \text{beyond the pole edges} \end{cases} \tag{8–15}$$

By using the facts that $A_P \approx \pi rl$ and $\phi = A_P B$, the torque expression reduces to

$$\tau_{\text{ind}} = \begin{cases} \dfrac{2}{\pi} \phi i & \text{under the pole faces} \\ 0 & \text{beyond the pole edges} \end{cases} \tag{8–16}$$

Thus, *the torque produced in the machine is the product of the flux in the machine and the current in the machine*, times some quantity representing the mechanical construction of the machine (the percentage of the rotor covered by pole faces). In general, the torque in *any* real machine will depend on the same three factors:

1. The flux in the machine
2. The current in the machine
3. A constant representing the construction of the machine

EXAMPLE 8–1

Figure 8–6 shows a simple rotating loop between curved pole faces connected to a battery and a resistor through a switch. The resistor shown models the total resistance of the battery and the wire in the machine. The physical dimensions and characteristics of this machine are

$$r = 0.5 \text{ m} \qquad l = 1.0 \text{ m}$$
$$R = 0.3 \text{ } \Omega \qquad B = 0.25 \text{ T}$$
$$V_B = 120 \text{ V}$$

a.   What happens when the switch is closed?

b.   What is the machine's maximum starting current? What is its steady-state angular velocity at no load?

c.   Suppose a load is attached to the loop, and the resulting load torque is 10 N·m. What would the new steady-state speed be? How much power is supplied to the shaft of the machine? How much power is being supplied by the battery? Is this machine a motor or a generator?

d.   Suppose the machine is again unloaded, and a torque of 7.5 N-m is applied to the shaft in the direction of rotation. What is the new steady-state speed? Is this machine now a motor or a generator?

e.   Suppose the machine is running unloaded. What would the final steady-state speed of the rotor be if the flux density were reduced to 0.20 T?

■ **Solution**

a.   When the switch in Figure 8–6 is closed, a current will flow in the loop. Since the loop is initially stationary, $e_{ind} = 0$. Therefore, the current will be given by

$$i = \frac{V_B - e_{ind}}{R} = \frac{V_B}{R}$$

This current flows through the rotor loop, producing a torque

$$\tau_{ind} = \frac{2}{\pi} \phi i \qquad \text{CCW}$$

This induced torque produces an angular acceleration in a counterclockwise direction, so the rotor of the machine begins to turn. But as the rotor begins to turn, an induced voltage is produced in the motor, given by

$$e_{ind} = \frac{2}{\pi} \phi \omega \uparrow$$

so the current i falls. As the current falls, $\tau_{ind} = (2/\pi)\phi i \downarrow$ decreases, the machine winds up in steady state with $\tau_{ind} = 0$, and the battery voltage $V_B = e_{ind}$.

b.   At *starting conditions,* the machine's current is

$$i = \frac{V_B}{R} = \frac{120 \text{ V}}{0.3 \text{ } \Omega} = 400 \text{ A}$$

At *no-load steady-state conditions,* the induced torque τ_{ind} must be zero. But $\tau_{ind} = 0$ implies that current i must equal zero, since $\tau_{ind} = (2/\pi)\phi i$, and the flux is nonzero. The fact that $i = 0$ A means that the battery voltage $V_B = e_{ind}$. Therefore, the speed of the rotor is

$$V_B = e_{ind} = \frac{2}{\pi}\phi\omega$$

$$\omega = \frac{V_B}{(2/\pi)\phi} = \frac{V_B}{2rlB}$$

$$\omega = \frac{120 \text{ V}}{2(0.5 \text{ m})(1.0 \text{ m})(0.25 \text{ T})} = 480 \text{ rad/s}$$

c.   If a load torque of 10 N-m is applied to the shaft of the machine, it will begin to slow down. But as ω decreases, $e_{ind} = (2/\pi)\phi\omega \downarrow$ decreases and the rotor current increases [$i = (V_B - e_{ind} \downarrow)/R$]. As the rotor current increases, $|\tau_{ind}|$ increases too, until $|\tau_{ind}| = |\tau_{load}|$ at a lower speed ω.

At steady state, $|\tau_{load}| = |\tau_{ind}| = (2/\pi)\phi i$. Therefore,

$$i = \frac{\tau_{ind}}{(2/\pi)\phi} = \frac{\tau_{ind}}{2rlB}$$

$$= \frac{10 \text{ N-m}}{(2)(0.5 \text{ m})(1.0 \text{ m})(0.25 \text{ T})} = 40 \text{ A}$$

By Kirchhoff's voltage law, $e_{ind} = V_B - iR$, so

$$e_{ind} = 120 \text{ V} - (40 \text{ A})(0.3 \text{ }\Omega) = 108 \text{ V}$$

Finally, the speed of the shaft is

$$\omega = \frac{e_{ind}}{(2/\pi)\phi} = \frac{e_{ind}}{2rlB}$$

$$= \frac{108 \text{ V}}{(2)(0.5 \text{ m})(1.0 \text{ m})(0.25 \text{ T})} = 432 \text{ rad/s}$$

The power supplied to the shaft is

$$P = \tau\omega$$
$$= (10 \text{ N-m})(432 \text{ rad/s}) = 4320 \text{ W}$$

The power out of the battery is

$$P = V_B i = (120 \text{ V})(40 \text{ A}) = 4800 \text{ W}$$

This machine is operating as a *motor,* converting electric power to mechanical power.

d.   If a torque is applied in the direction of motion, the rotor accelerates. As the speed to increases, the internal voltage e_{ind} increases and exceeds V_B, so the current flows out of the top of the bar and into the battery. This machine is now a *generator.* This current causes an induced torque opposite to the direction of motion. The induced torque opposes the external applied torque, and eventually $|\tau_{load}| = |\tau_{ind}|$ at a higher speed ω.

The current in the rotor will be

$$i = \frac{\tau_{ind}}{(2/\pi)\phi} = \frac{\tau_{ind}}{2rlB}$$

$$= \frac{7.5 \text{ N-m}}{(2)(0.5 \text{ m})(1.0 \text{ m})(0.25 \text{ T})} = 30 \text{ A}$$

The induced voltage e_{ind} is

$$e_{ind} = V_B + iR$$
$$= 120 \text{ V} + (30 \text{ A})(0.3 \text{ } \Omega)$$
$$= 129 \text{ V}$$

Finally, the speed of the shaft is

$$\omega = \frac{e_{ind}}{(2/\pi)\phi} = \frac{e_{ind}}{2rlB}$$

$$= \frac{129 \text{ V}}{(2)(0.5 \text{ m})(1.0 \text{ m})(0.25 \text{ T})} = 516 \text{ rad/s}$$

e.   Since the machine is initially unloaded at the original conditions, the speed ω = 480 rad/s. If the flux decreases, there is a transient. However, after the transient is over, the machine must again have zero torque, since there is still no load on its shaft. If τ_{ind} = 0, then the current in the rotor must be zero, and $V_B = e_{ind}$. The shaft speed is thus

$$\omega = \frac{e_{ind}}{(2/\pi)\phi} = \frac{e_{ind}}{2rlB}$$

$$= \frac{120 \text{ V}}{(2)(0.5 \text{ m})(1.0 \text{ m})(0.20 \text{ T})} = 600 \text{ rad/s}$$

Notice that when the flux in the machine is decreased, its speed increases. This is the same behavior seen in real DC motors. ∎

8.2 | COMMUTATION IN A SIMPLE FOUR-LOOP DC MACHINE

Commutation is the process of converting the AC voltages and currents in the rotor of a DC machine to DC voltages and currents at its terminals. It is the most critical part of the design and operation of any DC machine. A more detailed study is necessary to determine just how this conversion occurs and to discover the problems associated with it. In this section, the technique of commutation will be explained for a machine more complex than the single rotating loop in Section 8.1 but less complex than a real DC machine. Section 8.3 will continue this development and explain commutation in real DC machines.

A simple four-loop, two-pole DC machine is shown in Figure 8–7. This machine has four complete loops buried in slots carved in the laminated steel of its rotor. The

Figure 8–7 | (a) A four-loop two-pole DC machine shown at time $\omega t = 0°$.
(b) The voltages on the rotor conductors at this time. (c) A winding
diagram of this machine showing the interconnections of the rotor loops.

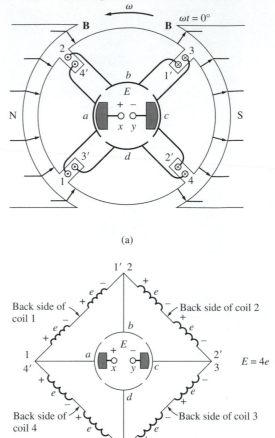

(a)

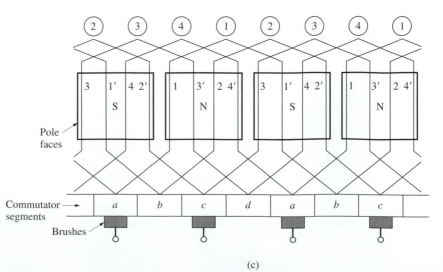

(b)

(c)

pole faces of the machine are curved to provide a uniform air-gap width and to give a uniform flux density everywhere under the faces.

The four loops of this machine are laid into the slots in a special manner. The "unprimed" end of each loop is the outermost wire in each slot, while the "primed" end of each loop is the innermost wire in the slot directly opposite. The winding's connections to the machine's commutator are shown in Figure 8–7b. Notice that loop 1 stretches between commutator segments a and b, loop 2 stretches between segments b and c, and so forth around the rotor.

At the instant shown in Figure 8–7, the 1, 2, 3', and 4' ends of the loops are under the north pole face, while the 1', 2', 3, and 4 ends of the loops are under the south pole face. The voltage in each of the 1, 2, 3', and 4' ends of the loops is given by

$$e_{ind} = (\mathbf{v} \times \mathbf{B}) \cdot \mathbf{l} \tag{1–45}$$
$$= vBl \qquad \text{positive out of the page} \tag{8–17}$$

The voltage in each of the 1', 2', 3, and 4 ends of the ends of the loops is given by

$$e_{ind} = (\mathbf{v} \times \mathbf{B}) \cdot \mathbf{l} \tag{1–45}$$
$$= vBl \qquad \text{positive into the page} \tag{8–18}$$

The overall result is shown in Figure 8–7b. In Figure 8–7b, each coil represents one side (or *conductor*) of a loop. If the induced voltage on any one side of a loop is called $e = vBl$, then the total voltage at the brushes of the machine is

$$\boxed{E = 4e \qquad \omega t = 0°} \tag{8–19}$$

Notice that there are two parallel paths for current through the machine. The existence of two or more parallel paths for rotor current is a common feature of all commutation schemes.

What happens to the voltage E of the terminals as the rotor continues to rotate? To find out, examine Figure 8–8. This figure shows the machine at time $\omega t = 45°$. At that time, loops 1 and 3 have rotated into the gap between the poles, so the voltage across each of them is zero. Notice that at this instant the brushes of the machine are shorting out commutator segments ab and cd. This happens just at the time when the loops between these segments have 0 V across them, so shorting out the segments creates no problem. At this time, only loops 2 and 4 are under the pole faces, so the terminal voltage E is given by

$$\boxed{E = 2e \qquad \omega t = 0°} \tag{8–20}$$

Now let the rotor continue to turn through another 45°. The resulting situation is shown in Figure 8–9. Here, the 1', 2, 3, and 4' ends of the loops are under the north pole face, and the 1, 2', 3', and 4 ends of the loops are under the south pole face. The voltages are still built up out of the page for the ends under the north pole face and into the page for the ends under the south pole face. The resulting voltage diagram is

Figure 8–8 | The same machine at time $\omega t = 45°$ showing the voltages on the conductors.

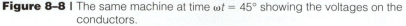

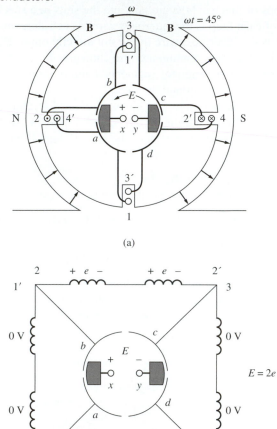

(a)

(b)

shown in Figure 8–8b. There are now four voltage-carrying ends in each parallel path through the machine, so the terminal voltage E is given by

$$E = 4e \quad \omega t = 90°$$ (8–21)

Compare Figure 8–7 to Figure 8–9. Notice that *the voltages on loops 1 and 3 have reversed between the two pictures, but since their connections have also reversed, the total voltage is still being built up in the same direction as before.* This fact is at the heart of every commutation scheme. Whenever the voltage reverses in a

Figure 8–9 | The same machine at time $\omega t = 90°$ showing the voltages on the conductors.

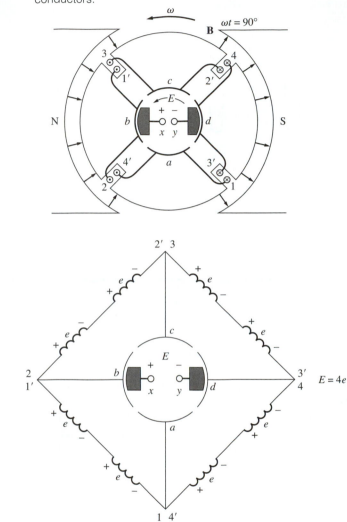

loop, the connections of the loop are also switched, and the total voltage is still built up in the original direction.

The terminal voltage of this machine as a function of time is shown in Figure 8–10. It is a better approximation to a constant DC level than the single rotating loop in Section 8–1 produced. As the number of loops on the rotor increases, the approximation to a perfect DC voltage continues to get better and better.

Figure 8–10 | The resulting output voltage of the machine in Figure 8–7.

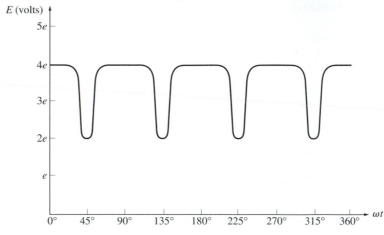

In summary,

Commutation is the process of switching the loop connections on the rotor of a DC machine just as the voltage in the loop switches polarity, in order to maintain an essentially constant DC output voltage.

As in the case of the simple rotating loop, the rotating segments to which the loops are attached are called *commutator segments*, and the stationary pieces that ride on top of the moving segments are called *brushes*. The commutator segments in real machines are typically made of copper bars. The brushes are made of a mixture containing graphite, so that they cause very little friction as they rub over the rotating commutator segments. Note that there is a small (and almost constant regardless of load) voltage drop across boundary between the brushes and the commutator segments.

Real DC machines have a construction similar to the four-loop example shown above, except that there are more loops and more commutation segments included on the rotor. For the details of the different possible connections of rotor loops, see Reference 1.

8.3 | PROBLEMS WITH COMMUTATION IN REAL MACHINES

The commutation process as described in Section 8.2 is not as simple in practice as it seems in theory, because two major effects occur in the real world to disturb it:

1. Armature reaction
2. *L di/dt* voltages

This section explores the nature of these problems and the solutions employed to mitigate their effects.

Armature Reaction

If the magnetic field windings of a DC machine are connected to a power supply and the rotor of the machine is turned by an external source of mechanical power, then a voltage will be induced in the conductors of the rotor. This voltage will be rectified into a DC output by the action of the machine's commutator.

Now connect a load to the terminals of the machine, and a current will flow in its armature windings. This current flow will produce a magnetic field of its own, which will distort the original magnetic field from the machine's poles. This distortion of the flux in a machine as the load is increased is called *armature reaction*. It causes two serious problems in real DC machines.

The first problem caused by armature reaction is *neutral-plane shift*. The *magnetic neutral plane* is defined as the plane within the machine where the velocity of the rotor wires is exactly parallel to the magnetic flux lines, so that e_{ind} in the conductors in the plane is exactly zero.

To understand the problem of neutral-plane shift, examine Figure 8–11. Figure 8–11a shows a two-pole DC machine. Notice that the flux is distributed uniformly under the pole faces. The rotor windings shown have voltages built up into the page for wires under the north pole face and out of the page for wires under the south pole face. The neutral plane in this machine is exactly vertical.

Now suppose a load is connected to this machine so that it acts as a generator. Current will flow out of the positive terminal of the generator, so current will be flowing into the page for wires under the north pole face and out of the page for wires under the south pole face. This current flow produces a magnetic field from the rotor windings, as shown in Figure 8–11c. This rotor magnetic field affects the original magnetic field from the poles that produced the generator's voltage in the first place. In some places under the pole surfaces, it subtracts from the pole flux, and in other places it adds to the pole flux. The overall result is that the magnetic flux in the air gap of the machine is skewed as shown in Figure 8–11d and e. Notice that the place on the rotor where the induced voltage in a conductor would be zero (the neutral plane) has shifted.

For the generator shown in Figure 8–11, the magnetic neutral plane shifted in the direction of rotation. If this machine had been a motor, the current in its rotor would be reversed and the flux would bunch up in the opposite corners from the bunches shown in the figure. As a result, the magnetic neutral plane would shift the other way.

In general, the neutral plane shifts in the direction of motion for a generator and opposite to the direction of motion for a motor. Furthermore, the amount of the shift depends on the amount of rotor current and hence on the load of the machine.

So what's the big deal about neutral-plane shift? It's just this: The commutator must short out commutator segments just at the moment when the voltage across them is equal to zero. If the brushes are set to short out conductors in the vertical plane, then the voltage between segments is indeed zero *until the machine is loaded*. When the machine is loaded, the neutral plane shifts, and the brushes short out

Figure 8–11 | The development of armature reaction in a DC generator. (a) Initially the pole flux is uniformly distributed, and the magnetic neutral plane is vertical. (b) The effect of the air gap on the pole flux distribution. (c) The armature magnetic field resulting when a load is connected to the machine. (d) Both rotor and pole fluxes are shown, indicating points where they add and subtract. (e) The resulting flux under the poles. The neutral plane has shifted in the direction of motion.

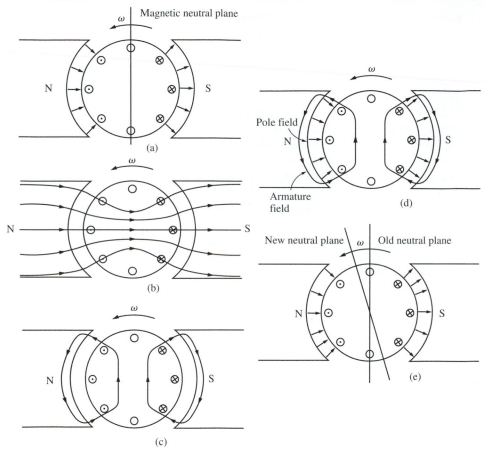

commutator segments with a finite voltage across them. The result is a current flow circulating between the shorted segments and large sparks at the brushes when the current path is interrupted as the brush leaves a segment. The end result is *arcing and sparking at the brushes*. This is a very serious problem, since it leads to drastically reduced brush life, pitting of the commutator segments, and greatly increased maintenance costs. Notice that this problem cannot be fixed even by placing the brushes over the full-load neutral plane, because then they would spark at no load.

In extreme cases, the neutral-plane shift can even lead to *flashover* in the commutator segments near the brushes. The air near the brushes in a machine is normally ionized as a result of the sparking on the brushes. Flashover occurs when the voltage of adjacent commutator segments gets large enough to sustain an arc in the ionized air above them. If flashover occurs, the resulting arc can even melt the commutator's surface.

The second major problem caused by armature reaction is called *flux weakening*. To understand flux weakening, refer to the magnetization curve shown in Figure 8–12. Most machines operate at flux densities near the saturation point. Therefore, at locations on the pole surfaces where the rotor magnetomotive force adds to the pole magnetomotive force, only a small increase in flux occurs. But at locations on the pole surfaces where the rotor magnetomotive force subtracts from the pole magneto-motive force, there is a larger decrease in flux. The net result is that *the total average flux under the entire pole face is decreased* (see Figure 8–13.)

Figure 8–12 | A typical magnetization curve shows the effects of pole saturation where armature and pole magnetomotive forces add.

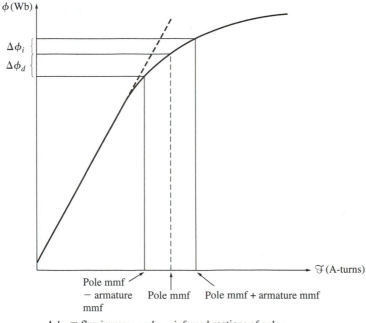

$\Delta\phi_i \equiv$ flux increase under reinforced sections of poles

$\Delta\phi_d \equiv$ flux decrease under subtracting sections of poles

Flux weakening causes problems in both generators and motors. In generators, the effect of flux weakening is simply to reduce the voltage supplied by the generator for any given load. In motors, the effect can be more serious. As Example 8–1 showed, when the flux in a motor is decreased, its speed increases. But increasing the speed of a motor can increase its load, resulting in more flux weakening. It is possible for some shunt DC motors to reach a runaway condition as a result of flux weakening, where the speed of the motor just keeps increasing until the machine is disconnected from the power line or until it destroys itself.

Figure 8–13 | The flux and magnetomotive force under the pole faces in a DC machine. At those points where the magnetomotive forces subtract, the flux closely follows the net magnetomotive force in the iron; but at those points where the magnetomotive forces add, saturation limits the total flux present. Note also that the neutral point of the rotor has shifted.

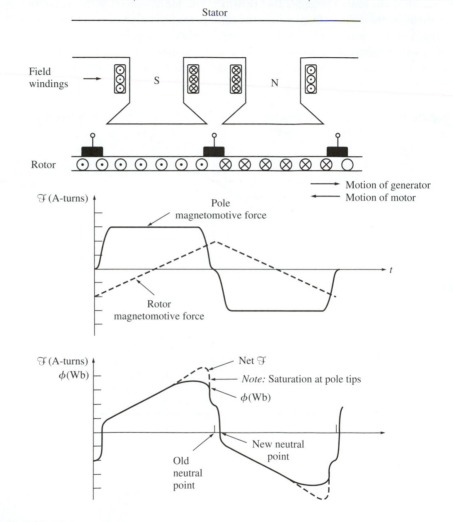

L di/dt Voltages

The second major problem is the *L di/dt* voltage that occurs in commutator segments being shorted out by the brushes, sometimes called *inductive kick*. To understand this problem, look at Figure 8–14. This figure represents a series of commutator segments and the conductors connected between them. Assuming that the current in the brush is 400 A, the current in each path is 200 A. Notice that when a commutator segment is shorted out, the current flow through that commutator segment must reverse. How fast must this reversal occur? Assuming that the machine is turning at 800 r/min and that there are 50 commutator segments (a reasonable number for a typical motor),

Figure 8–14 | (a) The reversal of current flow in a coil undergoing commutation. Note that the current in the coil between segments *a* and *b* must reverse direction while the brush shorts together the two commutator segments. (b) The current reversal in the coil undergoing commutation as a function of time for both ideal commutation and real commutation, with the coil inductance taken into account.

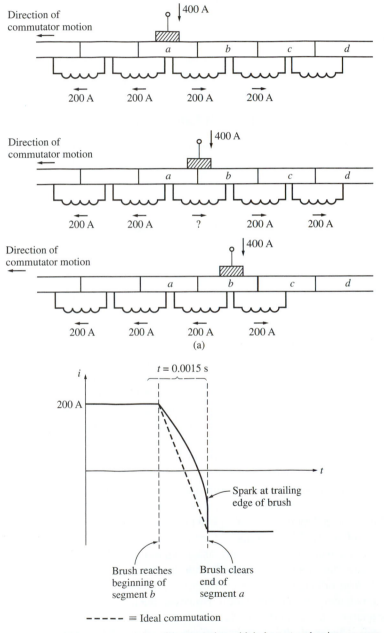

(a)

(b)

— — — — ≡ Ideal commutation

———————— ≡ Actual commutation with inductance taken into account

each commutator segments moves under a brush and clears it again in $t = 0.0015$ s. Therefore, the rate of change in current with respect to time in the shorted loop must *average*

$$\frac{di}{dt} = \frac{400 \text{ A}}{0.0015 \text{ s}} = 266,667 \text{ A/s} \tag{8–22}$$

With even a tiny inductance in the loop, a very significant inductive voltage kick $v = L\, di/dt$ will be induced in the shorted commutator segment. This high voltage naturally causes sparking at the brushes of the machine, resulting in the same arcing problems that the neutral-plane shift causes.

Solutions to the Problems with Commutation

Two approaches are currently used to partially or completely correct the problems of armature reaction and $L\, di/dt$ voltages:

1. Commutating poles or interpoles
2. Compensating windings

Each of these techniques is explained below, together with its advantages and disadvantages.

Commutating Poles or Interpoles   Because of the disadvantages noted above and especially because of the requirement that a person must adjust the brush positions of machines as their loads change, another solution to the problem of brush sparking was developed. The basic idea behind this new approach is that if the voltage in the wires undergoing commutation can be made zero, then there will be no sparking at the brushes. To accomplish this, small poles, called *commutating poles* or *interpoles*, are placed midway between the main poles. These commutating poles are located *directly over* the conductors being commutated. By providing a flux from the commutating poles, the voltage in the coils undergoing commutation can be exactly canceled. If the cancellation is exact, then there will be no sparking at the brushes.

The commutating poles do not otherwise change the operation of the machine, because they are so small that they affect only the few conductors about to undergo commutation. Notice that the *armature reaction* under the main pole faces is unaffected, since the effects of the commutating poles do not extend that far. This means that the flux weakening in the machine is unaffected by commutating poles.

How is cancellation of the voltage in the commutator segments accomplished for all values of loads? This is done by simply connecting the interpole windings in *series* with the windings on the rotor, as shown in Figure 8–15. As the load increases and the rotor current increases, the magnitude of the neutral-plane shift and the size of the $L\, di/dt$ effects increase too. Both these effects increase the voltage in the conductors undergoing commutation. However, the interpole flux increases too, producing a larger voltage in the conductors that opposes the voltage due to the neutral-plane shift. The net result is that their effects cancel over a broad range of loads. Note that interpoles work for both motor and generator operation, since when the machine changes from motor to generator, the current both in its rotor and in its interpoles reverses direction. Therefore, the voltage effects from them still cancel.

Figure 8–15 | A DC machine with interpoles.

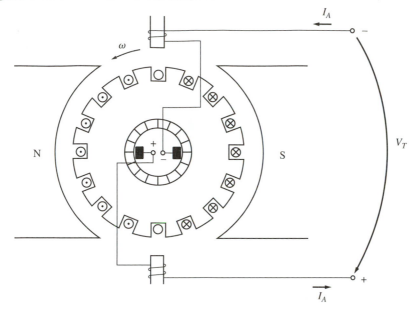

What polarity must the flux in the interpoles be? The interpoles must induce a voltage in the conductors undergoing commutation that is *opposite* to the voltage caused by neutral-plane shift and $L \, di/dt$ effects. In the case of a generator, the neutral plane shifts in the direction of rotation, meaning that the conductors undergoing commutation have the same polarity of voltage as the pole they just left (see Figure 8–16). To oppose this voltage, the interpoles must have the opposite flux, which is the flux of the upcoming pole. In a motor, however, the neutral plane shifts opposite to the direction of rotation, and the conductors undergoing commutation have the same flux as the pole they are approaching. In order to oppose this voltage, the interpoles must have the same polarity as the previous main pole. Therefore,

1. The interpoles must be of the same polarity as the next upcoming main pole in a generator.
2. The interpoles must be of the same polarity as the previous main pole in a motor.

The use of commutating poles or interpoles is very common, because they correct the sparking problems of DC machines at a fairly low cost. They are almost always found in any DC machine of 1 hp or larger. It is important to realize, though, that they do *nothing* for the flux distribution under the pole faces, so the flux-weakening problem is still present. Most medium-size, general-purpose motors correct for sparking problems with interpoles and just live with the flux-weakening effects.

Figure 8–16 | Determining the required polarity of an interpole. The flux from the interpole must produce a voltage that opposes the existing voltage in the conductor.

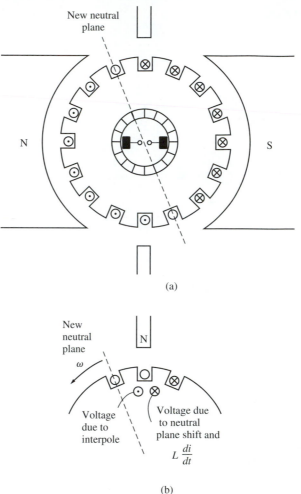

(a)

(b)

Compensating Windings For very heavy, severe duty cycle motors, the flux-weakening problem can be very serious. To completely cancel armature reaction and thus eliminate both neutral-plane shift and flux weakening, a different technique was developed. This third technique involves placing *compensating windings* in slots carved in the faces of the poles parallel to the rotor conductors, to cancel the distorting effect of armature reaction. These windings are connected in series with the rotor windings, so that whenever the load changes in the rotor, the current in the compensating windings changes, too. Figure 8–17 shows the basic concept. In Figure 8–17a, the pole flux is shown by itself. In Figure 8–17b, the rotor flux and the compensating winding flux are shown. Figure 8–17c represents the sum of these three fluxes, which is just equal to the original pole flux by itself.

Figure 8–17 | The effect of compensating windings in a DC machine. (a) The pole flux in the machine. (b) The fluxes from the armature and compensating windings. Notice that they are equal and opposite. (c) The net flux in the machine, which is just the original pole flux.

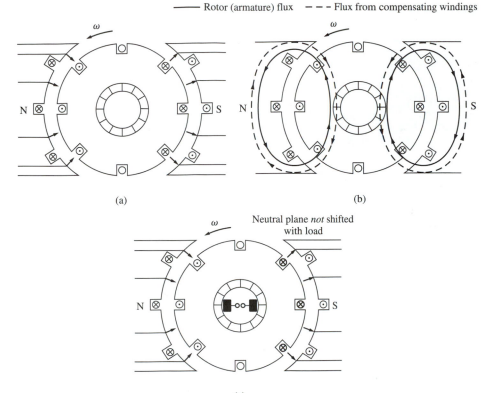

—— Rotor (armature) flux   – – – Flux from compensating windings

(a)   (b)

(c)

Figure 8–18 shows a more careful development of the effect of compensating windings on a DC machine. Notice that the magnetomotive force due to the compensating windings is equal and opposite to the magnetomotive force due to the rotor at every point under the pole faces. The resulting net magnetomotive force is just the magnetomotive force due to the poles, so the flux in the machine is unchanged regardless of the load on the machine. The stator of a large DC machine with compensating windings is shown in Figure 8–19.

The major disadvantage of compensating windings is that they are expensive, since they must be machined into the faces of the poles. Any motor that uses them must also have interpoles, since compensating windings do not cancel $L \, di/dt$ effects. The interpoles do not have to be as strong, though, since they are canceling only $L \, di/dt$ voltages in the windings, and not the voltages due to neutral-plane shifting. Because of the expense of having both compensating windings and interpoles on such a machine, these windings are used only where the extremely severe nature of a motor's duty demands them.

Figure 8–20 shows cutaway views of two real DC machines with compensating windings and interpoles.

Figure 8–18 | The flux and magnetomotive forces in a DC machine with compensating windings.

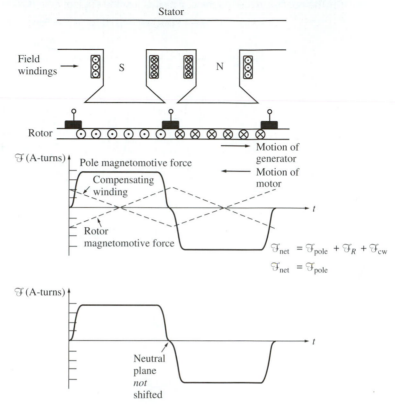

8.4 | POWER FLOW AND LOSSES IN DC MACHINES

DC motors take in electrical power and produce mechanical power. However, not all the power input to the machine appears in useful form at the other end—there is *always* some loss associated with the process.

The efficiency of a DC machine in percent is defined by the equation

$$\eta = \frac{P_{\text{out}}}{P_{\text{in}}} \times 100\% \qquad (8\text{–}23)$$

The difference between the input power and the output power of a machine is the losses that occur inside it. Therefore,

$$\eta = \frac{P_{\text{out}} - P_{\text{loss}}}{P_{\text{in}}} \times 100\% \qquad (8\text{–}24)$$

Figure 8–19 | The stator of a six-pole DC machine with interpoles and compensating windings. *(Courtesy of Westinghouse Electric Company.)*

The Losses in DC Machines

The losses that occur in DC machines can be divided into five basic categories:

1. Electrical or copper losses (I^2R losses)
2. Brush losses
3. Core losses
4. Mechanical losses
5. Stray load losses

Electrical or Copper Losses Copper losses are the losses that occur in the armature and field windings of the machine. The copper losses for the armature and field windings are given by

$$\text{Armature loss:} \quad P_A = I_A^2 R_A \qquad\qquad (8\text{–}25)$$
$$\text{Field loss:} \quad P_F = I_F^2 R_F \qquad\qquad (8\text{–}26)$$

where

P_A = armature loss
P_F = field circuit loss
I_A = armature current
I_F = field current
R_A = armature resistance
R_F = field resistance

Figure 8–20 I (a) A cutaway view of a 4000-hp, 700-V, 18-pole DC machine showing compensating windings, interpoles, equalizer, and commutator. *(Courtesy of General Electric Company.)* (b) A cutaway view of a smaller four-pole DC motor including interpoles but without compensating windings. *(Courtesy of MagneTek, Inc.)*

(a)

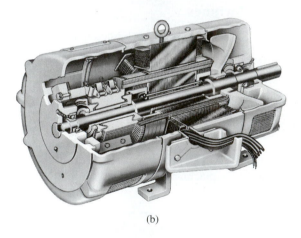

(b)

The resistance used in these calculations is usually the winding resistance at normal operating temperature.

Brush Losses   The brush drop loss is the power lost across the contact potential at the brushes of the machine. It is given by the equation

$$P_{BD} = V_{BD} I_A$$

(8–27)

where

P_{BD} = brush drop loss

V_{BD} = brush voltage drop

I_A = armature current

The reason that the brush losses are calculated in this manner is that the voltage drop across a set of brushes is approximately constant over a large range of armature currents. Unless otherwise specified, the brush voltage drop is usually assumed to be about 2 V.

Core Losses The core losses are the hysteresis losses and eddy current losses occurring in the metal of the motor. These losses are described in Chapter 1. These losses vary as the square of the flux density (B^2) and, for the rotor, as the 1.5th power of the speed of rotation ($n^{1.5}$).

Mechanical Losses The mechanical losses in a DC machine are the losses associated with mechanical effects. There are two basic types of mechanical losses: *friction* and *windage*. Friction losses are losses caused by the friction of the bearings in the machine, while windage losses are caused by the friction between the moving parts of the machine and the air inside the motor's casing. These losses vary as the cube of the speed of rotation of the machine.

Stray Losses (or Miscellaneous Losses) Stray losses are losses that cannot be placed in one of the previous categories. No matter how carefully losses are accounted for, some always escape inclusion in one of the above categories. All such losses are lumped into stray losses. For most machines, stray losses are taken by convention to be 1 percent of full load.

The Power-Flow Diagram

One of the most convenient techniques for accounting for power losses in a machine is the *power-flow diagram*. A power-flow diagram for a DC motor is shown in Figure 8–21. In this figure, electrical power is input to the machine, and the electrical I^2R losses and the brush losses must be subtracted. The remaining power is ideally converted from electrical to mechanical for at the point labeled P_{conv}. The electrical power that is converted is given by

$$P_{conv} = E_A I_A \tag{8–28}$$

and the resulting mechanical power is given by

$$P_{conv} = \tau_{ind}\omega_m \tag{8–29}$$

After the power is converted to mechanical form, the stray losses, mechanical losses, and core loses are subtracted, and the remaining mechanical power is output to the loads.

Figure 8–21 I Power-flow diagram for DC motor.

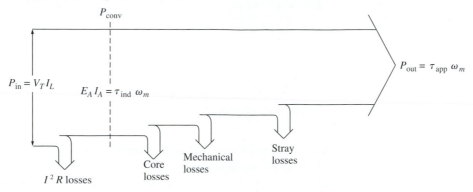

8.5 I TYPES OF DC MOTORS

There are five major types of DC motors in general use:

1. The separately excited DC motor
2. The shunt DC motor
3. The permanent-magnet DC motor
4. The series DC motor
5. The compounded DC motor

Each of these types will be examined in turn.

DC motors are, of course, driven from a DC power supply. Unless otherwise specified, *the input voltage to a DC motor is assumed to be constant*, because that assumption simplifies the analysis of motors and the comparison between different types of motors.

8.6 I THE EQUIVALENT CIRCUIT OF A DC MOTOR

The equivalent circuit of a DC motor is shown in Figure 8–23. In this figure, the armature circuit is represented by an ideal voltage source E_A and a resistor R_A. This representation is really the Thevenin equivalent of the entire rotor structure, including rotor coils, interpoles, and compensating windings, if present. The brush voltage drop is represented by a small battery V_{brush} opposing the direction of current flow in the machine. The field coils, which produce the magnetic flux in the generator, are represented by inductor L_F and resistor R_F. The separate resistor R_{adj} represents an external variable resistor used to control the amount of current in the field circuit.

There are a few variations and simplifications of this basic equivalent circuit. The brush drop voltage is often only a very tiny fraction of the generated voltage in a machine. Therefore, in cases where it is not too critical, the brush drop voltage may be left out or approximately included in the value of R_A. Also, the internal resistance

Figure 8–22 | Early DC motors. (a) A very early DC motor built by Elihu Thompson in 1886. It was rated at about 1/2 hp. *(Courtesy of General Electric Company.)* (b) A larger four-pole DC motor from about the turn of the century. *(Courtesy of General Electric Company.)*

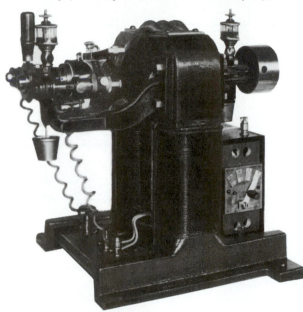

(a)

(b)

of the field coils is sometimes lumped together with the variable resistor, and the total is called R_F (see Figure 8–23b). A third variation is that some DC motors have more than one field coil, all of which will appear on the equivalent circuit.

Figure 8–23 | (a) The equivalent circuit of a DC motor. (b) A simplified equivalent circuit eliminating the brush voltage drop and combining R_{adj} with the field resistance.

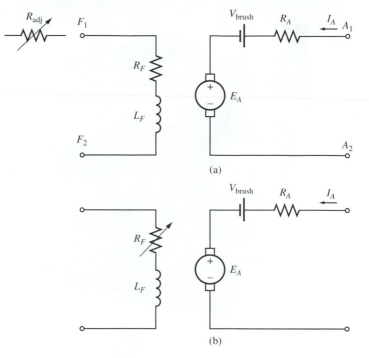

(a)

(b)

The internal generated voltage in this machine is given by the equation

$$E_A = K\phi\omega \qquad (8\text{–}30)$$

and the induced torque developed by the machine is given by

$$\tau_{ind} = K\phi I_A \qquad (8\text{–}31)$$

where K is a constant depending on the construction of a particular DC machine (the number of rotor coils, how they are interconnected, etc.). Note that these equations are similar to Equations (8–6) and (8–16) respectively for the single rotating loop. For the single rotating loop, the constant K is $2/\pi$.

These two equations, the Kirchhoff's voltage law equation of the armature circuit and the machine's magnetization curve, are all the tools necessary to analyze the behavior and performance of a DC motor.

8.7 | THE MAGNETIZATION CURVE OF A DC MACHINE

The internal generated voltage E_A of a DC motor or generator is given by Equation (8–30):

$$E_A = K\phi\omega \qquad\qquad (8\text{–}30)$$

Therefore, E_A is directly proportional to the flux in the machine and the speed of rotation of the machine. How is the internal generated voltage related to the field current in the machine?

The field current in a DC machine produces a field magnetomotive force given by $\mathcal{F} = N_F I_F$. This magnetomotive force produces a flux in the machine in accordance with its magnetization curve (Figure 8–24). Since the field current is directly proportional to the magnetomotive force and since E_A is directly proportional to the flux, it is customary to present the magnetization curve as a plot of E_A versus field current for a given speed ω_o (Figure 8–25).

Figure 8–24 | The magnetization curve of a ferromagnetic material (ϕ versus $\mathcal{F}$).

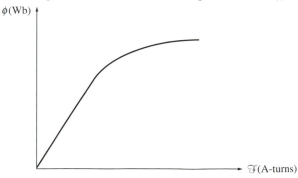

Figure 8–25 | The magnetization curve of a DC machine expressed as a plot of E_A versus I_F, for a fixed speed ω_o.

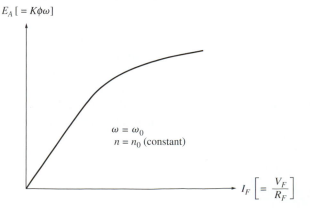

It is worth noting here that, to get the maximum possible power per pound of weight out of a machine, most motors and generators are designed to operate near the saturation point on the magnetization curve (at the knee of the curve). This implies that a fairly large increase in current is often necessary to get a small increase in E_A when operation is near full load.

The magnetization curves used in this book are also available in electronic form to simplify the solution of problems by MATLAB. Each magnetization curve is stored in a separate MAT file. Each MAT file contains three variables: `if_values`, containing the values of the field current; `ea_values`, containing the corresponding values of E_A; and `n_0`, containing the speed at which the magnetization curve was measured in units of revolutions per minute.

8.8 | SEPARATELY EXCITED AND SHUNT DC MOTORS

The equivalent circuit of a separately excited DC motor is shown in Figure 8–26a, and the equivalent circuit of a shunt DC motor is shown in Figure 8–26b. A separately excited DC motor is a motor whose field circuit is supplied from a separate constant-voltage power supply, while a shunt DC motor is a motor whose field circuit gets its power directly across the armature terminals of the motor. When the supply voltage to a motor is assumed constant, there is no practical difference in behavior between these two machines. Unless otherwise specified, whenever the behavior of a shunt motor is described, the separately excited motor is included, too.

The Kirchhoff's voltage law (KVL) equation for the armature circuit of these motors is

$$\boxed{V_T = E_A + I_A R_A} \qquad (8\text{–}32)$$

The Terminal Characteristic of a Shunt DC Motor

A terminal characteristic of a machine is a plot of the machine's output quantities versus each other. For a motor, the output quantities are shaft torque and speed, so the terminal characteristic of a motor is a plot of its output *torque versus speed*.

How does a shunt DC motor respond to a load? Suppose that the load on the shaft of a shunt motor is increased. Then the load torque τ_{load} will exceed the induced torque τ_{ind} in the machine, and the motor will start to slow down. When the motor slows down, its internal generated voltage drops ($E_A = K\phi\omega \downarrow$), so the armature current in the motor $I_A = (V_T - E_A \downarrow)/R_A$ increases. As the armature current rises, the induced torque in the motor increases ($\tau_{\text{ind}} = K\phi I_A \uparrow$), and finally the induced torque will equal the load torque at a lower mechanical speed of rotation ω.

The output characteristic of a shunt DC motor can be derived from the induced voltage and torque equations of the motor plus Kirchhoff's voltage law. The KVL equation for a shunt motor is

$$V_T = E_A + I_A R_A \qquad (8\text{–}32)$$

Figure 8–26 | (a) The equivalent circuit of a separately excited DC motor.
(b) The equivalent circuit of a shunt DC motor.

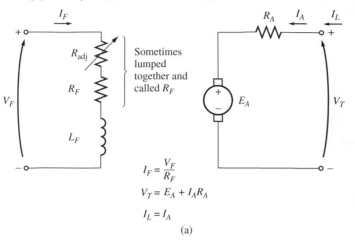

$$I_F = \frac{V_F}{R_F}$$

$$V_T = E_A + I_A R_A$$

$$I_L = I_A$$

(a)

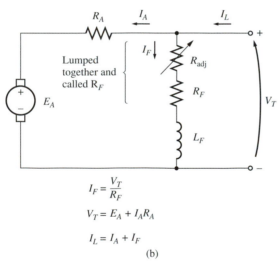

$$I_F = \frac{V_T}{R_F}$$

$$V_T = E_A + I_A R_A$$

$$I_L = I_A + I_F$$

(b)

The induced voltage $E_A = K\phi\omega$, so

$$V_T = K\phi\omega + I_A R_A \tag{8–33}$$

Since $\tau_{ind} = K\phi I_A$, current I_A can be expressed as

$$I_A = \frac{\tau_{ind}}{K\phi} \tag{8–34}$$

Combining Equations (8–33) and (8–34) produces

$$V_T = K\phi\omega + \frac{\tau_{ind}}{K\phi} R_A \tag{8–35}$$

Finally, solving for the motor's speed yields

$$\omega = \frac{V_T}{K\phi} - \frac{R_A}{(K\phi)^2}\tau_{\text{ind}}$$ (8–36)

This equation is just a straight line with a negative slope. The resulting torque–speed characteristic of a shunt DC motor is shown in Figure 8–27a.

Figure 8–27 | (a) Torque–speed characteristic of a shunt or separately excited DC motor with compensating windings to eliminate armature reaction. (b) Torque–speed characteristic of the motor with armature reaction present.

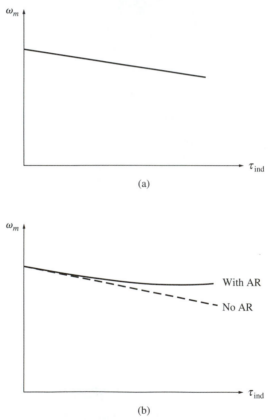

(a)

(b)

It is important to realize that, in order for the speed of the motor to vary linearly with torque, the other terms in this expression must be constant as the load changes. The terminal voltage supplied by the DC power source is assumed to be constant—if it is not constant, then the voltage variations will affect the shape of the torque–speed curve.

Another effect *internal to the motor* that can affect the shape of the torque–speed curve is armature reaction. If a motor has armature reaction, then as its load increases, the flux-weakening effects *reduce* its flux. As Equation (8–36) shows, the effect of a reduction in flux is to increase the motor's speed at any given load over the speed it would run at without armature reaction. The torque–speed characteristic of a shunt motor with armature reaction is shown in Figure 8–27b. If a motor has compensating windings, of course there will be no flux-weakening problems in the machine, and the flux in the machine will be constant.

If a shunt DC motor has compensating windings so that *its flux is constant regardless of load*, and the motor's speed and armature current are known at any one value of load, then it is possible to calculate its speed at any other value of load, as long as the armature current at that load is known or can be determined. Example 8–2 illustrates this calculation.

EXAMPLE 8–2

A 50-hp, 250-V, 1200 r/min, DC shunt motor with compensating windings has an armature resistance (including the brushes, compensating windings, and interpoles) of 0.06 Ω. Its field circuit has a total resistance $R_{adj} + R_F$ of 50 Ω, which produces a *no-load* speed of 1200 r/min. There are 1200 turns per pole on the shunt field winding (see Figure 8–28).

Figure 8–28 | The shunt motor in Example 8–2.

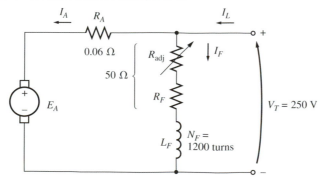

a. Find the speed of this motor when its input current is 100 A.
b. Find the speed of this motor when its input current is 200 A.
c. Find the speed of this motor when its input current is 300 A.
d. Plot the torque–speed characteristic of this motor.

■ **Solution**

The internal generated voltage of a DC machine with its speed expressed in revolutions per minute is given by

$$E_A = K\phi\omega \qquad\qquad (8\text{–}30)$$

Since the field current in the machine is constant (because V_T and the field resistance are both constant), and since there are no armature reaction effects, *the flux in this motor is constant*. The relationship between the speeds and internal generated voltages of the motor at two different load conditions is thus

$$\frac{E_{A2}}{E_{A1}} = \frac{K\phi\omega_2}{K\phi\omega_1} = \frac{n_2}{n_1} \tag{8-37}$$

The constant K cancels, since it is a constant for any given machine, and the flux ϕ cancels as described above. Therefore,

$$n_2 = \frac{E_{A2}}{E_{A1}} n_1 \tag{8-38}$$

At no load, the armature current is zero, so $E_{A1} = V_T = 250$ V, while the speed $n_1 = 1200$ r/min. If we can calculate the internal generated voltage at any other load, it will be possible to determine the motor speed at that load from Equation (8-38).

a. If $I_L = 100$ A, then the armature current in the motor is

$$I_A = I_L - I_F = I_L - \frac{V_T}{R_F}$$

$$= 100 \text{ A} - \frac{250 \text{ V}}{50 \ \Omega} = 95 \text{ A}$$

Therefore, E_A at this load will be

$$E_A = V_T - I_A R_A$$
$$= 250 \text{ V} - (95 \text{ A})(0.06 \ \Omega) = 244.3 \text{ V}$$

The resulting speed of the motor is

$$n_2 = \frac{E_{A2}}{E_{A1}} n_1 = \frac{244.3 \text{ V}}{250 \text{ V}} \ 1200 \text{ r/min} = 1173 \text{ r/min}$$

b. If $I_L = 200$ A, then the armature current in the motor is

$$I_A = 200 \text{ A} - \frac{250 \text{ V}}{50 \ \Omega} = 195 \text{ A}$$

Therefore, E_A at this load will be

$$E_A = V_T - I_A R_A$$
$$= 250 \text{ V} - (195 \text{ A})(0.06 \ \Omega) = 238.3 \text{ V}$$

The resulting speed of the motor is

$$n_2 = \frac{E_{A2}}{E_{A1}} n_1 = \frac{238.3 \text{ V}}{250 \text{ V}} \ 1200 \text{ r/min} = 1144 \text{ r/min}$$

c. If I_L, = 300 A, then the armature current in the motor is

$$I_A = I_L - I_F = I_L - \frac{V_T}{R_F}$$

$$= 300 \text{ A} - \frac{250 \text{ V}}{50 \text{ }\Omega} = 295 \text{ A}$$

Therefore, E_A at this load will be

$$E_A = V_T - I_A R_A$$
$$= 250 \text{ V} - (295 \text{ A})(0.06 \text{ }\Omega) = 232.3 \text{ V}$$

The resulting speed of the motor is

$$n_2 = \frac{E_{A2}}{E_{A1}} n_1 = \frac{232.3 \text{ V}}{250 \text{ V}} 1200 \text{ r/min} = 1115 \text{ r/min}$$

d. To plot the output characteristic of this motor, it is necessary to find the torque corresponding to each value of speed. At no load, the induced torque τ_{ind} is clearly zero. The induced torque for any other load can be found from the fact that power converted in a DC motor is

$$\boxed{P_{conv} = E_A I_A = \tau_{ind}\omega} \qquad \text{(8–28, 8–29)}$$

From this equation, the induced torque in a motor is

$$\tau_{ind} = \frac{E_A I_A}{\omega} \qquad \text{(8–39)}$$

Therefore, the induced torque when I_L = 100 A is

$$\tau_{ind} = \frac{(2443 \text{ V})(95 \text{ A})}{(1173 \text{ r/min})(1 \text{ min/60 s})(2\pi \text{ rad/r})} = 190 \text{ N-m}$$

The induced torque when I_L = 200 A is

$$\tau_{ind} = \frac{(2383 \text{ V})(195 \text{ A})}{(1144 \text{ r/min})(1 \text{ min/60})(2\pi \text{ rad/r})} = 388 \text{ N-m}$$

The induced torque when I_L = 300 A is

$$\tau_{ind} = \frac{(2323 \text{ V})(295 \text{ A})}{(1115 \text{ r/min})(1 \text{ min/60 s})(2\pi \text{ rad/r})} = 587 \text{ N-m}$$

The resulting torque-speed characteristic for this motor is plotted in Figure 8–29. ∎

Figure 8–29 I The torque–speed characteristic of the motor in Example 8–2.

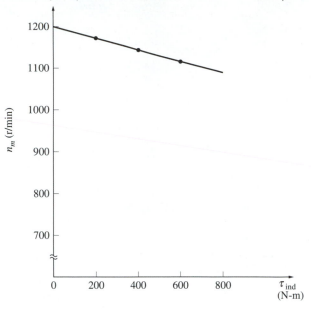

Nonlinear Analysis of a Shunt DC Motor

The flux ϕ and hence the internal generated voltage E_A of a DC machine are *non-linear* functions of its magnetomotive force. Therefore, anything that changes the magnetomotive force in a machine will have a nonlinear effect on the internal generated voltage of the machine. Since the change in E_A cannot be calculated analytically, the magnetization curve of the machine must be used to accurately determine its E_A for a given magnetomotive force. The two principal contributors to the magnetomotive force in the machine are its field current and its armature reaction, if present.

Since the magnetization curve is a direct plot of E_A versus I_F for a given speed ω_o, the effect of changing a machine's field current can be determined directly from its magnetization curve.

If a machine has armature reaction, its flux will be reduced with each increase in load. The total magnetomotive force in a shunt DC motor is the field circuit magnetomotive force less the magnetomotive force due to armature reaction (AR):

$$\mathcal{F}_{net} = N_F I_F - \mathcal{F}_{AR} \qquad (8\text{–}40)$$

Since magnetization curves are expressed as plots of E_A versus field current, it is customary to define an *equivalent field current* that would produce the same output voltage as the combination of all the magnetomotive forces in the machine. The resulting voltage E_A can then be determined by locating that equivalent field current on the magnetization curve. The equivalent field current of a shunt DC motor is given by

$$I_F^* = I_F - \frac{\mathcal{F}_{AR}}{N_F} \qquad (8\text{–}41)$$

One other effect must be considered when nonlinear analysis is used to determine the internal generated voltage of a DC motor. The magnetization curves for a machine are drawn for a particular speed, usually the rated speed of the machine. How can the effects of a given field current be determined if the motor is turning at other than rated speed?

The equation for the induced voltage in a DC machine is

$$E_A = K\phi\omega \qquad (8\text{-}30)$$

For a given effective field current, the flux in a machine is fixed, so the internal generated voltage is directly proportional to speed, and the internal generated voltage at any given speed can be related to the internal generated voltage at rated speed to speed by

$$\frac{E_A}{E_{A0}} = \frac{n}{n_0} \qquad (8\text{-}42)$$

where E_{A0} and n_0 represent the reference values of voltage and speed, respectively. If the reference conditions are known from the magnetization curve and the actual E_A is known from Kirchhoff's voltage law, then is possible to determine the actual speed n from Equation (8–42). The use of the magnetization curve and Equation (8–42) is illustrated in the following example, which analyzes a DC motor with armature reaction.

EXAMPLE 8–3

A 50-hp, 250-V, 1200 r/min, DC shunt motor *without* compensating windings has an armature resistance (including the brushes and interpoles) of 0.06 Ω. Its field circuit has a total resistance $R_F + R_{adj}$ of 50 Ω, which produces a no-load speed of 1200 r/min. There are 1200 turns per pole on the shunt field winding, and the armature reaction produces a demagnetizing magnetomotive force of 840 A-turns at a load current of 200 A. The magnetization curve of this machine is shown in Figure 8–30.

a. Find the speed of this motor when its input current is 200 A.

b. This motor is essentially identical to the one in Example 8–2 except for the absence of compensating windings. How does its speed compare to that of the previous motor at a load current of 200 A?

c. Calculate and plot the torque–speed characteristic for this motor.

■ Solution

a. If I_L = 200 A, then the armature current of the motor is

$$I_A = I_L - I_F = I_L - \frac{V_T}{R_F}$$

$$= 200 \text{ A} - \frac{250 \text{ V}}{50 \text{ Ω}} = 195 \text{ A}$$

Figure 8–30 | The magnetization curve of a typical 250-V DC motor, taken at a speed of 1200 r/min.

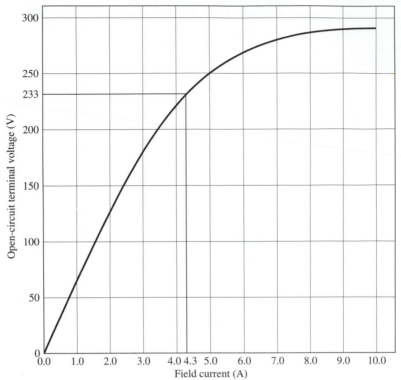

Therefore, the internal generated voltage of the machine is

$$E_A = V_T - I_A R_A$$
$$= 250 \text{ V} - (195 \text{ A})(0.06 \ \Omega) = 238.3 \text{ V}$$

At $I_L = 200$ A, the demagnetizing magnetomotive force due to armature reaction is 840 A-turns, so the effective shunt field current of the motor is

$$I_F^* = I_F - \frac{\mathcal{F}_{AR}}{N_F} \qquad (8\text{–}41)$$

$$= 5.0 \text{ A} - \frac{840 \text{ A-turns}}{1200 \text{ turns}} = 4.3 \text{ A}$$

From the magnetization curve, this effective field current would produce an internal generated voltage E_{A0} of 233 V at a speed n_0 of 1200 r/min.

We know that the internal generated voltage E_{A0} would be 233 V at a speed of 1200 r/min. Since the actual internal generated voltage E_A is 238.3 V, the actual operating speed of the motor must be

$$\frac{E_A}{E_{A0}} = \frac{n}{n_0} \qquad\qquad (8\text{--}42)$$

$$n = \frac{E_A}{E_{A0}} n_0 = \frac{2383 \text{ V}}{233 \text{ V}} (1200 \text{ r/min}) = 1227 \text{ r/min}$$

b. At 200 A of load in Example 8–2, the motor's speed was $n = 1144$ r/min. In this example, the motor's speed is 1227 r/min. *Notice that the speed of the motor with armature reaction is higher than the speed of the motor with no armature reaction.* This relative increase in speed is due to the flux weakening in the machine with armature reaction.

c. To derive the torque–speed characteristic of this motor, we must calculate the torque and speed for many different conditions of load. Unfortunately, the demagnetizing armature reaction magnetomotive force is only given for one condition of load (200 A). Since no additional information is available, we will assume that the strength of $\mathcal{F}_{AR}$ varies linearly with load current.

A MATLAB m-file that automates this calculation and plots the resulting torque–speed characteristic is shown below. It performs the same steps as part (a) to determine the speed for each load current, and then calculates the induced torque at that speed. Note that it reads the magnetization curve from a file called `fig8_30.mat`. This file and the other magnetization curves in this chapter are available for download from the book's World Wide Web site (see Preface for details).

```
% M-file: shunt_ts_curve.m
% M-file create a plot of the torque-speed curve of the
%   the shunt DC motor with armature reaction in
%   Example 8-3.

% Get the magnetization curve. This file contains the
% three variables if_value, ea_value, and n_0.
load fig8_30.mat

% First, initialize the values needed in this program.
v_t = 250;               % Terminal voltage (V)
r_f = 50;                % Field resistance (ohms)
r_a = 0.06;              % Armature resistance (ohms)
i_l = 10:10:300;         % Line currents (A)
n_f = 1200;              % Number of turns on field
f_ar0 = 600;             % Armature reaction @ 200 A (A-t/m)

% Calculate the armature current for each load.
i_a = i_l - v_t / r_f;

% Now calculate the internal generated voltage for
% each armature current.
e_a = v_t - i_a * r_a;
```

```
% Calculate the armature reaction MMF for each armature
% current.
f_ar = (i_a / 200) * f_ar0;

% Calculate the effective field current.
i_f = v_t / r_f - f_ar / n_f;

% Calculate the resulting internal generated voltage at
% 1200 r/min by interpolating the motor's magnetization
% curve.
e_a0 = interp1(if_values,ea_values,i_f,'spline');

% Calculate the resulting speed from Equation (8-42).
n = ( e_a ./ e_a0 ) * n_0;

% Calculate the induced torque corresponding to each
% speed from Equations (8-55) and (8-56).
t_ind = e_a .* i_a ./ (n * 2 * pi / 60);

% Plot the torque-speed curve
plot(t_ind,n,'Color','k','LineWidth',2.0);
hold on;
xlabel('\tau_{ind} (N-m)','Fontweight','Bold');
ylabel('\itn_{m} \rm\bf(r/min)','Fontweight','Bold');
title ('Shunt DC Motor Torque-Speed
Characteristic','Fontweight','Bold');
axis([ 0 600 1100 1300]);
grid on;
hold off;
```

The resulting torque–speed characteristic is shown in Figure 8–31. Note that for any given load, the speed of the motor with armature reaction is higher than the speed of the motor without armature reaction. ■

Speed Control of Shunt DC Motors

How can the speed of a shunt DC motor be controlled? There are two common methods to control the speed of the motor, and they have already been seen in the simple rotating loop earlier in this chapter. The two common ways in which the speed of a shunt DC machine can be controlled are by

1. Adjusting the field resistance R_F (and thus the field flux)
2. Adjusting the terminal voltage applied to the armature

Each of these methods is described in detail below.

Changing the Field Resistance   To understand what happens when the field resistor of a DC motor is changed, assume that the field resistor increases and observe the response. If the field resistance increases, then the field current decreases ($I_F = V_T/R_F \uparrow$), and as the field current decreases, the flux ϕ decreases with it. A decrease in flux causes an instantaneous decrease in the internal generated voltage

Figure 8–31 | The torque–speed characteristic of the motor with armature reaction in Example 8–3.

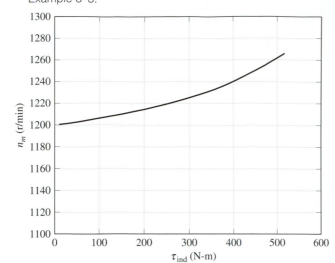

E_A ($= K\phi \downarrow \omega$), which causes a large increase in the machine's armature current, since

$$I_A \uparrow = \frac{V_T - E_A\downarrow}{R_A}$$

The induced torque in a motor is given by $\tau_{\text{ind}} = K\phi I_A$. Since the flux ϕ in this machine decreases while the current I_A increases, which way does the induced torque change? The easiest way to answer this question is to look at an example. Figure 8–32 shows a shunt DC motor with an internal resistance of 0.25 Ω. It is currently operating with a terminal voltage of 250 V and an internal generated voltage of 245 V. Therefore, the armature current flow is $I_A = (250 \text{ V} - 245 \text{ V})/0.25 \ \Omega = 20 \text{ A}$. What happens in this motor *if there is a 1 percent decrease in flux?* If the flux decreases by 1 percent, then E_A must decrease by 1 percent too, because $E_A = K\phi\omega$. Therefore, E_A will drop to

$$E_{A2} = 0.99 \ E_{A1} = 0.99(245 \text{ V}) = 242.55 \text{ V}$$

Figure 8–32 | A 250-V shunt DC motor with typical values of E_A and R_A.

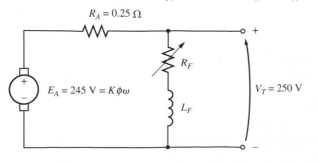

The armature current must then rise to

$$I_A = \frac{250 \text{ V} - 242.55 \text{ V}}{0.25 \ \Omega} = 29.8 \text{ A}$$

Thus a 1 percent decrease in flux produced a 49 percent increase in armature current.

So to get back to the original discussion, the increase in current predominates over the decrease in flux, and the induced torque rises:

$$\overset{\downarrow \ \Uparrow}{\tau_{\text{ind}} = K\phi I_A}$$

Since $\tau_{\text{ind}} > \tau_{\text{load}}$, the motor speeds up.

However, as the motor speeds up, the internal generated voltage E_A rises, causing I_A to fall. As I_A falls, the induced torque τ_{ind} falls too, and finally τ_{ind} again equals τ_{load} at a higher steady-state speed than originally.

To summarize the cause-and-effect behavior involved in this method of speed control:

1. Increasing R_F causes I_F ($= V_T/R_F$ ↑) to decrease.
2. Decreasing I_F decreases ϕ.
3. Decreasing ϕ lowers E_A ($= K\phi \downarrow \omega$).
4. Decreasing E_A increases $I_A = (V_T - E_A \downarrow)/R_A$.
5. Increasing I_A increases τ_{ind} ($= K\phi \downarrow I_A \Uparrow$), with the change in I_A dominant over the change in flux).
6. Increasing τ_{ind} makes $\tau_{\text{ind}} > \tau_{\text{load}}$, and the speed ω increases.
7. Increasing to increases $E_A = K\phi\omega$ ↑ again.
8. Increasing E_A decreases I_A.
9. Decreasing I_A decreases τ_{ind} until $\tau_{\text{ind}} = \tau_{\text{load}}$ at a higher speed ω.

The effect of increasing the field resistance on the output characteristic of a shunt motor is shown in Figure 8–33a. Notice that as the flux in the machine decreases, the no-load speed of the motor increases, while the slope of the torque–speed curve becomes steeper. Naturally, decreasing R_F would reverse the whole process, and the speed of the motor would drop.

A Warning about Field Resistance Speed Control   The effect of increasing the field resistance on the output characteristic of a shunt DC motor is shown in Figure 8–33. Notice that as the flux in the machine decreases, the no-load speed of the motor increases, while the slope of the torque–speed curve becomes steeper. This shape is a consequence of Equation (8–36), which describes the terminal characteristic of the motor. In Equation (8–36), the no-load speed is proportional to the reciprocal of the flux in the motor, while the slope of the curve is proportional to the reciprocal of the flux squared. Therefore, a decrease in flux causes the slope of the torque–speed curve to become steeper.

Figure 8–33a shows the terminal characteristic of the motor over the range from no-load to full-load conditions. Over this range, an increase in field resistance

Figure 8–33 | The effect of field resistance speed control on a shunt motor's torque–speed characteristic: (a) over the motor's normal operating range; (b) over the entire range from no-load to stall conditions.

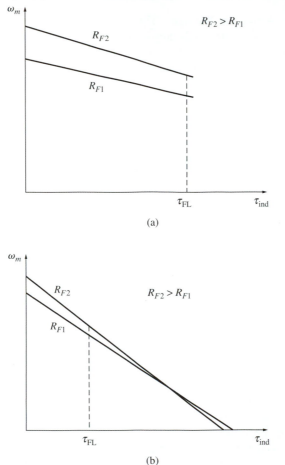

(a)

(b)

increases the motor's speed, as described in this section. For motors operating between no-load and full-load conditions, an increase in R_F may reliably be expected to increase operating speed.

Now examine Figure 8–33b. This figure shows the terminal characteristic of the motor over the full range from no-load to stall conditions. It is apparent from the figure that at *very slow* speeds an increase in field resistance will actually *decrease* the speed of the motor. This effect occurs because, at very low speeds, the increase in armature current caused by the decrease in E_A is no longer large enough to compensate for the decrease in flux in the induced torque equation. With the flux decrease actually larger than the armature current increase, the induced torque decreases, and the motor slows down.

Some small DC motors used for control purposes actually operate at speeds close to stall conditions. For these motors, an increase in field resistance might have no effect, or it might even decrease the speed of the motor. Since the results are not predictable, field resistance speed control should not be used in these types of DC motors. Instead, the armature voltage method of speed control should be employed.

Changing the Armature Voltage   The second form of speed control involves changing the voltage applied to the armature of the motor *without changing the voltage applied to the field.* A connection similar to that in Figure 8–34 is necessary for this type of control. In effect, the motor must be *separately excited* to use armature voltage control.

Figure 8–34 | Armature voltage control of a shunt (or separately excited) DC motor.

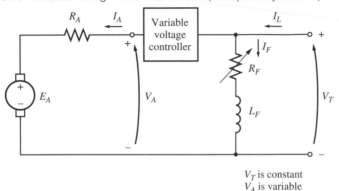

V_T is constant
V_A is variable

If the voltage V_A is increased, then the armature current in the motor must rise $[I_A = (V_A \uparrow - E_A)/R_A]$. As I_A increases, the induced torque $\tau_{ind} = K\phi I_A \uparrow$ increases, making $\tau_{ind} > \tau_{load}$, and the speed ω of the motor increases.

But as the speed ω increases, the internal generated voltage E_A ($= K\phi\omega \uparrow$) increases, causing the armature current to decrease. This decrease in I_A decreases the induced torque, causing τ_{ind} to equal τ_{load} at a higher rotational speed ω.

To summarize the cause-and-effect behavior in this method of speed control:

1. An increase in V_A increases I_A [$= (V_A \uparrow - E_A)/R_A$].
2. Increasing I_A increases τ_{ind} ($= K\phi I_A \uparrow$).
3. Increasing τ_{ind} makes $\tau_{ind} > \tau_{load}$, increasing ω.
4. Increasing ω increases E_A ($= K\phi\omega \uparrow$).
5. Increasing E_A decreases I_A [$= (V_A - E_A \uparrow)/R_A$].
6. Decreasing I_A decreases τ_{ind} until $\tau_{ind} = \tau_{load}$ at a higher ω.

The effect of an increase in V_A on the torque–speed characteristic of a separately excited motor is shown in Figure 8–35. Notice that the no-load speed of the motor is shifted by this method of speed control, but the slope of the curve remains constant.

The two methods of shunt motor speed control—field resistance variation and armature voltage variation—have different safe ranges of operation.

Figure 8–35 | The effect of armature voltage speed control on a shunt motor's torque–speed characteristic.

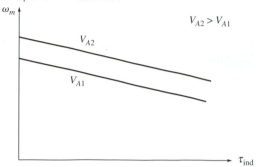

In field resistance control, the lower the field current in a shunt (or separately excited) DC motor, the faster it turns, and the higher the field current, the slower it turns. Since an increase in field current causes a decrease in speed, there is always a minimum achievable speed by field circuit control. This minimum speed occurs when the motor's field circuit has the maximum permissible current flowing through it.

If a motor is operating at its rated terminal voltage, power, and field current, then it will be running at rated speed, also known as *base speed*. Field resistance control can control the speed of the motor for speeds above base speed but not for speeds below base speed. To achieve a speed slower than base speed by field circuit control would require excessive field current, possibly burning up the field windings.

In armature voltage control, the lower the armature voltage on a separately excited DC motor, the slower it turns; and the higher the armature voltage, the faster it turns. Since an increase in armature voltage causes an increase in speed, there is always a maximum achievable speed by armature voltage control. This maximum speed occurs when the motor's armature voltage reaches its maximum permissible level.

If the motor is operating at its rated voltage, field current, and power, it will be turning at base speed. Armature voltage control can control the speed of the motor for speeds below base speed but not for speeds above base speed. To achieve a speed faster than base speed by armature voltage control would require excessive armature voltage, possibly damaging the armature circuit.

These two techniques of speed control are obviously complementary. Armature voltage control works well for speeds below base speed, and field resistance or field current control works well for speeds above base speed. By combining the two speed-control techniques in the same motor, it is possible to get a range of speed variations of up to 40 to 1 or more. Shunt and separately excited DC motors have excellent speed control characteristics.

There is a significant difference in the torque and power limits on the machine under these two types of speed control. The limiting factor in either case is the heating of the armature conductors, which places an upper limit on the magnitude of the armature current I_A.

For armature voltage control, *the flux in the motor is constant*, so the maximum torque in the motor is

$$\tau_{max} = K\phi I_{A,\,max} \tag{8–43}$$

This maximum torque is constant regardless of the speed of the rotation of the motor. Since the power out of the motor is given by $P = \tau\omega$, the maximum power of the motor at any speed under armature voltage control is

$$P_{max} = \tau_{max}\omega \tag{8–44}$$

Thus *the maximum power out of the motor is directly proportional to its operating speed* under armature voltage control.

On the other hand, when field resistance control is used, the flux does change. In this form of control, a speed increase is caused by a decrease in the machine's flux. In order for the armature current limit not to be exceeded, the induced torque limit must decrease as the speed of the motor increases. Since the power out of the motor is given by $P = \tau\omega$, and the torque limit decreases as the speed of the motor increases, *the maximum power out of a DC motor under field current control is constant*, while *the maximum torque varies as the reciprocal of the motor's speed*.

These shunt DC motor power and torque limitations for safe operation as a function of speed are shown in Figure 8–36.

Figure 8–36 | Power and torque limits as a function of speed for a shunt motor under armature volt and field resistance control.

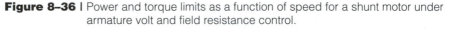

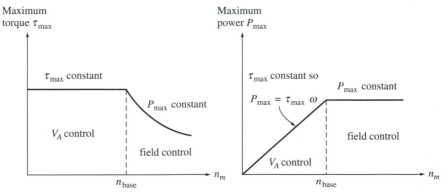

The following examples illustrate how to find the new speed of a DC motor if it is varied by field resistance or armature voltage control methods.

EXAMPLE 8–4

Figure 8–37a shows a 100-hp, 250-V, 1200 r/min shunt DC motor with an armature resistance of 0.03 Ω and a field resistance of 41.67 Ω. The motor has compensating windings, so armature reaction can be ignored. Mechanical and core losses may be assumed to be negligible for the purposes of this problem. The motor is assumed to be driving a load with a line current of 126 A and an initial speed of 1103 r/min. To simplify the problem, assume that the amount of armature current drawn by the motor remains constant.

Figure 8–37 | (a) The shunt motor in Example 8–4. (b) The separately excited DC motor in Example 8–4.

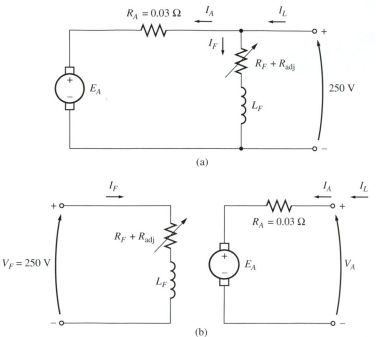

(a)

(b)

a. If the machine's magnetization curve is shown in Figure 8–30, what is the motor's speed if the field resistance is raised to 50 Ω?

b. Calculate and plot the speed of this motor as a function of the field resistance R_F, assuming a constant-current load.

■ **Solution**

a. The motor has an initial line current of 126 A, so the initial armature current is

$$I_{A1} = I_{L1} - I_{F1} = 126 \text{ A} - \frac{150 \text{ V}}{41.67 \text{ Ω}} = 120 \text{ A}$$

Therefore, the internal generated voltage is

$$E_{A1} = V_T - I_{A1}R_A = 250 \text{ V} - (120 \text{ A})(0.03 \text{ Ω})$$
$$= 246.4 \text{ V}$$

After the field resistance is increased to 50 Ω, the field current will become

$$I_{F2} = \frac{V_T}{R_F} = \frac{250 \text{ V}}{50 \text{ Ω}} = 5 \text{ A}$$

The ratio of the internal generated voltage at one speed to the internal generated voltage and another speed is given by the ratio of Equation (8–30) at the two speeds:

$$\frac{E_{A2}}{E_{A1}} = \frac{K\phi_2\omega_2}{K\phi_1\omega_1} = \frac{\phi_2 n_2}{\phi_1 n_1} \tag{8–45}$$

Because the armature current is assumed constant, $E_{A1} = E_{A2}$, and this equation reduces to

$$1 = \frac{\phi_2 n_2}{\phi_1 n_1}$$

or

$$n_2 = \frac{\phi_1}{\phi_2} n_1 \tag{8–46}$$

A magnetization curve is a plot of E_A versus I_F for a given speed. Since the values of E_A on the curve are directly proportional to the flux, the ratio of the internal generated voltages read off the curve is equal to the ratio of the fluxes within the machine. At $I_F = 5$ A, $E_{A0} = 250$ V, while at $I_F = 6$ A, $E_{A0} = 268$ V. Therefore, the ratio of fluxes is given by

$$\frac{\phi_1}{\phi_2} = \frac{268 \text{ V}}{250 \text{ V}} = 1.076$$

and the new speed of the motor is

$$n_2 = \frac{\phi_1}{\phi_2} n_1 = (1.076)(1103 \text{ r/min}) = 1187 \text{ r/min}$$

b. A MATLAB m-file that calculates the speed of the motor as a function of R_F is shown below.

```
% M-file: rf_speed_control.m
% M-file create a plot of the speed of a shunt DC
%    motor as a function of field resistance, assuming
%    a constant armature current (Example 8-4).

% Get the magnetization curve. This file contains the
% three variables if_value, ea_value, and n_0.
load fig3_30.mat

% First, initialize the values needed in this program.
v_t = 250;              % Terminal voltage (V)
r_f = 40:1:70;          % Field resistance (ohms)
r_a = 0.03;             % Armature resistance (ohms)
i_a = 120;              % Armature currents (A)
```

```
% The approach here is to calculate the e_a0 at the
% reference field current, and then to calculate the
% e_a0 for every field current. The reference speed is
% 1103 r/min, so by knowing the the e_a0 and reference
% speed, we will be able to calculate the speed at the
% other field current.

% Calculate the internal generated voltage at 1200 r/min
% for the reference field current (5 A) by interpolating
% the motor's magnetization curve. The reference speed
% corresponding to this field current is 1103 r/min.
e_a0_ref = interp1(if_values,ea_values,5,'spline');
n_ref = 1103;

% Calculate the field current for each value of field
% resistance.
i_f = v_t ./ r_f;

% Calculate the E_a0 for each field current by
% interpolating the motor's magnetization curve.
e_a0 = interp1(if_values,ea_values,i_f,'spline');

% Calculate the resulting speed from Equation (8-46):
% n2 = (phi1 / phi2) * n1 = (e_a0_1 / e_a0_2 ) * n1
n2 = ( e_a0_ref ./ e_a0 ) * n_ref;

% Plot the speed versus r_f curve.
plot(r_f,n2,'Color','k','LineWidth',2.0);
hold on;
xlabel('Field resistance, \Omega','Fontweight','Bold');
ylabel('\itn_{m} \rm\bf(r/min)','Fontweight','Bold');
title ('Speed vs \itR_{F} \rm\bf for a Shunt DC Motor', ...
       'Fontweight','Bold');
axis([40 70 0 1400]);
grid on;
hold off;
```

The resulting plot is shown in Figure 8–38.   ∎

Note that the assumption of a constant armature current as R_F changes is not a very good one for real loads. The current in the armature will vary with speed in a fashion dependent on the torque required by the type of load attached to the motor. These differences will cause a motor's speed versus R_F curve to be slightly different than the one shown in Figure 8–38, but it will have a similar shape.

EXAMPLE 8–5

The motor in Example 8–4 is now connected separately excited, as shown in Figure 8–37b. The motor is initially running with $V_A = 250$ V, $I_A = 120$ A, and $n = 1103$ r/min, while supplying a constant-torque load. What will the speed of this motor be if V_A is reduced to 200 V?

Figure 8–38 | Plot of speed versus field resistance for the shunt DC motor of Example 8–4.

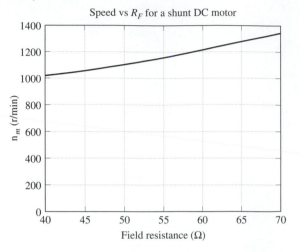

Speed vs R_F for a shunt DC motor

■ **Solution**

The motor has an initial line current of 120 A and an armature voltage V_A of 250 V, so the internal generated voltage E_A, is

$$E_A = V_T - I_A R_A = 250\ \text{V} - (120\ \text{A})(0.03\ \Omega) = 246.4\ \text{V}$$

By applying Equation (8–45) and realizing that the flux ϕ is constant, the motor's speed can be expressed as

$$\frac{E_{A2}}{E_{A1}} = \frac{\phi_2 n_2}{\phi_1 n_1} \qquad (8\text{–}45)$$

$$= \frac{n_2}{n_1}$$

$$n_2 = \frac{E_{A2}}{E_{A1}}\, n_1$$

To find E_{A2} use Kirchhoff's voltage law:

$$E_{A2} = V_T - I_{A2} R_A$$

Since the torque is constant and the flux is constant, I_A is constant. This yields a voltage of

$$E_{A2} = 200\ \text{V} - (120\ \text{A})(0.03\ \Omega) = 196.4\ \text{V}$$

The final speed of the motor is thus

$$n_2 = \frac{E_{A2}}{E_{A1}}\, n_1 = \frac{196.4\ \text{V}}{246.4\ \text{V}}\ 1103\ \text{r/min} = 879\ \text{r/min} \qquad ■$$

The Effect of an Open Field Circuit

The previous section of this chapter contained a discussion of speed control by varying the field resistance of a shunt motor. As the field resistance increased, the speed of the motor increased with it. What would happen if this effect were taken to the extreme, if the field resistor *really* increased? What would happen if the field circuit actually opened while the motor was running? From the previous discussion, the flux in the machine would drop drastically, all the way down to ϕ_{res}, and E_A ($= K\phi\omega$) would drop with it. This would cause a really enormous increase in the armature current, and the resulting induced torque would be quite a bit higher than the load torque on the motor. Therefore, the motor's speed starts to rise and just keeps going up.

The results of an open field circuit can be quite spectacular. When the author was an undergraduate at Louisiana State University, his laboratory group once made a mistake of this sort. The group was working with a small motor–generator set being driven by a 3-hp shunt DC motor. The motor was connected and ready to go, but there was just *one* little mistake—when the field circuit was connected, it was fused with a 0.3-A fuse instead of the 3-A fuse that was supposed to be used.

When the motor was started, it ran normally for about 3 s, and then suddenly there was a flash from the fuse. Immediately, the motor's speed skyrocketed. Someone turned the main circuit breaker off within a few seconds, but by that time the tachometer attached to the motor had pegged at 4000 r/min. The motor itself was only rated for 800 r/min.

Needless to say, that experience scared everyone present very badly and taught them to be *most* careful about field circuit protection. In DC motor starting and protection circuits, a *field loss relay* is normally included to disconnect the motor from the line in the event of a loss of field current.

A similar effect can occur in ordinary shunt DC motors operating with light fields if their armature reaction effects are severe enough. If the armature reaction on a DC motor is severe, an increase in load can weaken its flux enough to actually cause the motor's speed to rise. However, most loads have torque–speed curves whose torque *increases* with speed, so the increased speed of the motor increases its load, which increases its armature reaction, weakening its flux again. The weaker flux causes a further increase in speed, further increasing load etc. until the motor overspeeds. This condition is known as *runaway*.

In motors operating with very severe load changes and duty cycles, this flux-weakening problem can be solved by installing compensating windings. Unfortunately, compensating windings are too expensive for use on ordinary, run-of-the-mill motors. The solution to the runaway problem employed for less-expensive, less severe duty motors is to provide a turn or two of cumulative compounding to the motor's poles. As the load increases, the magnetomotive force from the series turns increases, which counteracts the demagnetizing magnetomotive force of the armature reaction. A shunt motor equipped with just a few series turns like this is called a *stabilized shunt* motor.

8.9 | THE PERMANENT-MAGNET DC MOTOR

A permanent-magnet DC (PMDC) *motor* is a DC motor whose poles are made of permanent magnets. Permanent-magnet DC motors offer a number of benefits compared with shunt DC motors in some applications. Since these motors do not require an external field circuit, they do not have the field circuit copper losses associated with shunt DC motors. Because no field windings are required, they can be smaller than corresponding shunt DC motors. PMDC motors are especially common in smaller fractional- and subfractional-horsepower sizes, where the expense and space of a separate field circuit cannot be justified.

However, PMDC motors also have disadvantages. Permanent magnets cannot produce as high a flux density as an externally supplied shunt field, so a PMDC motor will have a lower induced torque τ_{ind} per ampere of armature current I_A than a shunt motor of the same size and construction. In addition, PMDC motors run the risk of demagnetization. As mentioned earlier, the armature current I_A in a DC machine produces an armature magnetic field of its own. The armature mmf subtracts from the mmf of the poles under some portions of the pole faces and adds to the mmf of the poles under other portions of the pole faces (see Figures 8–11 and 8–13), reducing the overall net flux in the machine. This is the *armature reaction* effect. In a PMDC machine, the pole flux is just the residual flux in the permanent magnets. If the armature current becomes very large, there is some risk that the armature mmf may demagnetize the poles, permanently reducing and reorienting the residual flux in them. Demagnetization may also be caused by the excessive heating that can occur during prolonged periods of overload.

Figure 8–39a shows a magnetization curve for a typical ferromagnetic material. It is a plot of flux density **B** versus magnetizing intensity **H** (or equivalently, a plot of flux ϕ versus mmf $\mathcal{F}$). When a strong external magnetomotive force is applied to this material and then removed, a residual flux $\mathbf{B}_{res}$ will remain in the material. To force the residual flux to zero, it is necessary to apply a coercive magnetizing intensity $\mathbf{H}_C$ with a polarity opposite to the polarity of the magnetizing intensity **H** that originally established the magnetic field. For normal machine applications such as rotors and stators, a ferromagnetic material should be picked which has as small a $\mathbf{B}_{res}$ and $\mathbf{H}_C$ as possible, since such a material will have low hysteresis losses.

On the other hand, a good material for the poles of a PMDC motor should have *as large a residual flux density* $\mathbf{B}_{res}$ *as possible*, while simultaneously having *as large a coercive magnetizing intensity* $\mathbf{H}_C$ *as possible*. The magnetization curve of such a material is shown in Figure 8–39b. The large $\mathbf{B}_{res}$ produces a large flux in the machine, while the large $\mathbf{H}_C$ means that a very large current would be required to demagnetize the poles.

In the last 40 years, a number of new magnetic materials have been developed that have desirable characteristics for making permanent magnets. The major types of materials are the ceramic (ferrite) magnetic materials and the rare-earth magnetic materials. Figure 8–39c shows the second quadrant of the magnetization curves of some typical ceramic and rare-earth magnets, compared to the magnetization curve of a conventional ferromagnetic alloy (Alnico 5). It is obvious from the comparison that the best rare-earth magnets can produce the same residual flux as the best

Figure 8–39 | (a) The magnetization curve of a typical ferromagnetic material. Note the hysteresis loop. After a large magnetizing intensity **H** is applied to the core and then removed, a residual flux density $\mathbf{B}_{res}$ remains behind in the core. This flux can be brought to zero if a coercive magnetizing intensity $\mathbf{H}_C$ is applied to the core with the opposite polarity. In this case, a relatively small value of it will demagnetize the core. (b) The magnetization curve of a ferromagnetic material suitable for use in permanent magnets. Note the high residual flux density $\mathbf{B}_{res}$ and the relatively large coercive magnetizing intensity $\mathbf{H}_C$. (c) The second quadrant of the magnetization curves of some typical magnetic materials. Note that the rare-earth magnets combine both a high residual flux and a high coercive magnetizing intensity.

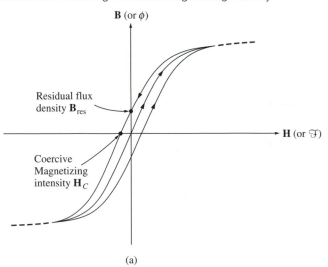

(a)

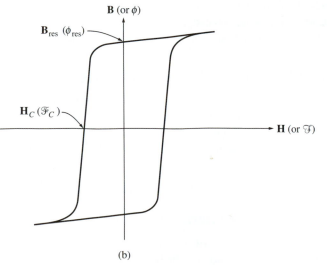

(b)

(continued)

Figure 8–39 | *(concluded)*

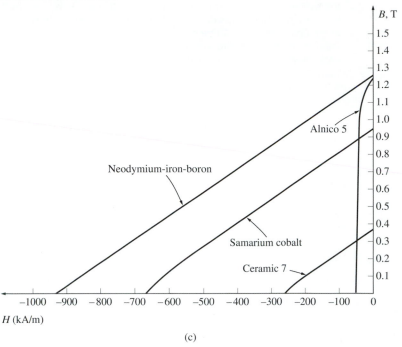

(c)

conventional ferromagnetic alloys, while simultaneously being largely immune to demagnetization problems due to armature reaction.

A permanent-magnet DC motor is basically the same machine as a shunt DC motor, except that *the flux of a PMDC motor is fixed*. Therefore, it is not possible to control the speed of a PMDC motor by varying the field current or flux. The only methods of speed control available for a PMDC motor are armature voltage control and armature resistance control.

For more information about PMDC motors, see References 4 and 10.

8.10 | THE SERIES DC MOTOR

A series DC motor is a DC motor whose field windings consist of a relatively few turns connected in series with the armature circuit. The equivalent circuit of a series DC motor is shown in Figure 8–40. In a series motor, the armature current, field current, and line current are all the same. The Kirchhoff's voltage law equation for this motor is

$$V_T = E_A + I_A(R_A + R_S) \tag{8–47}$$

Induced Torque in a Series DC Motor

The terminal characteristic of a series DC motor is very different from that of the shunt motor previously studied. The basic behavior of a series DC motor is due to the

Figure 8–40 | The equivalent circuit of a series DC motor.

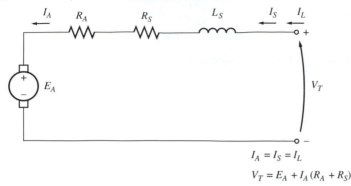

$$I_A = I_S = I_L$$
$$V_T = E_A + I_A(R_A + R_S)$$

fact that *the flux is directly proportional to the armature current*, at least until saturation is reached. As the load on the motor increases, its flux increases too. As seen earlier, an increase in flux in the motor causes a decrease in its speed. The result is that a series motor has a sharply drooping torque–speed characteristic.

The induced torque in this machine is given by Equation (8–31):

$$\tau_{ind} = K\phi I_A \qquad\qquad (8\text{–}31)$$

The flux in this machine is directly proportional to its armature current (at least until the metal saturates). Therefore, the flux in the machine can be given by

$$\phi = cI_A \qquad\qquad (8\text{–}48)$$

where c is a constant of proportionality. The induced torque in this machine is thus given by

$$\tau_{ind} = K\phi I_A = KcI_A^2 \qquad\qquad (8\text{–}49)$$

In other words, *the torque in the motor is proportional to the square of its armature current*. As a result of this relationship, it is easy to see that a series motor gives more torque per ampere than any other DC motor. It is therefore used in applications requiring very high torques. Examples of such applications are the starter motors in cars, elevator motors, and tractor motors in locomotives.

The Terminal Characteristic of a Series DC Motor

To determine the terminal characteristic of a series DC motor, an analysis will be done, based on the assumption of a linear magnetization curve, and then the effects of saturation will be considered in a graphical analysis.

The assumption of a linear magnetization curve implies that the flux in the motor will be given by Equation (8–48):

$$\phi = cI_A \qquad\qquad (8\text{–}48)$$

This equation will be used to derive the torque–speed characteristic curve for the series motor.

The derivation of a series motor's torque–speed characteristic starts with Kirchhoff's voltage law:

$$V_T = E_A + I_A(R_A + R_S) \tag{8–47}$$

From Equation (8–49), the armature current can be expressed as

$$I_A = \sqrt{\frac{\tau_{ind}}{Kc}}$$

Also, $E_A = K\phi\omega$. Substituting these expressions in Equation (8–47) yields

$$V_T = K\phi\omega + \sqrt{\frac{\tau_{ind}}{Kc}}\,(R_A + R_S) \tag{8–50}$$

If the flux can be eliminated from this expression, it will directly relate the torque of a motor to its speed. To eliminate the flux from the expression, notice that

$$I_A = \frac{\phi}{c}$$

and the induced torque equation can be rewritten as

$$\tau_{ind} = \frac{K}{c}\,\phi^2$$

Therefore, the flux in the motor can be rewritten as

$$\phi = \sqrt{\frac{c}{K}}\,\sqrt{\tau_{ind}} \tag{8–51}$$

Substituting Equation (8–51) into Equation (8–50) and solving for speed yields

$$V_T = K\sqrt{\frac{c}{K}}\,\sqrt{\tau_{ind}}\,\omega + \sqrt{\frac{\tau_{ind}}{Kc}}\,(R_A + R_S)$$

$$\sqrt{Kc}\,\sqrt{\tau_{ind}}\,\omega = V_T - \frac{R_A + R_S}{\sqrt{Kc}}\,\sqrt{\tau_{ind}}$$

$$\omega = \frac{V_T}{\sqrt{Kc}\,\sqrt{\tau_{ind}}} - \frac{R_A + R_S}{Kc}$$

The resulting torque–speed relationship is

$$\boxed{\omega = \frac{V_T}{\sqrt{Kc}}\,\frac{1}{\sqrt{\tau_{ind}}} - \frac{R_A + R_S}{Kc}} \tag{8–52}$$

Notice that for an unsaturated series motor, the speed of the motor varies as the reciprocal of the square root of the torque. That is quite an unusual relationship! This ideal torque–speed characteristic is plotted in Figure 8–41.

One disadvantage of series motors can be seen immediately from this equation. When the torque on this motor goes to zero, its speed goes to infinity. In practice, the torque can never go entirely to zero because of the mechanical, core, and stray losses

Figure 8–41 | The torque–speed characteristic of a series DC motor.

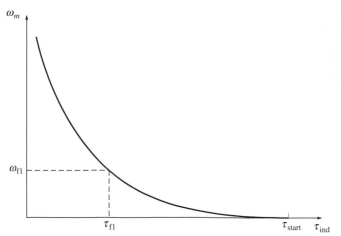

that must be overcome. However, if no other load is connected to the motor, it can turn fast enough to seriously damage itself. *Never* completely unload a series motor, and never connect one to a load by a belt or other mechanism that could break. If that were to happen and the motor were to become unloaded while running, the results could be serious.

The nonlinear analysis of a series DC motor with magnetic saturation effects, but ignoring armature reaction, is illustrated in Example 8–6.

EXAMPLE 8–6

Figure 8–40 shows a 250-V series DC motor with compensating windings, and a total series resistance $R_A + R_S$ of 0.08 Ω. The series field consists of 25 turns per pole, with the magnetization curve shown in Figure 8–42.

a. Find the speed and induced torque of this motor for when its armature current is 50 A.

b. Calculate and plot the torque–speed characteristic for this motor.

■ **Solution**

a. To analyze the behavior of a series motor with saturation, pick points along the operating curve and find the torque and speed for each point. Notice that the magnetization curve is given in units of magnetomotive force (ampere-turns) versus E_A for a speed of 1200 r/min, so calculated E_A values must be compared to the equivalent values at 1200 r/min to determine the actual motor speed.

For $I_A = 50$ A,

$$E_A = V_T - I_A(R_A + R_S) = 250 \text{ V} - (50 \text{ A})(0.08 \text{ }\Omega) = 246 \text{ V}$$

Figure 8–42 | The magnetization curve of the motor in Example 8–6. This curve was taken at speed $n_m = 1200$ r/min.

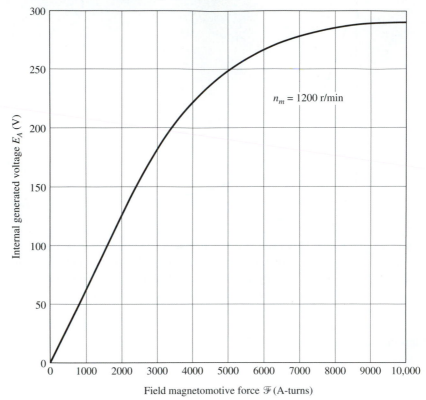

Since $I_A = I_F = 50$ A, the magnetomotive force is

$$\mathcal{F} = NI = (25 \text{ turns})(50 \text{ A}) = 1250 \text{ A-turns}$$

From the magnetization curve at $\mathcal{F} = 1250$ A-turns, $E_{A0} = 80$ V. To get the correct speed of the motor, remember that, from Equation (8–42),

$$n = \frac{E_A}{E_{A0}} n_0$$

$$= \frac{246 \text{ V}}{80 \text{ V}} 1200 \text{ r/min} = 3690 \text{ r/min}$$

To find the induced torque supplied by the motor at that speed, recall that $P_{conv} = E_A I_A = \tau_{ind}\omega$. Therefore,

$$\tau_{ind} = \frac{E_A I_A}{\omega}$$

$$= \frac{(246 \text{ V})(50 \text{ A})}{(3690 \text{ r/min})(1 \text{ min/60 s})(2\pi \text{ rad/r})} = 31.8 \text{ N-m}$$

b. To calculate the complete torque–speed characteristic, we must repeat the steps in *a* for many values of armature current. A MATLAB m-file that calculates the torque–speed characteristics of the series DC motor is shown below. Note that the magnetization curve used by this program works in terms of field magnetomotive force instead of effective field current.

```
% M-file: series_ts_curve.m
% M-file create a plot of the torque-speed curve of the
%    the series DC motor with armature reaction in
%    Example 8-6.

% Get the magnetization curve. This file contains the
% three variables mmf_value, ea_value, and n_0.
load fig8_42.mat

% First, initialize the values needed in this program.
v_t = 250;               % Terminal voltage (V)
r_a = 0.08;              % Armature + field resistance (ohms)
i_a = 10:10:300;         % Armature (line) currents (A)
n_s = 25;                % Number of series turns on field

% Calculate the MMF for each load
f = n_s * i_a;

% Calculate the internal generate voltage e_a.
e_a = v_t - i_a * r_a;

% Calculate the resulting internal generated voltage at
% 1200 r/min by interpolating the motor's magnetization
% curve.
e_a0 = interp1(mmf_values,ea_values,f,'spline');

% Calculate the motor's speed from Equation (8-42).
n = (e_a ./ e_a0) * n_0;

% Calculate the induced torque corresponding to each
% speed from Equations (8-55) and (8-56).
t_ind = e_a .* i_a ./ (n * 2 * pi / 60);

% Plot the torque-speed curve
plot(t_ind,n,'Color','k','LineWidth',2.0);
hold on;
xlabel('\tau_{ind} (N-m)','Fontweight','Bold');
ylabel('\itn_{m} \rm\bf(r/min)','Fontweight','Bold');
title ('Series DC Motor Torque-Speed Characteristic', ...
       'Fontweight','Bold');
axis([ 0 700 0 5000]);
grid on;
hold off;
```

The resulting motor torque–speed characteristic is shown in Figure 8–43. Notice the severe overspeeding at very small torques. ∎

Figure 8–43 | The torque–speed characteristic of the series DC motor in Example 8–6.

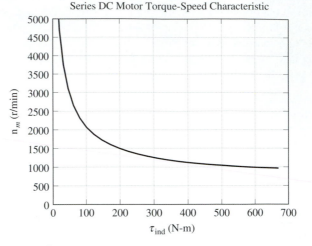

Speed Control of Series DC Motors

Unlike the shunt DC motor, there is only one efficient way to change the speed of a series DC motor. That method is to change the terminal voltage of the motor. If the terminal voltage is increased, the first term in Equation (8–52) is increased, resulting in a *higher speed for any given torque*.

Until the last 30 years or so, there was no convenient way to change V_T, so the only method of speed control available was the wasteful series resistance method. That has all changed today with the introduction of solid-state control circuits. It is now relatively easy to create a variable DC voltage to control the speed of a motor.

8.11 | THE COMPOUNDED DC MOTOR

A compounded DC motor is a motor with *both a shunt and a series field*. Such a motor is shown in Figure 8–44. The dots that appear on the two field coils have the same meaning as the dots on a transformer: *Current flowing into a dot produces a positive magnetomotive force*. If current flows into the dots on both field coils, the resulting magnetomotive forces add to produce a larger total magnetomotive force. This situation is known as *cumulative compounding*. If current flows into the dot on one field coil and out of the dot on the other field coil, the resulting magnetomotive forces subtract. In Figure 8–44 the round dots correspond to cumulative compounding of the motor, and the squares correspond to differential compounding.

The Kirchhoff's voltage law equation for a compounded DC motor is

$$V_T = E_A + I_A(R_A + R_S) \tag{8–53}$$

The currents in the compounded motor are related by

$$I_A = I_L - I_F \tag{8–54}$$

$$I_F = \frac{V_T}{R_F} \tag{8–55}$$

Figure 8–44 | The equivalent circuit of compounded DC motors: (a) long-shunt connection; (b) short-shunt connection.

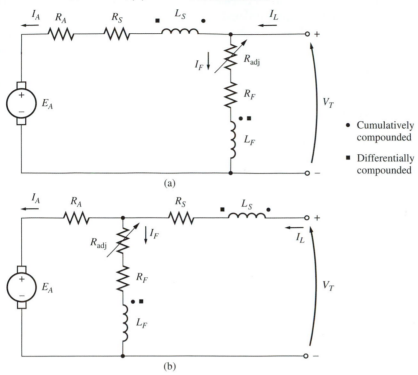

(a)

(b)

The net magnetomotive force and the effective shunt field current in the compounded motor are given by

$$\mathcal{F}_{\text{net}} = \mathcal{F}_F \pm \mathcal{F}_{\text{SE}} - \mathcal{F}_{\text{AR}} \qquad (8\text{–}56)$$

and

$$I_F^* = I_F + \frac{N_{\text{SE}}}{N_F} I_A - \frac{\mathcal{F}_{\text{AR}}}{N_F} \qquad (8\text{–}57)$$

where the positive sign in the equations is associated with a cumulatively compounded motor and the negative sign is associated with a differentially compounded motor.

The Torque–Speed Characteristic of a Cumulatively Compounded DC Motor

In the cumulatively compounded DC motor, there is a component of flux that is constant and another component that is proportional to its armature current (and thus to its load). Therefore, the cumulatively compounded motor has a higher starting torque than a shunt motor (whose flux is constant) but a lower starting torque than a series motor (whose entire flux is proportional to armature current).

In a sense, the cumulatively compounded DC motor combines the best features of both the shunt and the series motors. Like a series motor, it has extra torque for starting; like a shunt motor, it does not overspeed at no load.

At light loads, the series field has a very small effect, so the motor behaves approximately as a shunt DC motor. As the load gets very large, the series flux becomes quite important and the torque–speed curve begins to look like a series motor's characteristic. A comparison of the torque–speed characteristics of each of these types of machines is shown in Figure 8–45.

Figure 8–45 | (a) The torque–speed characteristic of a cumulatively compounded DC motor compared to series and shunt motors with the same full-load rating. (b) The torque–speed characteristic of it cumulatively compounded DC motor compared to a shunt motor with the same no-load speed.

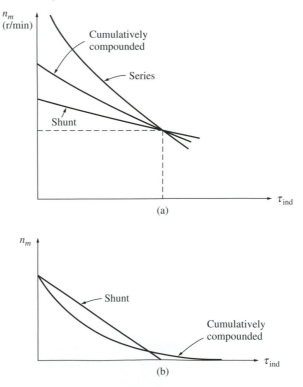

To determine the characteristic curve of a cumulatively compounded DC motor by nonlinear analysis, the approach is similar to that for the shunt and series motors seen before. Such an analysis will be illustrated in a later example.

The Torque–Speed Characteristic of a Differentially Compounded DC Motor

In a differentially compounded DC motor, *the shunt magnetomotive force and series magnetomotive force subtract from each other.* This means that, as the load on the motor increases, I_A increases and *the flux in the motor decreases.* But as the flux decreases, the speed of the motor increases. This speed increase causes another increase in load, which further increases I_A, further decreasing the flux, and increasing the speed again. The result is that a differentially compounded motor is unstable and tends to run away. This instability is *much* worse than that of a shunt motor with armature reaction. It is so bad that a differentially compounded motor is unsuitable for any application.

To make matters worse, it is impossible to start such a motor. At starting conditions the armature current and the series field current are very high. Since the series flux subtracts from the shunt flux, the series field can actually reverse the magnetic polarity of the machine's poles. The motor will typically remain still or turn slowly in the wrong direction while burning up, because of the excessive armature current. When this type of motor is to be started, its series field must be short circuited, so that it behaves as an ordinary shunt motor during the starting period.

Because of the stability problems of the differentially compounded DC motor, it is almost never *intentionally* used. However, a simple wiring error can create one by accident! A typical terminal characteristic for a differentially compounded DC motor is shown in Figure 8–46.

Figure 8–46 | The torque–speed characteristic of a differentially compounded DC motor.

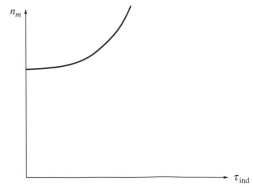

The Nonlinear Analysis of Compounded DC Motors

The determination of the torque and speed of a compounded DC motor is illustrated in Example 8–7.

EXAMPLE 8–7

A 100-hp, 250-V compounded DC motor with compensating windings has an internal resistance, including the series winding, of 0.04 Ω. There are 1000 turns per pole on the shunt field and 3 turns per pole on the series winding. The machine is shown in Figure 8–47, and its magnetization curve is shown in Figure 8–30. At no load, the field resistor has been adjusted to make the motor run at 1200 r/min. The core, mechanical, and stray losses may be neglected.

Figure 8–47 | The compounded DC motor in Example 8–7.

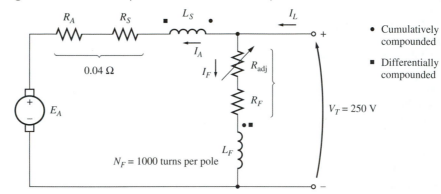

a.   What is the shunt field current in this machine at no load?
b.   If the motor is cumulatively compounded, find its speed when $I_A = 200$ A.
c.   If the motor is differentially compounded, find its speed when $I_A = 200$ A.

■ **Solution**
a.   At no load, the armature current is zero, so the internal generated voltage of the motor must equal V_T, which means that it must be 250 V. From the magnetization curve, a field current of 5 A will produce a voltage E_A of 250 V at 1200 r/min. Therefore, the shunt field current must be 5 A.

b.   When an armature current of 200 A flows in the motor, the machine's internal generated voltage is

$$E_A = V_T - I_A(R_A + R_S)$$
$$= 250 \text{ V} - (200 \text{ A})(0.04 \text{ }\Omega) = 242 \text{ V}$$

The effective field current of this cumulatively compounded motor is

$$I_F^* = I_F + \frac{N_{SE}}{N_F} I_A - \frac{\mathcal{F}_{AR}}{N_F} \qquad (8\text{–}57)$$

$$= 5 \text{ A} + \frac{3}{1000} \, 200 \text{ A} = 5.6 \text{ A}$$

From the magnetization curve, $E_{A0} = 262$ V at speed $n_0 = 1200$ r/min. Therefore, the motor's speed will be

$$n = \frac{E_A}{E_{A0}} n_0$$

$$= \frac{242 \text{ V}}{262 \text{ V}} 1200 \text{ r/min} = 1108 \text{ r/min}$$

c. If the machine is differentially compounded, the effective field current is

$$I_F^* = I_F - \frac{N_{SE}}{N_F} I_A - \frac{\mathcal{F}_{AR}}{N_F} \qquad (8\text{--}57)$$

$$= 5 \text{ A} - \frac{3}{1000} 200 \text{ A} = 4.4 \text{ A}$$

From the magnetization curve, $E_{A0} = 236$ V at speed $n_0 = 1200$ r/min. Therefore, the motor's speed will be

$$n = \frac{E_A}{E_{A0}} n_0$$

$$= \frac{242 \text{ V}}{236 \text{ V}} 1200 \text{ r/min} = 1230 \text{ r/min}$$

Notice that the speed of the cumulatively compounded motor decreases with load, while the speed of the differentially compounded motor increases with load. ∎

Speed Control in the Cumulatively Compounded DC Motor

The techniques available for the control of speed in a cumulatively compounded DC motor are the same as those available for a shunt motor:

1. Change the field resistance R_F.
2. Change the armature voltage V_A.

The arguments describing the effects of changing R_F or V_A are very similar to the arguments given earlier for the shunt motor.

Theoretically, the differentially compounded DC motor could be controlled in a similar manner. Since the differentially compounded motor is almost never used, that fact hardly matters.

8.12 | DC MOTOR STARTERS

In order for a DC motor to function properly on the job, it must have some special control and protection equipment associated with it. The purposes of this equipment are

1. To protect the motor against damage due to short circuits in the equipment
2. To protect the motor against damage from long-term overloads

3. To protect the motor against damage from excessive starting currents
4. To provide a convenient manner in which to control the operating speed of the motor

The first three functions will be discussed in this section.

DC Motor Problems on Starting

In order for a DC motor to function properly, it must be protected from physical damage during the starting period. At starting conditions, the motor is not turning, and so $E_A = 0$ V. Since the internal resistance of a normal DC motor is very low compared to its size (3 to 6 percent per unit for medium-size motors), a *very* high current flows.

Consider, for example, the 50-hp, 250-V motor in Example 8–2. This motor has an armature resistance R_A of 0.06 Ω, and a full-load current less than 200 A, but the current on starting is

$$I_A = \frac{V_T - E_A}{R_A}$$
$$= \frac{250 \text{ V} - 0 \text{ V}}{0.06 \text{ }\Omega} = 4167 \text{ A}$$

This current is over 20 times the motor's rated full-load current. It is possible for a motor to be severely damaged by such currents, even if they last for only a moment.

A solution to the problem of excess current during starting is to insert a *starting resistor* in series with the armature to limit the current flow until E_A can build up to do the limiting. This resistor must not be in the circuit permanently, because it would result in excessive losses and would cause the motor's torque–speed characteristic to drop off excessively with an increase in load.

Therefore, a resistor must be inserted into the armature circuit to limit current flow at starting, and it must be removed again as the speed of the motor builds up. In modern practice, a starting resistor is made up of a series of pieces, each of which is removed from the motor circuit in succession as the motor speeds up, in order to limit the current in the motor to a safe value while never reducing it to too low a value for rapid acceleration.

Figure 8–48 shows a shunt motor with an extra starting resistor that can be cut out of the circuit in segments by the closing of the 1A, 2A, and 3A contacts. Two actions are necessary in order to make a working motor starter. The first is to pick the size and number of resistor segments necessary in order to limit the starting current to its desired bounds. The second is to design a control circuit that shuts the resistor bypass contacts at the proper time to remove those parts of the resistor from the circuit.

Example 8–8 illustrates the selection of the size and number of resistor segments needed by an automatic starter circuit. The question of the timing required to cut the resistor segments out of the armature circuit will be examined later.

Figure 8–48 | A shunt motor with a starting resistor in series with its armature. Contacts 1A, 2A, and 3A short-circuit portions of the starting resistor when they close.

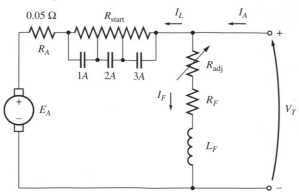

EXAMPLE 8–8

Figure 8–48 shows a 100-hp, 250-V, 350-A shunt DC motor with an armature resistance of 0.05 Ω. It is desired to design a starter circuit for this motor that will limit the maximum starting current to *twice* its rated value and which will switch out sections of resistance as the armature current falls to its rated value.

a.   How many stages of starting resistance will be required to limit the current to the range specified?

b.   What must the value of each segment of the resistor be? At what voltage should each stage of the starting resistance be cut out?

■ **Solution**

a.   The starting resistor must be selected so that the current flow equals twice the rated current of the motor when it is first connected to the line. As the motor starts to speed up, an internal generated voltage E_A will be produced in the motor. Since this voltage opposes the terminal voltage of the motor, the increasing internal generated voltage decreases the current flow in the motor. When the current flowing in the motor falls to rated current, a section of the starting resistor must be taken out to increase the starting current back up to 200 percent of rated current. As the motor continues to speed up, E_A continues to rise and the armature current continues to fall. When the current flowing in the motor falls to rated current again, another section of the starting resistor must be taken out. This process repeats until the starting resistance to be removed at a given stage is less than the resistance of the motor's armature circuit. At that point, the motor's armature resistance will limit the current to a safe value all by itself.

How many steps are required to accomplish the current limiting? To find out, define R_{tot} as the original resistance in the starting circuit. So R_{tot} is the sum of the resistance of each stage of the starting resistor together with the resistance of the armature circuit of the motor:

$$R_{tot} = R_1 + R_2 + \cdots + R_A \tag{8-58}$$

Now define $R_{tot,\, i}$ as the total resistance left in the starting circuit after stages 1 to i have been shorted out. The resistance left in the circuit after removing stages 1 through i is

$$R_{tot} = R_{i+1} + \cdots + R_A \tag{8-59}$$

Note also that the initial starting resistance must be

$$R_{tot} = \frac{V_T}{I_{max}}$$

In the first stage of the starter circuit, resistance R_1 must be switched out of the circuit when the current I_A falls to

$$I_A = \frac{V_T - E_A}{R_{tot}} = I_{min}$$

After switching that part of the resistance out, the armature current must jump to

$$I_A = \frac{V_T - E_A}{R_{tot,\, 1}} = I_{max}$$

Since $E_A\, (= K\phi\omega)$ is directly proportional to the speed of the motor, which cannot change instantaneously, the quantity $V_T - E_A$ must be constant at the instant the resistance is switched out. Therefore,

$$I_{min}R_{tot} = V_T - E_A = I_{max}R_{tot,\, 1}$$

or the resistance left in the circuit after the first stage is switched out is

$$R_{tot,\, 1} = \frac{I_{min}}{I_{max}} R_{tot} \tag{8-60}$$

By direct extension, the resistance left in the circuit after the nth stage is switched out is

$$R_{tot,\, n} = \left(\frac{I_{min}}{I_{max}}\right)^n R_{tot} \tag{8-61}$$

The starting process is completed when $R_{tot,\, n}$ for stage n is less than or equal to the internal armature resistance R_A of the motor. At that point, R_A can limit the current to the desired value all by itself. At the boundary where $R_A = R_{tot,\, n}$

$$R_A = R_{\text{tot}, n} = \left(\frac{I_{\min}}{I_{\max}}\right)^n R_{\text{tot}} \tag{8-62}$$

$$\frac{R_A}{R_{\text{tot}}} = \left(\frac{I_{\min}}{I_{\max}}\right)^n \tag{8-63}$$

Solving for *n* yields

$$n = \frac{\log(R_A/R_{\text{tot}})}{\log(I_{\min}/I_{\max})} \tag{8-64}$$

where *n* must be rounded up to the next integer value, since it is not possible to have a fractional number of starting stages. If *n* has a fractional part, then when the final stage of starting resistance is removed, the armature current of the motor will jump up to a value smaller than $I_{\max}$.

In this particular problem, the ratio $I_{\min}/I_{\max} = 0.5$, and R_{tot} is

$$R_{\text{tot}} - \frac{V_T}{I_{\max}} = \frac{250 \text{ V}}{700 \text{ A}} = 0.357 \ \Omega$$

so

$$n = \frac{\log(R_A/R_{\text{tot}})}{\log(I_{\min}/I_{\max})} = \frac{\log(0.05 \ \Omega/0.357 \ \Omega)}{\log(350 \text{ A}/700 \text{ A})} = 2.84$$

The number of stages required will be three.

b. The armature circuit will contain the armature resistor R_A and three starting resistors R_1, R_2, and R_3. This arrangement is shown in Figure 8–44.

At first, $E_A = 0$ V and $I_A = 700$ A, so

$$I_A = \frac{V_T}{R_A + R_1 + R_2 + R_3} = 700 \text{ A}$$

Therefore, the total resistance must be

$$R_A + R_1 + R_2 + R_3 = \frac{250 \text{ V}}{700 \text{ A}} = 0.357 \ \Omega \tag{8-65}$$

This total resistance will be placed in the circuit until the current falls to 350 A. This occurs when

$$E_A = V_T - I_A R_{\text{tot}} = 250 \text{ V} - (350 \text{ A})(0.357 \ \Omega) = 125 \text{ V}$$

When $E_A = 125$ V, I_A has fallen to 350 A and it is time to cut out the first starting resistor R_1. When it is cut out, the current should jump back to 700 A. Therefore,

$$R_A + R_2 + R_3 = \frac{V_T - E_A}{I_{\max}} = \frac{250 \text{ V} - 125 \text{ V}}{700 \text{ A}} = 0.1786 \ \Omega \tag{8-66}$$

This total resistance will be in the circuit until I_A again falls to 350 A. This occurs when E_A reaches

$$E_A = V_T - I_A R_{tot} = 250 \text{ V} - (350 \text{ A})(0.1786 \text{ }\Omega) = 187.5 \text{ V}$$

When $E_A = 187.5$ V, I_A has fallen to 350 A and it is time to cut out the second starting resistor R_2. When it is cut out, the current should jump back to 700 A. Therefore,

$$R_A + R_3 = \frac{V_T - E_A}{I_{max}} = \frac{250 \text{ V} - 187.5 \text{ V}}{700 \text{ A}} = 0.0893 \text{ }\Omega \qquad (8\text{--}67)$$

This total resistance will be in the circuit until I_A again falls to 350 A. This occurs when E_A reaches

$$E_A = V_T - I_A R_{tot} = 250 \text{ V} - (350 \text{ A})(0.0893 \text{ }\Omega) = 218.75 \text{ V}$$

When $E_A = 218.75$ V, I_A has fallen to 350 A and it is time to cut out the third starting resistor R_3. When it is cut out, only the internal resistance of the motor is left. By now, though, R_A alone can limit the motor's current to

$$I_A = \frac{V_T - E_A}{R_A} = \frac{250 \text{ V} - 218.75 \text{ V}}{0.05 \text{ }\Omega}$$

$$= 625 \text{ A} \qquad \text{(less than allowed maximum)}$$

From this point on, the motor can speed up by itself.

From Equations (8–63) to (8–65), the required resistor values can be calculated:

$R_3 = R_{tot,\,3} - R_A = 0.0893 \text{ }\Omega - 0.05 \text{ }\Omega = 0.0393 \text{ }\Omega$
$R_2 = R_{tot,\,2} - R_3 - R_A = 0.1786 \text{ }\Omega - 0.0393 \text{ }\Omega - 0.05 \text{ }\Omega = 0.0893 \text{ }\Omega$
$R_1 = R_{tot,\,1} - R_2 - R_3 - R_A = 0.357 \text{ }\Omega - 0.1786 \text{ }\Omega - 0.0393 \text{ }\Omega - 0.05 \text{ }\Omega = 0.1786 \text{ }\Omega$

And R_1, R_2, and R_3 are cut out when E_A reaches 125, 187.5, and 218.75 V, respectively. ∎

DC Motor Starting Circuits

Once the starting resistances have been selected, how can their shorting contacts be controlled to ensure that they shut at exactly the correct moment? Several different schemes are used to accomplish this switching, and two of the most common approaches will be examined in this section. Before that is done, though, it is necessary to introduce some of the components used in motor-starting circuits.

Figure 8–49 illustrates some of the devices commonly used in motor-control circuits. The devices illustrated are fuses, push button switches, relays, time delay relays, and overloads.

Figure 8–49a shows a symbol for a fuse. The fuses in a motor-control circuit serve to protect the motor against the danger of short circuits. They are placed in the power supply lines leading to motors. If a motor develops a short circuit, the fuses in

Figure 8–49 | (a) A fuse. (b) Normally open and normally closed push button switches. (c) A relay coil and contacts. (d) A time delay relay and contacts. (e) An overload and its normally closed contacts.

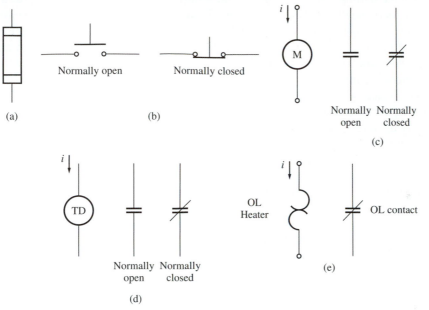

the line leading to it will burn out, opening the circuit before any damage has been done to the motor itself.

Figure 8–49b shows spring-type push button switches. There are two basic types of such switches—normally open and normally shut. *Normally open* contacts are open when the button is resting and closed when the button has been pushed, while *normally closed* contacts are closed when the button is resting and open when the button has been pushed.

A relay is shown in Figure 8–49c. It consists of a main coil and a number of contacts. The main coil is symbolized by a circle, and the contacts are shown as parallel lines. The contacts are of two types—normally open and normally closed. *A normally open* contact is one that is open when the relay is deenergized, and a *normally closed* contact is one that is closed when the relay is deenergized. When electric power is applied to the relay (the relay is energized), its contacts change state: The normally open contacts close, and the normally closed contacts open.

A time delay relay is shown in Figure 8–49d. It behaves exactly like an ordinary relay except that when it is energized there is an adjustable time delay before its contacts change state.

An overload is shown in Figure 8–49e. It consists of a heater coil and some normally shut contacts. The current flowing to a motor passes through the heater coils. If the load on a motor becomes too large, then the current flowing to the motor will heat up the heater coils, which will cause the normally shut contacts of the overload to open. These contacts can in turn activate some types of motor protection circuitry.

One common motor-starting circuit using these components is shown in Figure 8–50. In this circuit, a series of time delay relays shut contacts that remove each section of the starting resistor at approximately the correct time after power is applied to the motor. When the start button is pushed in this circuit, the motor's armature circuit is connected to its power supply, and the machine starts with all resistance in the circuit. However, relay 1TD energizes at the same time as the motor starts, so after some delay the 1TD contacts will shut and remove part of the starting resistance from the circuit. Simultaneously, relay 2TD is energized, so after another time delay the 2TD contacts will shut and remove the second part of the timing resistor. When the 2TD contacts shut, the 3TD relay is energized, so the process repeats again, and finally the motor runs at full speed with no starting resistance present in its circuit. If the time delays are picked properly, the starting resistors can be cut out at just the right times to limit the motor's current to its design values.

Another type of motor starter is shown in Figure 8–51. Here, a series of relays sense the value of E_A in the motor and cut out the starting resistance as E_A rises to preset levels. This type of starter is better than the previous one, since if the motor is loaded heavily and starts more slowly than normal, its armature resistance is still cut out when its current falls to the proper value.

Notice that both starter circuits have a relay in the field circuit labeled FL. This is a *field loss relay*. If the field current is lost for any reason, the field loss relay is deenergized, which turns off power to the M relay. When the M relay deenergizes, its normally open contacts open and disconnect the motor from the power supply. This relay prevents the motor from runaway if its field current is lost.

Notice also that there is an overload in each motor-starter circuit. If the power drawn from the motor becomes excessive, these overloads will heat up and open the OL normally shut contacts, thus turning off the M relay. When the M relay deenergizes, its normally open contacts open and disconnect the motor from the power supply, so the motor is protected against damage due to prolonged excessive loads.

8.13 | DC MOTOR EFFICIENCY CALCULATIONS

To calculate the efficiency of a DC motor, the following losses must be determined:

1. Copper losses
2. Brush drop losses
3. Mechanical losses
4. Core losses
5. Stray losses

The copper losses in the motor are the I^2R losses in the armature and field circuits of the motor. These losses can be found from a knowledge of the currents in the machine and the two resistances. To determine the resistance of the armature circuit in a machine, block its rotor so that it cannot turn and apply a *small* DC voltage to the armature terminals. Adjust that voltage until the current flowing in the armature is equal to the rated armature current of the machine. The ratio of the applied voltage to

Figure 8–50 | A DC motor starting circuit with rising time delay relays to cut out the starting resistor.

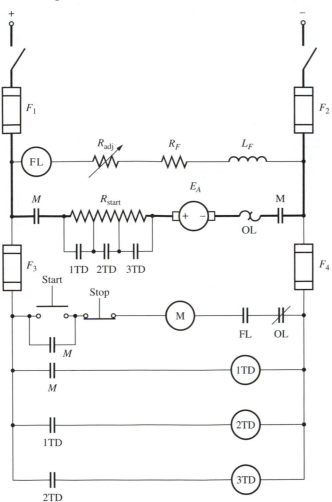

the resulting armature current flow is R_A. The reason that the current should be about equal to full-load value when this test is done is that R_A varies with temperature, and at the full-load value of the current, the armature windings will be near their normal operating temperature.

The resulting resistance will not be entirely accurate, because

1. The cooling that normally occurs when the motor is spinning will not be present.
2. Since there is an AC voltage in the rotor conductors during normal operation, they suffer from some amount of skin effect, which further raises armature resistance.

Figure 8–51 I (a) A DC motor starting circuit using countervoltage-sensing relays to cut out the starting resistor. (b) The armature current in a DC motor during starting.

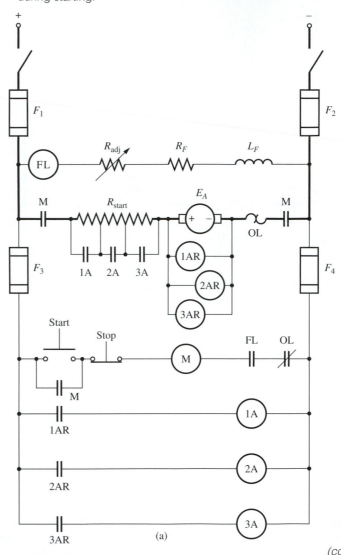

(a)

(continued)

IEEE Standard 113 deals with test procedures for DC machines. It gives a more accurate procedure for determining R_A, which can be used if needed.

The field resistance is determined by supplying the full-rated field voltage to the field circuit and measuring the resulting field current. The field resistance R_F is just the ratio of the field voltage to the field current.

Brush drop losses are often approximately lumped together with copper losses. If they are treated separately, they can be determined from a plot of contact potential

Figure 8–51 | *(concluded)*

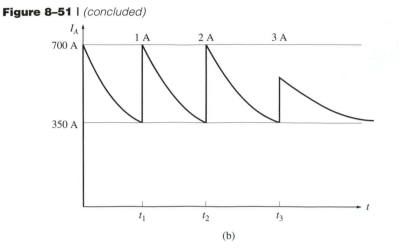

(b)

versus current for the particular type of brush being used. The brush drop losses are just the product of the brush voltage drop V_{BD} and the armature current I_A.

The core and mechanical losses are usually determined together. If a motor is allowed to turn freely at no load and at rated speed, then there is no output power from the machine. Since the motor is at no load, I_A is very small and the armature copper losses are negligible. Therefore, if the field copper losses are subtracted from the input power of the motor, the remaining input power must consist of the mechanical and core losses of the machine at that speed. These losses are called the *no-load rotational losses* of the motor. As long as the motor's speed remains nearly the same as it was when the losses were measured, the no-load rotational losses are a good estimate of mechanical and core losses under load in the machine.

An example of the determination of a motor's efficiency is given below.

EXAMPLE 8–9

A 50-hp, 250-V, 1200 r/min shunt DC motor has a rated armature current of 170 A and a rated field current of 5 A. When its rotor is blocked, an armature voltage of 10.2 V (exclusive of brushes) produces 170 A of current flow, and a field voltage of 250 V produces a field current flow of 5 A. The brush voltage drop is assumed to be 2 V. At no load with the terminal voltage equal to 240 V, the armature current is equal to 13.2 A, the field current is 4.8 A, and the motor's speed is 1150 r/min.

a.   How much power is output from this motor at rated conditions?

b.   What is the motor's efficiency?

■ **Solution**
The armature resistance of this machine is approximately

$$R_A = \frac{10.2 \text{ V}}{170 \text{ A}} = 0.06 \ \Omega$$

and the field resistance is

$$R_F = \frac{250 \text{ V}}{5 \text{ A}} = 50 \text{ }\Omega$$

Therefore, at full load the armature I^2R losses are

$$P_A = (170 \text{ A})^2(0.06 \text{ }\Omega) = 1734 \text{ W}$$

and the field circuit I^2R losses are

$$P_F = (5 \text{ A})^2(50 \text{ }\Omega) = 1250 \text{ W}$$

The brush losses at full load are given by

$$P_{\text{brush}} = V_{\text{BD}}I_A = (2 \text{ V})(170 \text{ A}) = 340 \text{ W}$$

The rotational losses at full load are essentially equivalent to the rotational losses at no load, since the no-load and full-load speeds of the motor do not differ too greatly. These losses may be ascertained by determining the input power to the armature circuit at no load and assuming that the armature copper and brush drop losses are negligible, meaning that the no-load armature input power is equal to the rotational losses.

$$P_{\text{tot}} = P_{\text{core}} + P_{\text{mech}} = (240 \text{ V})(13.2 \text{ A}) = 3168 \text{ W}$$

a.   The input power of this motor at the rated load is given by

$$P_{\text{in}} = V_T I_L = (250 \text{ V})(175 \text{ A}) = 43{,}750 \text{ W}$$

Its output power is given by

$$
\begin{aligned}
P_{\text{out}} &= P_{\text{in}} - P_{\text{brush}} - P_{\text{cu}} - P_{\text{core}} - P_{\text{mech}} - P_{\text{stray}} \\
&= 43{,}750 \text{ W} - 340 \text{ W} - 1734 \text{ W} - 1250 \text{ W} - 3168 \text{ W} - (0.01)(43{,}750 \text{ W}) \\
&= 36{,}820 \text{ W}
\end{aligned}
$$

where the stray losses are taken to be 1 percent of the input power.

b.   The efficiency of this motor at full load is

$$\eta = \frac{P_{\text{out}}}{P_{\text{out}}} \times 100\%$$

$$= \frac{36{,}820 \text{ W}}{43{,}750 \text{ W}} \times 100\% = 84.2\%$$ ∎

8.14 | SUMMARY

DC machines convert mechanical power to DC electric power, and vice versa. In this chapter, the basic principles of DC machine operation were explained first by looking at a machine consisting of a single rotating loop.

The concept of commutation as a technique for converting the AC voltage in rotor conductors to a DC output was introduced, and its problems were explored.

Next, we examined real DC motors. There are several different types of DC motors, differing in the manner in which their field fluxes are derived. These types of motors are separately excited, shunt, permanent-magnet, series, and compounded. The manner in which the flux is derived affects the way it varies with the load, which in turn affects the motor's overall torque–speed characteristic.

A shunt or separately excited DC motor has a torque–speed characteristic whose speed drops linearly with increasing torque. Its speed can be controlled by changing its field current, its armature voltage, or its armature resistance.

A permanent-magnet DC motor is the same basic machine except that its flux is derived from permanent magnets. Its speed can be controlled by any of the above methods except varying the field current.

A series motor has the highest starting torque of any DC motor but tends to over-speed at no load. It is used for very high torque applications where speed regulation is not important, such as a car starter.

A cumulatively compounded DC motor is a compromise between the series and the shunt motor, having some of the best characteristics of each. On the other hand, a differentially compounded DC motor is a complete disaster. It is unstable and tends to overspeed as load is added to it.

8.15 | QUESTIONS

8–1. What is commutation? How can a commutator convert AC voltages on a machine's armature to DC voltages at its terminals?

8–2. Why does curving the pole faces in a DC machine contribute to a smoother DC output voltage from it?

8–3. What is armature reaction? How does it affect the operation of a DC machine?

8–4. Explain the $L\,di/dt$ voltage problem in conductors undergoing commutation.

8–5. What are commutating poles or interpoles? How are they used?

8–6. What are compensating windings? What is their most serious disadvantage?

8–7. What types of losses are present in a DC machine?

8–8. How can the speed of a shunt DC motor be controlled? Explain in detail.

8–9. What is the practical difference between a separately excited and a shunt DC motor?

8–10. What effect does armature reaction have on the torque–speed characteristic of a shunt DC motor? Can the effects of armature reaction be serious? What can be done to remedy this problem?

8–11. What are the desirable characteristics of the permanent magnets in PMDC machines?

8–12. What are the principal characteristics of a series DC motor? What are its uses?

8–13. What are the characteristics of a cumulatively compounded DC motor?

8–14. What are the problems associated with a differentially compounded DC motor?

8–15. What happens in a shunt DC motor if its field circuit opens while it is running?

8–16. Why is a starting resistor used in DC motor circuits?

8–17. How can a DC starting resistor be cut out of a motor's armature circuit at just the right time during starting?

8.16 PROBLEMS

8–1. The following information is given about the simple rotating loop shown in Figure 8–6:

$$B = 0.4 \text{ T} \qquad V_B = 48 \text{ V}$$
$$l = 0.5 \text{ m} \qquad R = 0.4 \text{ }\Omega$$
$$r = 0.25 \text{ m} \qquad \omega = 500 \text{ rad/s}$$

   a. Is this machine operating as a motor or a generator? Explain.

   b. What is the current i flowing into or out of the machine? What is the power flowing into or out of the machine?

   c. If the speed of the rotor were changed to 550 rad/s, what would happen to the current flow into or out of the machine?

   d. If the speed of the rotor were changed to 450 rad/s, what would happen to the current flow into or out of the machine?

8–2. The power converted from one form to another within a DC motor was given by

$$P_{\text{conv}} = E_A I_A = \tau_{\text{ind}} \omega_m$$

Use the equations for E_A and τ_{ind} [Equations (8–30) and (8–31)] to prove that $E_A I_A = \tau_{\text{ind}} \omega_m$; that is, prove that the electric power disappearing at the point of power conversion is exactly equal to the mechanical power appearing at that point.

Problems 8–3 to 8–14 refer to the following DC motor:

$$P_{\text{rated}} = 30 \text{ hp} \qquad\qquad I_{L,\text{ rated}} = 110 \text{ A}$$
$$V_T = 240 \text{ V} \qquad\qquad N_F = 2700 \text{ turns per pole}$$
$$n_{\text{rated}} = 1200 \text{ r/min} \qquad N_{SE} = 14 \text{ turns per pole}$$
$$R_A = 0.19 \text{ }\Omega \qquad\qquad R_F = 75 \text{ }\Omega$$

$R_S = 0.02 \ \Omega$ $R_{adj} = 100 \ \text{to} \ 400 \ \Omega$
Rotational losses = 3550 W at full load
Magnetization curve as shown in Figure P8–1

In Problems 8–3 through 8–9, assume that the motor described on page 438 can be connected in shunt. The equivalent circuit of the shunt motor is shown in Figure P8–2.

8–3. If the resistor R_{adj} is adjusted to 175 Ω what is the rotational speed of the motor at no-load conditions?

8–4. Assuming no armature reaction, what is the speed of the motor at full load? What is the speed regulation of the motor?

8–5. If the motor is operating at full load and if its variable resistance R_{adj} is increased to 250 Ω, what is the new speed of the motor? Compare the full-load speed of the motor with $R_{adj} = 175 \ \Omega$ to the full-load speed with $R_{adj} = 250 \ \Omega$. (Assume no armature reaction, as in the previous problem.)

8–6. Assume that the motor is operating at full load and that the variable resistor R_{adj} is again 175 Ω. If the armature reaction is 1200 A-turns at full load, what is the speed of the motor? How does it compare to the result for Problem 8–5?

8–7. If R_{adj} can be adjusted from 100 to 400 Ω, what are the maximum and minimum no-load speeds possible with this motor?

8–8. What is the starting current of this machine if it is started by connecting it directly to the power supply V_T? How does this starting current compare to the full-load current of the motor?

8–9. Plot the torque–speed characteristic of this motor assuming no armature reaction, and again assuming a full-load armature reaction of 1200 A-turns.

For Problems 8–10 and 8–11, the shunt DC motor is reconnected separately excited, as shown in Figure P8–3. It has a fixed field voltage V_F of 240 V and an armature voltage V_A that can be varied from 120 to 240 V.

8–10. What is the no-load speed of this separately excited motor when $R_{adj} = 175 \ \Omega$ and (a) $V_A = 120 \ \text{V}$, (b) $V_A = 180 \ \text{V}$, (c) $V_A = 240 \ \text{V}$?

8–11. For the separately excited motor of Problem 8–10:

  a. What is the maximum no-load speed attainable by varying both V_A and R_{adj}?

  b. What is the minimum no-load speed attainable by varying both V_A and R_{adj}?

8–12. If the motor is connected cumulatively compounded as shown in Figure P8–4 and if $R_{adj} = 175 \ \Omega$, what is its no-load speed? What is its full-load speed? What is its speed regulation? Calculate and plot the torque–speed characteristic for this motor. (Neglect armature effects in this problem.)

Figure P8–1 | The magnetization curve for the DC motor in Problems 8–3 to 8–14. This curve was made at a constant speed of 1200 r/min.

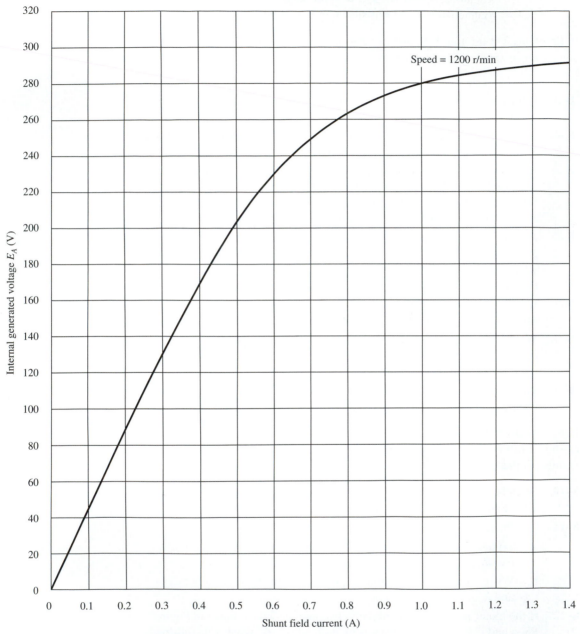

Speed = 1200 r/min

Internal generated voltage E_A (V)

Shunt field current (A)

8–13. The motor is connected cumulatively compounded and is operating at full load. What will the new speed of the motor be if R_{adj} is increased to 250 Ω? How does the new speed compared to the full-load speed calculated in Problem 8–12?

Figure P8–2 | The equivalent circuit of the shunt motor in Problems 8–3 to 8–9.

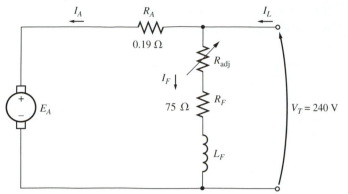

Figure P8–3 | The equivalent circuit of the separately excited motor in
Problems 8–10 and 8–11.

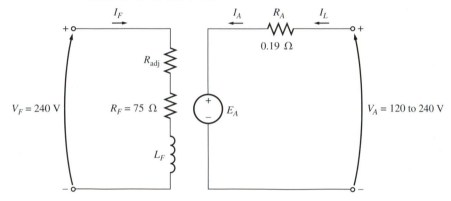

Figure P8–4 | The equivalent circuit of the compounded motor in Problems 8–12
to 8–14.

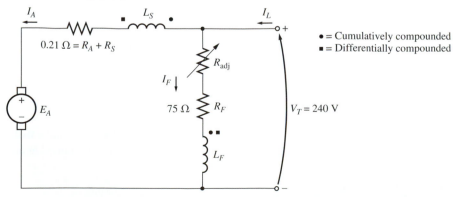

8–14. The motor is now connected differentially compounded.

    a.   If $R_{adj} = 175\ \Omega$, what is the no-load speed of the motor?

    b.   What is the motor's speed when the armature current reaches 40 A? 60 A?

    c.   Calculate and plot the torque–speed characteristic curve of this motor.

8–15. A 15-hp, 120-V series DC motor has an armature resistance of $0.1\ \Omega$ and a series field resistance of $0.08\ \Omega$. At full load, the current input is 115 A, and the rated speed is 1050 r/min. Its magnetization curve is shown in Figure P8–5. The core losses are 420 W, and the mechanical losses are 460 W at full load. Assume that the mechanical losses vary as the cube of the speed of the motor and that the core losses are constant.

    a.   What is the efficiency of the motor at full load?

    b.   What are the speed and efficiency of the motor if it is operating at an armature current of 70 A?

    c.   Plot the torque–speed characteristic for this motor.

8–16. A 20-hp, 240-V, 76-A, 900 r/min series motor has a field winding of 33 turns per pole. Its armature resistance is $0.09\ \Omega$, and its field resistance is $0.06\ \Omega$. The magnetization curve expressed in terms of magnetomotive force versus E_A at 900 r/min is given by the following table:

E_A, V	95	150	188	212	229	243
$\mathcal{F}$, A-turns	500	1000	1500	2000	2500	3000

Armature reaction is negligible in this machine.

    a.   Compute the motor's torque, speed, and output power at 33, 67, 100, and 133 percent of full-load armature current. (Neglect rotational losses.)

    b.   Plot the terminal characteristic of this machine.

8–17. A 300-hp, 440-V, 560-A, 863 r/min shunt DC motor has been tested, and the following data were taken:

Blocked-rotor test:

$$V_A = 16.3 \text{ V exclusive of brushes} \qquad V_F = 440 \text{ V}$$
$$I_A = 500 \text{ A} \qquad\qquad\qquad\qquad I_F = 8.86 \text{ A}$$

No-load operation:

$$V_A = 16.3 \text{ V including brushes} \qquad I_F = 8.76 \text{ A}$$
$$I_A = 23.1 \text{ A} \qquad\qquad\qquad\qquad n = 863 \text{ r/min}$$

What is this motor's efficiency at the rated conditions? [*Note*: Assume that (1) the brush voltage drop is 2 V; (2) the core loss is to be determined at an

Figure P8–5 | The magnetization curve for the series motor in Problem 8–15. This curve was taken at a constant speed of 1200 r/min.

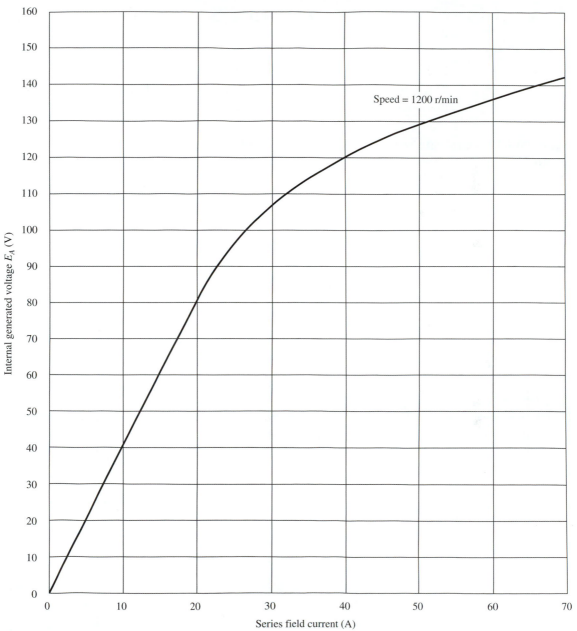

Internal generated voltage E_A (V)

Series field current (A)

Speed = 1200 r/min

armature voltage equal to the armature voltage under full load; and (3) stray load losses are 1 percent of full load.]

Problems 8–18 to 8–21 refer to a 240-V 100-A DC motor that has both shunt and series windings. Its characteristics are

$$R_A = 0.14 \ \Omega \qquad N_F = 1500 \text{ turns}$$
$$R_S = 0.05 \ \Omega \qquad N_{SE} = 15 \text{ turns}$$
$$R_F = 200 \ \Omega \qquad n_m = 1800 \text{ r/min}$$
$$R_{adj} = 0 \text{ to } 300 \ \Omega, \text{ currently set to } 120 \ \Omega$$

This motor has compensating windings and interpoles. The magnetization curve for this motor at 1800 r/min is shown in Figure P8–6.

Figure P8–6 | The magnetization curve for the DC motor in Problems 8–18 to 8–21.

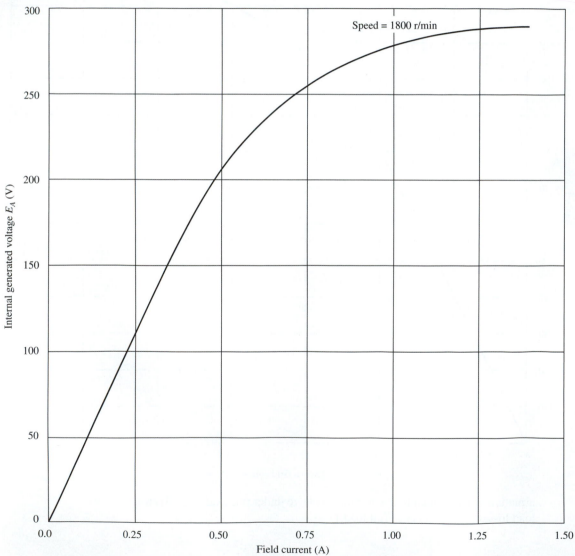

8–18. The motor described is connected in *shunt*.

    a. What is the no-load speed of this motor when $R_{adj} = 120\ \Omega$?

    b. What is its full-load speed?

    c. Under no-load conditions, what range of possible speeds can be achieved by adjusting R_{adj}?

8–19. This machine is now connected as a cumulatively compounded DC motor with $R_{adj} = 120\ \Omega$.

    a. What is the full-load speed of this motor?

    b. Plot the torque–speed characteristic for this motor.

    c. What is its speed regulation?

8–20. The motor is reconnected differentially compounded with $R_{adj} = 120\ \Omega$. Derive the shape of its torque–speed characteristic.

8–21. A series motor is now constructed from this machine by leaving the shunt field out entirely. Derive the torque–speed characteristic of the resulting motor.

8–22. An automatic starter circuit is to be designed for a shunt motor rated at 20 hp, 240 V, and 80 A. The armature resistance of the motor is $0.12\ \Omega$, and the shunt field resistance is $40\ \Omega$. The motor is to start with no more than 250 percent of its rated armature current, and as soon as the current falls to rated value, a starting resistor stage is to be cut out. How many stages of starting resistance are needed, and how big should each one be?

8.17 | REFERENCES

1. Chapman, Stephen J.: *Electric Machinery Fundamentals*, 3rd ed., McGraw-Hill, Burr Ridge, Ill., 1999.

2. Fitzgerald, A. E., and C. Kingsley, Jr.: *Electric Machinery*, McGraw-Hill, New York, 1952.

3. Fitzgerald, A. E., C. Kingsley, Jr., and S. D. Umans: *Electric Machinery*, 5th ed., McGraw-Hill, New York, 1990.

4. Heck, C.: *Magnetic Materials and Their Applications*, Butterworth, London, 1974.

5. IEEE Standard 113-1985, *Guide on Test Procedures for DC Machines*, IEEE, Piscataway, N.J., 1985. (Note that this standard has been officially withdrawn but is still available.)

6. Kloeffler, S. M., R. M. Kerchner, and J. L. Brenneman: *Direct Current Machinery*, rev. ed., Macmillan, New York, 1948.

7. Kosow, Irving L.: *Electric Machinery and Transformers*, Prentice-Hall, Englewood Cliffs, N.J., 1972.

8. McPherson, George: *An Introduction to Electrical Machines and Transformers*, Wiley, New York, 1981.

9. Siskind, Charles S.: *Direct-Current Machinery*, McGraw-Hill, New York, 1952.

10. Slemon, G. R., and A. Straughen: *Electric Machines*, Addison-Wesley, Reading, Mass., 1980.

CHAPTER 9

Transmission Lines

Generators and load are connected together through *transmission lines*, which transport electric power from the place where it is generated to the place where it is used. Efficient transmission lines are very important to modern power systems, because power generation is usually done at large electric generating stations located a long way from the cities and industries where the power is consumed. Most electric power consumers use small amounts of power, and they are scattered over wide areas. Transmissions lines must take the bulk power from the generators, transmit it to the locality where it will be used, and then distribute it to the individual homes and factories.

As a rule of thumb, the power handling capability of a transmission line is proportional to the square of the voltage on the line. Therefore, very high voltage transmission lines are used to transmit electric power over long distances to the areas where it will be used. Once the power reaches the area where it will be used, it is stepped down to a lower voltage in distribution substations, and then delivered to the individual customers on lower-voltage transmission lines called *distribution lines*.

There are two types of power transmission lines: overhead lines and buried cables. An overhead transmission line usually consists of three conductors or bundles of conductors containing the three phases of the power system. The conductors are usually *aluminum cable steel reinforced* (ACSR), with a steel core and aluminum conductors wrapped around the core. The steel core provides strength, while the aluminum conductors have a low resistance to minimize the losses in the transmission line. In an overhead transmission line, the conductors are suspended from a pole or tower via insulators. In addition to the phase conductors, a transmission line usually includes one or two steel wires known as *ground* or *shield* wires. These wires are electrically connected to the tower and the ground, and are thus at ground potential. In large transmission lines, these wires are located above the phase conductors, shielding them from lightning strikes, etc. Some examples of transmission and distribution lines are shown in Figure 9–1.

Figure 9–1 | Power lines come in a wide variety of constructions depending on voltage and local conditions: (a) Dual 345-kV transmission lines on a steel tower. (b) Dual 110-kV transmission lines on wooden poles. (c) A 13.8-kV distribution line in Louisiana. Because of the high incidence of lightning there, the ground wire is placed above the three phases. (d) A distribution line from South Australia with no ground wire. Because of a shortage of good wood in the state, these "Stobie" poles are constructed of concrete and steel.

(b)

(a)

(continued)

Cable lines are designed to be placed underground or under the water. In cables, the conductors are insulated from one another and surrounded by a protective sheath. Cable lines tend to be more expensive than overhead transmission lines, and they are harder to maintain and repair. Furthermore, cable lines have capacitance problems that prevent them from being used over very long distances. Nevertheless, they are increasingly popular in new urban areas where overhead transmission lines are considered an eyesore.

Transmission lines are characterized by a series resistance and inductance per unit length, and by a shunt capacitance per unit length. These values control the power-carrying capacity of each transmission line, and the voltage drop in the transmission line at full load. In this chapter, we will study the nature of each of these quantities, and learn how to calculate them in the simple case of a single-phase

Figure 9–1 | *(concluded)*

(c)   (d)

two-wire line. We will then use the equations for this line to explain the how conductor size and spacing affect resistance, inductance, and capacitance in any transmission line.

The more complex inductance and capacitance calculations for three-phase transmission lines with varying geometries and conductor bundle sizes will *not* be derived here. Most power engineers will never calculate the resistance, inductance, and capacitance per unit length of a transmission line—instead, they will simply look them up in a table provided by the line designers. Therefore, this chapter will concentrate more on how to *use* the transmission line characteristics, and not on how to calculate them. Refer to Reference 2 for more details on how to calculate transmission line values in complex situations.

9.1 | RESISTANCE

The DC resistance of a conductor is given by the equation

$$R_{DC} = \frac{\rho l}{A} \tag{9–1}$$

where

l = length of the conductor

A = cross-sectional area of the conductor

ρ = resistivity of the conductor

In SI units, length is measure in meters (m), area is measured in square meters (m^2), and resistivity is measured in ohm-meters (Ω-m). The DC *resistance per meter* of the conductor is given by the equation

$$r_{DC} = \frac{\rho}{A} \qquad (9\text{–}2)$$

where r_{DC} is measured in ohms per meter (Ω/m).

The resistivity of a conductor is a fundamental property of the material that the conductor is made from. It varies with both on the type and temperature of the material. For example, the resistivity of hard-drawn copper at 20° C is 1.77×10^{-8} Ω-m, and the resistivity of aluminum at 20° C is 2.83×10^{-8} Ω-m.

The resistivity of a conductor increases *linearly* with temperature over the normal range of temperatures found in transmission lines. If the resistivity of a conductor is known at one temperature, then it can be found at another temperature using the following equation:

$$\rho_{T2} = \frac{M + T_2}{M + T_1} \rho_{T1} \qquad (9\text{–}3)$$

where

T_1 = temperature 1, in °C

T_2 = temperature 2, in °C

ρ_{T1} = resistivity at temperature T_1

ρ_{T2} = resistivity at temperature T_2

M = temperature constant

The temperature constant M is the crossover temperature for the extrapolated resistivity of a particular conductor. Table 9–1 contains the resistivities at 20°C and temperature constants for several common types of conducting materials.

Table 9–1 | Resistivity and temperature constants of different materials

Material	Resistivity at 20°C (Ω-m)	Temperature constant (°C)
Annealed copper	1.72×10^{-8}	234.5
Hard-drawn copper	1.77×10^{-8}	241.5
Aluminum	2.83×10^{-8}	228.1
Iron	10.00×10^{-8}	180.0
Silver	1.59×10^{-8}	243.0

Notice from Table 9–1 that copper is a better conductor than aluminum, and that silver is an even better conductor than copper. If that is so, why are most transmission lines made out of aluminum? The answer is that aluminum is *much* cheaper than copper or silver. Since aluminum is cheap and relatively light, conductors made of aluminum can be bigger in diameter than the corresponding copper conductors, lowering the resistance of the line by having a larger area A to offset the higher resistivity ρ of the material.

The AC resistance of a conductor is always higher than its DC resistance because of *skin effect*. The AC current distribution in a conductor is not uniform. As frequency increases, more and more of the current is concentrated near the outer surface of the conductor. This concentration reduces the effective area of the conductor, raising its resistance in accordance with Equation (9–2),

The details of how to calculate the AC resistance of a conductor from its DC resistance and the frequency of the current can be found in Reference 2. As a practical matter, all wire vendors supply tables of the resistance per unit length for each conductor they manufacture at all common frequencies (50 and 60 Hz). Most power engineers will simply look up the values that they need from the vendor's tables.

9.2 | INDUCTANCE AND INDUCTIVE REACTANCE

The series inductance of a transmission line consists of two components—its internal inductance and its external inductance. The internal inductance is due to the magnetic flux inside the conductor, while the external inductance is due to the magnetic flux outside the conductor. We will consider the internal and external inductance separately.

The inductance of a transmission line is defined as the number of flux linkages produced per ampere of current flowing through the line.

$$L = \frac{\lambda}{I} \qquad (9\text{–}4)$$

The flux linkages are measured in weber-turns (Wb-turns). To calculate the internal and external inductances of a line, we must calculate the internal and external flux linkages and divide by the current flow.

Internal Inductance

A conductor of radius r carrying current I is shown in Figure 9–2. The magnetic field intensity at a distance x from the center of this conductor is given the symbol H_x. The magnitude of this magnetic intensity can be found from Ampere's law:

$$\oint \mathbf{H}_x \cdot d\mathbf{l} = I_x \qquad (9\text{–}5)$$

where $\mathbf{H}_x$ is the magnetic field intensity at each point along a closed path, $d\mathbf{l}$ is a vector of length dl along the path, and I_x is the net current enclosed in the path. For the circular path of radius x shown in Figure 9–2, the magnitude of $\mathbf{H}_x$ is constant, and the vector $d\mathbf{l}$ is always parallel to $\mathbf{H}_x$. Therefore, Equation (9–5) reduces to

$$2\pi x H_x = I_x$$

or
$$H_x = \frac{I_x}{2\pi x} \tag{9–6}$$

The units of magnetic intensity are henrys per meter (H/m).

Figure 9–2 | Cross section of a current-carrying conductor. The current flow in this conductor is out of the page.

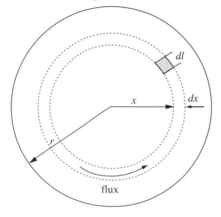

If we assume that the current is distributed uniformly in the conductor, then the net current enclosed in a circular path of radius x will be the total current in the conductor times the ratio of the area inside the path to the total area of the conductor.

$$I_x = \frac{\pi x^2}{\pi r^2} I \tag{9–7}$$

Therefore, the magnetic intensity at any radius x inside the conductor will be

$$H_x = \frac{x}{2\pi r^2} I \tag{9–8}$$

The flux density B_x at a distance x from the center of the conductor can be found from Equation (9–8) and Equation (1–21).

$$\mathbf{B} = \mu \mathbf{H} \tag{1–21}$$

Substituting Equation (1–21) into Equation (9–8) yields the desired expression.

$$B_x = \mu H_x = \frac{\mu x}{2\pi r^2} I \tag{9–9}$$

The units of flux density are webers per square meter, or teslas (T).

The differential magnetic flux contained in a circular tube of thickness dx and a distance x from the center of the conductor will be the flux density times the cross-sectional area of the element normal to the flux lines, which is dx times the axial length.

$$d\phi = \frac{\mu x I}{2\pi r^2} dx \quad \text{Wb/m} \tag{9-10}$$

The flux linkages $d\lambda$ per meter of length due to the flux in the tube are the product of the flux per meter and the fraction of the current linked.

$$d\lambda = \frac{\pi x^2}{\pi r^2} d\phi = \frac{\mu x^3 I}{2\pi r^4} dx \quad \text{Wb-turns/m} \tag{9-11}$$

We can find the total internal flux linkages per meter by integrating from the center of the conductor to the outside edge.

$$\lambda_{\text{int}} = \int d\lambda = \int_0^r \frac{\mu x^3 I}{2\pi r^4} dx = \frac{\mu I}{8\pi} \quad \text{Wb-turns/m} \tag{9-12}$$

Finally, we can substitute Equation (9–12) into Equation (9–4) to calculate the internal inductance per meter.

$$\boxed{l_{\text{int}} = \frac{\lambda_{\text{int}}}{I} = \frac{\mu}{8\pi} \quad \text{H/m}} \tag{9-13}$$

If the relative permeability of the conductor is 1 (as it would be for nonferromagnetic materials such as copper and aluminum), this inductance reduces to

$$l_{\text{int}} = \frac{\lambda_{\text{int}}}{I} = \frac{\mu_0}{8\pi} = \frac{4\pi \times 10^{-7}}{8\pi} = \frac{1}{2} \times 10^{-7} \text{ H/m} \tag{9-14}$$

External Inductance Between Two Points Outside of the Line

To calculate the inductance external to a conductor, we must be able to calculate the flux linkages of the conductor due only to the portion of flux between two points P_1 and P_2 that lie at distances D_1 and D_2 respectively from the center of the conductor (see Figure 9–3).

In the region external to the conductor, the magnetic intensity at a distance x from the center of the conductor is still given by Equation (9–6)

or
$$H_x = \frac{I_x}{2\pi x} \tag{9-6}$$

However, this time *all* of the current in the conductor lies within the tube, so I_x is equal to the total current I in the conductor.

$$H_x = \frac{I}{2\pi x} \tag{9-15}$$

Figure 9–3 | A conductor and two external points P_1 and P_2.

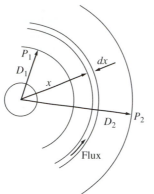

The flux density B_x at a distance x from the center of the conductor can be found from Equation (9–15) and Equation (1–21):

$$B_x = \mu H_x = \frac{\mu I}{2\pi x} \qquad (9\text{–}16)$$

The units of flux density are webers per square meter, or teslas (T).

The differential magnetic flux contained in a circular tube of thickness dx and a distance x from the center of the conductor will be the flux density times the cross-sectional area of the element normal to the flux lines, which is dx times the axial length.

$$d\phi = \frac{\mu I}{2\pi x}\, dx \quad \text{Wb/m} \qquad (9\text{–}17)$$

Since this flux links the full current carried by the conductor, $d\lambda = dx$, and

$$d\lambda = d\phi = \frac{\mu I}{2\pi x}\, dx \quad \text{Wb-turns/m} \qquad (9\text{–}18)$$

Therefore, the total flux linkage per meter between any two points P_1 and P_2 at distances D_1 and D_2 from the center of the conductor can be found as

$$\lambda_{\text{ext}} = \int_{D_1}^{D_2} d\lambda = \int_{D_1}^{D_2} \frac{\mu I}{2\pi x}\, dx = \frac{\mu I}{2\pi} \ln \frac{D_2}{D_1} \quad \text{Wb-turns/m} \qquad (9\text{–}19)$$

Finally, we can substitute Equation (9–19) into Equation (9–4) to calculate the external inductance per meter due only to the flux between points P_1 and P_2.

$$\boxed{l_{\text{ext}} = \frac{\lambda_{\text{ext}}}{I} = \frac{\mu}{2\pi} \ln \frac{D_2}{D_1} \quad \text{H/m}} \qquad (9\text{–}20)$$

Inductance of a Single-Phase Two-Wire Transmission Line

We now know enough to determine the series inductance of the single-phase two-wire transmission line shown in Figure 9–4. This figure shows the two conductors each of radius r and separated by a distance D, with current flowing into the page in the left-hand conductor and current flowing out of the page in the right-hand conductor. The magnitudes of the currents I flowing in the two conductors are equal an opposite.

Figure 9–4 | A single-phase two-wire transmission line. Each conductor has radius r, and the spacing between conductors is D.

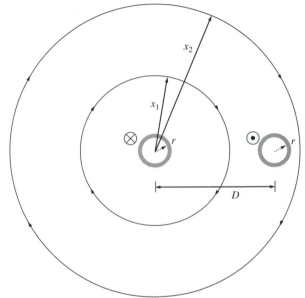

The figure also shows two circular paths around the left-hand conductor. The inner path of radius x_1 encloses only the left-hand conductor, so the line integral along that path will produce a net magnetic intensity, because there is a net current enclosed in the path.

$$\oint \mathbf{H}_x \cdot d\mathbf{l} = I_x \tag{9–5}$$

On the other hand, the outer path of radius x_2 encloses *both* conductors. Since the currents are equal and opposite in the two conductors, the *net* current enclosed by this path is 0! Therefore, there is no net contribution to the total inductance from the magnetic field at distances greater than D, the separation between the two conductors.

To find the total inductance per unit length of a phase in this transmission line, we must add up the internal inductance of the conductor and the external inductance between distance r (the surface of the conductor) and D (the largest distance that encloses a net current). These two components of inductance can be found from Equations (9–13) and (9–20). Therefore, the total inductance per unit length for a phase in this transmission line is

$$l = l_{int} + l_{ext} = \frac{\mu}{2\pi}\left(\frac{1}{4} + \ln\frac{D}{r}\right) \quad \text{H/m} \tag{9-21}$$

By symmetry, the total inductance of the other phase is the same, so the total inductance of the two-phase transmission line is

$$l = \frac{\mu}{\pi}\left(\frac{1}{4} + \ln\frac{D}{r}\right) \text{H/m} \tag{9-22}$$

where r is the radius of each conductor and D is the distance between conductors.

Understanding the Inductance of a Transmission Line

Equation (9–22) describes the inductance per unit length of a single-phase, two-wire transmission line. Similar equations can be derived for three-phase lines, and for lines using more than one wire in each phase. The mathematics involved is tedious, and depends on the geometry of the transmission line. It is not covered in this text. As a practical matter, most working engineers will find the inductance of a transmission line from lookup tables provided by the designers of the line. See Reference 2 if you would like to get more details about how to perform these calculations for arbitrary arrangements of conductors in two- and three-phase transmission lines.

Instead of spending a lot of time with various special cases, we will use Equation (9–22) to derive a general understanding about the behavior of inductance in all transmission lines. The key points are summarized below:

1. *The greater the spacing between the phases of a transmission line, the greater the inductance of the line.* This point is obvious from Equation (9–22). If the distance between the phases increases, the ratio D/r increases, and therefore the total inductance of the line increases. Since the phases of higher-voltage transmission lines must be spaced further apart for insulation purposes, *a single-conductor high-voltage line will tend to have a higher inductance than a single-conductor low-voltage line.*

Note also that the spacing between lines in cables is very small, so *the series inductance of buried cables will be much smaller than the inductance of overhead transmission lines.*

2. *The greater the radius of the conductors in a transmission line, the lower the inductance of the line.* If the radius of a conductor r is increased, the ratio D/r decreases, and therefore the total inductance of the line decreases. Practical transmission lines do not use conductors of extremely large radius, because they would be very heavy, inflexible, and expensive. However, they *do* approximate this behavior by bundling two, three or more conductors together in each phase, as shown in Figure 9–5. If a series of individual conductors is distributed along the circumference of a circle, then the behavior of the bundle is approximately like the behavior of a single conductor with a very large radius. The more conductors included in the bundle, the better this approximation becomes. This bundling of conductors in a phase can also be seen in the photographs in Figure 9–1.

Figure 9–5 | A bundle of conductors arranged along the circumference of a circle behaves approximately like a single conductor with the radius of the circle. The more conductors included in the bundle, the better the approximation becomes. (a) Three conductors in bundle. (b) Four conductors in bundle. (c) Photo of a transmission line with two conductors in bundle. (d) Photo of a transmission line with four conductors in bundle.

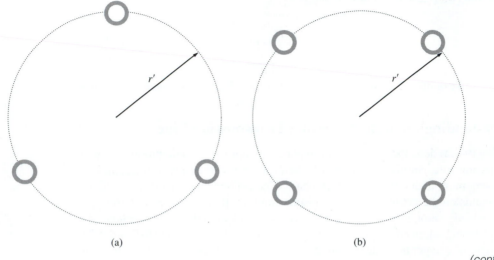

(a)   (b)

(continued)

Very high voltage transmission lines usually include two, three, or more parallel conductors per phase spaced at locations on the circumference of a circle. The parallel conductors reduce the series resistance of the transmission line, and the larger effective radius of the bundle reduces the series inductance of the transmission line, partially compensating for the increased inductance due to wider phase spacing.

Inductive Reactance

The series inductive reactance of a transmission line depends on both the inductance of the line and the frequency of the power system. If the inductance per unit length of a line is l, then the inductive reactance per unit length will be

$$x = j\omega l = j2\pi fl \tag{9–23}$$

where f is the frequency of the transmission line. To find the total series inductive reactance of a transmission line, we can just multiply the inductive reactance per unit length times the length of the line

$$X = xd \tag{9–24}$$

where d is the length of the line.

9.3 | CAPACITANCE AND CAPACITIVE REACTANCE

When a voltage is applied to a pair of conductors separated by a nonconducting dielectric medium, charges of equal magnitude but opposite sign accumulate on the

Figure 9–5 *(concluded)*

(c)

(d)

two conductors. The charge deposited on the conductors is proportional to the applied voltage, and the constant of proportionality is the capacitance C.

$$q = CV \qquad (9\text{–}25)$$

where

q = the charge on the conductors in coulombs

V = the voltage between the conductors in volts

C = the capacitance between the pair of conductors in farads

Traditional capacitors are designed as two parallel plates separated by a dielectric material, because this design maximizes capacitance. However, there is a capacitance between *any* two conductors, including the phases of a power system.

In AC power systems, a transmission line carries a time-varying voltage that differs in each phase. This alternating voltage causes the charges on individual phases to increase and decrease with time in accordance with Equation (9–25), and these changing charges produce a *charging current*. The charging current increases the current in the transmission line, and changes the power factor and voltage drop of the line. Note that charging current flows in a transmission line even when it is open circuited.

The capacitance of a transmission line can be determined using Gauss's law, just as the inductance of a transmission line was found from Ampere's law. Gauss's law states that the total electric charge within a closed surface is equal to the total electric flux emerging from the surface. In other words, the total charge within the closed surface is equal to the integral over the surface of the normal components of the electric flux density.

$$\oint_A \mathbf{D} \cdot d\mathbf{A} = q \tag{9-26}$$

where

q = the charge inside the surface, in coulombs
$\mathbf{D}$ = the electric flux density at the surface, in C/m^2
$d\mathbf{A}$ = the unit vector normal to the surface, in m^2

The electric flux density $\mathbf{D}$ at a point is given by the equation

$$\mathbf{D} = \varepsilon\mathbf{E} \tag{9-27}$$

where ε is the permittivity of the material surrounding the conductor, and $\mathbf{E}$ is the electric field intensity at that point. The permittivity of free space ε_0 is 8.85×10^{-12} farads per meter, and the permittivity of any other material is expressed as

$$\varepsilon = \varepsilon_r \varepsilon_0 \tag{9-28}$$

where ε_r is the *relative permittivity* of the material. The relative permittivity of air is almost exactly 1.0, so the permittivity of the air around an overhead transmission line is essentially equal to the free space permittivity ε_0.

The Electric Field Around a Long, Straight Conductor

The voltage or potential difference V between two points is equal to the work in joules needed to move a coulomb of charge from one point to the other one. The electric field intensity $\mathbf{E}$ is the force exerted on a coulomb of charge at a given point in the electric field. The units of potential difference are volts, and the units of electric field intensity are either newtons per coulomb or volts per meter (they are equivalent units).

Since capacitance is defined as the ratio of charge to voltage between two points, we must be able to calculate the voltage difference between those points in order to calculate capacitance. In this section, we will learn how to calculate the potential difference (voltage) between two points near an isolated conductor, and in the next section we will extend that calculation to determine the capacitance between a pair of conductors.

An isolated conductor with a positive charge on its surface is shown in Figure 9–6. Note that the lines of electric flux radiate uniformly outward from the surface of the conductor. For this nice, uniform situation, the flux density vector $\mathbf{D}$ is always

Figure 9–6 | Lines of electric flux originate on positive charges and terminate on negative charges. For a positively charged conductor in isolation, the lines of flux are uniformly distributed around the radius of the conductor.

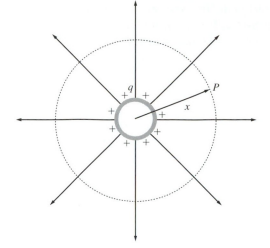

parallel to the normal vector $d\mathbf{A}$, and $\mathbf{D}$ is constant in magnitude at all points around a path of constant radius r from the center of the conductor. Therefore, Gauss's law reduces to

$$DA = Q \tag{9-29}$$
$$D(2\pi xl) = ql \tag{9-30}$$

where l is the length of the conductor being considered, q is the charge on the conductor per unit length, and Q is the total charge on the conductor. Therefore, the flux density D can be expressed in terms of the charge per unit length of the conductor and the distance from the conductor to the point P as

$$D = \frac{q}{2\pi x} \tag{9-31}$$

From Equations (9–27) and (9–31), the electric field intensity E at point P can be expressed as

$$E = \frac{q}{2\pi \varepsilon x} \tag{9-32}$$

Note that the direction of $\mathbf{E}$ is always radially outward from the conductor.

The potential difference between two points P_1 and P_2 is the integral of the electric field intensity $\mathbf{E}$ along the path of integration between the points.

$$V_{12} = \int_{P_1}^{P_2} \mathbf{E} \cdot d\mathbf{l} \tag{9-33}$$

where $d\mathbf{l}$ is a differential element tangential to the path of integration between P_1 and P_2. It is a property of electric fields that the path taken between the two points is irrelevant—the answer will always be the same regardless of the path taken.

By a clever choice of path between the points, we can make this calculation easy. Figure 9–7 shows a path between the two points that goes radially outward from P_1 to P_{int}, and then along a circle of constant radius from P_{int} to P_2. Note that the vectors **E** and $d\mathbf{l}$ are parallel in the first segment, and perpendicular in the second segment. Therefore, the dot product $\mathbf{E} \cdot d\mathbf{l}$ is just $E\, dx$ in the first segment, and zero in the second segment. Equation (9–33) thus reduces to

$$V_{12} = \int_{D_1}^{D_2} E\, dx \tag{9–34}$$

$$= \int_{D_1}^{D_2} \frac{q}{2\pi\varepsilon x}\, dx$$

$$= \frac{q}{2\pi\varepsilon} \ln\left(\frac{D_2}{D_1}\right) \tag{9–35}$$

Figure 9–7 | If the path of integration between points P_1 and P_2 is chosen to be two separate components, one radial and one circumferential, then the integration becomes very simple. The radial portion of the path is parallel to the **E** vector, and the circumferential portion of the path is perpendicular to the **E** vector.

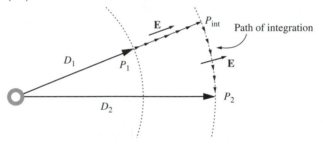

The Capacitance of a Single-Phase Two-Wire Transmission Line

To determine the capacitance per unit length of a single-phase two-wire transmission line, we must calculate the voltage difference between the two lines for a given amount of charge on the line.

Consider a transmission line consisting of two parallel conductors of equal radius separated by a distance D, as shown in Figure 9–8. To determine the potential difference between the two conductors, we must first determine the potential difference between a point at the surface of the first conductor and the location of the second conductor due to the charge on the conductor a. Then, we must determine the potential difference between a point at the surface of the second conductor and the location of the first conductor due to the charge on conductor b. Finally, we will sum up the two potential differences to get the total voltage between the conductors.

The potential difference due to the charge on conductor a can be found by applying Equation (9–35):

Figure 9–8 | A single-phase two-wire transmission line with a pair of conductors of radius r separated by a distance D.

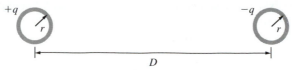

$$V_{ab,\,a} = \frac{q_a}{2\pi\varepsilon} \ln\left(\frac{D}{r}\right) \qquad (9\text{–}36)$$

Similarly, the potential difference due to the charge on conductor b is

$$V_{ba,\,b} = \frac{q_b}{2\pi\varepsilon} \ln\left(\frac{D}{r}\right) \qquad (9\text{–}37)$$

or

$$V_{ab,\,b} = -\frac{q_b}{2\pi\varepsilon} \ln\left(\frac{D}{r}\right) \qquad (9\text{–}38)$$

The total voltage between the lines is thus

$$V_{ab} = V_{ab,\,a} + V_{ab,\,b} = \frac{q_a}{2\pi\varepsilon} \ln\left(\frac{D}{r}\right) - \frac{q_b}{2\pi\varepsilon} \ln\left(\frac{D}{r}\right) \qquad (9\text{–}39)$$

Since $q_a = -q_b = q$, this equation reduces to

$$V_{ab} = \frac{q}{\pi\varepsilon} \ln\left(\frac{D}{r}\right) \qquad (9\text{–}40)$$

The capacitance per unit length between the two conductors of the transmission line is thus

$$c_{ab} = \frac{q}{V} = \frac{q}{\dfrac{q}{\pi\varepsilon} \ln\left(\dfrac{D}{r}\right)}$$

$$\boxed{c_{ab} = \frac{\pi\varepsilon}{\ln\left(\dfrac{D}{r}\right)}} \qquad (9\text{–}41)$$

The potential difference between each conductor and the ground or neutral is one-half the total potential difference between the two conductors, so the *capacitance to ground* of this single-phase transmission line will be

$$c_n = c_{an} = c_{bn} = \frac{2\pi\varepsilon}{\ln\left(\dfrac{D}{r}\right)} \qquad (9\text{–}42)$$

Understanding the Capacitance of a Transmission Line

Equation (9–41) describes the capacitance per unit length of a single-phase, two-wire transmission line. Similar equations can be derived for three-phase lines with various geometries. The mathematics involved is tedious, and is not covered in this text. As a practical matter, most working engineers will find the shunt capacitance of a transmission line from lookup tables provided by the designers of the line. See Reference 2 if you would like to get more details about how to perform these calculations for arbitrary arrangements of conductors in two- and three-phase transmission lines.

Instead of spending a lot of time with various special cases, we will use Equation (9–41) to derive a general understanding about the behavior of inductance in all transmission lines. The key points are summarized below:

1. *The greater the spacing between the phases of a transmission line, the lower the capacitance of the line.* This point is obvious from Equation (9–42). If the distance between the phases increases, the ratio D/r increases, and therefore the total capacitance of the line decreases. Since the phases of higher-voltage transmission lines must be spaced farther apart for insulation purposes, *a single-conductor high-voltage line will tend to have a lower capacitance than a single-conductor low-voltage line*.

Note also that the spacing between conductors in a cable is very small, so *the shunt capacitance of buried cables will be much larger than the shunt capacitance of overhead transmission lines*. This high capacitance means that a cable has a *much* higher shunt admittance than an overhead transmission line of equivalent length, and therefore a *much* higher charging current. This high charging current limits the maximum length of cables in power systems. Overhead transmission lines of several hundred kilometers are common, but an underground cable of that length would be unusable because of the high charging currents. Underground cables are normally used for short transmission lines in urban areas, where the short line length keeps the total capacitance down and the charging current under control.

2. *The greater the radius of the conductors in a transmission line, the higher the capacitance of the line.* If the radius of a conductor r is increased, the ratio D/r decreases, and therefore the total capacitance of the line increases. As we saw earlier in this chapter, several conductors are sometimes bundled together in high-voltage transmission lines to reduce the inductance of the lines. However, this bundling increases the effective radius of each phase, increasing the capacitance. Good transmission line design is a compromise among the conflicting requirements for low series inductance, low shunt capacitance, and a large enough separation to provide insulation between the phases.

Shunt Capacitive Admittance

The shunt capacitive admittance of a transmission line depends on both the capacitance of the line and the frequency of the power system. If the capacitance per unit length of a line is c, then the *shunt admittance* per unit length will be

$$y = j\omega c = j2\pi fc \tag{9–43}$$

where f is the frequency of the transmission line. To find the total shunt admittance of a transmission line, we can just multiply the admittance per unit length times the length of the line

$$Y = yd = j2\pi fcd \qquad (9\text{--}44)$$

where d is the length of the line. The corresponding capacitive reactance is the reciprocal of the admittance:

$$Z_C = 1/Y_C = -j\frac{1}{2\pi fcd} \qquad (9\text{--}45)$$

EXAMPLE 9–1

An 8000-V, 60-Hz, single-phase, two-wire transmission line consists of hard-drawn aluminum conductors with a radius of 2 cm. The two conductors are spaced 1.2 m apart, the transmission line is 30 km long, and the temperature of the conductors is 20°C. Answer the following questions about this transmission line:

a. What is series resistance per kilometer of this transmission line?
b. What is series inductance per kilometer of this transmission line?
c. What is shunt capacitance per kilometer of this transmission line?
d. What is total series reactance of this transmission line?
e. What is total shunt admittance of this transmission line?

■ **Solution**

a. The series resistance of this transmission line is given by Equation (9–1).

$$R = \frac{\rho l}{A} \qquad (9\text{--}1)$$

If we ignore skin effect, the resistivity of the line at 20°C will be 2.83×10^{-8} Ω-m. The resistance per kilometer of the line will be

$$r = \frac{\rho l}{A} = \frac{(2.83 \times 10^{-8}\ \Omega\text{-m})(1000\ \text{m})}{\pi (0.02)^2} = 0.0225\ \Omega/\text{km}$$

b. The series inductance per meter of this transmission line is given by Equation (9–22).

$$l = \frac{\mu}{\pi}\left(\frac{1}{4} + \ln\frac{D}{r}\right)\ \text{H/m} \qquad (9\text{--}22)$$

$$= \frac{\mu}{\pi}\left(\frac{1}{4} + \ln\frac{1.2}{0.02}\right) = 1.738 \times 10^{-6}\ \text{H/m}$$

Therefore the inductance per kilometer will be

$$l = 1.738 \times 10^{-3}\ \text{H/km}$$

c. The shunt capacitance per meter of this transmission line is given by Equation (9–41).

$$c_{ab} = \frac{\pi\varepsilon}{\ln\left(\dfrac{D}{r}\right)} \tag{9–41}$$

$$= \frac{\pi\,(8.854 \times 10^{-12}\ \text{F/m})}{\ln\left(\dfrac{1.2}{0.02}\right)} = 6.794 \times 10^{-12}\ \text{F/m}$$

Therefore the capacitance per kilometer will be

$$c_{ab} = 6.794 \times 10^{-9}\ \text{F/km}$$

d. The series impedance of this transmission line per kilometer will be.

$$z_{se} = r + jx = r + j2\pi fx = 0.0225 + j2\pi(60)(1.738 \times 10^{-3})$$
$$= 0.225 + j0.655\ \Omega/\text{km}$$

Therefore the total series impedance will be

$$Z_{se} = (0.225 + j0.655\ \Omega/\text{km})(30\ \text{km}) = 0.675 + j19.7\ \Omega$$

e. The shunt admittance of this transmission line per kilometer will be.

$$y_{sh} = j2\pi fc = j2\pi(60)(6.794 \times 10^{-9}\ \text{F/km}) = j2.561 \times 10^{-6}\ \text{S/m}$$

Therefore the total shunt admittance will be

$$Y_{sh} = (j2.561 \times 10^{-6}\ \text{S/km})(30\ \text{km}) = j7.684 \times 10^{-5}\ \text{S}$$

The corresponding shunt capacitive reactance is

$$Z_{sh} = 1/Y_{sh} = 1/(j7.684 \times 10^{-5}\ \text{S}) = -j13.0\ \text{k}\Omega \qquad \blacksquare$$

9.4 | TRANSMISSION LINE MODELS

Unlike generators, motors, or transformers, transmission lines are physically extended over tens or hundreds of kilometers. As a result, the resistance, inductance, and capacitance associated with the transmission line are also distributed along the length of the line (see Figure 9–9).

Figure 9–9 | A transmission line is characterized by a series resistance per unit length, a series inductance per unit length, and a shunt capacitance per unit length. This might be modeled as a repeating series of lumped constants, each representing the resistance, reactance, and capacitance of a small segment of the entire line.

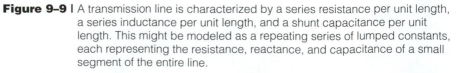
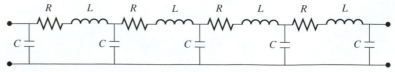

The distributed series and shunt elements of the transmission line make it harder to model than the transformers and motors that we have already studied. Such a distribution might be approximated by many small discrete lumped resistors, inductors, and capacitors, as shown in Figure 9–9, but the time required to calculate the voltages and currents flowing through the line would be excessive because of the need to solve for the voltages and currents at all the nodes in the transmission line. Alternately, we could solve the exact differential equations for a transmission line, but this is also not very practical in a large power system containing many transmission lines.

Fortunately, it is possible to make some simplifications of typical transmission line models without causing severe errors in calculations. Overhead transmission lines shorter than about 80 km (50 miles) can be modeled as a simple series resistance and inductance, since the shunt capacitance will be negligible over short distances. The per-phase model of a short transmission line is shown in Figure 9–10a. Note that the inductive reactance at 60 Hz for overhead lines will typically be much larger than the resistance of the lines, while the shunt capacitance will be negligible.

Medium-length lines are those between about 80 km (50 miles) and 240 km (150 miles). The shunt capacitance cannot be neglected for these lines, but it can be adequately modeled by two lumped capacitors, one before and one after the series impedance, each equal to one-half the line's total capacitance. The per-phase model of a medium-length transmission line is shown in Figure 9–10b.

Lines longer than about 240 km (150 miles) are long transmission lines; they will be discussed later in the chapter.

Figure 9–10 | (a) For short transmission lines, the shunt capacitance can be neglected. This produces in a per-phase model consisting of a series resistance and inductance only. (b) For medium-length transmission lines, the shunt capacitance can be divided into two lumped components, one before and one after the series impedances.

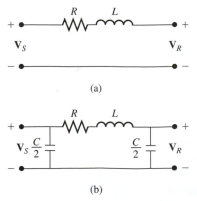

The total series resistance, series reactance, and shunt admittance of a transmission line can be calculated from the following equations

$$R = rd \tag{9-46}$$
$$X = xd \tag{9-47}$$
$$Y = yd \tag{9-48}$$

where r is the resistance per kilometer, x is the reactance per kilometer, y is the shunt admittance per kilometer, and d is the length of the transmission line in kilometers. The values r, x, and y can be computed from the geometry of the transmission line in accordance with Reference 2, but most practicing engineers will simply look up the values for a specific transmission line geometry from reference tables. The units of R and X are ohms (Ω), and the units of Y are siemens (S).

9.5 | THE SHORT TRANSMISSION LINE

The per-phase equivalent circuit for a short transmission line is shown in Figure 9–11. In this figure, $\mathbf{V}_S$ is the sending end voltage, $\mathbf{V}_R$ is the receiving end voltage, $\mathbf{I}_S$ is the sending end current, and $\mathbf{I}_R$ is the receiving end current. Note that because of the assumption of no shunt admittance, $\mathbf{I}_S = \mathbf{I}_R$.

By Kirchhoff's voltage law, the sending voltage $\mathbf{V}_S$ is related to the receiving voltage $\mathbf{V}_R$ by the equation

$$\mathbf{V}_S = \mathbf{V}_R + \mathbf{ZI} = \mathbf{V}_R + R\mathbf{I} + jX_L\mathbf{I} \qquad (9\text{–}49)$$

or
$$\boxed{\mathbf{V}_R = \mathbf{V}_S - R\mathbf{I} - jX_L\mathbf{I}} \qquad (9\text{–}50)$$

This last equation has *exactly the same form* as the per-phase equivalent circuit of a synchronous generator [Equation (5–9)], so many of the things that we have learned about synchronous generators will also apply to transmission lines.

Figure 9–11 | The per-phase equivalent circuit for a short transmission line. The values R and L are the total resistance and inductance for the entire line.

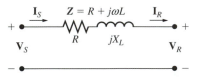

The Phasor Diagram of a Short Transmission Line

Because the voltages and currents in an AC transmission line are AC, they are usually expressed as phasors. Since phasors have both a magnitude and an angle, the relationship between them must be expressed by a two-dimensional plot called a phasor diagram. Figure 9–12 shows these relationships when the line is supplying loads at lagging power factor, at unity power factor, and leading power factor. From Equation (9–50), the source voltage $\mathbf{V}_S$ differs from the received voltage $\mathbf{V}_R$ by the resistive and inductive voltage drops. All voltages and currents are referenced to $\mathbf{V}_R$, which is arbitrarily assumed to be at an angle of 0°.

These three phasor diagrams can be compared in the figure. Note that *for a given source voltage V_S and magnitude of line current, the received voltage is lower for lagging loads and higher for leading loads.*

Figure 9–12 | The phasor diagram of a transmission line at (a) lagging, (b) unity, and (c) leading power factor.

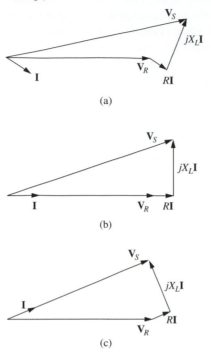

(a)

(b)

(c)

9.6 | TRANSMISSION LINE CHARACTERISTICS

In real overhead transmission lines, the line reactance X_L is normally much larger than the line resistance R, so R is often neglected in the *qualitative* study of transmission line behavior. We will now do so to study the general characteristics of transmission lines. (In some of the numerical examples, the resistance R will be included.)

The Effect of Load Changes on a Transmission Line

To understand the operating characteristics of a transmission line operating alone, examine an ideal generator supplying a load. A diagram of a single generator supplying a load is shown in Figure 9–13. What happens when we increase the load drawn from this generator?

Figure 9–13 | An ideal generator supplying a load through a transmission line.

An increase in the load is an increase in the real and/or reactive power drawn from the generator. Such a load increase increases the load current drawn from the generator and through the transmission line. Because the generator is ideal, its voltage and frequency will remain constant regardless of load.

The output of the ideal generator is the source voltage of the transmission line $\mathbf{V}_S$, so the magnitude of this voltage will not change with load. If $\mathbf{V}_S$ is constant, just what does vary with a changing load? The way to find out is to construct phasor diagrams showing an increase in the load, keeping the constraints on $\mathbf{V}_S$ in mind.

First, examine a load operating at a lagging power factor. If more load is added at the *same power factor*, then $|\mathbf{I}|$ increases but remains at the same angle θ with respect to $\mathbf{V}_R$ as before. Therefore, the voltage drop across the reactance $jX_L\mathbf{I}$ is larger than before but at the same angle. Since the line resistance is assumed to be zero,

$$\mathbf{V}_S = \mathbf{V}_R + jX_L\mathbf{I} \tag{9–49}$$

and $jX_L\mathbf{I}$ must stretch between $\mathbf{V}_R$ at an angle of $0°$ and $\mathbf{V}_S$, which is constrained to be of the same magnitude as before the load increase. If these constraints are plotted on a phasor diagram, there is one and only one point at which the armature reaction voltage can be parallel to its original position while increasing in size. The resulting plot is shown in Figure 9–14a.

If the constraints are observed, then we can see that as the load increases, the voltage at the end of the transmission line $\mathbf{V}_R$ decreases rather sharply.

Now suppose the transmission line is loaded with unity-power-factor loads. What happens if new loads are added at the same power factor? With the same constraints as before, we can see that this time $\mathbf{V}_R$ decreases only slightly (see Figure 9–14b).

Figure 9–14 | The effect of an increase in transmission line loads at constant power factor upon its terminal voltage: (a) Lagging power factor. (b) Unity power factor. (c) Leading power factor.

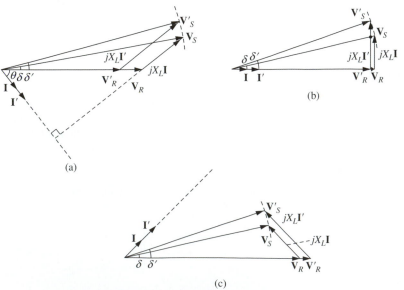

Finally, let the transmission line be loaded with leading-power-factor loads. If new loads are added at the same power factor this time, the armature reaction voltage lies outside its previous value, and V_R actually *rises* (see Figure 9–14c). In this last case, an increase in the load in the transmission line produced an increase in the terminal voltage. Such a result is not something one would expect based on intuition alone.

General conclusions from this discussion of transmission line behavior are

1. If lagging loads ($+Q$ or inductive reactive power loads) are added at the end of a line, the voltage at the end of the line V_R decreases significantly.
2. If unity-power-factor loads (no reactive power) are added at the end of a line, there is a slight decrease in V_R.
3. If leading loads ($-Q$ or capacitive reactive power loads) are added at the end of a line, V_R will rise.

A convenient way to compare the voltage behavior of a transmission line is by its *voltage regulation*. The voltage regulation (VR) of a transmission line is defined by the equation

$$VR = \frac{V_{nl} - V_{fl}}{V_{fl}} \times 100\% \qquad (4\text{–}56)$$

where V_{nl} is the no-load voltage of the transmission line and V_{fl} is the full-load voltage of the transmission line. A transmission line operating at a lagging power factor has a fairly large positive voltage regulation, a transmission line operating at a unity power factor has a small positive voltage regulation, and a transmission line operating at a leading power factor often has a negative voltage regulation.

Power Flows in a Transmission Line

The real power input to a transmission can be calculated from the equation

$$P_{in} = 3V_S I_S \cos \theta_S \qquad (9\text{–}51)$$

or

$$P_{in} = \sqrt{3} V_{LL,S} I_S \cos \theta_S \qquad (9\text{–}52)$$

where V_S is the magnitude of the source line-to-neutral voltage and $V_{LL,S}$ is the magnitude of the source line-to-line voltage. Note that the per-phase equivalent circuit implicitly assumes a wye connection, so the current I_S is the same in either case. Similarly, the real output power from a transmission line can be calculated from

$$P_{out} = 3V_R I_R \cos \theta_R \qquad (9\text{–}53)$$

or

$$P_{out} = \sqrt{3} V_{LL,R} I_R \cos \theta_R \qquad (9\text{–}54)$$

The reactive power input to a transmission can be calculated from the equation

$$Q_{\text{in}} = 3V_S I_S \sin \theta_S \qquad (9\text{--}55)$$

or

$$Q_{\text{in}} = \sqrt{3}V_{\text{LL}, S} I_S \sin \theta_S \qquad (9\text{--}56)$$

and the reactive output power from a transmission line can be calculated from

$$Q_{\text{out}} = 3V_R I_R \sin \theta_R \qquad (9\text{--}57)$$

or

$$Q_{\text{out}} = \sqrt{3}V_{\text{LL}, R} I_R \sin \theta_R \qquad (9\text{--}58)$$

The apparent power input to a transmission can be calculated from the equation

$$S_{\text{in}} = 3V_S I_S \qquad (9\text{--}59)$$

or

$$S_{\text{in}} = \sqrt{3}V_{\text{LL}, S} I_S \qquad (9\text{--}60)$$

and the apparent output power from a transmission line can be calculated from

$$S_{\text{out}} = 3V_R I_R \qquad (9\text{--}61)$$

or

$$S_{\text{out}} = \sqrt{3}V_{\text{LL},R} I_R \qquad (9\text{--}62)$$

If the line resistance R is ignored (since $X_L \gg R$), then a very useful equation can be derived to approximate the output power of the transmission line. To derive this equation, examine the phasor diagram in Figure 9–15. This figure shows a simplified phasor diagram of a transmission line with the line resistance ignored, and with $I_S = I_R = I$. Notice that the vertical segment bc can be expressed as either $V_S \sin \delta$ or $X_L I \cos \theta$. Therefore,

$$I \cos \theta = \frac{V_S \sin \delta}{X_L}$$

and substituting into Equation (9–51),

$$P = \frac{3V_S V_R \sin \delta}{X_L} \qquad (9\text{--}63)$$

Since the resistances are assumed to be zero in Equation (9–63), there are no electrical losses in this transmission line, and the P in this equation is both P_{in} and P_{out}.

Equation (9–63) shows that the power supplied by a transmission line depends on the angle δ between $\mathbf{V}_S$ and $\mathbf{V}_R$. Notice that the maximum power that the transmission line can supply occurs when $\delta = 90°$. At $\delta = 90°$, $\sin \delta = 1$, and

Figure 9–15 | Simplified phasor diagram of a transmission line with line resistance ignored.

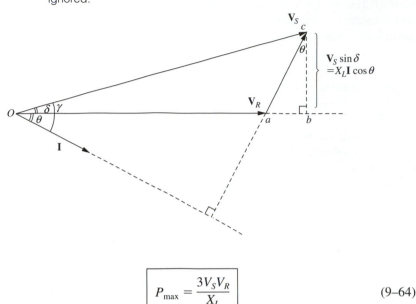

$$P_{\text{max}} = \frac{3V_S V_R}{X_L} \qquad\qquad (9\text{–}64)$$

The maximum power indicated by this equation is called the *steady-state stability limit* of the transmission line. Normally, real transmission lines never even come close to that limit. They would normally overheat because of excessive resistive losses long before that point, and power system protective devices such as circuit breakers are designed to disconnect transmission lines under extreme loads before any permanent damage can be done. Full-load δ angles of 25° are more typical of real transmission lines.

Let's examine Equations (9–63) and (9–64) more closely. They reveal a number of interesting properties of transmission lines.

1.   Notice from Equation (9–64) that the maximum power handling capability of a transmission line is a function of the *square* of its voltage. Thus if all other things are equal, a 220-kV transmission line will have 4 times the power handling capability of a 110-kV transmission line. This fact has led to a strong push for higher voltage transmission lines over the years.

This push to higher voltages is ultimately limited by other constraints. Very high voltages and currents produce very strong time-varying electric and magnetic fields. These strong fields can cause interference with nearby radios and TV sets. In addition, very high voltage transmission lines can produce unacceptable audio noise as the conductors vibrate with the expanding and collapsing fields. Both the electromagnetic interference and audio noise must be kept to reasonable levels.

Another problem is caused by the intense electric fields around conductors in very high voltage lines. When the electric field intensities around the wires get high enough, they can actually strip electrons out of atoms in the air. When this happens, an ionized current path forms, producing a glowing effect known as *corona*. Corona substantially increases the power losses in a transmission line, especially in damp weather. The voltage level at which corona losses become serious varies, depending on the construction of transmission line—lines with bundled conductors have much lower losses than lines with single conductors for a given voltage level.

2.   The maximum power handling capability of a transmission line is inversely related to its series reactance. This is not normally a problem for short and medium-length transmission lines, but long transmission lines can have quite high reactances. Recall that the series reactance *per kilometer* of a line is a function of its construction only. Thus, the longer a line with a given construction is, the higher its total series reactance will be. Some very long transmission lines include *series capacitors* to reduce the total series reactance and thus increase the total power-handling capability of the line.

3.   In normal power system operation, the magnitudes of voltages V_S and V_R do not change very much, so the angle δ between the two basically controls the power flowing through the transmission line. Some power systems take advantage of this fact by placing a phase-shifting transformer at one end of a transmission line, and varying phase of the voltage to control the power flow along the line.

Transmission Line Efficiency

The efficiency of a transmission line is the ratio of the output power from the line to the input power supplied to the line:

$$\eta = \frac{P_{\text{out}}}{P_{\text{in}}} \times 100\% \qquad (4\text{–}51)$$

Transmission Line Ratings

One of the limiting factors in transmission line operation is the resistive heating in the transmission line. This heating is a function of the square of the current flowing in the transmission line, regardless of the phase angle at which it flows. Therefore, transmission lines are typically rated at a nominal voltage and apparent power, similar to the ratings used for transformers.

Transmission Line Limits

There are several practical constraints that control the maximum real and reactive power that a transmission line can supply. The most important practical constraints are:

1.   The maximum steady-state line current must be limited to prevent overheating in the transmission lines. The power lost in a transmission line is approximately given by the equation

$$P_{\text{loss}} = 3I_L^2 R \qquad (9\text{–}65)$$

so the greater the current flow, the greater the resistive heating losses are.

2. The voltage drop in a practical transmission line should be limited to about 5 percent. In other words, the ratio of the magnitude of the receiving end voltage to the magnitude of the sending end voltage should be $|\mathbf{V}_R|/|\mathbf{V}_S| \leq 0.95$. This limit prevents excessive voltage variations in a power system.

3. The angle δ in a transmission line should typically be $\leq 30°$. This limitation ensures that the power flow in the transmission line is well below the static stability limit, ensuring that the power system can handle transients in a stable manner.

Any one of these three limitations can be the limiting factor in different circumstances. In short transmission lines where X and Z are relatively small, the resistive heating of the line is usually the limiting factor on the power that the transmission line can supply. In longer lines running at *lagging* power factor, the voltage drop across the line is usually the limiting factor, and in longer lines running at *leading* power factor, the maximum angle δ can be the limiting factor.

Example Problem

The following problems illustrate simple calculations involving voltages, currents, power flows, voltage regulation, and efficiency in transmission lines.

EXAMPLE 9–2

A 220-kV, 150 MVA, 60-Hz, three-phase transmission line is 140 km long. The characteristic parameters of the transmission line are

$$r = 0.09 \ \Omega/\text{km} \qquad x = 0.88 \ \Omega/\text{km} \qquad y = 4.1 \times 10^{-6} \ \text{S/km}$$

The voltage at the receiving end of the transmission line is 210 kV. Although this transmission line would normally be considered a medium-length transmission line, we will treat it as "short" for the purpose of this example.[1] Answer the following questions about this transmission line.

a. What is series impedance and shunt admittance of this transmission line?

b. What is the sending end voltage if the line is supplying rated voltage and apparent power at 0.85 PF lagging? at unity PF? at 0.85 PF leading?

c. What is the voltage regulation of the transmission line for each of the cases in (b)?

d. What is the efficiency of the transmission line when it is supplying rated apparent power at 0.85 PF lagging?

■ **Solution**

a. The series resistance, series reactance, and shunt admittance of this transmission line are:

---

[1] In the next two examples, we will treat this line as a medium-length and long line, and compare the results produced by each assumption.

$$R = rd = (0.12\ \Omega/\text{km})(140\ \text{km}) = 16.8\ \Omega \tag{9–46}$$
$$X = xd = (0.88\ \Omega/\text{km})(140\ \text{km}) = 123.2\ \Omega \tag{9–47}$$
$$Y = yd = (4.1 \times 10^{-6}\ \text{S/km})(140\ \text{km}) = 5.74 \times 10^{-4}\ \text{S} \tag{9–48}$$

b.   The current out of this transmission line will be given by

$$S_{\text{out}} = \sqrt{3}V_{\text{LL},\,R}I_R \tag{9–62}$$

or
$$I_R = \frac{S_{\text{out}}}{\sqrt{3}V_{\text{LL},\,R}} = \frac{150\ \text{MVA}}{\sqrt{3}(210\ \text{kV})} = 412\ \text{A}$$

and the line-to-neutral voltage out of the line will be

$$V_R = \frac{210\ \text{kV}}{\sqrt{3}} = 121\ \text{kV}$$

Since we are treating this transmission line as "short," the admittance may be ignored. The line-to-neutral voltage $\mathbf{V}_S$ at the sending end of the line when the power factor is 0.85 lagging will be

$$\mathbf{V}_S = \mathbf{V}_R + \mathbf{I}R + \mathbf{I}X_L \tag{9–49}$$
$$= 121\angle 0°\ \text{kV} + (412\angle{-31.8°}\ \text{A})(16.8\ \Omega + j123.2\ \Omega)$$
$$= 121\angle 0°\ \text{kV} + (412\angle{-31.8°}\ \text{A})(124\angle 82.2°\ \Omega)$$
$$= 158.6\angle 14.4°\ \text{kV}$$

The resulting line-to-line voltage at the sending end will be

$$\boxed{\mathbf{V}_{\text{LL},\,S} = \sqrt{3}(158.6\ \text{kV}) = 275\ \text{kV} \qquad \text{PF} = 0.85\ \text{lagging}}$$

The line-to-neutral voltage $\mathbf{V}_S$ at the sending end of the line when the power factor is 1.0 will be

$$\mathbf{V}_S = \mathbf{V}_R + \mathbf{I}R + \mathbf{I}X_L \tag{9–49}$$
$$= 121\angle 0°\ \text{kV} + (412\angle 0°\ \text{A})(16.8\ \Omega + j123.2\ \Omega)$$
$$= 121\angle 0°\ \text{kV} + (412\angle 0°\ \text{A})(124\angle 82.2°\ \Omega)$$
$$= 137.6\angle 21.6°\ \text{kV}$$

The resulting line-to-line voltage at the sending end will be

$$\boxed{\mathbf{V}_{\text{LL},\,S} = \sqrt{3}(137.6\ \text{kV}) = 238\ \text{kV} \qquad \text{PF} = 1.0}$$

The line-to-neutral voltage at the sending end of the line when the power factor is 0.85 leading will be

$$\mathbf{V}_S = \mathbf{V}_R + \mathbf{I}R + \mathbf{I}X_L \tag{9–49}$$
$$= 121\angle 0°\ \text{kV} + (412\angle 31.8°\ \text{A})(16.8\ \Omega + j123.2\ \Omega)$$
$$= 121\angle 0°\ \text{kV} + (412\angle 31.8°\ \text{A})(124\angle 82.2°\ \Omega)$$
$$= 110.5\angle 25.0°\ \text{kV}$$

The resulting line-to-line voltage at the sending end will be

$$\mathbf{V}_{LL,\,S} = \sqrt{3}(110.5 \text{ kV}) = 191 \text{ kV} \qquad PF = 0.85 \text{ leading}$$

c. The voltage regulation of a transmission line is given by Equation (4–56)

$$VR = \frac{V_{nl} - V_{fl}}{V_{fl}} \times 100\% \qquad (4\text{–}56)$$

The no-load received voltage will be the same as the source voltage, since there will be no current flowing in the line at no-load conditions. The full-load received voltage will be 210 kV. The resulting voltage regulation at 0.85 power factor lagging is

$$VR = \frac{275 - 210}{210} \times 100\% = 31.1\%$$

The resulting voltage regulation at 1.00 power factor is

$$VR = \frac{238 - 210}{210} \times 100\% = 13.7\%$$

The resulting voltage regulation at 0.85 power factor leading is

$$VR = \frac{191 - 210}{210} \times 100\% = -8.7\%$$

d. The output power from the transmission line at a power factor of 0.85 lagging is

$$P_{out} = 3V_R I_R \cos \theta_R \qquad (9\text{–}53)$$
$$= 3(121 \text{ kV})(412 \text{ A})(0.85) = 127 \text{ kW}$$

The input power from the transmission line is

$$P_{in} = 3V_S I_S \cos \theta_S \qquad (9\text{–}51)$$
$$= 3(158.6 \text{ kV})(412 \text{ A}) \cos[14.4 - (-31.8)] = 135.7 \text{ kW}$$

The resulting transmission line efficiency at full load and 0.85 power factor lagging is

$$\eta = \frac{P_{out}}{P_{in}} \times 100\% \qquad (4\text{–}51)$$
$$= \frac{127 \text{ kW}}{135.7 \text{ kW}} \times 100\% = 93.6\% \qquad \blacksquare$$

MATLAB can greatly simplify these transmission line calculations by eliminating the tedious conversions from rectangular to polar form and back again, and also by automatically performing the complex additions and multiplications. For example, the following MATLAB statements calculate the source line-to-neutral and line-to-line voltages when the power factor is 0.85 lagging, 1.0, and 0.85 leading. They use the functions **r2p** and **p2r** to perform the rectangular-to-polar and polar-to-rectangular conversions respectively.[2]

```
% M-file: ex9_2.m
% M-file to calculate the sending end line-to-neutral voltage,
% line-to-line voltage, and voltage regulation for a short
% transmission line.

% Initialize received voltage and line impedance
vr = 121000;
z = 16.8 + j*123.2;

% Perform sending-end phase voltage calculation for 0.85 pf lagging
il = p2r(412,-31.8);
vs = vr + il * z;
[mag,phase] = r2p(vs);
voltage_regulation = (abs(vs) - abs(vr)) / abs(vr) * 100;
disp(['0.85 PF lagging: Vs = ' num2str(mag) ' /_ ' num2str(phase)]);
disp(['                Vll,s = ' num2str(mag * sqrt(3))]);
disp(['                   VR = ' num2str(voltage_regulation) '%']);

% Perform sending-end phase voltage calculation for 1.0 pf
il = p2r(412,0);
vs = vr + il * z;
[mag,phase] = r2p(vs);
voltage_regulation = (abs(vs) - abs(vr)) / abs(vr) * 100;
disp(['1.0 PF:          Vs = ' num2str(mag) ' /_ ' num2str(phase)]);
disp(['                Vll,s = ' num2str(mag * sqrt(3))]);
disp(['                   VR = ' num2str(voltage_regulation) '%']);

% Perform sending-end phase voltage calculation for 0.85 pf leading
il = p2r(412,31.8);
vs = vr + il * z;
[mag,phase] = r2p(vs);
voltage_regulation = (abs(vs) - abs(vr)) / abs(vr) * 100;
disp(['0.85 PF leading: Vs = ' num2str(mag) ' /_ ' num2str(phase)]);
disp(['                Vll,s = ' num2str(mag * sqrt(3))]);
disp(['                   VR = ' num2str(voltage_regulation) '%']);
```

When this code is executed, the results are:

```
» ex9_2
0.85 PF lagging: Vs = 158624.7014 /_ 14.4162
                Vll,s = 274746.0422
                   VR = 31.0948%
```

---

[2]These MATLAB functions are documented fully in the supporting software package available at the book's website.

```
1.0 PF:          Vs  = 137623.9475 /_ 21.6428
              Vll,s = 238371.6694
                 VR = 13.7388%
0.85 PF leading: Vs  = 110526.1883 /_ 25.0436
              Vll,s = 191436.9737
                 VR = -8.656%
```

9.7 | TWO-PORT NETWORKS AND THE *ABCD* MODEL

A transmission line is an example of a two-port network. A *two-port network* is a network that can be isolated from the outside world by two connections, or ports, as shown in Figure 9–16.

Figure 9–16 | A two-port network.

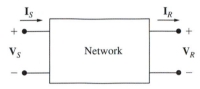

If this network is linear, then an elementary circuit theorem (analogous to Thevenin's theorem) states that the relationship between the sending and receiving end voltages and currents will be given by the following equations:

$$\mathbf{V}_S = A\mathbf{V}_R + B\mathbf{I}_R$$
$$\mathbf{I}_S = C\mathbf{V}_R + D\mathbf{I}_R$$

(9–66)

where constants A and D are dimensionless, constant B has units of ohms (Ω), and constant C has units of siemens (S). These constants are sometimes referred to as *generalized circuit constants*, or *ABCD constants.*

The *ABCD* constants have a simple physical interpretation. Constant A represents the effect of a change in the receiving end voltage on the sending end voltage, and constant D represents the effect of a change in the receiving end current on the sending end current. Both of these constants are dimensionless. Constant B represents the effect of a change in the receiving end current on the sending end voltage. Constant C represents the effect of a change in the receiving end voltage on the sending end current.

Transmission lines are two-port linear networks, and they are often represented by *ABCD* models. For the short transmission line model that we have already studied, $\mathbf{I}_S = \mathbf{I}_R = \mathbf{I}$, and the *ABCD* constants become

$$A = 1 \qquad B = Z$$
$$C = 0 \qquad D = 1$$

(9–67)

9.8 | THE MEDIUM-LENGTH TRANSMISSION LINE

The shunt admittance is included in calculations for lines of medium length, but the total admittance distributed along the line is treated as though it were divided into two lumped capacitors placed at the sending and receiving ends of the line. The admittance of each capacitor is equal to half of the total line admittance. The resulting circuit is known as a π circuit, and it is shown in Figure 9–17.

Figure 9–17 | The per-phase equivalent circuit for a medium-length transmission line. The values R and L are the total resistance and inductance for the entire line, and the total admittance Y is the shunt admittance distributed over the entire line.

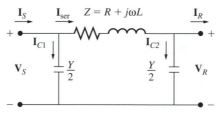

To calculate the relationship between the sending and receiving voltage and currents in this transmission line, examine Figure 9–17. Note that the current in the receiving end capacitance is given by

$$\mathbf{I}_{C2} = \mathbf{V}_R \frac{Y}{2} \tag{9–68}$$

and the current in the series impedance elements will be

$$\mathbf{I}_{ser} = \mathbf{V}_R \frac{Y}{2} + \mathbf{I}_R \tag{9–69}$$

Therefore, by Kirchhoff's voltage law, the voltage on the sending end of the transmission line will be

$$\mathbf{V}_S = Z\mathbf{I}_{ser} + \mathbf{V}_R = Z(\mathbf{I}_{C2} + \mathbf{I}_R) + \mathbf{V}_R$$
$$= Z\left(\mathbf{V}_R \frac{Y}{2} + \mathbf{I}_R\right) + \mathbf{V}_R$$
$$= \left(\frac{YZ}{2} + 1\right)\mathbf{V}_R + Z\mathbf{I}_R \tag{9–70}$$

The current flowing in the source will be

$$\mathbf{I}_S = \mathbf{I}_{C1} + \mathbf{I}_{ser} = \mathbf{I}_{C1} + \mathbf{I}_{C2} + \mathbf{I}_R$$
$$= \mathbf{V}_S \frac{Y}{2} + \mathbf{V}_R \frac{Y}{2} + \mathbf{I}_R \tag{9–71}$$

Substituting Equation (9–70) into Equation (9–71), we get

$$\mathbf{I}_S = Y\left(\frac{ZY}{4} + 1\right)\mathbf{V}_R + \left(\frac{ZY}{2} + 1\right)\mathbf{I}_R \tag{9–72}$$

Equations (9–70) and (9–72) look like the two-port relationships in Equation (9–66), so we can determine the *ABCD* constants of a medium-length transmission line by inspection to be

$$A = \frac{ZY}{2} + 1 \qquad B = Z$$

$$C = Y\left(\frac{ZY}{4} + 1\right) \qquad D = \frac{ZY}{2} + 1 \qquad (9\text{–}73)$$

Note that if we ignore the shunt capacitance of a transmission line, the shunt admittance $Y = 0$, and the *ABCD* constants in Equations (9–73) reduce to the short transmission line constants given in Equations (9–67).

EXAMPLE 9–3

The 220-kV, 150 MVA, 60-Hz three-phase transmission line of Example 9–2 is now to be treated as a "medium-length" transmission line. Answer the following questions about this line.

a. What is series impedance and shunt admittance of this transmission line?
b. What is the sending end voltage if the line is supplying rated voltage and apparent power at 0.85 PF lagging?
c. What is the voltage regulation of the transmission line for the case in (b)?
d. What is the efficiency of the transmission line when it is supplying rated apparent power at 0.85 PF lagging?
e. How much error was caused in Example 9–2 by treating the transmission line as a "short" line?

■ **Solution**

a. The series resistance, series reactance, and shunt admittance of this transmission line are the same as before:

$$R = rd = (0.12 \ \Omega/\text{km})(140 \text{ km}) = 16.8 \ \Omega \qquad (9\text{–}46)$$
$$X = xd = (0.88 \ \Omega/\text{km})(140 \text{ km}) = 123.2 \ \Omega \qquad (9\text{–}47)$$
$$Y = yd = (4.1 \times 10^{-6} \text{ S/km})(140 \text{ km}) = 5.74 \times 10^{-4} \text{ S} \qquad (9\text{–}48)$$

b. The current out of this transmission line will be given by

$$S_{\text{out}} = \sqrt{3} V_{\text{LL}, R} I_R \qquad (9\text{–}62)$$

or

$$I_R = \frac{S_{\text{out}}}{\sqrt{3} V_{\text{LL}, R}} = \frac{150 \text{ MVA}}{\sqrt{3}(210 \text{ kV})} = 412 \text{ A}$$

and the phase (line-to-neutral) voltage out of the line will be

$$V_R = \frac{210 \text{ kV}}{\sqrt{3}} = 121 \text{ kV}$$

The *ABCD* constants for this line are:

$$A = D = \frac{ZY}{2} + 1 = \frac{(16.8 + j123.2 \ \Omega)(j5.74 \times 10^{-4} \ S)}{2} + 1$$

$$= 0.9646 + j0.0048$$

$$B = 16.8 + j123.2 \ \Omega$$

$$C = Y\left(\frac{ZY}{4} + 1\right)$$

$$= (j5.74 \times 10^{-4} \ S)\left[\frac{(16.8 + j123.2 \ \Omega)(j5.74 \times 10^{-4} \ S)}{4} + 1\right]$$

$$= -1.3838 \times 10^{-6} + j5.6385 \times 10^{-4} \ S$$

Since we are treating this transmission line as a medium-length line, the line-to-neutral voltage $\mathbf{V}_S$ at the sending end of the line when the power factor is 0.85 lagging can be found from the *ABCD* constants and Equation (9–66):

$$\mathbf{V}_S = A\mathbf{V}_R + B\mathbf{I}_R$$

$$= (0.9646 + j0.0048)(121\angle 0° \ kV)$$

$$+ (16.8 + j123.2 \ \Omega)(412\angle -31.8° \ A)$$

$$= 154.6\angle 15° \ kV$$

The resulting line-to-line voltage at the sending end will be

$$\mathbf{V}_{LL, \ S} = \sqrt{3}(154.6 \ kV) = 268 \ kV \qquad PF = 0.85 \ \text{lagging}$$

c.   The voltage regulation of a transmission line is given by Equation (4–56)

$$\text{VR} = \frac{V_{nl} - V_{fl}}{V_{fl}} \times 100\% \tag{4–56}$$

The no-load received voltage will be the same as the source voltage since there will be no current flowing in the line, and the full-load received voltage will be 210 kV. The resulting voltage regulation at 0.85 power factor lagging is

$$\text{VR} = \frac{268 - 210}{210} \times 100\% = 27.6\%$$

d.   To calculate the efficiency of this transmission line, we must first know the sending end (input) current. This current can be calculated from the *ABCD* constants and Equation (9–66):

$$\mathbf{I}_S = C\mathbf{V}_R + D\mathbf{I}_R$$

$$= (-1.3838 \times 10^{-6} + j5.6385 \times 10^{-4} \ S)(121\angle 0° \ kV)$$

$$+ (0.9646 + j0.0048)(412\angle -31.8° \ A)$$

$$= 366\angle -22.4° \ A$$

The output power from the transmission line is

$$P_{out} = 3V_R I_R \cos \theta_R \tag{9–53}$$

$$= 3(121 \ kV)(412 \ A)(0.85) = 127 \ kW$$

The input power from the transmission line is

$$P_{in} = 3 V_S I_S \cos \theta_S \qquad (9\text{-}51)$$
$$= 3(154.6 \text{ kV})(366 \text{ A}) \cos[15.0 - (-22.4)] = 134.9 \text{ kW}$$

The resulting transmission line efficiency at full load and 0.85 power factor lagging is

$$\eta = \frac{P_{out}}{P_{in}} \times 100\% \qquad (4\text{-}51)$$

$$= \frac{127 \text{ kW}}{134.9 \text{ kW}} \times 100\% = 94.1\%$$

e. The sending end voltages calculated from the short transmission line and medium-length transmission line models were

$$\mathbf{V}_S = 158.6\angle 14.4° \text{ kV} \qquad \text{Short-line model}$$
$$= 154.6\angle 15° \text{ kV} \qquad \text{Medium-length-line model}$$

The sending end currents calculated from the short transmission line and medium-length transmission line models were

$$\mathbf{I}_S = 412\angle -31.8° \text{ A} \qquad \text{Short-line model}$$
$$= 366\angle -22.4° \text{ A} \qquad \text{Medium-length-line model}$$

Note that the sending end current is less and the sending end power factor is higher for the medium-length transmission line model. This is true because the shunt admittance of the transmission line itself provides a significant fraction of the reactive power being consumed by the load! The sending line current calculation is in error by about 11% if the shunt capacitance is ignored.

It is clear from these calculations that the shunt capacitance should *not* be ignored when working with this 140 km transmission line. ∎

MATLAB can also greatly simplify these calculations. For example, the following m-file calculates the sending end phase voltage and current for the transmission line of Example 9–3.

```
% M-file: ex9_3.m
% M-file to calculate the sending end voltage and current
%   for a medium-length transmission line.

% Initialize received voltage, impedance, and admittance
vr = 121000;
z  = 16.8 + j*123.2;
y  = j*5.7400e-4;

% Calculate ABCD constants
A = z*y/2 + 1;
B = z;
C = y * (z*y/4 + 1);
D = z*y/2 + 1;
```

```
% Calculate the sending end voltage
ir = p2r(412,-31.8);
vs = A * vr + B * ir;
[mag,phase] = r2p(vs);
disp(['Vs = ' num2str(mag) ' /_ ' num2str(phase) ' V']);
disp(['Vll,s = ' num2str(mag * sqrt(3))]);

% Calculate the sending end current
ir = p2r(412,-31.8);
is = C * vr + D * ir;
[mag,phase] = r2p(is);
disp(['Is = ' num2str(mag) ' /_ ' num2str(phase) ' A']);
```

When this code is executed, the results are:

```
» ex9_3
Vs = 154634.891 /_ 15.0202 V
Vll,s = 267835.4879
Is = 366.2664 /_ -22.3901 A
```

9.9 | THE LONG TRANSMISSION LINE

For very long transmission lines, it is no longer accurate enough to approximate the line's shunt admittance as two lumped constants of value $Y/2$ at either end of the transmission line. Instead, the shunt capacitance and series impedance must be treated as distributed quantities, and the voltages and currents on the line must be found by solving the differential equations of the transmission line. We will not attempt to solve the exact differential equations of a transmission line here. It is beyond the scope of this survey text—see Sections 6.4 through 6.7 of Reference 2 for a detailed solution.

However, it *is* possible to model a long transmission line as a π model with a *modified* series impedance Z' and a *modified* shunt admittance Y', and to perform calculations on that model using *ABCD* constants (see Figure 9–18). The modified values Z' and Y' can be determined by solving the differential equations of a long transmission line. The resulting values are[3]

$$Z' = Z \frac{\sinh \gamma d}{\gamma d} \tag{9–74}$$

and

$$Y' = Y \frac{\tanh (\gamma d/2)}{\gamma d/2} \tag{9–75}$$

where

Z = series impedance of the line in ohms

Y = shunt admittance of the line in siemens

---
[3]Adapted from Reference 2, Section 6.7.

Figure 9–18 | The per-phase equivalent circuit for a long transmission line.

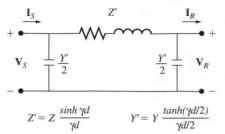

$$Z' = Z \frac{\sinh \gamma d}{\gamma d} \qquad Y' = Y \frac{\tanh(\gamma d/2)}{\gamma d/2}$$

γ = *propagation constant* of the transmission line ($\gamma = \sqrt{yz}$), where y is the shunt admittance per kilometer and z is the series impedance per kilometer

d = length of the transmission line, in kilometers

Note that as γd gets small, the quantities $\sinh(\gamma d)/(\gamma d)$ and $\tanh(\gamma d/2)/(\gamma d/2)$ both approach 1.0. Therefore, for shorter lines, $Z' \approx Z$ and $Y' \approx Y$. This is consistent with our model for a medium-length transmission line.

The *ABCD* constants for a long transmission line are given by

$$A = \frac{Z'Y'}{2} + 1 \qquad B = Z'$$
$$C = Y'\left(\frac{Z'Y'}{4} + 1\right) \qquad D = \frac{Z'Y'}{2} + 1 \qquad (9\text{–}76)$$

The calculations for a long transmission line are similar to the calculations for a medium-length transmission line, except that Z' and Y' are used instead of Z and Y. However, the mathematics gets too messy to do easily on a hand calculator.

EXAMPLE 9–4

The 220-kV, 150 MVA, 60-Hz three-phase transmission line of Example 9–2 is to be treated as a long transmission line. Answer the following questions about this line.

a. What is modified series impedance and shunt admittance of this transmission line?

b. What is the sending end voltage if the line is supplying rated voltage and apparent power at 0.85 PF lagging?

c. What is the voltage regulation of the transmission line for the case in (b)?

d. What is the efficiency of the transmission line when it is supplying rated apparent power at 0.85 PF lagging?

e. How much error was caused in Examples 9–2 and 9–3 by treating the transmission line as a "short" line? As a "medium-length" line?

■ **Solution**

a. The series resistance, series reactance, and shunt admittance of this
    transmission line are the same as before:

$$R = rd = (0.12 \ \Omega/\text{km})(140 \ \text{km}) = 16.8 \ \Omega \qquad (9\text{–}46)$$
$$X = xd = (0.88 \ \Omega/\text{km})(140 \ \text{km}) = 123.2 \ \Omega \qquad (9\text{–}47)$$
$$Y = yd = (4.1 \times 10^{-6} \ \text{S/km})(140 \ \text{km}) = 5.74 \times 10^{-4} \ \text{S} \qquad (9\text{–}48)$$

The propagation constant of this line is

$$\gamma = \sqrt{yz} = \sqrt{(j4.1 \times 10^{-6} \ \text{S/km})(0.12 + j0.88 \ \Omega/\text{km})}$$
$$= 0.000129 + j0.001903 \ \text{km}^{-1}$$

and the quantity γd is

$$\gamma d = (0.000129 + j0.001903 \ \text{km}^{-1})(140 \ \text{km}) = 0.0181 + j0.2665$$

The modified series impedance is given by

$$Z' = Z \sinh \frac{\sinh \gamma d}{\gamma d} \qquad (9\text{–}74)$$
$$= (16.8 + j123.2 \ \Omega) \frac{\sinh (0.0181 + j0.2665)}{(0.0181 + j0.2665)}$$
$$= 16.406 + j121.8 \ \Omega$$

The modified shunt admittance is given by

$$Y' = Y \frac{\tanh (\gamma d/2)}{\gamma d/2} \qquad (9\text{–}75)$$
$$= (j5.74 \ \Omega \times 10^{-4} \ \text{S}) \frac{\tanh [(0.0181 + j0.2665)/2]}{[(0.0181 + j0.2665)/2]}$$
$$= 4.68 \times 10^{-7} + j5.77 \times 10^{-4} \ \text{S}$$

b. The current out of this transmission line will be given by

$$S_{\text{out}} = \sqrt{3} V_{\text{LL}, R} I_R \qquad (9\text{–}62)$$

or

$$I_R = \frac{S_{\text{out}}}{\sqrt{3} V_{\text{LL}, R}} = \frac{150 \ \text{MVA}}{\sqrt{3}(210 \ \text{kV})} = 412 \ \text{A}$$

and the line-to-neutral voltage out of the line will be

$$V_R = \frac{210 \ \text{kV}}{\sqrt{3}} = 121 \ \text{kV}$$

The *ABCD* constants for this line are:

$$A = D = \frac{Z'Y'}{2} + 1 = \frac{(16.406 + j121.8 \ \Omega)(4.68 \times 10^{-10} + j5.77 \times 10^{-4} \ \text{S})}{2} + 1$$
$$= 0.9648 + j0.0048$$
$$B = 16.406 + j121.8 \ \Omega$$

$$C = Y'\left(\frac{Z'Y'}{4} + 1\right)$$
$$= (4.68 \times 10^{-7} + j5.77 \times 10^{-4}\ \text{S})$$
$$\left[\frac{(16.406 + j121.8\ \Omega)(4.68 \times 10^{-10} + j5.77 \times 10^{-4}\ \text{S})}{2} + 1\right]$$
$$= -9.16 \times 10^{-7} + j5.67 \times 10^{-4}\ \text{S}$$

Since we are treating this transmission line as a long line, the line-to-neutral voltage $\mathbf{V}_S$ at the sending end of the line when the power factor is 0.85 lagging can be found from the *ABCD* constants and Equation (9–66):

$$\mathbf{V}_S = A\mathbf{V}_R + B\mathbf{I}_R$$
$$= (0.9648 + j0.0048)(121\angle 0°\ \text{kV})$$
$$\quad + (16.406 + j121.8\ \Omega)(412\angle{-31.8°}\ \text{A})$$
$$= 154.1\angle 14.9°\ \text{kV}$$

The resulting line-to-line voltage at the sending end will be

$$\mathbf{V}_{LL,S} = \sqrt{3}(154.1\ \text{kV}) = 267\ \text{kV} \qquad \text{PF} = 0.85\ \text{lagging}$$

c.   The voltage regulation of a transmission line is given by Equation (4–56)

$$\text{VR} = \frac{V_{nl} - V_{fl}}{V_{fl}} \times 100\% \tag{4–56}$$

The no-load received voltage will be the same as the source voltage since there will be no current flowing in the line, and the full-load received voltage will be 210 kV. The resulting voltage regulation at 0.85 power factor lagging is

$$\text{VR} = \frac{267 - 210}{210} \times 100\% = 27.1\%$$

d.   To calculate the efficiency of this transmission line, we must first know the sending end (input) current. This current can be calculated from the *ABCD* constants and Equation (9–66):

$$\mathbf{I}_S = C\mathbf{V}_R + D\mathbf{I}_R$$
$$= (-9.16 \times 10^{-7} + j5.67 \times 10^{-4}\ \text{S})(121\angle 0°\ \text{kV})$$
$$\quad + (0.9648 + j0.0048)(412\angle{-31.8°}\ \text{A})$$
$$= 366.2\angle{-22.3°}\ \text{A}$$

The output power from the transmission line is

$$P_{out} = 3V_R I_R \cos\theta_R \tag{9–53}$$
$$= 3(121\ \text{kV})(412\ \text{A})(0.85) = 127\ \text{kW}$$

The input power from the transmission line is

$$P_{in} = 3V_S I_S \cos\theta_S \tag{9–51}$$
$$= 3(154.1\ \text{kV})(366.2\ \text{A})\cos[14.9 - (-22.3)] = 134.8\ \text{kW}$$

The resulting transmission line efficiency at full load and 0.85 power factor lagging is

$$\eta = \frac{P_{out}}{P_{in}} \times 100\% \qquad\qquad (4\text{--}51)$$

$$= \frac{127 \text{ kW}}{134.8 \text{ kW}} \times 100\% = 94.2\%$$

e. The sending end voltages calculated from the short transmission line, medium-length transmission line, and long transmission line models were

$\mathbf{V}_S = 158.6\angle 14.4° \text{ kV}$	Short-line model
$\mathbf{V}_S = 154.6\angle 15° \text{ kV}$	Medium-length-line model
$\mathbf{V}_S = 154.1\angle 14.9° \text{ kV}$	Long-line model

The sending end currents calculated from the short transmission line, medium-length transmission line, and long transmission line models were

$\mathbf{I}_S = 412\angle -31.8° \text{ A}$	Short-line model
$\mathbf{I}_S = 366\angle -22.4° \text{ A}$	Medium-length-line model
$\mathbf{I}_S = 366.2\angle -22.3° \text{ A}$	Long-line model

Note that there is very little difference between the voltages and currents calculated with the medium-length transmission line model and the long transmission line mode. There is no need to do all of the extra calculations for a long transmission line in this case. This result is consistent with the use of the medium-length transmission line model for line lengths of up to 240 km. ∎

MATLAB can also greatly simplify these calculations. For example, the following m-file calculates the sending end line-to-neutral voltage and current for the transmission line of Example 9–4.

```
% M-file: ex8_3.m
% M-file to calculate the sending end voltage and current
%   for a long transmission line.

% Initialize received voltage and per-length values
vr = 121000;
z  = (0.12 + j*0.88);       % Impedance per km
y  = j*4.1e-6;              % Admittance per km
l  = 140;                   % Length (km)

% Series reactance and shunt admittance
Z  = z * l;                 % Total series impedance (ohms)
Y  = y * l;                 % Total shunt admittance (S)

% Calculate gamma * length
gamma = sqrt(y*z);
gl = gamma * l;
```

```
% Calculate ZP and YP
ZP = Z * sinh(gl) / gl;
YP = Y * tanh(gl/2) / (gl/2);

% Calculate ABCD constants
A = ZP*YP/2 + 1;
B = ZP;
C = YP * (ZP*YP/4 + 1);
D = ZP*YP/2 + 1;

% Calculate the sending end voltage
ir = p2r(412,-31.8);
vs = A * vr + B * ir;
[mag,phase] = r2p(vs);
disp(['vs = ' num2str(mag) ' /_ ' num2str(phase) ' V']);
disp(['Vll,s = ' num2str(mag * sqrt(3))]);

% Calculate the sending end current
ir = p2r(412,-31.8);
is = C * vr + D * ir;
[mag,phase] = r2p(is);
disp(['is = ' num2str(mag) ' /_ ' num2str(phase) ' A']);
```

When this code is executed, the results are:

```
» ex9_4
vs = 154119.5015 /_ 14.9106 V
Vll,s = 266942.807
is = 366.241 /_ -22.3329 A
```

9.10 | SUMMARY

Generators and loads are connected by transmission lines. Unlike other power system components, transmission lines are physically extended over tens or hundreds of kilometers. As a result, the resistance, inductance, and capacitance associated with the transmission line are also distributed along the length of the line. Transmission lines are characterized by a series resistance per unit length, a series inductance per unit length, and a shunt capacitance per unit length.

In general, the series inductance of a transmission line *increases* as the spacing between phases increases, and decreases as the radius of the individual conductor (or bundle of conductors) in each phase increases. The shunt capacitance of a transmission line *decreases* as the spacing between phases increases, and increases as the radius of the individual conductor (or bundle of conductors) in each phase increases.

The distributed resistance, inductance, and capacitance of real transmission lines is modeled using lumped resistors, inductors and capacitors. In the short transmission line model, the series resistance and inductive reactance of transmission line is modeled, and the shunt capacitance is ignored. In the medium-length transmission line model, the shunt capacitance is split into two pieces and included on either side of the series elements, producing a π circuit. The long transmission line model is also a π circuit, but the values of the series impedance and the shunt capacitance are adjusted to account for the effects of their distributed nature. Transmission lines are typically analyzed by using *ABCD* constants.

The maximum power handling capability of a transmission line is a function of the square of its voltage, so power transmission lines tend to be built at high voltages. However, there is an upper limit to the voltage, set by audio noise, radio and TV interference, and corona losses.

The maximum power handling capability of a transmission line is inversely related to its series reactance. This is not normally a problem for short and medium-length transmission lines, but long transmission lines can have quite high reactances. Series capacitors are sometimes used to reduce the series reactance of long transmission lines.

9.11 | QUESTIONS

9–1. How can the resistance of a transmission line be calculated?

9–2. What is skin effect? How does skin effect change the resistance of an AC transmission line?

9–3. What factors influence the inductance of a transmission line?

9–4. Suppose that there are two transmission lines using a single conductor per phase, and that the conductors in the two transmission lines are identical (same material and same diameter). If one transmission line is rated at 34.5 kV and the other one is rated at 238 kV, which transmission line is likely to have the highest inductance? Why?

9–5. Suppose someone asked you to evaluate a proposal for an underwater power cable between North Carolina and the Bahamas. What would you say about the probability of success of this proposal? Why?

9–6. What line lengths are generally considered to be short transmission lines? Medium-length transmission lines? Long transmission lines?

9–7. What approximations are made in a short transmission line model?

9–8. What approximations are made in a medium-length transmission line model?

9–9. What happens to the receiving end voltage as the load on a transmission line is increased if the load has a lagging power factor? Sketch a phasor diagram showing the resulting behavior.

9–10. What happens to the receiving end voltage as the load on a transmission line is increased if the load has unity power factor? Sketch a phasor diagram showing the resulting behavior.

9–11. What happens to the receiving end voltage as the load on a transmission line is increased if the load has a leading power factor? Sketch a phasor diagram showing the resulting behavior.

9–12. What is the significance of the angle δ between $\mathbf{V}_S$ and $\mathbf{V}_R$ in a transmission line?

9–13. What is the significance of the constants *A*, *B*, *C*, and *D* in an *ABCD* representation of a two-port network?

9.12 | PROBLEMS

9–1. Calculate the DC resistance in ohms per kilometer for an aluminum conductor with a 3-cm diameter.

9–2. Calculate the DC resistance in ohms per *mile* for a hard-drawn copper conductor with a 1-inch diameter. (Note that 1 mile = 1.609 km.)

Problems 9–3 through 9–7 refer to a single-phase, 8-kV, 50-Hz, 50-km-long transmission line consisting of two aluminum conductors with a 3-cm diameter separated by a spacing of 4 m.

9–3. Calculate the inductive reactance of this line in ohms.

9–4. Assume that the 50-Hz AC resistance of the line is 5 percent greater than its DC resistance, and calculate the series impedance of the line in ohms per kilometer.

9–5. Calculate the shunt admittance of the line in siemens per kilometer.

9–6. The single-phase transmission line is operating with the receiving side of the line open circuited. The sending end voltage is 8 kV at 50 Hz. How much charging current is flowing in the line?

9–7. The single-phase transmission line is now supplying 8 kV to an 800 kVA, 0.9 PF lagging single-phase load.

    a. What is the sending end voltage and current of this transmission line?

    b. What is the efficiency of the transmission line under these conditions?

    c. What is the voltage regulation of the transmission line under these conditions?

Problems 9–8 through 9–10 refer to a single-phase, 8-kV, 50-Hz, 50-km-long underground cable consisting of two aluminum conductors with a 3-cm diameter separated by a spacing of 15 cm.

9–8. The single-phase transmission line referred to in Problems 9–3 through 9–7 is to be replaced by an underground cable. The cable consists of two aluminum conductors with a 3-cm diameter, separated by a center-to-center spacing of 15 cm. As before, assume that the 50-Hz AC resistance of the line is 5 percent greater than its DC resistance, and calculate the series impedance and shunt admittance of the line in ohms per kilometer. Also, calculate the total impedance and admittance for the entire line.

9–9. The underground cable is operating with the receiving side of the line open circuited. The sending end voltage is 8 kV at 50 Hz. How much charging current is flowing in the line? How does this charging current in the cable compare to the charging current of the overhead transmission line?

9–10. The underground cable is now supplying 8 kV to an 800-kVA, 0.9 PF lagging, single-phase load.

    a. What is the sending end voltage and current of this transmission line?

    b. What is the efficiency of the transmission line under these conditions?

    c. What is the voltage regulation of the transmission line under these conditions?

9–11. A 138-kV, 200-MVA, 60-Hz, three-phase power transmission line is 100 km long and has the following characteristics:

$$r = 0.103 \ \Omega/\text{km}$$
$$x = 0.525 \ \Omega/\text{km}$$
$$y = 3.3 \times 10^{-6} \ \text{S/km}$$

    a.   What is per phase series impedance and shunt admittance of this transmission line?

    b.   Should it be modeled as a short, medium, or long transmission line?

    c.   Calculate the *ABCD* constants of this transmission line.

    d.   Sketch the phasor diagram of this transmission line when the line is supplying rated voltage and apparent power at a 0.90 power factor lagging.

    e.   Calculate the sending end voltage if the line is supplying rated voltage and apparent power at 0.90 PF lagging.

    f.   What is the voltage regulation of the transmission line for the conditions in (e)?

    g.   What is the efficiency of the transmission line for the conditions in (e)?

9–12. If the series resistance and shunt admittance of the transmission line in Problem 9–11 are ignored, what would the value of the angle δ be at rated conditions and 0.90 PF lagging?

9–13. If the series resistance and shunt admittance of the transmission line in Problem 9–11 are *not* ignored, what would the value of the angle δ be at rated conditions and 0.90 PF lagging?

9–14. Assume that the transmission line of Problem 9–11 is to supply a load at 0.90 PF lagging with no more than a 5 percent voltage drop and a torque angle $\delta \leq 30°$. Treat the line as a medium-length transmission line. What is the maximum power that this transmission line can supply without violating one of these constraints? Which constraint is violated first?

9–15. The transmission line of Problem 9–11 is connected between two infinite buses, as shown in Figure P9–1. Answer the following questions about this transmission line.

Figure P9–1 I A three-phase transmission line connecting two infinite buses together.

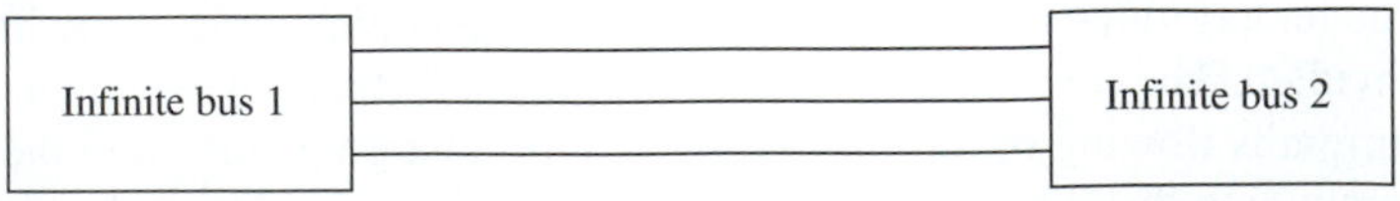

    a.   If the per-phase (line-to-neutral) voltage on the sending infinite bus is $80\angle 10° $ kV and the per-phase voltage on the receiving infinite bus is $76\angle 0°$ kV, how much real and reactive power are being supplied by the transmission line to the receiving bus?

b. If the per-phase voltage on the sending infinite bus is changed to
    $82\angle 10°$ kV, how much real and reactive power are being supplied by
    the transmission line to the receiving bus? Which changed more, the
    real or the reactive power supplied to the load?

c. If the per-phase voltage on the sending infinite bus is changed to
    $80\angle 15°$ kV, how much real and reactive power are being supplied by
    the transmission line to the receiving bus? Compared to the conditions
    in part (a), which changed more, the real or the reactive power supplied
    to the load?

d. From the above results, how could real power flow be controlled in a
    transmission line? How could reactive power flow be controlled in
    a transmission line?

9–16. A 50-Hz, three-phase transmission line is 300 km long. It has a total series
        impedance of $23 + j75\ \Omega$ and a shunt admittance of $j500\ \mu S$. It delivers
        50 MW at 220 kV, with a power factor of 0.88 lagging. Find the voltage at
        the receiving end using (a) the short line approximation. (b) The medium-
        length-line approximation. (c) The long line equation. How accurate are
        the short- and medium-length line approximations for this case?

9–17. A 60-Hz, three-phase, 110-kV transmission line has a length of 100 miles
        and a series impedance of $0.20 + j0.85\ \Omega$/mile and a shunt admittance of
        6×10^{-5} S/mile. The transmission line is supplying 120 MW at a power
        factor of 0.85 lagging, and the receiving end voltage is 110 kV.

a. What are the voltage, current, and power factor at the receiving end of
    this line?

b. What are the voltage, current, and power factor at the sending end of
    this line?

c. How much power is being lost in this transmission line?

d. What is the current angle δ of this transmission line? How close is the
    transmission line to its steady-state stability limit?

9.13 | REFERENCES

1. Elgerd, Olle I.: *Electric Energy Systems Theory: An Introduction*, McGraw-Hill, New
   York, 1971.

2. Granger, John J., and William D. Stevenson, Jr.: *Power Systems Analysis*, McGraw-Hill,
   New York, 1994.

3. Gross, Charles A.: *Power Systems Analysis*, John Wiley and Sons, New York, 1979.

4. Yamayee, Zia A., and Juan L. Bala, Jr.: *Electromechanical Energy Devices and Power
   Systems*, John Wiley and Sons, New York, 1994.

Power System Representation and Equations

In this chapter, we will start to combine the components that we have studied earlier in the book into complete power systems. This chapter introduces the standard symbols used in one-line representations of power systems, discusses the application of the per-unit system to power systems, and shows how to create the basic equations used in power system analysis.

In this and subsequent chapters, there is a slight change in notation. In earlier chapters, an impedance or admittance was shown in bold face, and the *magnitude* of the impedance or admittance was shown in italics. From here on, we need to distinguish between matrices of impedances, single impedance values, and the magnitudes of single impedance values. Therefore, *matrices of impedances or admittances* will be shown in bold face ($\mathbf{Z}_{\text{bus}}$), *individual impedances or admittances* will be shown in italics (Z_{11}), and the *magnitudes of individual impedances or admittances* will be shown with absolute value bars ($|Z_{11}|$).

10.1 | ONE-LINE OR SINGLE-LINE DIAGRAMS

As we learned in Chapter 2, almost all modern power systems are three-phase systems, with similar phases equal in amplitude and shifted in phase from each other by 120°. Because the three phases are all similar, it is customary to sketch power systems in a simple form with a single line representing all three phases of the real power system. When combined with a standard set of symbols for electrical components, these one-line diagrams provide a compact way to represent a great deal of information. Some of the standard symbols used in one-line diagrams are shown in Figure 10–1. These symbols are defined in Reference 4.

Figure 10–2 shows a one-line diagram of a simple power system. This power system contains two synchronous machines, two loads, two busses, two transformers, plus a transmission line to connect the busses together. All of the devices are

Figure 10–1 | Symbols used in power system one-line diagrams.

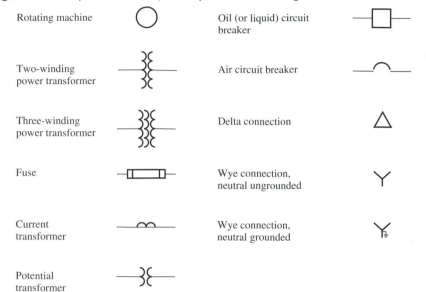

Rotating machine		Oil (or liquid) circuit breaker
Two-winding power transformer		Air circuit breaker
Three-winding power transformer		Delta connection
Fuse		Wye connection, neutral ungrounded
Current transformer		Wye connection, neutral grounded
Potential transformer		

Figure 10–2 | A one-line diagram of a simple power system.

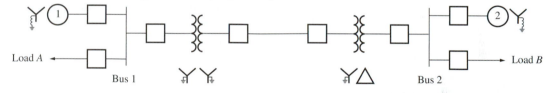

Load A   Bus 1   Bus 2   Load B

protected by oil circuit breakers (OCBs). Note that this diagram indicates the type of connection used for each machine and transformer, and also indicates the points on the power system that are connected to ground.

It is important to know where a power system is connected to ground, since such connections affect the current that flows in nonsymmetrical faults. These connections are shown on the diagram as a wye connection with a ground symbol in the middle. Sometimes, the connections to ground are direct, and sometimes they are through resistors or inductors. If a resistor or inductor is included in the ground connection, it can help reduce the fault current that flows in unsymmetrical faults, while having no impact on the performance of the system in steady-state operations. (Recall from Chapter 2 that the current flowing in the neutral of a balanced three-phase system is zero. Therefore, inserting an impedance into the neutral line has no impact on the balanced steady-state operation of the power system.)

One-line diagrams such as this usually include additional information, such as the ratings of machines and transformers, the power being consumed or supplied by all of the loads in the system, and the impedances of the various devices in the system.

10.2 | PER-PHASE, PER-UNIT EQUIVALENT CIRCUITS

As we learned in Chapter 2, the easiest way to analyze a balanced three-phase circuit is to produce a *per-phase equivalent circuit,* which represents one of the three phases with all Δ connections converted to their equivalent Y connections by the Δ–Y transform. We can solve for the voltages and currents in that single phase, knowing that the voltages and currents in the other two phases would be exactly the same except for a 120° phase shift. Example 2–2 illustrates this sort of analysis.

As we learned in Chapter 3, the *per-unit system* has a major advantage for power system analysis. When values are expressed in the per-unit system, the voltage level changes caused by transformers disappear from the solution. Example 3–3 demonstrates how the per-unit system simplifies the solution of a circuit containing transformers.

Real power systems are three-phase systems containing a mixture of Y and Δ connections, and also containing very, very many transformers. Because the systems are three-phase, it is convenient to analyze them with a per-phase equivalent circuit. Because there are so many transformers, it is impractical to refer the voltages, currents, and impedances through all of them to get to a common level for analysis purposes. Therefore, we need to create *per-phase, per-unit equivalent circuits* to solve power system problems.

As we saw in Chapter 3, the per-unit system applies just as well to three-phase systems as to single-phase systems. The single-phase base Equations (3–60) to (3–63) apply to three-phase systems on a *per-phase* basis. The per-phase base voltage, current, apparent power, and impedance of a circuit are related by the following equations

$$I_{\text{base}} = \frac{S_{1\phi,\,\text{base}}}{V_{\text{LN,\,base}}} \tag{10–1}$$

$$Z_{\text{base}} = \frac{V_{\text{LN,\,base}}}{I_{\text{base}}} \tag{10–2}$$

$$Z_{\text{base}} = \frac{(V_{\text{LN,\,base}})^2}{S_{1\phi,\,\text{base}}} \tag{10–3}$$

Here, $V_{\text{LN,\,base}}$ is the base voltage from line to neutral in the three-phase circuit, which is the same as the base phase voltage of a Y-connected circuit. $S_{1\phi,\,\text{base}}$ is the base apparent power of a single phase in the circuit.

The base current and impedance in a per-unit system can also be expressed in terms of the *three-phase* apparent power and line-to-line voltages. Remember that the apparent power of a complete three-phase circuit is 3 times the apparent power of a single phase, and that the line-to-line voltage of a three-phase circuit is $\sqrt{3}$ times the line-to-neutral voltage of the circuit. According to these relationships, equations relating the base quantities become

$$I_{\text{base}} = \frac{S_{3\phi,\,\text{base}}}{\sqrt{3}V_{\text{LL},\,\text{base}}} \tag{10-4}$$

$$Z_{\text{base}} = \frac{V_{\text{LL},\,\text{base}}}{\sqrt{3}I_{\text{base}}} \tag{10-5}$$

$$Z_{\text{base}} = \frac{(V_{\text{LL},\,\text{base}})^2}{S_{3\phi,\,\text{base}}} \tag{10-6}$$

In the per-unit system, all voltages, currents, and powers in a circuit are represented as a fraction of the base value for that unit.

$$\text{Quantity in per unit} = \frac{\text{actual value}}{\text{base value of quantity}} \tag{10-7}$$

If any two of the four base quantities are known (voltage, apparent power, current, and impedance), then the other base values can be calculated from equations (10–1) to (10–6). Traditionally, the base apparent power and base voltage are specified at a point in the circuit, and the other values are calculated from them. Note that the base voltage varies by the voltage ratio of each transformer in the circuit. However, the base apparent power remains constant throughout a circuit, because the base current decreases by the same amount as the base voltage increases when passing through a transformer.

As we noted earlier in the book, many transformers and machines have their internal impedances specified as per-unit resistances and reactances, using the voltage and apparent power ratings of the device itself as the base quantities. If these impedances are expressed in per-unit to a base other than the one selected as the base for a power system, we must covert the impedances to per-unit on the new base. This conversion could be done by using the original base impedance to convert the impedances back into ohms, and then using the power system base impedance to convert the value on ohms to per-unit on the new base.

Alternatively, we could combine the two steps into a single equation. The single equation to convert a per-unit impedance on a given base to a per-unit impedance on a new base is

$$\text{Per-unit } Z_{\text{new}} = \text{per-unit } Z_{\text{given}} \left(\frac{V_{\text{given}}}{V_{\text{new}}}\right)^2 \left(\frac{S_{\text{new}}}{S_{\text{given}}}\right) \tag{10-8}$$

EXAMPLE 10–1

A 13.8-kV, 100-MVA, 60-Hz, three-phase synchronous generator has a nameplate resistance R of 10 percent or (0.10 per unit) and a reactance X of 80 percent (or 0.80 per unit). These values are specified on the base of the machine's rating. The base quantities of the power system it is connected to are $V_{\text{LL, base}}$ = 14.4 kV and S_{base} = 500 MVA. Find the per-unit impedance of the generator on the base of the power system.

■ Solution

Note that the per-unit values on a machine's nameplate could be expressed as either a percent or a fraction. A per-unit resistance of 10 percent or 0.10 are two ways of representing the same value, and you may encounter both at one time or another.

The per-unit resistance and reactance can be found either by first converting the given per-unit values to ohms and then converting them back to per-unit on the new base, or else by calculating directly from Equation (10–8).

To convert to ohms first and then back to per unit on the new base, we will need to calculate both the given base impedance of the generator and the new base impedance of the power system.

$$Z_{base, \, gen} = \left(\frac{V_{LL, \, base}}{S_{3\phi, \, base}}\right)^2 = \frac{(13.8 \text{ kV})^2}{100 \text{ MVA}} = 1.904 \; \Omega \qquad (10\text{–}6)$$

$$Z_{base, \, system} = \frac{(V_{LL, \, base})^2}{S_{3\phi, \, base}} = \frac{(14.4 \text{ kV})^2}{500 \text{ MVA}} = 0.4147 \; \Omega \qquad (10\text{–}6)$$

Then,

$$R = R_{pu} Z_{base, \, gen} = (0.10)(1.904 \; \Omega) = 0.190 \; \Omega$$
$$X = X_{pu} Z_{base, \, gen} = (0.80)(1.904 \; \Omega) = 1.524 \; \Omega$$
$$R_{pu} = \frac{R}{Z_{base, \, system}} = \frac{0.190 \; \Omega}{0.4147 \; \Omega} = 0.459 \text{ per unit}$$
$$X_{pu} = \frac{X}{Z_{base, \, system}} = \frac{1.524 \; \Omega}{0.4147 \; \Omega} = 3.67 \text{ per unit}$$

Alternatively, the calculation can be done in a single step with Equation 10–8:

$$\text{Per-unit } Z_{new} = \text{per-unit } Z_{given} \left(\frac{V_{given}}{V_{new}}\right)^2 \left(\frac{S_{new}}{S_{given}}\right) \qquad (10\text{–}8)$$

$$R_{pu, \, system} = (0.1)\left(\frac{13.8 \text{ kV}}{14.4 \text{ kV}}\right)^2 \left(\frac{500 \text{ MVA}}{100 \text{ MVA}}\right) = 0.459 \text{ per unit}$$

$$X_{pu, \, system} = (0.8)\left(\frac{13.8 \text{ kV}}{14.4 \text{ kV}}\right)^2 \left(\frac{500 \text{ MVA}}{100 \text{ MVA}}\right) = 3.67 \text{ per unit} \qquad ■$$

EXAMPLE 10–2

A simple power system consisting of one synchronous generator and one synchronous motor connected by two transformers and a transmission line is shown in Figure 10–3. Create a per-phase, per-unit equivalent circuit for this power system using a base apparent power of 100 MVA and a base line voltage at generator G_1 of 13.8 kV.

Figure 10–3 | A one-line diagram of the simple power system in Example 10–2.

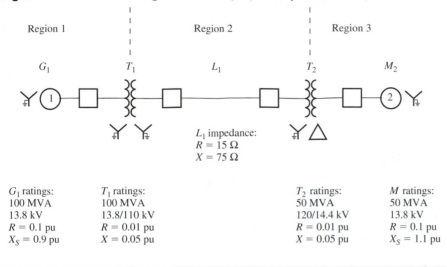

| Region 1 | | Region 2 | | Region 3 |

G_1 T_1 L_1 T_2 M_2

L_1 impedance:
$R = 15\ \Omega$
$X = 75\ \Omega$

G_1 ratings:	T_1 ratings:	T_2 ratings:	M ratings:
100 MVA	100 MVA	50 MVA	50 MVA
13.8 kV	13.8/110 kV	120/14.4 kV	13.8 kV
$R = 0.1$ pu	$R = 0.01$ pu	$R = 0.01$ pu	$R = 0.1$ pu
$X_S = 0.9$ pu	$X = 0.05$ pu	$X = 0.05$ pu	$X_S = 1.1$ pu

■ **Solution**

To create a per-phase, per-unit equivalent circuit, we must first calculate the impedances of each component in the power system in per-unit to the system base. The system base apparent power is S_{base} = 100 MVA everywhere in the power system. The base voltage in the three regions will vary as the voltage ratios of the transformers that delineate the regions. The base voltages are:

$$V_{base,\ 1} = 13.8\ \text{kV} \qquad\qquad \text{Region 1}$$

$$V_{base,\ 2} = V_{base,\ 1}\left(\frac{110\ \text{kV}}{13.8\ \text{kV}}\right) = 110\ \text{kV} \qquad \text{Region 2} \qquad (10\text{–}9)$$

$$V_{base,\ 3} = V_{base,\ 2}\left(\frac{14.4\ \text{kV}}{120\ \text{kV}}\right) = 13.2\ \text{kV} \qquad \text{Region 3}$$

The corresponding base impedances in each region are:

$$Z_{base,\ 1} = \frac{(V_{LL,\ base})^2}{S_{3\phi,\ base}} = \frac{(13.8\ \text{kV})^2}{100\ \text{MVA}} = 1.904\ \Omega \qquad \text{Region 1}$$

$$Z_{base,\ 2} = \frac{(V_{LL,\ base})^2}{S_{3\phi,\ base}} = \frac{(110\ \text{kV})^2}{100\ \text{MVA}} = 121\ \Omega \qquad \text{Region 2} \qquad (10\text{–}10)$$

$$Z_{base,\ 3} = \frac{(V_{LL,\ base})^2}{S_{3\phi,\ base}} = \frac{(13.2\ \text{kV})^2}{100\ \text{MVA}} = 1.743\ \Omega \qquad \text{Region 3}$$

The impedances of G_1 and T_1 are specified in per-unit on a base of 13.8 kV and 100 MVA, which is the same as the system base in Region 1. Therefore, the per-unit resistances and reactances of these components on the system base are unchanged:

$$R_{G1,\,pu} = 0.1 \text{ per unit}$$
$$X_{G1,\,pu} = 0.9 \text{ per unit}$$
$$R_{T1,\,pu} = 0.01 \text{ per unit}$$
$$X_{T1,\,pu} = 0.05 \text{ per unit}$$

The transmission line appears in Region 2 of the power system. The impedance of the transmission line is specified in ohms, and the base impedance in that region is 121 Ω. Therefore, the per-unit resistance and reactance of the transmission line on the system base are:

$$R_{\text{line, system}} = \left(\frac{15\ \Omega}{121\ \Omega}\right) = 0.124 \text{ per unit} \tag{10–11}$$

$$X_{\text{line, system}} = \left(\frac{75\ \Omega}{121\ \Omega}\right) = 0.620 \text{ per unit}$$

The impedance of T_2 is specified in per-unit on a base of 14.4 kV and 50 MVA in Region 3. Therefore, the per-unit resistances and reactances of this component on the system base are:

$$\text{Per-unit } Z_{\text{new}} = \text{per-unit } Z_{\text{given}} \left(\frac{V_{\text{given}}}{V_{\text{new}}}\right)^2 \left(\frac{S_{\text{new}}}{S_{\text{given}}}\right) \tag{10–8}$$

$$R_{T2,\,pu} = (0.01)\left(\frac{14.4\ \text{kV}}{13.2\ \text{kV}}\right)^2 \left(\frac{100\ \text{MVA}}{50\ \text{MVA}}\right) = 0.238 \text{ per unit} \tag{10–12}$$

$$X_{T2,\,pu} = (0.05)\left(\frac{14.4\ \text{kV}}{13.2\ \text{kV}}\right)^2 \left(\frac{100\ \text{MVA}}{50\ \text{MVA}}\right) = 0.119 \text{ per unit}$$

The impedance of M_2 is specified in per-unit on a base of 13.8 kV and 50 MVA in Region 3. Therefore, the per-unit resistances and reactances of this component on the system base are:

$$\text{Per-unit } Z_{\text{new}} = \text{per-unit } Z_{\text{given}} \left(\frac{V_{\text{given}}}{V_{\text{new}}}\right)^2 \left(\frac{S_{\text{new}}}{S_{\text{given}}}\right) \tag{10–8}$$

$$R_{M2,\,pu} = (0.1)\left(\frac{14.8\ \text{kV}}{13.2\ \text{kV}}\right)^2 \left(\frac{100\ \text{MVA}}{50\ \text{MVA}}\right) = 0.219 \text{ per unit} \tag{10–13}$$

$$X_{M2,\,pu} = (1.1)\left(\frac{14.8\ \text{kV}}{13.2\ \text{kV}}\right)^2 \left(\frac{100\ \text{MVA}}{50\ \text{MVA}}\right) = 2.405 \text{ per unit} \tag{10–14}$$

Therefore, the per-phase, per-unit equivalent circuit of this transmission system is as shown in Figure 10–4. ∎

10.3 | WRITING NODE EQUATIONS FOR EQUIVALENT CIRCUITS

Once the per-phase, per-unit equivalent circuit of a power system is created, it can be used to find the voltages, currents, and powers present at various points in a power system. The most common technique used to solve such circuits is *nodal analysis*.

Figure 10–4 | Per-phase, per-unit equivalent circuit of the simple power system in Example 10–2.

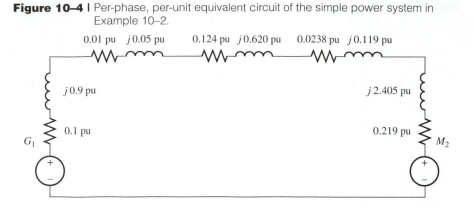

In nodal analysis, we use Kirchhoff's current law equations to determine the voltages at each node (each bus) in the power system, and then use the resulting voltages to calculate the currents and power flows at various points in the system.

Figure 10–5 shows a simple three-phase power system containing three busses connected by three transmission lines. The system includes a generator connected to bus 1, a load connected to bus 2, and a motor connected to bus 3.

Figure 10–5 | A simple three-phase power system.

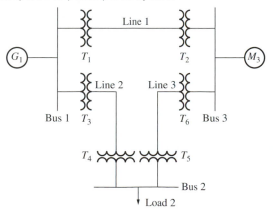

The per-phase, per-unit equivalent circuit of this power system is shown in Figure 10–6. Here, the busses have been labeled as nodes ①, ②, and ③, while the neutral has been labeled as node ⓝ. The per-unit series impedances of the transformers and transmission lines between each pair of busses have been added up, and the resulting impedances have been expressed as admittances ($Y = 1/Z$) for ease of use in nodal analysis. In addition, a shunt admittance at each bus is shown between the bus and the neutral. This admittance can include the shunt admittance of the transmission line models, as well as the shunt admittance associated with any generators or loads on a bus.

Figure 10–6 | Per-phase, per-unit equivalent circuit for the power system shown in Figure 10–5.

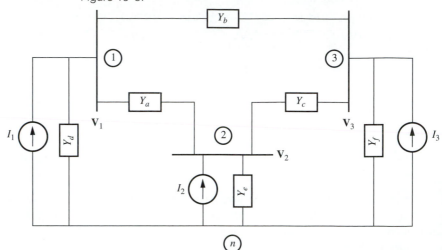

The voltages between each bus and the neutral in this equivalent circuit are represented with single subscripts, while the voltages between any two busses are represented with double subscripts. Thus the voltage V_1 is the voltage between bus 1 and the neutral, while the voltage V_{12} is the voltage between bus 1 and bus 2.

Also, note that the generators and loads on the system are represented by current sources injecting currents into the appropriate node. This convention makes nodal analysis easier, and it creates no difficulties for the user, since we can always convert a synchronous machine equivalent circuit containing voltage sources into one containing current sources using Norton's theorem (see Appendix B).

The convention used in Figure 10–6 is that the current sources always flow *into* a node, which means that the power flow for generators will turn out to be positive, while the power flow for motors and loads will turn out to be negative.

Kirchhoff's current law (KCL) states that *the sum of all currents entering any node is equal to the sum of all currents leaving the node.* We can use this law to create a system of simultaneous equations with the node voltages as unknowns, and we can solve that system of equations to find the voltages present in the power system.

If we assume that the current from the current sources are entering each node, and that all other currents are leaving the node, then applying Kirchhoff's current law at node ① yields the equation:

Sum of currents out of the node = sum of currents into the node

$$(\mathbf{V}_1 - \mathbf{V}_2)Y_a + (\mathbf{V}_1 - \mathbf{V}_3)Y_b + \mathbf{V}_1Y_d = \mathbf{I}_1 \qquad (10\text{–}15)$$

Similarly, applying KCL to nodes ② and ③ produces the equations

$$(\mathbf{V}_2 - \mathbf{V}_1)Y_a + (\mathbf{V}_2 - \mathbf{V}_3)Y_c + \mathbf{V}_2Y_e = \mathbf{I}_2 \qquad (10\text{–}16)$$
$$(\mathbf{V}_3 - \mathbf{V}_1)Y_b + (\mathbf{V}_3 - \mathbf{V}_2)Y_c + \mathbf{V}_3Y_f = \mathbf{I}_3 \qquad (10\text{–}17)$$

Rearranging these equations to collect the terms in each voltage produces the following result:

$$\begin{aligned}
(Y_a + Y_b + Y_d)\mathbf{V}_1 - Y_a\mathbf{V}_2 - Y_b\mathbf{V}_3 &= \mathbf{I}_1 \\
-Y_a\mathbf{V}_1 + (Y_a + Y_c + Y_e)\mathbf{V}_2 - Y_c\mathbf{V}_3 &= \mathbf{I}_2 \\
-Y_b\mathbf{V}_1 - Y_c\mathbf{V}_2 + (Y_b + Y_c + Y_f)\mathbf{V}_3 &= \mathbf{I}_3
\end{aligned} \tag{10–18}$$

Equation (10–18) can be expressed in matrix form as

$$\begin{bmatrix} Y_a + Y_b + Y_d & -Y_a & -Y_b \\ -Y_a & Y_a + Y_c + Y_e & -Y_c \\ -Y_b & -Y_c & Y_b + Y_c + Y_f \end{bmatrix}\begin{bmatrix} \mathbf{V}_1 \\ \mathbf{V}_2 \\ \mathbf{V}_3 \end{bmatrix} = \begin{bmatrix} \mathbf{I}_1 \\ \mathbf{I}_2 \\ \mathbf{I}_3 \end{bmatrix} \tag{10–19}$$

This is an equation of the form

$$\mathbf{Y}_{\text{bus}}\mathbf{V} = \mathbf{I} \tag{10–20}$$

where $\mathbf{Y}_{\text{bus}}$ is the *bus admittance matrix* of a system. The bus admittance matrix of a system has the form

$$\mathbf{Y}_{\text{bus}} = \begin{bmatrix} Y_{11} & Y_{12} & Y_{13} \\ Y_{21} & Y_{22} & Y_{23} \\ Y_{31} & Y_{32} & Y_{33} \end{bmatrix} \tag{10–21}$$

As we can see from Equation (10–19), the bus admittance matrix has a regular form that is easy to calculate:

■ The diagonal elements Y_{ii} are equal to the *sum* of all admittances directly connected to node *i*.

■ The off-diagonal elements Y_{ij} are equal to the *negative* of the admittances directly connected to node *i* and node *j*.

The diagonal admittances of $\mathbf{Y}_{\text{bus}}$ are known as the *self-admittances* or *driving-point admittances* of the nodes, and the off-diagonal admittances of $\mathbf{Y}_{\text{bus}}$ are known as the *mutual admittances* or *transfer admittances* of the nodes.

The matrix $\mathbf{Y}_{\text{bus}}$ is extremely useful in load flow calculations, as we shall see in the next chapter. Inverting this matrix yields the *bus impedance matrix* $\mathbf{Z}_{\text{bus}}$, which is very important in fault current studies.

$$\mathbf{Z}_{\text{bus}} = \mathbf{Y}_{\text{bus}}^{-1} \tag{10–22}$$

Note that this simple technique for constructing $\mathbf{Y}_{\text{bus}}$ works only for components that are not mutually coupled. The construction techniques that apply for mutually coupled components are covered in Reference 2.

Once $\mathbf{Y}_{\text{bus}}$ has been calculated, the solution to Equation (10–22) becomes

$$\mathbf{V} = \mathbf{Y}_{\text{bus}}^{-1}\mathbf{I} \tag{10–23}$$

or
$$\mathbf{V} = \mathbf{Z}_{\text{bus}}\mathbf{I} \tag{10–24}$$

10.4 | SOLVING POWER SYSTEM NODE EQUATIONS WITH MATLAB

There are many different techniques available for solving systems of simultaneous linear equations, such as substitution, gaussian elimination, LU factorization, and so forth. In most power system textbooks, substantial time is devoted to discussing the different techniques available for such solutions. Fortunately for us, MATLAB has very efficient equation solvers built directly into it, so the solution of a system of nodal equations will be a simple matter.

If a system of n simultaneous linear equations in n unknowns can be expressed in the form

$$Ax = b \tag{10–25}$$

where A is an $n \times n$ matrix, and b is an n-element column vector, then the solution to the system of equations can be expressed as

$$x = A^{-1}b \tag{10–26}$$

where A^{-1} is the matrix inverse of A.

In MATLAB, the solution of Equation (10–25) can be accomplished either by directly evaluating Equation (10–26), or by using the left division (\) symbol. For example, consider the system of equations

$$\begin{aligned}
1.0x_1 + 0.5x_2 - 0.5x_3 &= 1.0 \\
0.5x_1 + 1.0x_2 + 0.25x_3 &= 2.0 \\
-0.5x_1 + 0.25x_2 + 1.0x_3 &= 1.0
\end{aligned} \tag{10–27}$$

Here matrices A and b are:

$$A = \begin{bmatrix} 1.0 & 0.5 & -0.5 \\ 0.5 & 1.0 & 0.25 \\ -0.5 & 0.25 & 1.0 \end{bmatrix} \tag{10–28}$$

$$b = \begin{bmatrix} 1 \\ 2 \\ 1 \end{bmatrix} \tag{10–29}$$

This system of equations can be solved using MATLAB in either of the following ways:

```
» A = [1 .5 -.5;.5 1 .25; -.5 .25 1];
» b = [1;2;1];
» x = inv(A) * b
x =
    1.0000
    1.2000
    1.2000
» x = A\b
x =
    1.0000
    1.2000
    1.2000
```

EXAMPLE 10–3

A simple power system consisting of four busses interconnected by five transmission lines is shown in Figure 10–7. The power system includes one generator attached to

bus 1 and one synchronous motor attached to bus 3. The per-phase, per-unit equivalent circuit of this power system is shown in Figure 10–8. Note that all impedances are being treated as pure reactances for simplicity in this case. (Recall that

Figure 10–7 | A simple three-phase power system containing four busses and
five transmission lines.

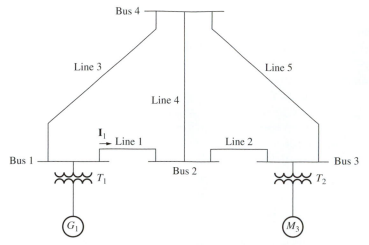

Figure 10–8 | The per-phase, per-unit equivalent circuit of the power system in
Figure 10–7. Note that the values shown are per-unit *impedances.*

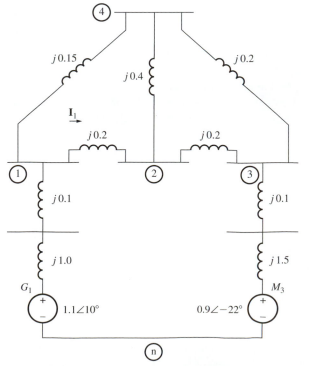

reactance is much larger than resistance in typical transformers, synchronous machines, and overhead transmission lines, so this sort of simplification makes sense.) Find the per-unit voltage at each bus in this power system, and the per-unit current flow in line 1.

■ **Solution**

The first step in solving for the bus voltages in this circuit is to convert the voltage sources in the circuit into equivalent current sources by using Norton's theorem. Then, we must convert all of the impedance values into admittances and build up the admittance matrix $\mathbf{Y}_{bus}$. Next, we will use $\mathbf{Y}_{bus}$ to solve for the bus voltages, and then use the voltages on buses 1 and 2 to calculate the current in line 1.

First, we need to find the Norton equivalent circuits for the combination of G_1 and T_1, and for the combination of M_3 and T_2. The combination of G_1 and T_1 is shown in Figure 10–9a. The Thevenin impedance of this combination is $Z_{TH} = j1.1$, and the short-circuit current is

$$\mathbf{I}_{SC} = \frac{\mathbf{V}_{OC}}{Z_{TH}} = \frac{1.1\angle 10°}{j1.1} = 1.0\angle -80°$$

The resulting Norton equivalent circuit is shown in Figure 10–9b.

The combination of M_3 and T_2 is shown in Figure 10–9c. The Thevenin impedance of this combination is $Z_{TH} = j1.6$, and the short-circuit current is

$$\mathbf{I}_{SC} = \frac{\mathbf{V}_{OC}}{Z_{TH}} = \frac{0.9\angle -22°}{j1.6} = 0.563\angle -112°$$

The resulting Norton equivalent circuit is shown in Figure 10–9d.

The per-phase, per-unit circuit with the current sources included is shown in Figure 10–10, and the same circuit with the impedance values converted to admittances is shown in Figure 10–11. We can now use that figure to construct $\mathbf{Y}_{bus}$ in the manner described in this section. The resulting admittance matrix is

$$\mathbf{Y}_{bus} = \begin{bmatrix} -j12.576 & j5.0 & 0 & j6.667 \\ j5.0 & -j12.5 & j5.0 & j2.5 \\ 0 & j5.0 & -10.625 & j5.0 \\ j6.667 & j2.5 & j5.0 & -j14.167 \end{bmatrix} \tag{10–30}$$

The current vector for this circuit is

$$\mathbf{I} = \begin{bmatrix} 1.0\angle -80° \\ 0 \\ 0.563\angle -112° \\ 0 \end{bmatrix} \tag{10–31}$$

so the equation describing this power system is

$$\mathbf{Y}_{bus}\mathbf{V} = \mathbf{I} \tag{10-20}$$

$$\begin{bmatrix} -j12.576 & j5.0 & 0 & j6.667 \\ j5.0 & -j12.5 & j5.0 & j2.5 \\ 0 & j5.0 & -j10.625 & j5.0 \\ j6.667 & j2.5 & j5.0 & -j14.167 \end{bmatrix} \mathbf{V} = \begin{bmatrix} 1.0\angle -80° \\ 0 \\ 0.563\angle -112° \\ 0 \end{bmatrix} \tag{10-32}$$

The solution to this equation is

$$\mathbf{V} = \mathbf{Y}_{bus}^{-1}\mathbf{I}$$

$$= \begin{bmatrix} -j12.576 & j5.0 & 0 & j6.667 \\ j5.0 & -j12.5 & j5.0 & j2.5 \\ 0 & j5.0 & -j10.625 & j5.0 \\ j6.667 & j2.5 & j5.0 & -j14.167 \end{bmatrix}^{-1} \begin{bmatrix} 1.0\angle -80° \\ 0 \\ 0.563\angle -112° \\ 0 \end{bmatrix} \tag{10-33}$$

This solution is easy to calculate in MATLAB, since matrix inversion and matrix multiplication are built into the language. The results are:

$$\mathbf{V} = \begin{bmatrix} 0.989\angle -0.60° \\ 0.981\angle -1.58° \\ 0.974\angle -2.62° \\ 0.982\angle -1.48° \end{bmatrix} \text{V} \tag{10-34}$$

Next the current in line 1 can be calculated from the equation

$$\mathbf{I}_1 = (\mathbf{V}_1 - \mathbf{V}_1)Y_{line\ 1} \tag{10-35}$$
$$= (0.989\angle -0.60° - 0.981\angle -1.58°)(-j5.0)$$
$$= 0.092\angle -25.16°$$

A MATLAB program to perform these calculations is shown below. Note that this program uses the utility functions **r2p** and **p2r** to convert complex numbers between rectangular and polar forms.

```
% M-file: ex10_3.m
% M-file to calculate the per-unit bus voltage and the
%   current in line 1 in the power system of Example 10-3.

% Create Y-bus
Ybus = [-j*12.576     j*5.0          0          j*6.667; ...
         j*5.0       -j*12.5        j*5.0       j*2.5; ...
            0         j*5.0        -j*10.625    j*5.0; ...
         j*6.667      j*2.5         j*5.0      -j*14.167];

% Create the current vector I
I = [ p2r(1.0,-80); ...
      0; ...
      p2r(0.563,-112); ...
      0];

% Calculate Vbus
Vbus = Ybus \ I;
```

```
% Display results
for ii = 1:4
   [mag, phase] = r2p(Vbus(ii));
   str = ['The voltage at bus ' int2str(ii) ' = ' ...
          num2str(mag) '/' num2str(phase)];
   disp(str);
end

% Now calculate the current in line 1
i1 = (Vbus(1) - Vbus(2)) * (-j*5.0);
[mag, phase] = r2p(i1);
str = ['The current in line 1 = ' num2str(mag) '/' num2str(phase)];
disp(str);
```

Figure 10–9 | (a) The combination of G_1 and T_1. (b) The Norton equivalent circuit for the combination of G_1 and T_1. (c) The combination of M_3 and T_2. (d) The Norton equivalent circuit for the combination of M_3 and T_2.

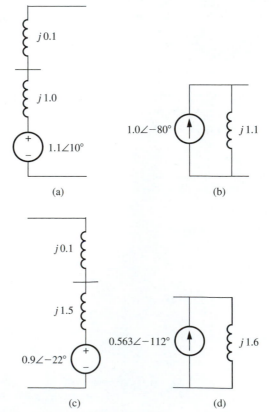

When this program is executed, the results are:

```
» ex10_3
The voltage at bus 1 = 0.98886/-0.60171
The voltage at bus 2 = 0.98136/-1.5782
The voltage at bus 3 = 0.9738/-2.6195
The voltage at bus 4 = 0.9821/-1.4799
The current in line 1 = 0.091941/-25.1562
```

∎

Figure 10–10 | The per-phase, per-unit equivalent circuit of the power system after converting the voltage sources to current sources. Note that the values shown are per-unit *impedances*.

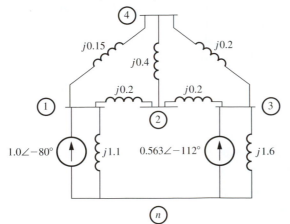

Figure 10–11 | The per-phase, per-unit equivalent circuit of the power system after converting the impedances to *admittances*.

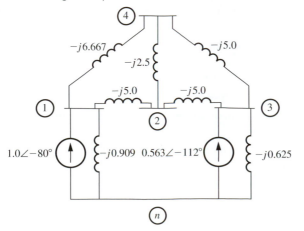

10.5 | SUMMARY

Power systems are usually represented by one-line diagrams, in which a single line represents the three phases connecting power system components together. One-line diagrams are drawn with a standard set of symbols for machines, circuit breakers, transformers, and so forth. They usually contain information such as voltage levels, device ratings, and connection types (wye or delta), although the information on the diagram will vary depending on the purpose for which it is used.

Power systems are usually analyzed by using per-phase, per-unit equivalent circuits. This analysis is done on a per-phase basis because the magnitudes of the voltages and currents in each phase of a balanced three-phase system are identical. The per-unit system is used because it eliminates changes of voltage level within a system.

The voltages in power systems are usually calculated by nodal analysis. All sources are represented as current sources by using Norton's theorem, and all impedances are represented as admittances. Then, nodal equations are written at every bus in the system, and the resulting set of simultaneous equations is solved to determine the bus voltages.

The system of equations to solve takes the form of the equation

$$\mathbf{Y}_{bus}\mathbf{V} = \mathbf{I} \qquad (10\text{--}20)$$

where $\mathbf{Y}_{bus}$ is the bus admittance matrix of a system. $\mathbf{Y}_{bus}$ can usually be constructed by inspection once a power system is represented with current sources and admittances, according to the following rules:

- The diagonal elements Y_{ii} are equal to the *sum* of all admittances directly connected to node i.
- The off-diagonal elements Y_{ij} are equal to the *negative* of the admittances directly connected to node i and node j.

The resulting system of equations can be solved easily by using the built-in functions in MATLAB to invert $\mathbf{Y}_{bus}$ and multiply it by the current vector $\mathbf{I}$.

10.6 | QUESTIONS

10–1. Why are one-line diagrams used to represent power systems?

10–2. What symbols can appear on a one-line diagram?

10–3. What is Norton's theorem? How is it used in solving systems of power system equations?

10–4. How is the bus admittance matrix $\mathbf{Y}_{bus}$ created?

10–5. What is the relationship between the bus admittance matrix $\mathbf{Y}_{bus}$ and the bus impedance matrix $\mathbf{Z}_{bus}$?

10–6. What steps are required to solve for the bus voltages in a power system?

Figure P10–1 | One-line diagram of the power system in Problem 10–3.

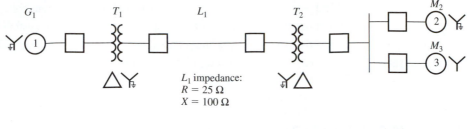

G_1 ratings:	T_1 ratings:		T_2 ratings:	M_2 ratings:	M_3 ratings:
30 MVA	35 MVA		30 MVA	20 MVA	10 MVA
13.8 kV	13.2/115 kV		120/12.5 kV	12.5 kV	12.5 kV
$R = 0.1$ pu	$R = 0.01$ pu		$R = 0.01$ pu	$R = 0.1$ pu	$R = 0.1$ pu
$X_S = 1.0$ pu	$X = 0.10$ pu		$X = 0.08$ pu	$X_S = 1.1$ pu	$X_S = 1.1$ pu

10.7 | PROBLEMS

10–1. Sketch the per-phase, per-unit equivalent circuit of the power system in Figure 10–2. (Treat each load on the system as a resistance in series with a reactance.) Note that you do not have enough information to actually calculate the values of components in the equivalent circuit.

10–2. A 20,000-kVA, 110/13.8-kV, Y–Δ three-phase transformer has a series impedance of $0.02 + j0.08$ pu. Find the per-unit impedance of this transformer in a power system with a base apparent power of 500 MVA and a base voltage on the high side of 120 kV.

10–3. Find the per-phase equivalent circuit of the power system shown in Figure P10–1.

10–4. Two 4.16-kV, three-phase synchronous motors are connected to the same bus. The motor ratings are:

> Motor 1: 5000 hp, 0.8 PF lagging, 95% efficiency, $R = 3\%$, $X_S = 90\%$
> Motor 2: 3000 hp, 1.0 PF, 95% efficiency, $R = 3\%$, $X_S = 90\%$

Calculate the per-unit impedances of these motors to a base of 20 MVA, 4.16 kV. (*Note:* To calculate these values, you will first have to determine the rated apparent power of each motor considering its rated output power, efficiency, and power factor.)

10–5. A Y-connected synchronous generator rated 100 MVA, 13.2 kV has a rated impedance of $R = 5\%$ and $X_S = 80\%$ per unit. It is connected to a 50-Ω transmission line through a 13.8/120-kV, 100-MVA Δ–Y transformer with a rated impedance of $R = 2\%$ and $X = 8\%$ per unit. The base for the power system is 200 MVA, 120 kV at the transmission line.

a. Sketch the one-line diagram of this power system, with symbols labeled appropriately.

b. Find per-unit impedance of the generator on the system base.

c. Find per-unit impedance of the transformer on the system base.

d. Find per-unit impedance of the transmission line on the system base.

e. Find the per-phase, per-unit equivalent circuit of this power system.

10–6. Assume that the power system of the previous problem is connected to a resistive Y-connected load of 16 Ω per phase. If the internal generated voltage of the generator is $\mathbf{E}_A = 1.10\angle 20°$ per unit, what is the terminal voltage of the generator? How much power is being supplied to the load?

10–7. Figure P10–2 shows a one-line diagram of a three-phase power system. The ratings of the various components in the system are:

Synchronous generator 1:	40 MVA, 13.8 kV, $R = 3\%$, $X_S = 80\%$
Synchronous motor 2:	20 MVA, 13.8 kV, $R = 3\%$, $X_S = 80\%$
Synchronous motor 3:	10 MVA, 13.2 kV, $R = 3\%$, $X_S = 100\%$
Y–Δ transformers:	20 MVA, 13.8/138 kV, $R = 2\%$, $X = 10\%$
Y–Y transformers:	10 MVA, 13.8/138 kV, $R = 2\%$, $X = 10\%$
Line 1:	$R = 10\ \Omega$, $X = 50\ \Omega$
Line 2:	$R = 5\ \Omega$, $X = 30\ \Omega$
Line 3:	$R = 5\ V$, $X = 30\ \Omega$

Figure P10–2 | One-line diagram of the power system in Problem 10–7.

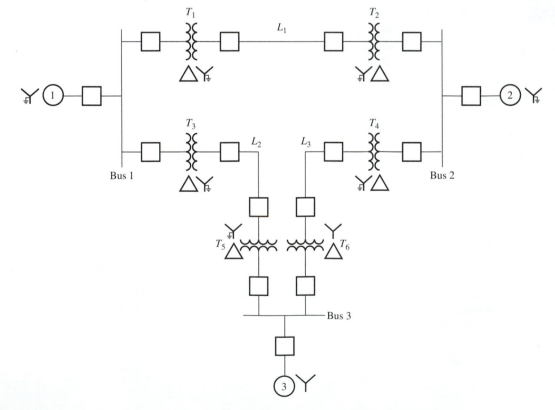

The per-unit system base for this power system is 40 MVA, 128 kV in transmission line 1. Create the per-phase, per-unit equivalent circuit for this power system.

10–8. Calculate the bus admittance matrix $\mathbf{Y}_{\text{bus}}$ and the bus impedance matrix $\mathbf{Z}_{\text{bus}}$ for the power system shown in Figure P10–2.

10–9. Assume that internal generated voltages of the generators and motors in the per-unit equivalent circuit of the previous problem have the following values:

$$\mathbf{E}_{A1} = 1.15\angle 22°$$
$$\mathbf{E}_{A2} = 1.00\angle -20°$$
$$\mathbf{E}_{A3} = 0.95\angle -15°$$

 a. Find the per-unit voltages on each bus in the power system.

 b. Find the actual voltages on each bus in the power system.

 c. Find the current flowing in each transmission line in the power system.

 d. Determine the magnitude and direction of the real and reactive power flowing in each transmission line.

 e. Are any of the components in the power system overloaded?

10.8 | REFERENCES

 1. Elgerd, Olle I.: *Electric Energy Systems Theory: An Introduction*, McGraw-Hill, New York, 1971.

 2. Granger, John J., and William D. Stevenson, Jr.: *Power Systems Analysis*, McGraw-Hill, New York, 1994.

 3. Gross, Charles A.: *Power Systems Analysis*, John Wiley and Sons, New York, 1979.

 4. IEEE Standard 315-1975: *IEEE Graphic Symbols for Electrical and Electronics Diagrams*, Institute of Electrical and Electronics Engineers, 1975 (reaffirmed 1993).

 5. Yamayee, Zia A., and Juan L. Bala, Jr.: *Electromechanical Energy Devices and Power Systems*, John Wiley and Sons, New York, 1994.

Introduction to Power-Flow Studies

A power system engineer must be able to analyze the performance of power systems both in normal operating conditions and under fault (short-circuit) conditions. The analysis of a power system in normal steady-state operation is known as a *power-flow study* or *load-flow study*. Power-flow studies attempt to determine the voltages, currents, and real and reactive power flows in a power system under a given set of load conditions. They are the topic of this chapter. The analyses of power systems under fault (short-circuit) conditions will be postponed until subsequent chapters.

An engineer who understands the operation of a power system under normal conditions will be able to plan ahead for its safe operation and future growth. A good understanding of steady-state conditions allows the engineer to play "What if?" games.

For example, suppose that a new customer intends to open an industrial plant that will require 100 MW of power on the outskirts of some city. Will there be enough power-handling capacity in the system for this new load, or will the additional load cause some components to be overloaded? Will it be necessary to build new transmission lines, or to increase the number or size of installed transformers? Will it be necessary to build new generation capacity?

As another example, suppose that a power system is properly supplying loads, and a transmission line within the system must be taken off line for maintenance. Can the remaining lines in the power system handle the required loads without exceeding their ratings? These sorts of questions are asked on a regular basis.

Furthermore, the central planning departments of power companies try to anticipate the power needs 10 to 20 years into the future, and to simulate systems serving those needs. These studies help them to identify the need for additional transmission lines and generation capability early enough to plan properly for the future.

11.1 | BASIC TECHNIQUES FOR POWER-FLOW STUDIES

A power-flow study is an analysis of the voltages, currents, and power flows in a power system under steady-state conditions. In a power-flow study, the power systems engineer *makes an assumption* about either the voltage at a bus or the power being supplied to the bus for each bus in the power system, and then determines the magnitude and phase angles of the bus voltages, line currents, etc. that result from the assumed combination of voltages and power flows.

The simplest way to perform power-flow calculations is by iteration. We create a bus admittance matrix $\mathbf{Y}_{bus}$ for the power system, and make an initial estimate for the voltages at each bus in the system. We can then update the voltage estimate for each bus in the system one at a time, based on the estimates for the voltages and power flows at every other bus and the values of the bus admittance matrix. Since the voltage at a given bus depends on the voltages at all of the other busses in the system, and the voltages at the other busses are just estimates, the updated voltage will not be correct. However, it will usually be closer to the correct answer than the original guess. If this process is repeated, the voltages at each bus will get closer and closer to the correct answer. When the voltages at each bus no longer change significantly during an iteration, the solution has converged to the correct answer.

The equations used to update the estimates of voltage at a bus differ for different types of busses, as we will describe below. Each bus in a power system is classified as one of three types:

1. *Load bus.* A *load bus* is a bus at which the real and reactive power are specified, and for which the bus voltage will be calculated. Real and reactive power supplied to a power system are defined to be positive quantities, while real and reactive power consumed from the power system are defined to be negative quantities. A load bus is also known as *PQ bus*, because the real power P_i and reactive power Q_i at the bus are specified. All busses that do not have generators attached are load busses.

2. *Generator bus.* A *generator bus* is a bus at which the magnitude of the voltage is kept constant by adjusting the field current of a synchronous generator tied to the bus. As we learned in Chapter 6, when a generator is connected to a power system, increasing the field current of the generator increases both the reactive power supplied by the generator and the terminal voltage of the power system. At a generator bus, we assume that the field current is being adjusted to maintain a constant terminal voltage V_T. We also learned in Chapter 6 that the generator's prime mover controls the real power supplied by the generator to the power system. Increasing the prime mover's governor set points increases the power that the generator supplies to the power system. Thus at a generator bus, we can control and specify the *magnitude of the bus voltage* $|\mathbf{V}_i|$ and *real power P_i supplied*. A generator bus is also known as *PV bus*, because the real power P_i and magnitude of the bus voltage $|\mathbf{V}_i|$ at the bus are specified.

3. *Slack bus or swing bus.* The *slack bus* or *swing bus* is a special generator bus that serves as the reference bus for the power system. Its voltage is assumed to be fixed in both magnitude and phase—most often it is taken to be $1\angle 0°$ pu. The real and reactive power of the slack bus is uncontrolled; it supplies whatever real or reactive power is necessary to make the power flows in the system balance.

These categories generally correspond to the ways in which real power systems operate. Load busses are busses that supply power to loads, and the amount of power supplied will be whatever the loads demand. The voltage on a load bus in a real power system will go up and down with changing loads, unless the power company compensates for these changes with tap-changing under load (TCUL) transformers or switched capacitors. Thus, load busses have specified values of P and Q, while **V** varies with load conditions.

Real generators operate most efficiently when they are running at full load, so *power systems try to keep all but one (or a few) generator running at 100 percent capacity*, while allowing the remaining generator (called the *swing* generator) to handle increases and decreases in load demand. Thus, most busses with generators attached to them will be supplying a *fixed* amount of power (the rated full load power of the generators), and the magnitude of their voltages will be maintained constant by field circuits of the generators. These generator busses have specified values of P and $|\mathbf{V}_i|$.

Finally, the controls on the swing generator will be adjusted to maintain a constant power system voltage and frequency, allowing real and reactive power to increase or decrease as required whenever loads change. The bus that the swing generator is connected to is called the slack bus.

In this section, we will learn how to perform power-flow studies on a power system that has only load busses and a slack bus. In the next section, we will learn how to handle generator busses.

Constructing $\mathbf{Y}_{\text{bus}}$ for Power-Flow Analysis

The most common approach to power-flow analysis is based on the bus admittance matrix $\mathbf{Y}_{\text{bus}}$. We learned how to create the bus admittance matrix in Chapter 10. The form of the bus admittance matrix used for load flow studies is slightly different than the one we studied in Chapter 10, because the internal impedances of generators and loads connected to the power system are not included in $\mathbf{Y}_{\text{bus}}$. Instead, they are accounted for as specified real and reactive powers input and output from the busses.

For example, consider the simple power system shown in Figure 11–1. This power system has four busses, five transmission lines, one generator, and three loads. The series per-unit impedances of the five transmission lines are specified in Table 11–1, while the shunt admittances of the transmission lines are ignored. In this case, the Y_{ii} terms of the bus admittance matrix can be constructed by summing the admittances of all transmission lines connected to each bus, and the Y_{ij} ($i \neq j$) terms are just the negative of the line admittances stretching between busses i and j. For example, the term Y_{11} will be the sum of the admittances of all transmission lines connected to bus 1, which are transmission lines 1 and 5, so

$Y_{11} = 1.7647 - j7.0588$ pu.[1] Similarly, the term Y_{12} will be the *negative* of all the admittances stretching between bus 1 and bus 2, which will be the negative of the admittance of transmission line 1, so $Y_{12} = -0.5882 + j2.3529$ pu.

Figure 11–1 | A simple power system with four busses, five transmission lines, one generator, and three loads.

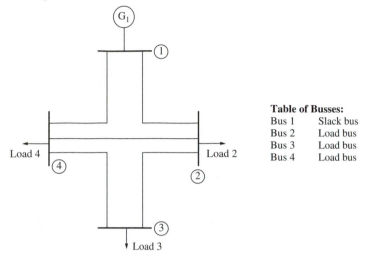

Table of Busses:
Bus 1   Slack bus
Bus 2   Load bus
Bus 3   Load bus
Bus 4   Load bus

Table 11–1 | Per-unit series impedances and the corresponding admittances of the transmission lines in Figure 11–1

Transmission line number	From/to (bus to bus)	Series impedance Z (pu)	Series admittance Y (pu)
1	1–2	$0.10 + j0.40$	$0.5882 - j2.3529$
2	2–3	$0.10 + j0.50$	$0.3846 - j1.9231$
3	2–4	$0.10 + j0.40$	$0.5882 - j2.3529$
4	3–4	$0.05 + j0.20$	$1.1765 - j4.7059$
5	4–1	$0.05 + j0.20$	$1.1765 - j4.7059$

The complete bus admittance matrix $\mathbf{Y}_{bus}$ can be derived by repeating these calculations for every term in the matrix. It is:

$$\mathbf{Y}_{bus} = \begin{bmatrix} 1.7647 - j7.0588 & -0.5882 + j2.3529 & 0 & -1.1765 + j4.7059 \\ -0.5882 + j2.3529 & 1.5611 - j6.6290 & -0.3846 + j1.9231 & -0.5882 + j2.3529 \\ 0 & -0.3846 + j1.9231 & 1.5611 - j6.6290 & -1.1765 + j4.7059 \\ -1.1765 + j4.7059 & -0.5882 + j2.3529 & -1.1765 + j4.7059 & 2.9412 - j11.7647 \end{bmatrix} \tag{11–1}$$

---

[1]If the shunt admittances of the transmission lines are *not* ignored, then we will treat each transmission line as a π circuit, with half of the shunt admittance placed at each end of the line. In that case, the self admittance Y_{ii} at each bus would also include half of the shunt admittance of each transmission line connected to the bus.

Power-Flow Analysis Equations

The basic equation for power-flow analysis is derived from the nodal analysis equations for the power system:

$$\mathbf{Y}_{bus}\,\mathbf{V} = \mathbf{I} \tag{10–20}$$

For the four-bus power system shown in Figure 11–1, Equation (10–20) becomes

$$\begin{bmatrix} Y_{11} & Y_{12} & Y_{13} & Y_{14} \\ Y_{21} & Y_{22} & Y_{23} & Y_{24} \\ Y_{31} & Y_{32} & Y_{33} & Y_{34} \\ Y_{41} & Y_{42} & Y_{43} & Y_{44} \end{bmatrix} \begin{bmatrix} \mathbf{V}_1 \\ \mathbf{V}_2 \\ \mathbf{V}_3 \\ \mathbf{V}_4 \end{bmatrix} = \begin{bmatrix} \mathbf{I}_1 \\ \mathbf{I}_2 \\ \mathbf{I}_3 \\ \mathbf{I}_4 \end{bmatrix} \tag{11–2}$$

where Y_{ij} are the elements of the bus admittance matrix, $\mathbf{V}_i$ are bus voltages, and $\mathbf{I}_i$ are the currents injected at each node. For bus 2 in the power system, this equation reduces to

$$Y_{21}\mathbf{V}_1 + Y_{22}\mathbf{V}_2 + Y_{23}\mathbf{V}_3 + Y_{24}\mathbf{V}_4 = \mathbf{I}_2 \tag{11–3}$$

Unfortunately, the loads on a real power system are specified in terms of real and reactive power, *not* as currents. The relationship between the per-unit real and reactive power supplied to the power system at bus i and the per-unit current injected into the system at that bus can be found from the complex power Equations (1–60) and (1–61):

$$\mathbf{S} = \mathbf{V}\mathbf{I}^* = P + jQ \tag{11–4}$$

where $\mathbf{V}$ is the per-unit voltage at the bus, $\mathbf{I}^*$ is the complex conjugate of the per-unit current injected at the bus, and P and Q are the per-unit real and reactive powers, respectively. Note that P is a positive number for generators and a negative number for loads.

Therefore, the current injected at bus 2 can be found as

$$\mathbf{V}_2\mathbf{I}_2^* = P_2 + jQ_2$$

$$\mathbf{I}_2^* = \frac{P_2 + jQ_2}{\mathbf{V}_2}$$

or

$$\mathbf{I}_2 = \frac{P_2 - jQ_2}{\mathbf{V}_2^*} \tag{11–5}$$

If Equation (11–5) is substituted into Equation (11–3), the result is

$$Y_{21}\mathbf{V}_1 + Y_{22}\mathbf{V}_2 + Y_{23}\mathbf{V}_3 + Y_{24}\mathbf{V}_4 = \frac{P_2 - jQ_2}{\mathbf{V}_2^*} \tag{11–6}$$

Solving this equation for $\mathbf{V}_2$ yields the equation

$$\mathbf{V}_2 = \frac{1}{Y_{22}}\left[\frac{P_2 - jQ_2}{\mathbf{V}_2^*} - (Y_{21}\mathbf{V}_1 + Y_{23}\mathbf{V}_3 + Y_{24}\mathbf{V}_4) \right] \tag{11–7}$$

Similar equations can be created for each load bus in the power system.

Equation (11–7) gives an *updated estimate* for $\mathbf{V}_2$ based on the specified values of real and reactive power, plus the current estimates of all of the bus voltages in the power system. Note that the updated estimate for $\mathbf{V}_2$ will *not* be the same as the original estimate of $\mathbf{V}_2{}^*$ that was used to derive it. If the newly calculated value for $\mathbf{V}_2$ were substituted back into the equation and used to calculate a new $\mathbf{V}_2$ several times, the differences between the estimates would converge to a single value of $\mathbf{V}_2$ based on the estimated values of the other bus voltages in the system. However, this voltage would *not* be the correct bus voltage for the entire power system, because the voltages at the other nodes also need to be updated. Since the voltages at each bus in the system depend on the voltages at all the other busses, the only way to find the correct answer is to update all of the bus voltages in the power system during each iteration.

The bus voltages in a power system are found by updating the voltage estimates for each bus in the power system one after another (except for the slack bus), and then repeating the process until the voltage values no longer change much between iterations. This method is known as the *Gauss-Siedel iterative method*. The basic procedure for the Gauss-Siedel iterative method is:

1. *Calculate the bus admittance matrix* $\mathbf{Y}_{bus}$. Include the admittances of all transmission lines, transformers, etc., *between* busses, but exclude the admittances of the loads or generators themselves.

2. *Select a slack bus.* One of the busses in the power system should be chosen as the slack bus, and its voltage will arbitrarily be assumed to be $1.0\angle 0°$.

3. *Select initial estimates for all bus voltages.* It is necessary to make an initial estimate for the voltages at every bus before the Gauss-Siedel iteration is started. It is very important that the estimates be reasonable, because very poor choices can result in convergence to incorrect values. Most of the time, we assume that the voltage at every load bus is $1.0\angle 0°$ to start with. This choice is commonly called a *flat start;* it usually results in good convergence. (Erroneous convergence can occur if initial estimates of voltage differ widely in magnitude and/or phase.)

4. *Write voltage equations for every other bus in the system.* The voltage equations for every load bus will be analogous to Equation (11–7). The generic form of these voltage equations is

$$\mathbf{V}_i = \frac{1}{Y_{ii}} \left(\frac{P_i - jQ_i}{\mathbf{V}_i^*} - \sum_{\substack{k=1 \\ (k \neq i)}}^{N} Y_{ik}\mathbf{V}_k \right) \tag{11–8}$$

5. *Calculate an updated estimate of the voltage at each load bus in succession using Equation (11–8).* Calculate an updated estimate of the voltage at bus 1, bus 2, etc., except for the slack bus.

6. *Compare the differences between the old voltage estimates and the new voltage estimates.* If the difference between the old voltage estimates and the new ones are less than some specified tolerance for all busses, then we are done. Otherwise, repeat step 5.

7. *Confirm that the resulting solution is reasonable.* Examine the bus voltages to determine that the resulting solution is not erroneous. A valid solution typically has bus voltages whose phase range is less than 45°—larger ranges of phase

may indicate that the system converged to an incorrect solution. If the solution appears to be erroneous, change the initial conditions and try again.

The following two examples illustrate this procedure. Example 11–1 demonstrates the procedure using simple hand calculations on a two-bus power system, while Example 11–2 uses MATLAB to solve for the voltages and power flows in the four-bus power system shown in Figure 11–1.

EXAMPLE 11–1

Figure 11–2 shows a simple two-bus power system, with a generator attached to bus 1 and loads attached to bus 2. A single transmission line connects the busses. The series impedance of the line is $Z_1 = 0.10 + j0.50$ pu. The shunt admittance of the transmission line may be neglected. Assume that bus 1 is the slack bus, and that it has a voltage $\mathbf{V}_1 = 1.0\angle 0°$ pu. The per-unit real and reactive power *supplied to the loads* from the power system at bus 2 is $P_2 = 0.30$ pu, $Q_2 = 0.20$ pu. (Since these values are flowing out of the power system, the real and reactive power *supplied to the power system* at each of these busses will be the negative of the above numbers.) Determine the voltage at each bus in this power system for the specified load conditions.

Figure 11–2 | A simple two-bus power system, with a single transmission line between the busses.

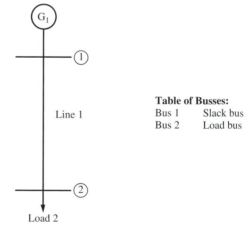

Line 1

Table of Busses:
Bus 1   Slack bus
Bus 2   Load bus

Load 2

■ Solution
We will follow the seven steps outlined above to solve this problem.

1. *Calculate the bus admittance matrix* $\mathbf{Y}_{bus}$. The Y_{ii} terms bus admittance matrix can be constructed by summing the admittances of all transmission lines connected to each bus, and the Y_{ij} ($i \neq j$) terms are just the negative of the line admittances stretching between busses i and j. For example, the term Y_{11} will be the sum of the admittances of all transmission lines

connected to bus 1 (a single line in this case). The series admittance of line 1 is

$$Y_{\text{Line 1}} = \frac{1}{Z_{\text{Line 1}}} = \frac{1}{0.10 + j0.50} = 0.3846 - j1.9231 \quad (11\text{--}9)$$

Therefore, the term Y_{11} will be

$$Y_{11} = 0.3846 - j1.9231 \quad (11\text{--}10)$$

A similar calculation applies to Y_{22}, etc. The complete bus admittance matrix $\mathbf{Y}_{\text{bus}}$ is

$$\mathbf{Y}_{\text{bus}} = \begin{bmatrix} 0.3846 - j1.9231 & -0.3846 + j1.9231 \\ -0.3846 + j1.9231 & 0.3846 - j1.9231 \end{bmatrix} \quad (11\text{--}11)$$

2. *Select a slack bus.* We will select bus 1 as the slack bus, since it is the only bus in the system connected to a generator. The voltage at bus 1 will arbitrarily be assumed to be $1.0\angle 0°$.

3. *Select initial estimates for all bus voltages.* We will make a flat start, making the initial voltage estimates at every bus $1.0\angle 0°$.

4. *Write voltage equations for every other bus in the system.* The voltage equation for bus 2 is:

$$\mathbf{V}_2 = \frac{1}{Y_{22}} \left[\frac{P_2 - jQ_2}{\mathbf{V}_{2,\,\text{old}}^*} - (Y_{21}\mathbf{V}_1) \right] \quad (11\text{--}12)$$

Since the real power and reactive power *supplied* to the system at Bus 2 are $P_2 = -0.30$ pu and $Q_2 = -0.20$ pu, since $Y_{22} = 0.3846 - j1.9231$ and $Y_{21} = -0.3846 + j1.9231$, and since $\mathbf{V}_1 = 1\angle 0°$, Equation (11–12) reduces to

$$\mathbf{V}_2 = \frac{1}{0.3846 - j1.9231} \left[\frac{-0.3 - j0.2}{\mathbf{V}_{2,\,\text{old}}^*} - [(-0.3846 + j1.9231)\mathbf{V}_1] \right] \quad (11\text{--}13)$$

$$\mathbf{V}_2 = \frac{1}{1.9612\angle -78.8°} \left[\frac{0.3603\angle -146.3°}{\mathbf{V}_{2,\,\text{old}}^*} - (1.9612\angle 101.3°)(1\angle 0°) \right] \quad (11\text{--}14)$$

5. *Calculate an updated estimate of the voltage at each load bus in succession using Equation (11–8).* In this simple problem, we only have to calculate updated voltages for bus 2, since the voltage at the slack bus (bus 1) is assumed constant. We will repeat this calculation until the voltage converges to a constant value.

The initial estimate for the voltage at bus 2 is $\mathbf{V}_{2,\,0} = 1\angle 0°$. With this starting point, the new estimate for the voltage at bus 2 becomes

$$\mathbf{V}_{2,\,1} = \frac{1}{1.9612\angle -78.7°} \left[\frac{0.3603\angle -146.3°}{\mathbf{V}_{2,\,\text{old}}^*} - (1.9612\angle 101.3°)(1\angle 0°) \right]$$

$$= \frac{1}{1.9612\angle -78.7°} \left[\frac{0.3603\angle -146.3°}{1\angle 0°} - (1.9612\angle 101.3°) \right]$$

$$= 0.8797\angle -8.499°$$

If this new estimate of $\mathbf{V}_2$ is plugged back into Equation (11–14), a second estimate can be calculated:

$$\mathbf{V}_{2,2} = \frac{1}{1.9612\angle -78.7°}\left[\frac{0.3603\angle -146.3°}{(0.8797\angle -8.499°)^*} - (1.9612\angle 101.3°)(1\angle 0°)\right]$$
$$= 0.8412\angle -8.499°$$

Similarly, the third iteration would be:

$$\mathbf{V}_{2,3} = \frac{1}{1.9612\angle -78.7°}\left[\frac{0.3603\angle -146.3°}{(0.8412\angle -8.499°)^*} - (1.9612\angle 101.3°)(1\angle 0°)\right]$$
$$= 0.8345\angle -8.962°$$

The fourth iteration would be:

$$\mathbf{V}_{2,4} = \frac{1}{1.9612\angle 78.7°}\left[\frac{0.3603\angle -146.3°}{(0.8345\angle -8.962°)^*} - (1.9612\angle 101.3°)(1\angle 0°)\right]$$
$$= 0.8320\angle -8.962°$$

The fifth iteration would be:

$$\mathbf{V}_{2,5} = \frac{1}{1.9612\angle -78.7°}\left[\frac{0.3603\angle -146.3°}{(0.8320\angle -8.962°)^*} - (1.9612\angle 101.3°)(1\angle 0°)\right]$$
$$= 0.8315\angle -8.994°$$

At this point, the magnitude of the voltage is barely changing, so we might consider this value to be close enough to the correct answer, and stop iterating. This power system converged to an answer in five iterations. The voltages at each bus in the power system are:

$$\mathbf{V}_1 = 1.0\angle 0°$$
$$\mathbf{V}_2 = 0.8315\angle -8.994°$$

7. *Confirm that the resulting solution is reasonable.* These results appear reasonable, since the phase angles of the various bus voltages in the system differ by only 10°. The current flowing in the transmission line from bus 1 to bus 2 is

$$\mathbf{I}_1 = \frac{(\mathbf{V}_1 - \mathbf{V}_2)}{Z_{\text{Line 1}}} = \frac{1\angle 0° - 0.8315\angle -8.994°}{0.10 + j0.50} \qquad (11\text{–}15)$$
$$= 0.4333\angle -42.65°$$

and the power supplied by the transmission line to bus 2 is

$$\mathbf{S} = \mathbf{VI}^* = (0.8315\angle -8.994°)(0.4333\angle -42.65°)^* \quad (11\text{–}16)$$
$$= 0.2999 + j0.1997$$

This is just the amount of power being consumed by the loads, so this solution appears to be correct. ∎

The results of Example 11–1 should be interpreted as follows: *if* the real and reactive power supplied by bus 2 is $0.3 + j0.2$ pu, and *if* the voltage on the slack bus is $1\angle 0°$ pu, then the voltage at bus 2 will be $\mathbf{V}_2 = 0.8315\angle -8.994°$. Note that this voltage is correct only for the assumed conditions; if a different amount of power were supplied by bus 2, a different voltage $\mathbf{V}_2$ will be calculated.

In power flow studies, we postulate some reasonable combination of powers being supplied to loads, and determine the resulting voltages at all of the busses in the power system. Once the voltages at the busses are known, we can calculate the current flowing in each transmission line in the system, and then check to see that none of the transmission lines are overloaded in that particular situation.

EXAMPLE 11–2

Refer to the power system shown in Figure 11–1. Assume that bus 1 is the slack bus, and that it has a voltage $\mathbf{V}_1 = 1.0\angle 0°$ pu. The per-unit real and reactive power *supplied to the loads* from the power system at busses 2, 3, and 4 is $P_2 = 0.20$ pu, $Q_2 = 0.15$ pu, $P_3 = 0.20$ pu, $Q_3 = 0.15$ pu, $P_4 = 0.20$ pu, and $Q_4 = 0.15$ pu. (Since these values are flowing out of the power system, the real and reactive power *supplied to the power system* at each of these busses will be the negative of the above numbers.)

Determine the voltage at each bus in this power system for the specified load conditions.

■ Solution

We will follow the seven steps outlined above to solve this problem. The problem must be solved as a MATLAB program, because the time required to solve it by hand calculations would be excessive. We will build a MATLAB program and execute it to calculate the bus voltages.

1. *Calculate the bus admittance matrix* $\mathbf{Y}_{bus}$. The bus admittance matrix was calculated earlier. It is

$$\mathbf{Y}_{bus} = \begin{bmatrix} 1.7647 - j7.0558 & -0.5882 + j2.3529 & 0 & -1.1765 + j4.7059 \\ -0.5882 + j2.3529 & 1.5661 - j6.6290 & -0.3846 + j1.9231 & -0.5882 + j2.3529 \\ 0 & -0.3846 + j1.9231 & 1.5611 - j6.6290 & -1.1765 + j4.7059 \\ -1.1765 + j4.7059 & -0.5882 + j2.3529 & -1.1765 + j4.7059 & 2.9412 - j11.7647 \end{bmatrix} \quad (11\text{--}1)$$

The MATLAB statement required to initialize $\mathbf{Y}_{bus}$ to the values in Equation (11–1) is

```
% Create Y-bus
Ybus = ...
[ 1.7647-j*7.0558  -0.5882+j*2.3529   0                 -1.1765+j*4.7059; ...
 -0.5882+j*2.3529   1.5611-j*6.6290  -0.3846+j*1.9231   -0.5882+j*2.3529; ...
  0                -0.3846+j*1.9231   1.5611-j*6.6290   -1.1765+j*4.7059; ...
 -1.1765+j*4.7059  -0.5882+j*2.3529  -1.1765+j*4.7059    2.9412-j*11.7647 ];
```

2. *Select a slack bus.* We will select bus 1 as the slack bus, since it is the only bus in the system connected to a generator. The voltage at bus 1 will be arbitrarily be assumed to be $1.0\angle 0°$.

3. *Select initial estimates for all bus voltages.* We will make a flat start, making the initial voltage estimates at every bus be $1.0\angle 0°$.

The MATLAB statements required to initialize the bus voltages are:

```
% Initialize the bus voltages to 1.0 at 0 degrees
for ii = 1:n_bus
   Vbus(ii) = 1;
end
```

4. *Write voltage equations for every other bus in the system.* The voltage equations for busses 2, 3, and 4 are:

$$\mathbf{V}_2 = \frac{1}{Y_{22}} \left[\frac{P_2 - jQ_2}{\mathbf{V}_{2,\,\text{old}}^*} - (Y_{21}\mathbf{V}_1 + Y_{23}\mathbf{V}_3 + Y_{24}\mathbf{V}_4) \right] \qquad (11\text{–}17)$$

$$\mathbf{V}_3 = \frac{1}{Y_{33}} \left[\frac{P_3 - jQ_3}{\mathbf{V}_{3,\,\text{old}}^*} - (Y_{31}\mathbf{V}_1 + Y_{32}\mathbf{V}_2 + Y_{34}\mathbf{V}_4) \right] \qquad (11\text{–}18)$$

$$\mathbf{V}_4 = \frac{1}{Y_{44}} \left[\frac{P_4 - jQ_4}{\mathbf{V}_{4,\,\text{old}}^*} - (Y_{41}\mathbf{V}_1 + Y_{42}\mathbf{V}_2 + Y_{43}\mathbf{V}_3) \right] \qquad (11\text{–}19)$$

5. *Calculate an updated estimate of the voltage at each load bus in succession using Equation (11–8).* We will calculate an updated estimate of the voltages at bus 2, bus 3, and bus 4 using MATLAB. The statements required to perform this calculation are given below.

To evaluate these voltage equations, we need to create a nested MATLAB loop. The outer loop increments over each nonslack bus, and while the inner loop sums up the current contributions from every other bus at this bus. The code to evaluate Equations (11–17) to (11–19) is:

```
% Calculate the updated bus voltage
for ii = 1:n_bus

   % Skip the slack bus!
   if ii ~= swing_bus

      % Calculate updated voltage at bus 'ii'. First, sum
      % up the current contributions at bus 'ii' from all
      % other busses.
      temp = 0;
      for jj = 1:n_bus
         if ii ~= jj
            temp = temp - Ybus(ii,jj) * Vbus_old(jj);
         end
      end

      % Add in the current injected at this node
      temp = (P(ii) - j*Q(ii)) / conj(Vbus_old(ii)) + temp;

      % Get updated estimate of Vbus at 'ii'
      Vbus(ii) = 1/Ybus(ii,ii) * temp;
   end
end
```

6. *Compare the differences between the old voltage estimates and the new voltage estimates.* We will consider the answers to have converged if both the real and the imaginary parts of every voltage in the solution change by less than 0.0001 during an iteration. Once convergence has been achieved, we will write out the bus voltages in magnitude/phase format.

The code to compare the old voltage estimates to the new voltage estimates must cause the program to break out of the execution loop if *every* value is within tolerance. This can be done by first clearing an "out of tolerance" flag, and then turning it on if either the real or the imaginary component of *any* bus voltage exceeds the specified tolerance. If not, the loop should exit. The MATLAB code to perform this step is shown below:

```
% Compare the old and new estimate of the voltages.
% Note that we will compare the real and the imag parts
% separately, and both must be within tolerances.
test = 0;
for ii = 1:n_bus

   % Compare real parts
   if abs( real(Vbus(ii)) - real(Vbus_old(ii)) ) > eps
      test = 1;
   end

   % Compare imaginary parts
   if abs( imag(Vbus(ii)) - imag(Vbus_old(ii)) ) > eps
      test = 1;
   end
end

% Did we converge? If so, get out if the loop.
if test == 0
   break;
end
```

Finally, the bus voltage must be written out after the loop finishes executing, and we must confirm that a reasonable solution was reached.

The final MATLAB program is shown below. Note that it counts the iterations performed before the solution converges to give us a feel for how much computation is occurring.

```
% M-file: ex11_2.m
% M-file to solve the power-flow problem of Example 11-2.

% Set problem size and initial conditions
n_bus = 4;
swing_bus = 1;
```

```
% Create Y-bus
 Ybus = ...
[ 1.7647-j*7.0588 -0.5882+j*2.3529  0                -1.1765+j*4.7059; ...
 -0.5882+j*2.3529  1.5611-j*6.6290 -0.3846+j*1.9231 -0.5882+j*2.3529; ...
  0               -0.3846+j*1.9231  1.5611-j*6.6290 -1.1765+j*4.7059; ...
 -1.1765+j*4.7059 -0.5882+j*2.3529 -1.1765+j*4.7059  2.9412-j*11.7647 ];

% Initialize the real and reactive power supplied to the
% power system at each bus. Note that the power at the
% swing bus doesn't matter.
P(2) = -0.2;
P(3) = -0.3;
P(4) = -0.2;
Q(2) = -0.15;
Q(3) = -0.15;
Q(4) = -0.10;

% Initialize the bus voltages to 1.0 at 0 degrees
for ii = 1:n_bus
    Vbus(ii) = 1;
end

% Set convergence criterion
eps = 0.0001;

% Initialize the iteration counter
n_iter = 0;

% Create an infinite loop
while (1)

   % Increment the iteration count
   n_iter = n_iter + 1;

   % Save old bus voltages for comparison purposes
   Vbus_old = Vbus;

   % Calculate the updated bus voltage
   for ii = 1:n_bus

      % Skip the slack bus!
      if ii ~= swing_bus

         % Calculate updated voltage at bus 'ii'. First, sum
         % up the current contributions at bus 'ii' from all
         % other busses.
         temp = 0;
         for jj = 1:n_bus
            if ii ~= jj
               temp = temp - Ybus(ii,jj) * Vbus_old(jj);
            end
         end
```

```
        % Add in the current injected at this node
        temp = (P(ii) - j*Q(ii)) / conj(Vbus_old(ii)) + temp;

        % Get updated estimate of Vbus at 'ii'
        Vbus(ii) = 1/Ybus(ii,ii) * temp;
      end
   end

   % Compare the old and new estimate of the voltages.
   % Note that we will compare the real and the imag parts
   % separately, and both must be within tolerances.
   test = 0;
   for ii = 1:n_bus

      % Compare real parts
      if abs( real(Vbus(ii)) - real(Vbus_old(ii)) ) > eps
         test = 1;
      end

      % Compare imaginary parts
      if abs( imag(Vbus(ii)) - imag(Vbus_old(ii)) ) > eps
         test = 1;
      end
   end

   % Did we converge? If so, get out of the loop.
   if test == 0
      break;
   end
end

% Display results
for ii = 1:n_bus
   [mag, phase] = r2p(Vbus(ii));
   str = ['The voltage at bus ' int2str(ii) ' = ' ...
          num2str(mag) '/' num2str(phase)];
   disp(str);
end

% Display the number of iterations
str = ['Number of iterations = ' int2str(n_iter) ];
disp(str);
```

When this program is executed, the results are:

```
» ex11_2
The voltage at bus 1 = 1/0
The voltage at bus 2 = 0.89837/-5.3553
The voltage at bus 3 = 0.87191/-7.7492
The voltage at bus 4 = 0.91346/-4.8938
Number of iterations = 25
```

The resulting bus voltages are

$$\mathbf{V}_{bus}(1) = 1.000\angle 0° \text{ pu}$$
$$\mathbf{V}_{bus}(2) = 0.898\angle -5.35° \text{ pu} \qquad (11-20)$$
$$\mathbf{V}_{bus}(3) = 0.872\angle -7.75° \text{ pu}$$
$$\mathbf{V}_{bus}(4) = 0.913\angle -4.89° \text{ pu}$$

Step 7 requires that, after executing the program, we confirm that the answers appear consistent, and that the program did not converge on an erroneous solution. The consistency of the answers can be proved by summing up the currents at one or more busses to ensure that the sum comes out to zero, as required by Kirchhoff's current law. The solution is likely to not be erroneous because the spread of phase angles in the answers is less than 45°. ∎

A Warning about Network Equation Solutions

Note that relationship between voltage and current at a load bus is given by Equation (11–5):

$$\mathbf{I}_2 = \frac{P_2 - jQ_2}{\mathbf{V}_2^*} \qquad (11-5)$$

This is a *nonlinear relationship*—when the voltage at the bus goes up, the current at the bus goes down. Therefore, the power system network equations that we are solving are fundamentally *nonlinear equations*. From mathematics, we know that sets of nonlinear equations can sometimes have multiple solutions, or even no solution at all. As a result, a power-flow study can sometimes converge to an erroneous solution, or even keep running forever in a "limit cycle" without ever converging to a solution. The more complex the power system being analyzed, and the more heavily loaded it is, the more likely this is to occur.

In general, the solution that a nonlinear system of equations converges to depends on the starting point for the analysis. The closer we start to the correct answer, the more likely the equations are to converge to it. In general, a flat start (all bus voltages assumed to be $1\angle 0°$) is pretty close to the correct answer for most power systems, because power systems are designed to keep voltage reasonably constant for the user. However, sometimes we will fail to converge to the proper answer even with a flat start. In that case, trying different starting conditions might produce convergence to the correct answer.

Also, since some sets of power system equations and starting conditions can iterate forever without coming to a solution, a good power network analysis program should include a maximum number of iterations, and should stop running if that number of iterations is reached without converging to a solution.

Speeding Up Network Equation Solutions

Note from the results of Example 11–2 that it took *25* iterations before the voltages in the load flow problem converged to within a tolerance 0.0001. A major advantage of the Gauss-Siedel iterative method is that it is relatively stable and usually results in the right answer. Unfortunately, the method converges to a solution relatively slowly, requiring a lot of computer time. Fortunately, there are a couple of simple tricks that can be applied to speed up the convergence process. These tricks are described below.

1. *Applying updated bus voltage values immediately.* The first trick is to begin using the new estimates of bus voltage as soon as they are calculated, instead of waiting to the beginning of the next iteration to apply them. This change has the effect of applying the better estimates of bus voltage at a given bus to the estimation of the voltage at the other busses in the system, resulting in faster convergence. For example, consider the bus voltage update code in Example 11–2. Here, the values in **Vbus_old** are used to calculate updated values in **Vbus** for all the load busses in the system.

```
% Save old bus voltages for comparison purposes
Vbus_old = Vbus;

% Calculate the updated bus voltage
for ii = 1:n_bus

    % Skip the slack bus!
    if ii ~= swing_bus

        % Calculate updated voltage at bus 'ii'. First, sum
        % up the current contributions at bus 'ii' from all
        % other busses.
        temp = 0;
        for jj = 1:n_bus
            if ii ~= jj
                temp = temp - Ybus(ii,jj) * Vbus_old(jj);
            end
        end

        % Add in the current injected at this node
        temp = (P(ii) - j*Q(ii)) / conj(Vbus_old(ii)) + temp;

        % Get updated estimate of Vbus at 'ii'
        Vbus(ii) = 1/Ybus(ii,ii) * temp;
    end
end
```

If we modify this code slightly so that array Vbus is used in the calculation, then the updates at bus 2 will be applied immediately to calculation at bus 3, etc., resulting in faster convergence. Note that we must still keep copies of Vbus_old to test for convergence, even if we are not otherwise using that array. The slightly modified code is shown below.

```
% Save old bus voltages for comparison purposes
Vbus_old = Vbus;
```

```
% Calculate the updated bus voltage
for ii = 1:n_bus

    % Skip the slack bus!
    if ii ~= swing_bus

        % Calculate updated voltage at bus 'ii'. First, sum
        % up the current contributions at bus 'ii' from all
        % other busses.
        temp = 0;
        for jj = 1:n_bus
            if ii ~= jj
                temp = temp - Ybus(ii,jj) * Vbus(jj);
            end
        end

        % Add in the current injected at this node
        temp = (P(ii) - j*Q(ii)) / conj(Vbus(ii)) + temp;

        % Get updated estimate of Vbus at 'ii'
        Vbus(ii) = 1/Ybus(ii,ii) * temp;
    end
end
```

When the code is executed with this slight modification, the results are:

```
» ex11_2a
The voltage at bus 1 = 1/0
The voltage at bus 2 = 0.89821/-5.3595
The voltage at bus 3 = 0.87168/-7.7564
The voltage at bus 4 = 0.91327/-4.8986
Number of iterations = 16
```

This simple change results in convergence in 16 iterations instead of 25, a great improvement in computational efficiency.

2. *Applying acceleration factors.* The second trick that can be applied to the Gauss-Siedel method is called *acceleration factors*. Studies have shown that the Gauss-Siedel method moves bus voltage estimates slowly toward optimum values, with each step tending to be in the same direction as the previous one. If the *difference* between the new voltage estimate and the old voltage estimate is multiplied by an "acceleration factor" and added to the old voltage estimate, then in a single step we can jump closer to the true solution than would be possible with the ordinary Gauss-Siedel method. This acceleration factor must be greater than 1.0 to speed the solution, but if it is too large, the factor can actually hurt, because the update will "jump over" the correct point. In fact, a very large acceleration factor will make the equations unstable. A suitable acceleration factor for systems that also apply immediate voltage updates might be 1.3 to 1.4. An acceleration factor higher than 2.0 is unstable, and convergence will never occur.

A modified version of the MATLAB code that applies both immediate updates and an acceleration factor is shown below:

```
% Save old bus voltages for comparison purposes
Vbus_old = Vbus;

% Calculate the updated bus voltage
for ii = 1:n_bus

   % Skip the slack bus!
   if ii ~= swing_bus

      % Calculate updated voltage at bus 'ii'. First, sum
      % up the current contributions at bus 'ii' from all
      % other busses.
      temp = 0;
      for jj = 1:n_bus
         if ii ~= jj
            temp = temp - Ybus(ii,jj) * Vbus(jj);
         end
      end

      % Add in the current injected at this node
      temp = (P(ii) - j*Q(ii)) / conj(Vbus(ii)) + temp;

      % Get updated estimate of Vbus at 'ii'
      Vnew = 1/Ybus(ii,ii) * temp;

      % Apply an acceleration factor to the new voltage estimate
      Vbus(ii) = Vbus_old(ii) + acc_fac * (Vnew - Vbus_old(ii));
   end
end
```

When the code is executed with this slight modification, the results are:

```
» ex11_2b
The voltage at bus 1 = 1/0
The voltage at bus 2 = 0.89812/-5.3619
The voltage at bus 3 = 0.87158/-7.7605
The voltage at bus 4 = 0.91321/-4.9005
Number of iterations = 8
```

The combination of these two improvements improved the speed of convergence by a factor of 3, from 25 to 8 iterations!

Additional speed-ups are possible by using more advanced solution techniques such as the *Newton-Raphson method*. These techniques perform the same function as the Gauss-Siedel iterative method, but are more computationally efficient. Unfortunately, they are also more complex to explain, and we do not have time to cover them in this survey text. An explanation of the Newton-Raphson method may be found in Section 9.3 of Reference 2.

11.2 | ADDING GENERATOR BUSSES TO POWER-FLOW STUDIES

At a generator bus, the real power P_i and the magnitude of the bus voltage $|\mathbf{V}_i|$ are specified. Since we do not know the reactive power Q_i for that bus, we must estimate it before applying Equation (11–8) to get an updated estimate of bus voltage. The value of reactive power at the generator bus can be estimated by solving Equation (11–8) for Q_i:

$$\mathbf{V}_i = \frac{1}{Y_{ii}}\left(\frac{P_i - jQ_i}{\mathbf{V}_i^*} - \sum_{\substack{k=1 \\ (k \neq i)}}^{N} Y_{ik}\mathbf{V}_k \right) \tag{11–8}$$

$$Y_{ii}\mathbf{V}_i = \frac{P_i - jQ_i}{\mathbf{V}_i^*} - \sum_{\substack{k=1 \\ (k \neq i)}}^{N} Y_{ik}\mathbf{V}_k$$

$$\frac{P_i - jQ_i}{\mathbf{V}_i^*} = Y_{ii}\mathbf{V}_i + \sum_{\substack{k=1 \\ (k \neq i)}}^{N} Y_{ik}\mathbf{V}_k$$

$$P_i - jQ_i = \mathbf{V}_i^*\left(Y_{ii}\mathbf{V}_i + \sum_{\substack{k=1 \\ (k \neq i)}}^{N} Y_{ik}\mathbf{V}_k \right)$$

If we bring the case $k = i$ into the summation, this equation becomes

$$P_i - jQ_i = \mathbf{V}_i^* \sum_{k=1}^{N} Y_{ik}\mathbf{V}_k$$

or

$$Q_i = -\text{Im}\left(\mathbf{V}_i^* \sum_{k=1}^{N} Y_{ik}\mathbf{V}_k \right) \tag{11–21}$$

Once the reactive power at the bus has been estimated, we can update the bus voltage at the generator bus using P_i and Q_i just as we would a load bus. However, the *magnitude* of the generator bus voltage is also forced to remain constant. To keep it constant, we must multiply the new voltage estimate by the *ratio* of the magnitude of the old voltage estimate to the magnitude of the new voltage estimate. This multiplication has the effect of updating the phase angle of the voltage while keeping the magnitude of the voltage constant.

In other words, the steps required to update the voltage at a generator bus are:

1. Estimate the reactive power Q_i at the bus using Equation (11–21).
2. Update the estimated voltage at the bus using Equation (11–8), as though the bus were a load bus.
3. Force the magnitude of the estimated voltage to be constant by multiplying the new voltage estimate by the ratio of the magnitude of the original estimate to the magnitude of the new estimate. This has the effect of updating the voltage phase estimate without changing the voltage amplitude.

EXAMPLE 11-3

Figure 11–3 shows a simple power system with four busses, five transmission lines, two generators, and two loads. Since this power system has generators connected to two busses, it will have one slack bus, one generator bus, and two load busses. Assume that bus 1 is the slack bus in the power system, and that it has a voltage $V_1 = 1.0\angle 0°$ pu. Bus 3 is a generator bus. The generator is supplying a real power $P_3 = 0.30$ pu to the power system, with a voltage magnitude of 1.0 pu. The per-unit real and reactive power loads on the power system at busses 2 and 4 are $P_2 = 0.30$ pu, $Q_2 = 0.15$ pu, $P_4 = 0.20$ pu, and $Q_4 = 0.15$ pu. Since these values are flowing *out* of the power system, the real and reactive power *supplied* to the power system at each of these busses will be the negative of the above numbers. The series impedances of each of the transmission lines in this system are the same as for Example 11–2; they are given in Table 11–1.

Determine the voltage at each bus in this power system for the specified load conditions.

Figure 11–3 | A simple power system with four busses, five transmission lines, two generators, and two loads. This power system includes a generator bus.

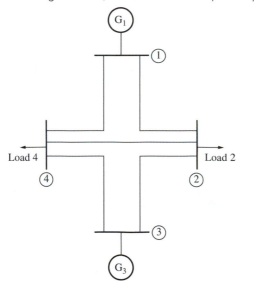

Table of Buses:

Bus 1	Slack bus
Bus 2	Load bus
Bus 3	Generator bus
Bus 4	Load bus

■ Solution

This problem is similar to Example 11–2, except that it includes a generator bus. We will build a MATLAB program and execute it to calculate the bus voltages.

Since this power system has the same topology as the one in Example 11–2, and the impedances of each transmission line are identical, Y_{bus} will be the same as in the previous example. It is:

$$\mathbf{Y}_{bus} = \begin{bmatrix} 1.7647 - j7.0588 & -0.5882 + j2.3529 & 0 & -1.1765 + j4.7059 \\ -0.5882 + j2.3529 & 1.5611 - j6.6290 & -0.3846 + j1.9231 & -0.5882 + j2.3529 \\ 0 & -0.3846 + j1.9231 & 1.5611 - j6.6290 & -1.1765 + j4.7059 \\ -1.1765 + j4.7059 & -0.5882 + j2.3529 & -1.1765 + j4.7059 & 2.9412 - j11.7647 \end{bmatrix} \quad (11\text{--}1)$$

The MATLAB statement required to initialize $\mathbf{Y}_{bus}$ to the values in Equation (11–1) is

```
% Create Y-bus
Ybus = ...
[ 1.7647-j*7.0588  -0.5882+j*2.3529   0                -1.1765+j*4.7059; ...
 -0.5882+j*2.3529   1.5611-j*6.6290  -0.3846+j*1.9231  -0.5882+j*2.3529; ...
  0                -0.3846+j*1.9231   1.5611-j*6.6290  -1.1765+j*4.7059; ...
 -1.1765+j*4.7059  -0.5882+j*2.3529  -1.1765+j*4.7059   2.9412-j*11.7647 ];
```

As stated in the problem, bus 1 is the slack bus. The voltage at bus 1 will arbitrarily be assumed to be $1.0\angle 0°$. Bus 3 is a generator bus, meaning that we will have to estimate the reactive power at that bus before calculating the bus voltage, and then force the *magnitude* of the voltage to remain constant after calculating the bus voltage. We will once again make a flat start, making the initial voltage estimates at every bus to be $1.0\angle 0°$.

The voltage equations for busses 2, 3, and 4 are the same as before. However, bus 3 is a now generator bus, so we must first estimate the reactive power Q_3 at that bus before estimating $\mathbf{V}_3$, and then we must update the estimate of $\mathbf{V}_3$ after it is calculated to maintain a constant voltage magnitude. The resulting voltage equations for the three busses are:

$$\mathbf{V}_2 = \frac{1}{Y_{22}}\left[\frac{P_2 - jQ_2}{\mathbf{V}_{2,\,old}^*} - (Y_{21}\mathbf{V}_1 + Y_{23}\mathbf{V}_3 + Y_{24}\mathbf{V}_4)\right] \quad (11\text{--}22)$$

$$Q_3 = -\mathrm{Im}\left(\mathbf{V}_3^* \sum_{k=1}^{N} Y_{ik}\mathbf{V}_k\right) \quad (11\text{--}23)$$

$$\mathbf{V}_3 = \frac{1}{Y_{33}}\left[\frac{P_3 - jQ_3}{\mathbf{V}_{3,\,old}^*} - (Y_{31}\mathbf{V}_1 + Y_{32}\mathbf{V}_2 + Y_{34}\mathbf{V}_4)\right] \quad (11\text{--}24)$$

$$\mathbf{V}_3 = \mathbf{V}_3\frac{|\mathbf{V}_{3,\,old}|}{|\mathbf{V}_3|} \quad (11\text{--}25)$$

$$\mathbf{V}_4 = \frac{1}{Y_{44}}\left[\frac{P_4 - jQ_4}{\mathbf{V}_{4,\,old}^*} - (Y_{41}\mathbf{V}_1 + Y_{42}\mathbf{V}_2 + Y_{43}\mathbf{V}_3)\right] \quad (11\text{--}26)$$

To evaluate these equations, we need to create a nested MATLAB loop. The outer loop increments over each nonslack bus, and the inner loop sums up the current contributions from every other bus at this bus. Tests are added to identify generator busses so that the reactive power can be calculated before the voltage is updated, and the magnitude of the voltage can be corrected after the voltage is updated. The code to evaluate Equations (11–22) to (11–26) is shown on page 533, with the modifications to support generator busses highlighted.

```
% Calculate the updated bus voltage
for ii = 1:n_bus

   % Skip the swing bus!
   if ii ~= swing_bus

      % If this is a generator bus, update the reactive
      % power estimate.
      if gen_bus(ii)
         temp = 0;
         for jj = 1:n_bus
            temp = temp + Ybus(ii,jj) * Vbus(jj);
         end
         temp = conj(Vbus(ii)) * temp;
         Q(ii) = -imag(temp);
      end

      % Calculate updated voltage at bus 'ii'. First, sum
      % up the current contributions at bus 'ii' from all
      % other busses.
      temp = 0;
      for jj = 1:n_bus
         if ii ~= jj
            temp = temp - Ybus(ii,jj) * Vbus(jj);
         end
      end

      % Add in the current injected at this node
      temp = (P(ii) - j*Q(ii)) / conj(Vbus(ii)) + temp;

      % Get updated estimate of Vbus at 'ii'
      Vnew = 1/Ybus(ii,ii) * temp;

      % Apply an acceleration factor to the new voltage estimate
      Vbus(ii) = Vbus_old(ii) + acc_fac * (Vnew - Vbus_old(ii));

      % If this is a generator bus, update the magnitude of the
      % voltage to keep it constant.
      if gen_bus(ii)
         Vbus(ii) = Vbus(ii) * abs(Vbus_old(ii)) / abs(Vbus(ii));
      end
   end
end
```

The final MATLAB program is shown on page 534, with the changes to accommodate generator busses highlighted. Note that it counts the iterations performed before the solution converges to give us a feel for how much computation is occurring. Also, this program is designed to stop executing after a maximum number (100) of iterations is reached, so it will not run in an infinite loop if the solutions fail to converge!

```
% M-file: ex11_3.m
% M-file to solve the power-flow problem of Example 11-3.
% This set of equations includes a generator bus.

% Set problem size and initial conditions
n_bus = 4;
swing_bus = 1;
acc_fac = 1.3;

% Specify which busses are generator busses with flags.
% Note that 1 is "true" and 0 is "false" in MATLAB. In
% this example, Bus 3 is a generator bus, and the others
% are not.
gen_bus = [0 0 1 0];

% Create Y-bus
 Ybus = ...
[ 1.7647-j*7.0588 -0.5882+j*2.3529  0                -1.1765+j*4.7059; ...
 -0.5882+j*2.3529  1.5611-j*6.6290 -0.3846+j*1.9231 -0.5882+j*2.3529; ...
  0               -0.3846+j*1.9231  1.5611-j*6.6290 -1.1765+j*4.7059; ...
 -1.1765+j*4.7059 -0.5882+j*2.3529 -1.1765+j*4.7059  2.9412-j*11.7647 ];

% Initialize the real and reactive power supplied to the
% power system at each bus. Note that the power at the
% swing bus doesn't matter, and the reactive power at the
% generator bus will be recomputed dynamically.
P(2) = -0.2;
P(3) = 0.3;
P(4) = -0.2;
Q(2) = -0.15;
Q(3) = 0.0;
Q(4) = -0.10;

% Initialize the bus voltages to 1.0 at 0 degrees
for ii = 1:n_bus
   Vbus(ii) = 1;
end

% Set convergence criterion
eps = 0.0001;

% Initialize the iteration counter
n_iter = 0;

% Set a maximum number of iterations here so that
% the program will not run in an infinite loop if
% it fails to converge to a solution.
for iter = 1:100

   % Increment the iteration count
   n_iter = n_iter + 1;
```

```
% Save old bus voltages for comparison purposes
Vbus_old = Vbus;

% Calculate the updated bus voltage
for ii = 1:n_bus

    % Skip the swing bus!
    if ii ~= swing_bus

        % If this is a generator bus, update the reactive
        % power estimate.
        if gen_bus(ii)
            temp = 0;
            for jj = 1:n_bus
                temp = temp + Ybus(ii,jj) * Vbus(jj);
            end
            temp = conj(Vbus(ii)) * temp;
            Q(ii) = -imag(temp);
        end

        % Calculate updated voltage at bus 'ii'. First, sum
        % up the current contributions at bus 'ii' from all
        % other busses.
        temp = 0;
        for jj = 1:n_bus
            if ii ~= jj
                temp = temp - Ybus(ii,jj) * Vbus(jj);
            end
        end

        % Add in the current injected at this node
        temp = (P(ii) - j*Q(ii)) / conj(Vbus(ii)) + temp;

        % Get updated estimate of Vbus at 'ii'
        Vnew = 1/Ybus(ii,ii) * temp;

        % Apply an acceleration factor to the new voltage estimate
        Vbus(ii) = Vbus_old(ii) + acc_fac * (Vnew - Vbus_old(ii));

        % If this is a generator bus, update the magnitude of the
        % voltage to keep it constant.
        if gen_bus(ii)
            Vbus(ii) = Vbus(ii) * abs(Vbus_old(ii)) / abs(Vbus(ii));
        end
    end
end

% Compare the old and new estimate of the voltages.
% Note that we will compare the real and the imag parts
% separately, and both must be within tolerances.
test = 0;
for ii = 1:n_bus
```

```
      % Compare real parts
      if abs( real(Vbus(ii)) - real(Vbus_old(ii)) ) > eps
         test = 1;
      end

      % Compare imaginary parts
      if abs( imag(Vbus(ii)) - imag(Vbus_old(ii)) ) > eps
          test = 1;
      end
   end

   % Did we converge? If so, get out of the loop.
   if test == 0
      break;
   end
end

% Did we exceed the maximum number of iterations?
if iter == 100
   disp('Max number of iterations exceeded!');
end

% Display results
for ii = 1:n_bus
   [mag, phase] = r2p(Vbus(ii));
   str = ['The voltage at bus ' int2str(ii) ' = ' ...
         num2str(mag) '/' num2str(phase)];
   disp(str);
end

% Display the number of iterations
str = ['Number of iterations = ' int2str(n_iter) ];
disp(str);
```

When this program is executed, the results are:

```
» ex11_3
The voltage at bus 1 = 1/0
The voltage at bus 2 = 0.96361/-0.97462
The voltage at bus 3 = 1/1.8443
The voltage at bus 4 = 0.98019/-0.27034
Number of iterations = 8
```

The resulting bus voltages are

$$\mathbf{V}_{bus}(1) = 1.000\angle 0° \text{ pu}$$
$$\mathbf{V}_{bus}(2) = 0.964\angle -0.97° \text{ pu}$$
$$\mathbf{V}_{bus}(3) = 1.000\angle 1.84° \text{ pu}$$
$$\mathbf{V}_{bus}(4) = 0.980\angle -0.27° \text{ pu} \tag{11-27}$$

This solution looks reasonable; the spread of bus voltage phase angles in the answers is less than 45°. ∎

11.3 | THE INFORMATION DERIVED FROM POWER-FLOW STUDIES

After the bus voltages are calculated at all busses in a power system, a power-flow program can be set up to provide alerts if the voltage at any given bus exceeds ± 5 percent of the nominal value. This information is important to the engineer, since power is supposed to be supplied at a constant voltage level. These high or low voltages indicate spots where some sort of compensatory work is required, either with switched capacitors or with TCUL transformers. If the voltage is too low at a bus, additional capacitors can be switched on to the bus and the problem can be re-solved to determine the effect of this action on the bus voltage. By solving such problems repeatedly, it is possible to define the amount of switched capacitance needed to ensure that the voltage on a bus remains in tolerance under all load conditions.

In addition, it is possible to determine the net real and reactive power either supplied to or removed from each bus by generators or loads connected to it. To calculate the real and reactive power at a bus, we first calculate the net current injected at the bus. The net current injected at a bus will be equal to the sum of all the currents *leaving* the bus through transmission lines. The current leaving the bus on each transmission line will be equal to the difference between the voltages at either end of the transmission line multiplied by the admittance of the line, so the total current injected at the now will be

$$\mathbf{I}_i = \sum_{\substack{k=1 \\ k \neq i}}^{N} Y_{ik}(V_i - V_k) \tag{11–28}$$

The resulting real and reactive power injected at the bus can be found from the equation

$$\mathbf{S}_i = -\mathbf{V}_i \mathbf{I}_i^* = P_i + jQ_i \tag{11–29}$$

where the minus sign takes into account the fact that current is assumed to be injected instead of leaving the node.

Similarly, the power-flow study can show the real and reactive power flowing in every transmission line in the system. The current flow out of a node along a particular transmission line between bus i and bus j can be calculated from the equation

$$\mathbf{I}_{ij} = Y_{ij}(\mathbf{V}_i - \mathbf{V}_j) \tag{11–30}$$

where Y_{ij} is the admittance of the transmission line between those two busses. The resulting real and reactive power can be calculated from the equation

$$\mathbf{S}_{ij} = -\mathbf{V}_i \mathbf{I}_{ij}^* = P_{ij} + jQ_{ij} \tag{11–31}$$

Furthermore, by comparing the real and reactive power flows at either end of the transmission line, we can determine the real and reactive power losses on each line!

In modern professional power-flow programs, this information is displayed graphically on a computer screen, with real and reactive power flows into and out of each bus and transmission line being shown as arrows. The display uses color to highlight areas where the power system is overloaded. This graphical representation makes it very easy to locate "hot spots."

With all of this information available, the user of a power-flow study can determine whether any of the components in a power system will be overloaded by the particular conditions of the study. If the ratings of any components are exceeded, then the conditions of this study are an unacceptable steady-state operating condition, and something else must be tried.

This is where power-flow studies aid the engineer in playing "What if?" games. A power engineer usually begins by analyzing the power system in its normal operating conditions, called the *base case*. Then, the engineer can project possible future load conditions to see if any parts of the power system get into trouble. If a transmission line appears to be overloaded in a power-flow study representing a possible future load condition, the engineer can add a proposed new transmission line to the power system and re-solve the power-flow problem to see how much the new transmission line has helped. Lines of varying capacities can be added to the system at various points until a good solution is found for supplying the anticipated loads without overloading any components, and with a good margin of reserve for uncertainty in future load growth. With this information in hand, engineers can plan to add capacity over the next 5–10 years to always stay ahead of the need.

Another important type of "What if?" game is: What happens if a transmission line or other component is removed from the current system because of failure or maintenance? Can the remaining components of the power system supply all required loads without being overloaded? If not, then some loads must be shed to keep the power system safe if it ever winds up in this configuration. The system should be designed with enough redundancy that it can function properly under all load conditions with the loss of a few transmission lines, and these "What if?" studies help the engineers to determine just how much redundancy is required, and where the components should be located.

Finally, power-flow studies help a power system to operate more efficiently. A typical power system has many generators in many different geographical locations, and of course the loads on the system are distributed in many different locations. The loads on the power system can be supplied by the generators in many different ways, with the generators at some power stations supplying full power while the generators at other stations serve as swing generators. If power-flow studies are performed for different combinations of generation, the engineer can determine which combination produces the minimum transmission losses, and thus provides the power most efficiently. The study of the efficient operation of power systems is called *economic dispatch*. It is a major subdiscipline within power systems engineering.

11.4 | A SIMPLE POWER-FLOW PROGRAM

The software accompanying this book includes a simple MATLAB power-flow analysis program, called `power_flow.m`. This program calculates the voltages and powers at every bus, plus the real and reactive power flowing on each transmission line. It also provides alerts if any of the bus voltages are out of tolerance, or if any of the transmission lines are overloaded.

The source code for this program is available for download at the book's website. It can be used to find the voltages and power flows in a power system, where the voltages, impedances, and admittances are specified in per-unit, and the real and reactive power flows are specified in MW and MVAR.

The input data for program `power_flow` is placed in an input file, which can be created using any available editor. The file must contain three types of lines—a `SYSTEM` line to define the system name and base MVA, `BUS` lines to define the initial voltage estimates and the power flows at each bus, and `LINE` lines to define the transmission lines connecting the various busses together. It may also contain comment lines.

A typical input file is shown in Figure 11–4. This file contains `SYSTEM`, `BUS`, and `LINE` lines, plus comment lines that begin with a % symbol in column 1.

Figure 11–4 | A sample input file for the MATLAB program `power_flow`. This input file solves the problem in Example 11–3.

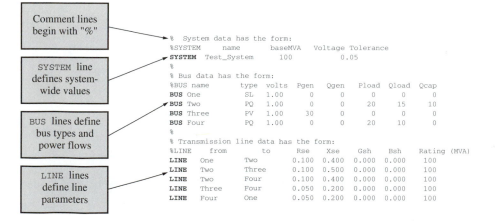

The `SYSTEM` card specifies the name of the power system case being studied, the base apparent power of the power system (in MVA), and the allowable per-unit voltage tolerance on the busses. The format of a `SYSTEM` line is shown below (note that the data is free format—the values do not have to appear in any particular columns), and the values in the line are defined in Table 11–2.

```
%SYSTEM      name       baseMVA    Voltage Tolerance    Max Iterations
SYSTEM   Test_System    100           0.05                  100
```

Table 11–2 | Fields defined on the **SYSTEM** line

Field	Description
SYSTEM	Line identifier
Name	System study name (spaces are not allowed within the name)
Voltage Tolerance	The acceptable voltage tolerance on busses, in per-unit. For example, a tolerance of 0.05 means that voltages between 0.95 and 1.05 pu are acceptable.
Max Iterations	The maximum number of iterations to perform in trying to converge to a solution. This value is optional; if it is not present, the maximum number of iterations is 200.

The format of a BUS line is shown below, and the values in the line are defined in Table 11–3.

```
%BUS name    type    volts    Pgen    Qgen    Pload    Qload    Qcap
BUS One      SL      1.00     0       0       0        0        0
```

Table 11–3 | Fields defined on the **BUS** line

Field	Description
BUS	Line identifier
Name	Bus name (spaces are not allowed within the name)
Type	Bus type, which is one of:
	SL Slack bus
	PV Generator bus
	PQ Load bus
Pgen	Generator power supplied to bus, in MW
Qgen	Generator reactive power supplied to bus, in MVAR
Pload	Real power supplied to loads at bus, in MW
Qload	Reactive power supplied to loads at bus, in MVAR
Qcap	Capacitive reactive power at bus, in MVAR

The format of a LINE line is shown below, and the values in the line are defined in Table 11–4.

```
%LINE    from    to    Rse      Xse      Gsh      Bsh      Rating(MVA)
LINE     One     Two   0.100    0.400    0.000    0.000    100
```

Table 11–4 | Fields defined on the **LINE** line

Field	Description
LINE	Line identifier
From Bus Name	Bus where transmission line starts.
To Bus Name	Bus where transmission line ends.
Rse	Line series resistance, in per-unit
Xse	Line series reactance, in per-unit
Gsh	Line shunt conductance, in per-unit
Bsh	Line shunt susceptance, in per-unit
Rating	Line power rating, in MVA

Figure 11-5 | Result of executing power_flow with the input file shown in Figure 11–4. Note that the bus voltages match the values that we calculated in Example 11–3.

```
» power_flow ex11_3_system
Input summary statistics:
18 lines in system file
1 SYSTEM lines
4 BUS lines
5 LINE lines

                           Results for Case Test_System
```

			=====Bus Information=====		|--Generation--|	|----Load----|	|--Cap--|	=====Line Information=====		
Bus no.	Bus Name	Type	Volts / angle (pu)	angle (deg)	(MW)	(MVAR)	(MW)	(MVAR)	(MVAR)	To Bus	---Line Flow--- (MW)	(MVAR)
1	One	SL	1.000/	0.00	10.47	16.42	0.00	0.00	0.00	Two	5.99	7.63
										Four	4.48	8.79
2	Two	PQ	0.964/	-0.97	0.00	0.00	20.00	15.00	0.00	One	-5.89	-7.25
										Three	-10.42	-4.69
										Four	-3.66	-3.06
3	Three	PV	1.000/	1.85	30.00	10.77	0.00	0.00	0.00	Two	10.56	5.40
										Four	19.46	5.38
4	Four	PQ	0.980/	-0.27	0.00	0.00	20.00	10.00	0.00	Two	3.69	3.16
										Three	-19.25	-4.56
										One	-4.44	-8.59
				Totals	40.47	27.19	40.00	25.00	0.00			

```
                              Line Losses
```

Line no.	From Bus	To Bus	Ploss (MW)	Qloss (MVAR)
1	One	Two	0.09	0.38
2	Two	Three	0.14	0.70
3	Two	Four	0.02	0.10
4	Three	Four	0.20	0.81
5	Four	One	0.05	0.19
		Totals	0.51	2.19

```
                                 Alerts
```

NONE

Done in 6 iterations.

Figure 11–6 | The Ozzie Outback Electric Power Company system.

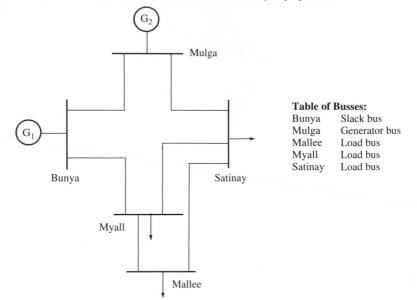

Table of Busses:

Bunya	Slack bus
Mulga	Generator bus
Mallee	Load bus
Myall	Load bus
Satinay	Load bus

To solve a power-flow problem by this program, you must first calculate the per-unit impedances and admittances of all components to a common system base, and then create a file containing those input parameters. Note that each "line" in the power system should include *all* of the impedances in the line, whether they come from the line itself or from transformers at the end of the line. In general, the initial voltage at load busses should be assumed to be 1.0.

When program `power_flow` is executed, the results are as shown in Figure 11–5.

Program `power_flow` is simple and easy to use, but lacks many features found in commercial power-flow programs. For example, commercial programs support tap-changing and phase-shifting transformers in power lines, and they usually have interactive graphical displays. Another significant limitation of this simple program is that it uses the Gauss-Siedel iterative method to solve for bus voltages. This method is very effective for small systems of equations, but would become prohibitively time-consuming for power systems with large numbers of busses.

EXAMPLE 11–4

The Ozzie Outback Electric Power (OOEP) Company maintains the power system shown in Figure 11–6. This power system contains five busses, with generators attached to two of them and loads attached to the remaining ones. The power system has six transmission lines connecting the busses together, with the characteristics shown in Table 11–5. Per-unit values are given to a system base apparent power of 100 MVA. The generators at busses Bunya and Mulga are each capable of supplying 200 MW at 0.85 power factor lagging.

Table 11–5 | Transmission lines in the OOEP system.

From	To	R_{SE} (pu)	X_{SE} (pu)	Rating (MVA)
Bunya	Mulga	0.001	0.051	100
Mulga	Satinay	0.007	0.035	250
Bunya	Myall	0.007	0.035	200
Myall	Satinay	0.007	0.035	100
Myall	Mallee	0.011	0.051	60
Mallee	Satinay	0.011	0.051	60

You are an engineer in the company, and you have been informed that some new customers intend to build facilities that will draw extra power. You need to perform a power-flow study to see if the new load conditions will cause trouble for the power system. The new anticipated load conditions are shown in Table 11–6. Determine if the bus voltages will remain within a 5 percent tolerance, and if any transmission line will be overloaded by these new load conditions.

Table 11–6 | Anticipated loads on the OOEP system

Bus	Real load (MW)	Reactive load (MVAR)
Mallee	100	75
Myall	100	70
Satinay	120	100

■ Solution

To solve this problem, we must create an input file and execute program `power_flow`. The input file must contain the system, bus, and line data describing this system.

This power system contains two generators. We will assume that the generator at Mulga is supplying full rated power (200 MW), while the generator at Bunya is the swing generator, supplying whatever additional power is needed. Having the generator at Mulga running at full capacity should improve the overall efficiency of the power system.

Note that it is necessary to make an assumption about the voltages at the slack bus and generator bus. Since we don't have any better information, we will assume that the magnitude of both voltages is 1.0 pu. Also, we will assume an initial estimate of 1.0 pu for the voltages of each load bus in the system.

An appropriate input file for this system is shown below.

```
% File describing a possible future condition of the
% Ozzie Outback Electric Power system.
%
% System data has the form:
%SYSTEM    name        baseMVA     Voltage Tolerance
SYSTEM  OOEP_Future1    100            0.05
%
% Bus data has the form:
%BUS name      type  volts  Pgen   Qgen   Pload   Qload   Qcap
BUS Bunya      SL    1.00    0      0       0       0       0
BUS Mulga      PV    1.00   200     0       0       0       0
BUS Mallee     PQ    1.00    0      0      100      75      0
BUS Myall      PQ    1.00    0      0      100      70      0
BUS Satinay    PQ    1.00    0      0      120     100      0
%
% Transmission line data has the form:
%LINE    from      to      Rse    Xse    Gsh    Bsh    Rating(MVA)
LINE    Bunya    Mulga    0.011  0.051  0.000  0.000     100
LINE    Mulga    Satinay  0.007  0.035  0.000  0.000     250
LINE    Bunya    Myall    0.007  0.035  0.000  0.000     200
LINE    Myall    Satinay  0.007  0.035  0.000  0.000     100
LINE    Myall    Mallee   0.011  0.051  0.000  0.000      60
LINE    Mallee   Satinay  0.011  0.051  0.000  0.000      60
```

When program `power_flow` is executed, the results are as shown in Figure 11–7. As you can see, the bus voltages are out of tolerance for busses Mallee, Myall, and Satinay, and transmission lines 3, 5, and 6 are overloaded! This would certainly not be an acceptable future configuration for this power system. ∎

Figure 11-7 | Result of executing `power_flow` for the OOEP power system. Note the alerts on out-of-tolerance bus voltages and overloaded power lines.

```
>> power_flow OOEP_Future1
Input summary statistics:
23 lines in system file
1 SYSTEM lines
5 BUS lines
6 LINE lines

                          Results for Case OOEP_Future1

|===============Bus Information===============|============Line Information=====|
Bus   Bus   Volts / angle  |--Generation--|-----Load----|--Cap--|   To   |---Line Flow---|
no.   Name  Type  (pu)  (deg)   (MW)  (MVAR)   (MW)  (MVAR) (MVAR)   Bus     (MW)  (MVAR)
|-----------------------------------------------------------------------------------------|
1   Bunya   SL  1.000/  0.00  127.09 143.28    0.00   0.00   0.00  Mulga   -22.56    5.01
                                                                   Myall   149.65  138.27

2   Mulga   PV  0.964/ -0.69  200.00 139.60    0.00   0.00   0.00  Bunya    22.62   -4.74
                                                                   Satinay 177.59  144.39

3   Mallee  PQ  0.913/ -3.97    0.00   0.00  100.00  75.00   0.00  Myall   -49.12  -40.94
                                                                   Satinay -50.67  -34.05

4   Myall   PQ  0.942/ -2.60    0.00   0.00  100.00  70.00   0.00  Bunya  -146.74 -123.74
                                                                   Satinay  -2.75   10.30
                                                                   Mallee   49.66   43.45

5   Satinay PQ  0.938/ -2.49    0.00   0.00  120.00 100.00   0.00  Mulga  -173.92 -126.06
                                                                   Myall     2.75  -10.26
                                                                   Mallee   51.16   36.33
|=========================================================================================|

          Totals          327.09 282.88  320.00 245.00   0.00
|=========================================================================================|
```

(continued)

Figure 11-7 | *(concluded)*

					Line Losses
Line no.	From Bus	To Bus	Ploss (MW)	Qloss (MVAR)	
1	Bunya	Mulga	0.06	0.27	
2	Mulga	Satinay	3.67	18.34	
3	Bunya	Myall	2.91	14.53	
4	Myall	Satinay	0.01	0.04	
5	Myall	Mallee	0.54	2.50	
6	Mallee	Satinay	0.49	2.28	
		Totals	7.67	37.96	

Alerts

```
ALERT: Voltage on bus Mallee out of tolerance.
ALERT: Voltage on bus Myall out of tolerance.
ALERT: Voltage on bus Satinay out of tolerance.
ALERT: Rating on line 3 exceeded: 203.75 MVA > 200.00 MVA.
ALERT: Rating on line 5 exceeded: 65.98 MVA > 60.00 MVA.
ALERT: Rating on line 6 exceeded: 61.05 MVA > 60.00 MVA.
```

Done in 16 iterations.

EXAMPLE 11–5

Since the future load conditions for the OOEP system are unacceptable, you have been given the job of making a recommendation for changes to solve the problem. You could propose any number of different solutions, such as adding additional generation and a new bus or adding additional transmission lines, but these solutions are *very* expensive. Before embarking on them you must ensure that there is no cheaper way to solve the problem.

One possible way to improve the situation is to add capacitor banks to one or more busses in the power system. Capacitor banks supply reactive power to the system at the point where they are connected, so that the reactive power does not have to be supplied by the generators through the transmission lines. This action both reduces the current flow in the lines and increases the voltage at nearby busses.

Determine the effect on the power system of adding a 125-MVAR capacitor bank at the Mallee bus.

■ **Solution**

This problem is similar to the previous one, except that we have added a capacitor bank to the Mallee bus. The input file is shown below:

```
% File describing a possible future condition of the
% Ozzie Outback Electric Power system. This version
% adds a capacitor bank at the Mallee bus.
%
% System data has the form:
%SYSTEM     name        baseMVA    Voltage Tolerance
SYSTEM  OOEP_Future2    100              0.05
%
% Bus data has the form:
%BUS name       type   volts  Pgen   Qgen   Pload   Qload   Qcap
BUS Bunya        SL    1.00     0      0       0       0       0
BUS Mulga        PV    1.00    200     0       0       0       0
BUS Mallee       PQ    1.00     0      0      100      75     125
BUS Myall        PQ    1.00     0      0      100      70       0
BUS Satinay      PQ    1.00     0      0      120     100       0
%
% Transmission line data has the form:
%LINE     from        to      Rse     Xse     Gsh     Bsh     Rating(MVA)
LINE    Bunya      Mulga    0.011   0.051   0.000   0.000      100
LINE    Mulga      Satinay  0.007   0.035   0.000   0.000      250
LINE    Bunya      Myall    0.007   0.035   0.000   0.000      200
LINE    Myall      Satinay  0.007   0.035   0.000   0.000      100
LINE    Myall      Mallee   0.011   0.051   0.000   0.000       60
LINE    Mallee     Satinay  0.011   0.051   0.000   0.000       60
```

When program `power_flow` is executed, the results are as shown in Figure 11–8. The bus voltages are now in tolerance at all busses, and the power flows in all transmission lines are within limits. This is certainly an improvement over the previous situation. ■

Figure 11-8 | Result of executing power_flow for the OOEP power system with a capacitor bank added. Note the improvements in bus voltages and reduced power flows in transmission lines.

```
» power_flow OOEP_Future2
Input summary statistics:
24 lines in system file
1 SYSTEM lines
5 BUS lines
6 LINE lines
```

                         Results for Case OOEP_Future2

Bus no.	Bus Name	Type	Volts / (pu)	angle (deg)	Generation (MW)	(MVAR)	Load (MW)	(MVAR)	Cap (MVAR)	To Bus	Line Flow (MW)	(MVAR)
1	Bunya	SL	1.000/	0.00	124.62	75.12	0.00	0.00	0.00	Mulga	-23.15	5.14
										Myall	147.78	69.98
2	Mulga	PV	1.000/	0.71	200.00	70.88	0.00	0.00	0.00	Bunya	23.22	-4.86
										Satinay	177.00	75.79
3	Mallee	PQ	0.972/	-4.44	0.00	0.00	100.00	75.00	125.00	Myall	-48.94	21.36
										Satinay	-50.81	28.63
4	Myall	PQ	0.966/	-2.78	0.00	0.00	100.00	70.00	0.00	Bunya	-145.91	-60.62
										Satinay	-3.18	10.43
										Mallee	49.27	-19.82
5	Satinay	PQ	0.963/	-2.66	0.00	0.00	120.00	100.00	0.00	Mulga	-174.40	-62.81
										Myall	3.19	-10.39
										Mallee	51.21	-26.79
		Totals			324.62	146.00	320.00	245.00	125.00			

                Line Losses

Line no.	From Bus	To Bus	Ploss (MW)	Qloss (MVAR)
1	Bunya	Mulga	0.06	0.29
2	Mulga	Satinay	2.60	12.98
3	Bunya	Myall	1.87	9.36
4	Myall	Satinay	0.01	0.04
5	Myall	Mallee	0.33	1.54
6	Mallee	Satinay	0.40	1.84
		Totals	5.27	26.04

```
Alerts
NONE

Done in 16 iterations.
```

Although we have shown that adding a capacitor bank at Mallee helps the situation, we are not done in planning for the future growth of the power system. For example, we now know that the busses are in tolerance at full load, but what happens at no load? It is possible that the bus voltages may now be too *high* at low loads—we must check that. If so, then the capacitor bank might have to be switchable instead of being permanently connected to the power system. In addition, we should try other combinations to see if the power system functions better if the capacitors are distributed among several busses instead of being lumped together at a single bus. Several more cases will have to be studied before you (the engineer) can make a recommendation to management about how to handle the new loads.

11.5 | ADDITIONAL FEATURES OF POWER-FLOW PROGRAMS

In this chapter, we have introduced the basics of power-flow analysis, and solved a number of problems with a simple power-flow analysis program. Advanced power-flow analysis programs such as those actually used in industry include a number of additional features that we have not considered. For example, advanced programs can incorporate tap-changing transformers, phase-shifting transformers, and mutually coupled components such as two transmission lines sharing the same poles. They also make it easier to enter values by accepting them in volts, ohms, etc. and doing the per-unit conversions internally. In addition, advanced programs almost always use the Newton-Raphson technique to solve power-flow equations.

These additional features are beyond the scope of this survey text. Many of them are described in Chapter 9 of Reference 2.

11.6 | SUMMARY

The analysis of a power system in normal steady-state operation is known as a *power-flow study* or *load flow study*. Power-flow studies attempt to determine the voltages, currents, and real and power flows in a power system under a given set of load conditions. An engineer who understands the operation of a power system under normal conditions will be able to plan ahead for its safe operation and future growth. A good understanding of steady-state conditions allows the engineer to play "What if?" games concerning future growth conditions.

The power-flow study techniques that we considered in this chapter are based on the admittance bus matrix $\mathbf{Y}_{bus}$. The busses on a power system are classified as load busses (only loads), generator busses (a generator is attached), or the slack bus (a generator that supplies all remaining power requirements). Real and reactive power flows are specified at load busses, real power flow and $|\mathbf{V}_{bus}|$ are specified at generator busses, and both the voltage magnitude and angle are specified at the slack bus (usually $1\angle 0°$).

The voltages in the power system are then solved using iterative methods. The simplest such method, and the only one described in this book, is the Gauss-Siedel iterative method. In this method, the voltage at each bus is estimated from a knowledge of power flows and an estimate of all of the other bus voltages in the system. As long as the initial estimates for bus voltage are reasonably close to the correct answers, this iterative process will usually converge to a correct answer. However, since the network system equations are fundamentally nonlinear, they can converge to erroneous solutions, or even fail to converge at all.

We also introduced the MATLAB-based program `power_flow`, which can be used to solve power-flow problems in simple power systems.

11.7 | QUESTIONS

11–1. What is the purpose of a power-flow study?

11–2. What categories of busses are used in a power-flow study? What are the characteristics of each type? What values are specified *a priori* for each type of bus?

11–3. Write the equation used to update bus voltage estimates in a power-flow study.

11–4. What additional equations and assumptions are required to include generator busses in a power-flow study?

11–5. What is the Gauss-Siedel iterative method?

11–6. What techniques can be used to speed up convergence to a solution with the Gauss-Siedel iterative method?

11.8 | PROBLEMS

11–1. Find the per-unit currents and the real and reactive power flows in each transmission line in the power system of Example 11–2. Also, calculate the losses in each transmission line.

11–2. Assume that a capacitor bank is added to bus 4 in the power system of Example 11–2. The capacitor bank *consumes* a reactive power of $Q_{cap} = -0.25$ pu, or alternatively, the capacitor *supplies* a reactive power of $+0.25$ pu to bus 4. Determine the per-unit bus voltages in the power system after the capacitor is added.

11–3. Calculate the per-unit currents and the real and reactive power flows in each transmission line in the power system of Problem 11–2 after the capacitor has been added. Did the currents in the lines go up or down as a result of the addition of the capacitor?

Problems 11–4 through 11–8 refer to the simple power system with five busses and six transmission lines shown in Figure P11–1. The base apparent power of this power system is 100 MVA, and the tolerance on each bus voltage is 5 percent. The bus data for the power system is given in Table 11–7, and the line data for the power system is given in Table 11–8.

Figure P11–1 | The power system of Problems 11–4 to 11–7.

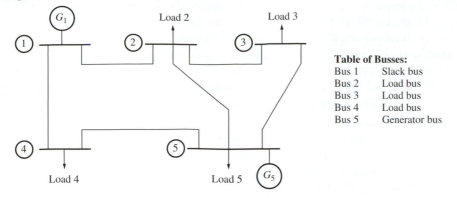

Table of Busses:
Bus 1   Slack bus
Bus 2   Load bus
Bus 3   Load bus
Bus 4   Load bus
Bus 5   Generator bus

Table 11–7 | Bus data for the power system in Figure P11–1

Bus name	Type	V (pu)	Generation		Loads	
			P (MW)	Q (MVAR)	P (MW)	Q (MVAR)
1	SL	$1\angle 0°$				
2	PQ	$1\angle 0°$			60	35
3	PQ	$1\angle 0°$			70	40
4	PQ	$1\angle 0°$			80	50
5	PV	$1\angle 0°$	190		40	30

Table 11–8 | Line data for the power system in Figure P11–1
(the shunt admittances of all lines are negligible)

Transmission line number	From/to (bus to bus)	Series impedance Z (pu)	Rated MVA
1	1–2	$0.0210 + j0.1250$	50
2	1–4	$0.0235 + j0.0940$	100
3	2–3	$0.0250 + j0.1500$	50
4	2–5	$0.0180 + j0.0730$	100
5	3–5	$0.0220 + j0.1100$	70
6	4–5	$0.0190 + j0.0800$	100

11–4. Answer the following questions about this power system:

a. Find the voltages at each bus in this power system.

b. Find the real and reactive power flows in each transmission line.

c. Are any of the bus voltages out of tolerance in this power system?

d. Are any of the transmission lines overloaded?

11–5. Suppose that transmission line 3 in the previous problem (between busses 2 and 3) is open circuited for maintenance. Find the bus voltages and transmission line powers in the power system with the line removed. Are any of the voltages out of tolerance? Are any of the transmission lines overloaded?

11–6. Suppose that a 40-MVAR capacitor bank is added to bus 3 of the power system in Problem 11–5. What happens to the bus voltages in this power system? What happens to the apparent powers of the transmission lines? Is this situation better or worse than the one in Problem 11–5?

11–7. Assume that the power system is restored to its original configuration. A new plant consuming 20 MW at 0.95 PF lagging is to be added to bus 4. Will this new load cause any problems for this power system? If this new load will cause problems, what solution could you recommend?

11–8. Write your own program to solve for the voltages and currents in the power system of Problem 11–4 instead of using program `power_flow`. You may write the program in any programming language with which you are familiar. How do your answers compare to the ones produced by `power_flow`?

11–9. Figure P11–2 shows a one-line diagram of a simple power system. Assume that the generator G_1 is supplying a constant 13.8 kV at bus 1, that load M_2 is consuming 20 MVA at 0.85 PF lagging, and that load M_3 is consuming 10 MVA at 0.90 PF leading. Calculate the bus admittance matrix $\mathbf{Y}_{bus}$ for this system, and then use it to determine the voltage at bus 2 in this power system. (*Note:* assume that the system base apparent power is 30 MVA.)

11–10. Suppose that a 10-MVAR capacitor bank were added to bus 2 in the power system of Figure P11–2. What is the voltage of bus 2 now?

Problems 11–11 through 11–15 refer to the simple power system with six busses and six transmission lines shown in Figure P11–3. The base apparent power of this power system is 100 MVA, and the tolerance on each bus voltage is 5 percent. Typical daytime bus data for the power system is given in Table 11–9, typical nighttime bus data for the power system is given in Table 11–10. The line data for the power system is given in Table 11–11.

Figure P11–2 | One-line diagram of the power system in Problems 11–9 and 11–10.

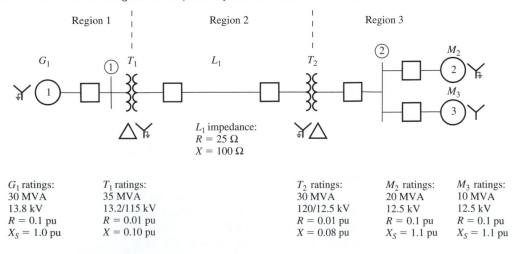

Region 1   Region 2   Region 3

G_1 ratings:	T_1 ratings:		T_2 ratings:	M_2 ratings:	M_3 ratings:
30 MVA	35 MVA		30 MVA	20 MVA	10 MVA
13.8 kV	13.2/115 kV		120/12.5 kV	12.5 kV	12.5 kV
$R = 0.1$ pu	$R = 0.01$ pu		$R = 0.01$ pu	$R = 0.1$ pu	$R = 0.1$ pu
$X_S = 1.0$ pu	$X = 0.10$ pu		$X = 0.08$ pu	$X_S = 1.1$ pu	$X_S = 1.1$ pu

L_1 impedance:
$R = 25\ \Omega$
$X = 100\ \Omega$

Figure P11–3 I The power system of Problems 11–11 to 11–15.

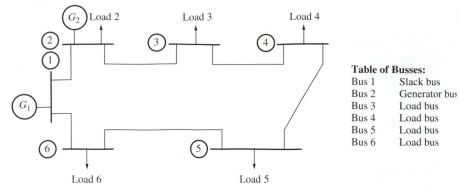

Table of Busses:

Bus 1	Slack bus
Bus 2	Generator bus
Bus 3	Load bus
Bus 4	Load bus
Bus 5	Load bus
Bus 6	Load bus

Table 11–9 I Typical daytime bus data for the power system in Figure P11–3

Bus name	Type	V (pu)	Generation		Loads	
			P (MW)	Q (MVAR)	P (MW)	Q (MVAR)
1	SL	$1\angle 0°$				
2	PV	$1\angle 0°$	100		60	35
3	PQ	$1\angle 0°$			40	25
4	PQ	$1\angle 0°$			60	40
5	PQ	$1\angle 0°$			30	10
6	PQ	$1\angle 0°$			40	25

Table 11–10 I Typical nighttime bus data for the power system in Figure P11–3

Bus name	Type	V (pu)	Generation		Loads	
			P (MW)	Q (MVAR)	P (MW)	Q (MVAR)
1	SL	$1\angle 0°$				
2	PV	$1\angle 0°$	100		60	35
3	PQ	$1\angle 0°$			40	25
4	PQ	$1\angle 0°$			60	40
5	PQ	$1\angle 0°$			30	10
6	PQ	$1\angle 0°$			40	25

Table 11–11 I Line data for the power system in Figure P11–3
(the shunt admittances of all lines are negligible)

Transmission line number	From/to (bus to bus)	Series impedance Z (pu)	Rated MVA
1	1–2	$0.010 + j0.080$	200
2	2–3	$0.010 + j0.080$	200
3	3–4	$0.008 + j0.064$	200
4	4–5	$0.020 + j0.150$	100
5	5–6	$0.008 + j0.064$	200
6	6–1	$0.010 + j0.080$	200

11–11. The bus data given in Table 11–9 represent typical daytime loads on the power system.

    a.   Find the voltages at each bus in this power system under typical daytime conditions.

    b.   Find the voltages at each bus in this power system under typical nighttime conditions.

    c.   Find the real and reactive power flows in each transmission line in each case.

    d.   Are any of the bus voltages out of tolerance in this power system?

    e.   Are any of the transmission lines overloaded?

11–12. If there are any problems with the bus voltages under typical daytime loads, propose a solution for this problem. Define a set of capacitors at various busses in the system to compensate for out-of-tolerance voltage variations.

11–13. What happens to the bus voltages under nighttime conditions if the capacitors proposed in Problem 11–12 are left attached to the system at night? Is it OK to permanently connect the capacitors to the power system, or must they be switched?

11–14. Suppose that the power system is operating under typical daytime loads, and the transmission line between bus 4 and bus 5 open-circuits. What happens to the voltages on each bus now? Are any of the transmission lines overloaded?

11–15. Suppose that a new transmission line is to be added to the power system between bus 1 and bus 4. The line is rated at 200 MVA, and its series impedance is $0.008 + j0.064$ pu. Assume that no capacitors have been added to the system, and determine the daytime and nighttime voltages at every bus. Did the new transmission line resolve the voltage problems?

11.9 | REFERENCES

1. Elgerd, Olle I.: *Electric Energy Systems Theory: An Introduction*, McGraw-Hill, New York, 1971.

2. Granger, John J., and William D. Stevenson, Jr.: *Power Systems Analysis*, McGraw-Hill, New York, 1994.

3. Gross, Charles A.: *Power Systems Analysis*, John Wiley and Sons, New York, 1979.

4. Stagg, Glen W., and Ahmed H. El-Abiad: *Computer Methods in Power Systems Analysis*, McGraw-Hill, New York, 1968.

Symmetrical Faults

A *fault* in a circuit is any failure that interferes with the normal flow of current to the loads. In most faults, a current path forms between two or more phases, or between one or more phases and the neutral (ground). This current path has relatively low impedance, resulting in excessive current flows.

High-voltage transmission lines have strings of insulators between each phase and the supporting towers that carry the transmission line. These insulators are normally large enough to prevent *flashover*, which is a condition in which the voltage difference between the phase and ground is large enough to ionize the air around the insulators, creating a current path from the phase to the structure of the tower, which is grounded. If flashover does occur on a phase of the transmission line, the ionized air provides a current path between the phase and ground, producing an *arc*. Such faults involving a single phase and ground are called *single line-to-ground faults*. The short-circuit path has a low impedance, so very high currents flow through the faulted line, into the ground, and back into the power system through the grounded wye connections of generators on the system. Faults involving ionized current paths are also known as *transient faults*, because the fault will disappear if the ionized current path is cleared. Transient faults usually clear if power is removed from the transmission line for a short time and then restored.

Single line-to-ground faults can also occur if one phase of a transmission line breaks and comes into direct contact with the ground, or if a string of insulators supporting the line breaks, allowing the line to sag to the ground. This type of fault is called a *permanent fault*, because it will remain even if power is removed from the transmission line for a short time and then restored. In most power systems, about three-quarters of all faults are transient or permanent single line-to-ground faults.

There are several other types of faults. Sometimes, all three phases of a transmission line are shorted together. These types of faults are known as *symmetrical three-phase faults*. At other times, two phases of a transmission line touch, or flashover occurs between two phases, resulting in a *line-to-line fault*. If two lines touch and also touch the ground, a *double line-to-ground fault* occurs.

Lightning causes most faults on high-voltage transmission lines. When lightning strikes a phase of a transmission line, it produces a very high transient voltage that can far exceed the rated voltage of the line. This high voltage will usually cause flashover between the phase and the grounded transmission towers, creating an arc. Once the current starts flowing through the arc, it continues to do so even after the lightning disappears.

When a fault occurs, the high current flows in the line are detected by protective circuitry, and the circuit breakers on the affected transmission line are automatically opened for a brief period (about $\frac{1}{3}$ second). This small time allows the ionized air forming the fault to dissipate and deionize. The circuit breakers are then reclosed. If the power system fault was a transient fault caused by flashover, normal operation should be restored because the arc conducting the current has been extinguished. Many transient faults are thus cleared automatically. If the fault is still present on the line after the circuit breakers reclose, they will open again, isolating the transmission line until the permanent fault can be located and cleared by a repair crew.

The circuit breakers used to protect a power system must be designed to operate properly and without damage even when the highest possible fault current is flowing in the system. To select the proper circuit breaker sizes, we must know the maximum current flow that could occur during a fault. In addition, we must know the magnitude of fault currents so that we can adjust the power system's protective circuitry to open circuit breakers when faults occur, but not under normal full-load operating conditions. The calculation of current flows under fault conditions is known as *fault analysis*. It is the topic of the next two chapters.

The types of faults described above are summarized in Table 12–1. They can be divided into two broad categories: *symmetrical faults*, in which the magnitudes of the AC currents in each phase are the same; and *unsymmetrical faults*, in which the magnitudes of the AC currents in each phase differ. Symmetrical faults are covered in this chapter, and unsymmetrical faults are covered in Chapter 13.

Table 12–1 | Types of power system faults, in decreasing frequency of occurrence

Type of fault	Abbreviation	Type
Single line-to-ground	SLG	Unsymmetrical
Line-to-line	LL	Unsymmetrical
Double line-to-ground	LLG	Unsymmetrical
Symmetrical three-phase	3P	Symmetrical

12.1 | FAULT CURRENT TRANSIENTS IN MACHINES

When a symmetrical three-phase fault occurs at the terminals of a synchronous generator, the resulting current flow in the phases of the generator can appear as shown in Figure 12–1. The current in each phase shown in Figure 12–1 can be represented as a transient DC component added on top of a symmetrical AC component. The symmetrical AC component by itself is shown in Figure 12–2.

Figure 12-1 | The total fault currents as a function of time during a symmetrical three-phase fault on a synchronous generator.

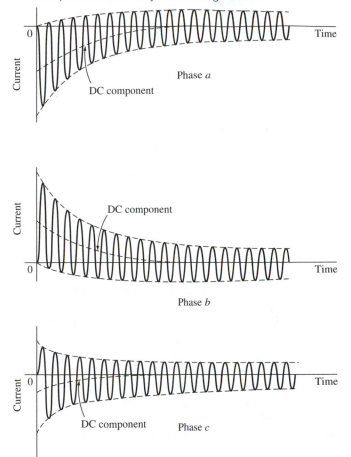

Before the fault, only AC voltages and currents were present within the generator, while immediately after the fault, both AC and DC currents are present. Where did the DC currents come from? Remember that a synchronous generator is basically an inductive circuit—a synchronous machine is modeled by an internal generated voltage in series with an inductance. Also, recall that *a current cannot change instantaneously in an inductor.* When the fault occurs, the AC component of current jumps to a very large value, but the total current cannot change at that instant. The transient DC component of current is just large enough that the *sum* of the AC and DC components just after the fault equals the AC current flowing just before the fault. Since the instantaneous values of current at the moment of the fault are different in each phase, the magnitude of the DC component of current will be different in each phase.

These DC components of current decay fairly quickly, but they initially average about 50 or 60 percent of the AC current flow the instant after the fault occurs. The

Figure 12–2 | The symmetrical AC component of the fault current.

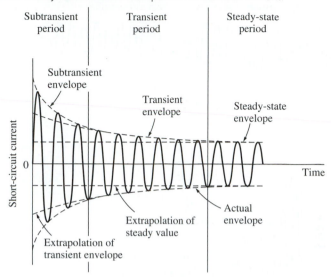

total initial current is therefore typically 1.5 or 1.6 times the AC component taken alone.

The AC symmetrical component of current is shown in Figure 12–2. It can be divided into roughly three periods. During the first cycle or so after the fault occurs, the AC current is very large and falls very rapidly. This period of time is called the *subtransient period*. After it is over, the current continues to fall at a slower rate, until at last it reaches a steady state. The period of time during which it falls at a slower rate is called the *transient period*, and the time after it reaches steady state is known as the *steady-state period*.

If the rms magnitude of the AC component of current is plotted as a function of time on a semilogarithmic scale, it is possible to observe the three periods of fault current. Such a plot is shown in Figure 12–3. It is possible to determine the time constants of the decays in each period from such a plot.

The AC rms current flowing in the generator during the subtransient period is called the *subtransient current* and is denoted by the symbol I''. This current is caused by the amortisseur or damper windings of synchronous machines. The time constant of the subtransient current is given the symbol T'', and it can be determined from the slope of the subtransient current in the plot in Figure 12–3. This current can often be 10 times the size of the steady-state fault current.

The rms current flowing in the generator during the transient period is called the *transient current* and is denoted by the symbol I'. It is caused by a transient DC component of current induced in *the field circuit* of a synchronous generator at the time of the fault. This transient field current increases the internal generated voltage of a synchronous generator, and so causes an increased fault current. Since the time constant of the DC field circuit is much longer than the time constant of the damper windings, the transient period lasts much longer than the subtransient period. This

Figure 12-3 | A semilogarithmic plot of the rms magnitude of the AC component of fault current as a function of time. The subtransient and transient time constants of the current can be determined from such a plot.

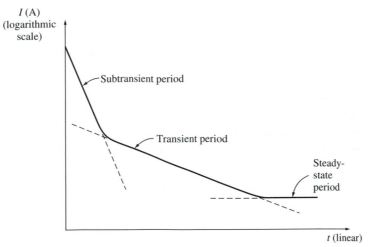

time constant is given the symbol T'. The average rms current during the transient period is often as much as 5 times the steady-state fault current.

After the transient period, the fault current reaches a steady-state condition. The steady-state rms current during a fault is denoted by the symbol I_{ss}. It is given approximately by the fundamental frequency component of the internal generated voltage E_A within the machine divided by its synchronous reactance:

$$I_{ss} = \frac{E_A}{X_S} \qquad \text{steady state} \qquad (12\text{--}1)$$

The rms magnitude of the AC fault current in a synchronous generator varies continuously as a function of time. If I'' is the subtransient component of current at the instant of the fault, I' is the transient component of current at the instant of the fault, and I_{ss} is the steady-state fault current, then the rms magnitude of the current at any time after a fault occurs at the terminals of the generator is

$$I(t) = (I'' - I')e^{-t/T''} + (I' - I_{ss})e^{-t/T'} + I_{ss} \qquad (12\text{--}2)$$

It is customary to define subtransient and transient reactances for a synchronous generator as a convenient way to describe the subtransient and transient components of fault current. The *subtransient reactance* of a power system is defined as the ratio of the fundamental component of the internal generated voltage E_A to the sub-transient component of current at the beginning of the fault. It is given by

$$X'' = \frac{E_A}{I''} \qquad \text{subtransient} \qquad (12\text{--}3)$$

Similarly, the *transient reactance* of a power system is defined as the ratio of the fundamental component of E_A to the transient component of current I' at the beginning of the fault. This value of current is found by extrapolating the subtransient region in Figure 12–3 back to time zero:

$$X' = \frac{E_A}{I'} \qquad \text{transient} \qquad\qquad (12\text{–}4)$$

For the purposes of sizing protective equipment, the subtransient current is often assumed to be E_A/X'', and the transient current is assumed to be E_A/X', since these are the maximum values that the respective currents take on.

Note that the above discussion of faults assumes that all three phases were shorted out simultaneously. If the fault does not involve all three phases equally, then more complex methods of analysis are required to understand it. These methods (known as symmetrical components) are discussed in Chapter 13.

Fault Transients in Synchronous Motors

When a short circuit develops on a power system with a synchronous motor, the motor becomes a temporary generator, converting its energy of rotation back into electrical power, which is supplied to the power system. Since a synchronous motor is physically the same machine as a synchronous generator, it also has a subtransient reactance X'' and transient reactance X' that must be considered in determining the total fault currents flowing in the power system.

Fault Transients in Induction Motors

As we learned in Chapter 7, an induction motor is simply an AC machine that has only amortisseur or damper windings on its rotor. Since damper windings are the major source of current during the subtransient period, the induction motors attached to a power system should be considered when calculating the subtransient currents flowing in fault.

Since the currents in a damper winding are of negligible importance during the transient and steady-state periods of faults, induction motors may be ignored in analyzing fault currents after the subtransient period ends.

EXAMPLE 12–1

A 100-MVA, 13.8-kV, Y-connected, three-phase, 60-Hz synchronous generator is operating at the rated voltage and no load when a three-phase fault develops at its terminals. Its reactances in per unit to the machine's own base are

$$X_S = 1.00 \qquad X' = 0.25 \qquad X'' = 0.12$$

and its time constants are

$$T' = 1.10 \text{ s} \qquad T'' = 0.04 \text{ s}$$

The initial DC component in this machine averages 50 percent of the initial AC component.

a. What is the AC component of current in this generator the instant after the fault occurs?
b. What is the total current (AC plus DC) flowing in the generator right after the fault occurs?
c. What will the AC component of the current be after two cycles? After 5 s?

■ **Solution**
The base current of this generator is given by the equation

$$I_{L, \text{base}} = \frac{S_{\text{base}}}{\sqrt{3} V_{L, \text{base}}} \tag{10–4}$$

$$= \frac{100 \text{ MVA}}{\sqrt{3}(13.8 \text{ kV})} = 4184 \text{ A}$$

The subtransient, transient, and steady-state currents in per unit and in amperes are

$$I'' = \frac{E_A}{X''} = \frac{1.0}{0.12} = 8.333$$
$$= (8.333)(4184 \text{ A}) = 34,900 \text{ A}$$
$$I' = \frac{E_A}{X'} = \frac{1.0}{0.25} = 4.00$$
$$= (4.00)(4184 \text{ A}) = 16,700 \text{ A}$$
$$I_{\text{ss}} = \frac{E_A}{X'} = \frac{1.0}{1.0} = 1.00$$
$$= (1.00)(4184 \text{ A}) = 4184 \text{ A}$$

a. The initial AC component of current is $I'' = 34{,}900$ A.
b. The total current (AC plus DC) at the beginning of the fault is

$$I_{\text{tot}} = 1.5 I'' = 52{,}350 \text{ A}$$

c. The AC component of current as a function of time is given by Equation (12–2):

$$I(t) = (I'' - I')e^{-t/T''} + (I' - I_{\text{ss}})e^{-t/T'} + I_{\text{ss}} \tag{12–2}$$
$$= 18{,}200e^{-t/0.04 \text{ s}} + 12{,}516e^{-t/1.1 \text{ s}} + 4184 \text{ A}$$

At two cycles, $t = 1/30$ s, the total current is

$$I\left(\frac{1}{30}\right) = 7910 \text{ A} + 12{,}142 \text{ A} + 4184 \text{ A} = 24{,}236 \text{ A}$$

After two cycles, the transient component of current is clearly the largest one and this time is in the transient period of the short circuit. At 5 s, the current is down to

$$I(5) = 0 \text{ A} + 133 \text{ A} + 4184 \text{ A} = 4317 \text{ A}$$

This is part of the steady-state period of the short circuit. ■

EXAMPLE 12–2

Two generators are connected in parallel to the low-voltage side of a unit transformer, as shown in Figure 12–4. Generators G_1 and G_2 are each rated at 50 MVA, 13.8 kV, with a subtransient reactance of 0.20 pu. Transformer T_1 is rated at 100 MVA, 13.8/115 kV, with a series reactance of 0.08 pu. (Note that the resistances of all components are being ignored in this problem.)

Assume that initially the voltage on the high voltage side of the transformer is 120 kV, that the transformer is unloaded, and that there are no circulating currents between the generators. Calculate the subtransient fault current that will flow if a three-phase fault occurs at the high-voltage side of transformer.

Figure 12–4 | One-line diagram for the power system in Example 12–2.

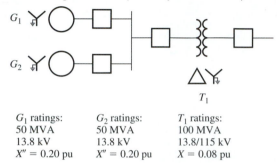

G_1 ratings:
50 MVA
13.8 kV
$X'' = 0.20$ pu

G_2 ratings:
50 MVA
13.8 kV
$X'' = 0.20$ pu

T_1 ratings:
100 MVA
13.8/115 kV
$X = 0.08$ pu

■ Solution

If we choose the per-unit base values for this power system to be 100 MVA and 115 kV at the high-voltage side of the transformer, then the base voltage at the low-voltage side of the transformer will be 13.8 kV.

The subtransient reactance of the two generators to the system base will be

$$\text{per-unit } Z_{new} = \text{per-unit } Z_{given} \left(\frac{V_{given}}{V_{new}}\right)^2 \left(\frac{S_{new}}{S_{given}}\right) \qquad (10\text{–}8)$$

$$X_1'' = X_2'' = \left(\frac{13.8 \text{ kV}}{13.8 \text{ kV}}\right)^2 \left(\frac{100 \text{ MVA}}{50 \text{ MVA}}\right) = 0.40 \text{ pu}$$

The reactance of the transformer is already given on the system base, so it will not have to change.

$$X_T = 0.08 \text{ pu}$$

The per-unit voltage on the high-voltage side of the transformer is

$$\text{Quantity in per unit} = \frac{\text{actual value}}{\text{base value of quantity}} \qquad (10\text{–}7)$$

$$V = \left(\frac{120 \text{ kV}}{115 \text{ kV}}\right) = 1.044 \text{ pu}$$

Since there is no load on the power system, the voltage at the terminals of each generator (and the internal generated voltage E_A of each generator) must also be 1.044 pu. The per-phase equivalent circuit of this power system is shown in Figure 12–5a. Note that we have arbitrarily chosen the internal generated voltage E_A of each generator to be at a phase angle of 0°. (We know that the phase angles of both voltages are the same, because there were no circulating currents before the fault occurred.)

If we can find the voltage at bus 1 in this power system, then we can solve for the subtransient fault current. To find the voltage $\mathbf{V}_1$, we must convert the per-unit impedances to admittances, and the voltage sources to equivalent current sources. The Thevenin impedance of each generator is $Z_{TH} = j0.40$, so the short-circuit current of each generator is

$$\mathbf{I}_{SC} = \frac{\mathbf{V}_{OC}}{Z_{TH}} = \frac{1.044\angle 0°}{j0.40} = 2.61\angle -90°$$

The resulting equivalent circuit is shown in Figure 12–5b.

Figure 12–5 | (a) Per-phase equivalent circuit for the power system in Example 12–2 with values expressed as impedances. (b) Per-phase equivalent circuit with values expressed as admittances and voltage sources converted to current sources.

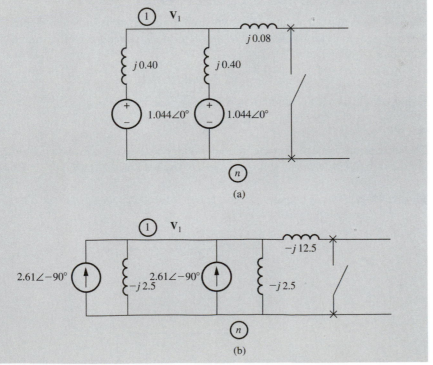

We can now write a node equation for voltage $\mathbf{V}_1$:

Sum of currents out of the node = sum of currents into the node

$$\mathbf{V}_1(-j2.5) + \mathbf{V}_1(-j2.5) + \mathbf{V}_1(-j12.5) = 2.61\angle -90° + 2.61\angle -90°$$

$$\mathbf{V}_1(-j17.5) = 5.22\angle -90°$$

$$\mathbf{V}_1 = \frac{5.22\angle -90°}{-j17.5} = 0.298\angle 0° \text{ pu}$$

Therefore, the subtransient current in the fault is

$$\mathbf{I}_F = \mathbf{V}_1(-j12.5) = (0.298\angle 0°)(-j12.5) = 3.729\angle -90° \text{ pu}$$

The base current at the high-voltage side of the transformer is

$$I_{base} = \frac{S_{3\phi,\,base}}{\sqrt{3}V_{LL,\,base}} = \frac{100\text{ MVA}}{\sqrt{3}(115\text{ KV})} = 502\text{ A} \qquad (10\text{--}4)$$

Therefore, the subtransient fault current will be

$$I_F = I_{F,\,pu}I_{base} = (3.729)(502\text{ A}) = 1872\text{ A} \qquad \blacksquare$$

12.2 | INTERNAL GENERATED VOLTAGES OF LOADED MACHINES UNDER TRANSIENT CONDITIONS

The per-phase equivalent circuit of a synchronous generator is shown in Figure 12–6. The internal generated voltage of this synchronous generator can be found from Kirchhoff's Voltage Law:

$$\mathbf{E}_A = \mathbf{V}_\phi + jX_S\mathbf{I}_A + R_A\mathbf{I}_A \qquad (12\text{--}5)$$

If the series resistance is ignored, this equation reduces to

$$\mathbf{E}_A = \mathbf{V}_\phi + jX_S\mathbf{I}_A \qquad (12\text{--}6)$$

These equations indicate that the internal voltage within a synchronous generator is not the same as its terminal voltage if there is any load on the machine.

Figure 12–6 | The per-phase equivalent circuit of a synchronous generator under steady-state conditions.

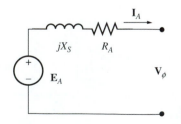

A similar equation exists for the relationship between internal and terminal voltages in a synchronous generator under subtransient and transient conditions. The per-phase equivalent circuit for a synchronous generator under *subtransient* conditions is shown in Figure 12–7. The Kirchhoff's voltage law equation for this circuit is

$$\mathbf{E}_A'' = \mathbf{V}_\phi + jX''\mathbf{I}_A \tag{12–7}$$

Therefore, the internal voltage $\mathbf{E}_A''$ under subtransient conditions can be calculated from a knowledge of the load current $\mathbf{I}_A$ and terminal voltage $\mathbf{V}_\phi$ just before the fault occurs. This voltage will be the voltage driving the subtransient fault current flow from the generator. It is sometimes called the *voltage behind subtransient reactance.*

Since the voltage behind subtransient reactance varies as a function of the load on the generator before the fault occurs, *the subtransient current flow in a fault will vary depending on the prefault load conditions of the power system.* This variation will typically be less than 10 percent for different load conditions. In fact, we can calculate the approximate fault currents in a power system by ignoring the voltage behind subtransient reactance and treating that voltage as equal to the prefault phase voltage of the generator.

Figure 12–7 | The per-phase equivalent circuit of a synchronous generator under subtransient conditions.

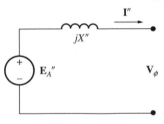

Similarly, the per-phase equivalent circuit for a synchronous generator under *transient* conditions is shown in Figure 12–8. The Kirchhoff's Voltage law equation for this circuit is

$$\mathbf{E}_A' = \mathbf{V}_\phi + jX'\mathbf{I}_A \tag{12–8}$$

Therefore, the internal voltage $\mathbf{E}_A'$ under transient conditions can be calculated from a knowledge of the load current $\mathbf{I}_A$ and terminal voltage $\mathbf{V}_\phi$ just before the fault occurs. This voltage will be the voltage driving the transient fault current flow from the generator. It is sometimes called the *voltage behind transient reactance.*

Figure 12–8 | The per-phase equivalent circuit of a synchronous generator under transient conditions.

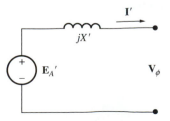

Since the voltage behind transient reactance varies as a function of the load on the generator before the fault occurs, *the transient current flow in a fault will vary, depending on the prefault load conditions of the power system.* This variation will typically be less than 10 percent for different load conditions. As with the sub-transient case, we can calculate the approximate transient fault currents in a power system by ignoring the voltage behind transient reactance and treating that voltage as equal to the prefault phase voltage of the generator.

Internal Generated Voltages of Synchronous Motors

Since synchronous motors are the same physical machines as synchronous generators, they have the same types of subtransient and transient reactances. When a motor is short-circuited, it no longer receives power from the line, but its field circuit remains energized, and the inertia of the machine and its loads keep the motor turning. The motor acts as a generator, supplying power to the fault. The equivalent circuit of a synchronous motor is the same as the equivalent circuit of a synchronous generator, except that the assumed direction of current flow is reversed. As a result, the equations for the internal generated voltage, voltage behind subtransient reactance, and voltage behind transient reactance become:

$$\mathbf{E}_A = \mathbf{V}_\phi - jX_S\mathbf{I}_A \qquad \text{Internal generated voltage} \qquad (12\text{--}9)$$
$$\mathbf{E}_A'' = \mathbf{V}_\phi - jX''\mathbf{I}_A \qquad \text{Voltage behind subtransient reactance} \qquad (12\text{--}10)$$
$$\mathbf{E}_A' = \mathbf{V}_\phi - jX'\mathbf{I}_A \qquad \text{Voltage behind transient reactance} \qquad (12\text{--}11)$$

These voltages would be used in subtransient and transient fault current analysis in the same manner as the one for synchronous generators.

EXAMPLE 12–3

A 100-MVA, 13.8-kV, 0.9 PF lagging, Y-connected, three-phase, 60-Hz synchronous generator is operating at rated voltage and full load when a symmetrical three-phase fault develops at its terminals. Its reactances in per unit to the machine's own base are

$$X_S = 1.00 \qquad X' = 0.25 \qquad X'' = 0.12$$

a. If this generator is operating at full load when the fault develops, what is the subtransient fault current produced by this generator?

b. If this generator is operating at no load and rated voltage when the fault develops, what is the subtransient fault current produced by this generator? (*Note:* This calculation is equivalent to ignoring the effects of prefault load on fault currents.)

c. How much difference does calculating the voltage behind subtransient reactance make in the fault current calculation?

■ Solution

The base current of this generator is given by the equation

$$I_{L, \text{base}} = \frac{S_{\text{base}}}{\sqrt{3} V_{L, \text{base}}} \tag{10-4}$$

$$= \frac{100 \text{ MVA}}{\sqrt{3}(13.8 \text{ kV})} = 4184 \text{ A}$$

a. Before the fault, the generator was operating at rated conditions, so the per-unit current was

$$\mathbf{I}_A = 1.0\angle -25.84° \text{ pu} \tag{12-12}$$

The voltage behind subtransient reactance is

$$\mathbf{E}_A'' = \mathbf{V}_\phi + jX''\mathbf{I}_A \tag{12-7}$$
$$= 1\angle 0° + (j0.12)(1.0\angle -25.84°)$$
$$= 1.058\angle 5.86°$$

Therefore, the per-unit fault current when the terminals are shorted is

$$\mathbf{I}_F = \frac{1.058\angle 5.86°}{j0.12} = 8.815\angle -84.1°$$

The fault current in amperes is

$$I_F = (8.815)(4184 \text{ A}) = 36,880 \text{ A}$$

b. Before the fault, the generator was assumed to be at no-load conditions, so the per-unit current was

$$\mathbf{I}_A = 0.0\angle 0° \text{ pu} \tag{12-13}$$

The voltage behind subtransient reactance is

$$\mathbf{E}_A'' = \mathbf{V}_\phi + jX''\mathbf{I}_A \tag{12-7}$$
$$= 1\angle 0° + (j0.12)(0.0\angle 0°)$$
$$= 1.0\angle 0°$$

Therefore, the per-unit fault current when the terminals are shorted is

$$\mathbf{I}_F = \frac{1.0\angle 0°}{j0.12} = 8.333\angle -90°$$

The fault current in amperes is

$$I_F = (8.333)(4184 \text{ A}) = 34,870 \text{ A}$$

c. The difference in fault current between these two cases is

$$\text{Difference} = \frac{36,880 - 34,870}{34,870} \times 100\% = 5.76\%$$

> The difference in the fault current when the voltage behind subtransient reactance is considered and when it is ignored is only about 5.8 percent. Since the actual load in a power system cannot always be predicted before a fault occurs, analyses generally ignore this small effect and treat power systems as though they were initially unloaded. ∎

12.3 | FAULT CURRENT CALCULATIONS USING THE BUS IMPEDANCE MATRIX

The discussion of fault calculations so far has been limited to simple circuits. To determine the fault current flowing in a system, we have created a per-phase, per-unit equivalent circuit of the power system using either subtransient reactances if we wish to solve for subtransient currents, or transient reactances if we wish to solve for transient currents. Then, a short circuit was added between one node in the equivalent circuit and the neutral, and the current flow through that short was calculated by standard circuit analysis techniques.

This approach always works, but it can get rather complex if we are dealing with power systems involving many busses, transmission lines, generators, and loads. To handle really large power systems, we need to have a systematic approach to solve for the voltages and currents in the power system. We will do this by adapting the nodal analysis technique introduced in Chapter 10.

To solve for the fault currents in a power system, we will introduce a new voltage source in the system to represent the effects of a fault at a bus. By solving for the currents introduced by this additional voltage source, we will automatically be solving for the fault currents that flow at that bus.

To understand this approach, consider the simple power system from Example 10–3, which is repeated in Figure 12–9. If we wish to solve for the *subtransient* fault current at some node in this power system, then we must create a per-phase, per-unit equivalent circuit using *subtransient* reactances X'' instead of the synchronous

Figure 12–9 | The simple three-phase power system of Example 10–3.

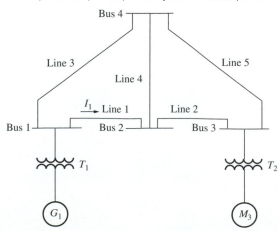

reactance X_S that we used in Example 10–3. In addition, we will assume that the power system is initially unloaded, making the voltages behind subtransient reactance $\mathbf{E}''_{A1}$ and $\mathbf{E}''_{A2}$ both equal to $1\angle 0°$ pu. The resulting equivalent circuit is shown in Figure 12–10.

Suppose that we wish to determine the subtransient fault current at bus 2 when a symmetrical three-phase fault occurs on that bus. Before the fault, the voltage at bus 2 is called $\mathbf{V}_f$. If we introduce a voltage source of value $\mathbf{V}_f$ between bus 2 and the neutral, nothing will have changed, so there will be no net current $\mathbf{I}''_f$ flowing through the source (see Figure 12–11).

Now suppose that we create a short circuit on bus 2, which brings the voltage on bus 2 down to 0. This is equivalent to inserting an *additional* voltage source $-\mathbf{V}_f$ in series with the existing voltage source, making the total voltage at bus 2 become 0 (see Figure 12–12). With this additional voltage source inserted, there will now be a fault current $\mathbf{I}''_f$. *This fault current is entirely due to the insertion of the new voltage source*, so we can use superposition to analyze the effects of only the new voltage source on the system. The resulting current flow $\mathbf{I}''_f$ will be the current for the *entire* power system, because the other sources in the system produced a net zero current.

Figure 12–10 I The per-phase, per-unit equivalent circuit of the power system in Figure 12–9. Note that the values shown are per-unit *impedances*, and the subtransient reactances are used instead of the synchronous reactances for each synchronous machine.

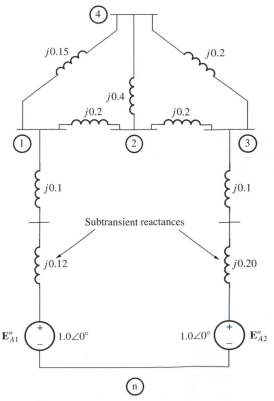

Figure 12–11 | The equivalent circuit with a voltage source $\mathbf{V}_f$ added at bus 2. Since the voltage at bus 2 was already $\mathbf{V}_f$, nothing has changed, and the fault current $\mathbf{I}''_f$ will be zero.

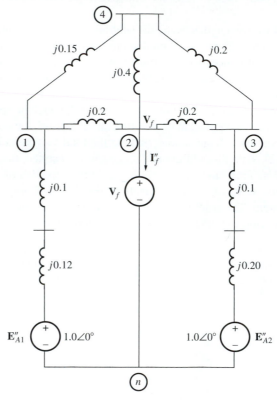

If all of the voltage sources except $-\mathbf{V}_f$ are set to zero and the impedances are converted to admittances, the power system appears as shown in Figure 12–13. We can now use that figure to construct the bus admittance matrix $\mathbf{Y}_{\text{bus}}$ in the manner described in Chapter 10. The resulting admittance matrix is

$$\mathbf{Y}_{\text{bus}} = \begin{bmatrix} -j\,16.212 & j\,5.0 & 0 & j\,6.667 \\ j\,5.0 & -j\,12.5 & j\,5.0 & j\,2.5 \\ 0 & j\,5.0 & -j\,13.333 & j\,5.0 \\ j\,6.667 & j\,2.5 & j\,5.0 & -j\,14.167 \end{bmatrix} \qquad (12\text{–}14)$$

The nodal equation describing this power system is

$$\mathbf{Y}_{\text{bus}}\,\mathbf{V} = \mathbf{I} \qquad (10\text{–}20)$$

With all other voltage sources set to zero, the voltage at bus 2 is $-\mathbf{V}_f$, and the current *entering* bus 2 is $-\mathbf{I}''_f$. Therefore, the nodal equation becomes

$$\begin{bmatrix} Y_{11} & Y_{12} & Y_{13} & Y_{14} \\ Y_{21} & Y_{22} & Y_{23} & Y_{24} \\ Y_{31} & Y_{32} & Y_{33} & Y_{34} \\ Y_{41} & Y_{42} & Y_{43} & Y_{44} \end{bmatrix} \begin{bmatrix} \Delta\mathbf{V}_1 \\ -\mathbf{V}_f \\ \Delta\mathbf{V}_3 \\ \Delta\mathbf{V}_4 \end{bmatrix} = \begin{bmatrix} 0 \\ -\mathbf{I}''_f \\ 0 \\ 0 \end{bmatrix} \qquad (12\text{–}15)$$

Figure 12–12 | When a voltage source $-\mathbf{V}_f$ is added, the voltage at bus 2 drops to zero and a fault current $\mathbf{I}_f''$ flows. Since there was no fault current before inserting the voltage source, the current now flowing is entirely due to the effects of adding this source.

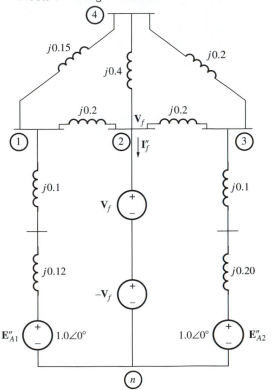

Figure 12–13 | The power system with all sources set to zero except for $-\mathbf{V}_f$, and with the impedance values converted to admittances.

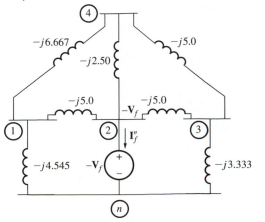

where $\Delta\mathbf{V}_1$, $\Delta\mathbf{V}_3$, and $\Delta\mathbf{V}_4$ are the changes in the voltages at those busses due to the current $-\mathbf{I}_f''$ injected at bus 2 by the fault.

The solution to Equation (12–15) is

$$\mathbf{V} = \mathbf{Y}_{bus}^{-1}\mathbf{I} = \mathbf{Z}_{bus}\mathbf{I} \qquad (10–24)$$

$$\begin{bmatrix} \Delta\mathbf{V}_1 \\ -\mathbf{V}_f \\ \Delta\mathbf{V}_3 \\ \Delta\mathbf{V}_4 \end{bmatrix} = \begin{bmatrix} Z_{11} & Z_{12} & Z_{13} & Z_{14} \\ Z_{21} & Z_{22} & Z_{23} & Z_{24} \\ Z_{31} & Z_{32} & Z_{33} & Z_{34} \\ Z_{41} & Z_{42} & Z_{43} & Z_{44} \end{bmatrix} = \begin{bmatrix} 0 \\ -\mathbf{I}_f'' \\ 0 \\ 0 \end{bmatrix} \qquad (12–16)$$

where $\mathbf{Z}_{bus} = \mathbf{Y}_{bus}^{-1}$. Since only bus 2 has current injected at it, we can see that Equation (12–16) reduces to

$$\begin{aligned} \Delta\mathbf{V}_1 &= -Z_{12}\mathbf{I}_f'' \\ -\mathbf{V}_f &= -Z_{22}\mathbf{I}_f'' \\ \Delta\mathbf{V}_3 &= -Z_{32}\mathbf{I}_f'' \\ \Delta\mathbf{V}_4 &= -Z_{42}\mathbf{I}_f'' \end{aligned} \qquad (12–17)$$

In other words, *the fault current at bus 2 is just the prefault voltage* $\mathbf{V}_f$ *at bus 2 divided by* Z_{22}, *the driving point impedance at bus 2.*

$$\mathbf{I}_f'' = \frac{\mathbf{V}_f}{Z_{22}} \qquad (12–18)$$

The voltage difference at each of the nodes due to the fault current can be calculated by substituting Equation (12–18) in Equations (12–17):

$$\begin{matrix} \Delta\mathbf{V}_1 \\ \Delta\mathbf{V}_2 \\ \Delta\mathbf{V}_3 \\ \Delta\mathbf{V}_4 \end{matrix} = \begin{matrix} -Z_{12}\mathbf{I}_f'' \\ -\mathbf{V}_f \\ -Z_{32}\mathbf{I}_f'' \\ -Z_{42}\mathbf{I}_f'' \end{matrix} = \begin{bmatrix} -\dfrac{Z_{12}}{Z_{22}}\mathbf{V}_f \\ -\mathbf{V}_f \\ -\dfrac{Z_{32}}{Z_{22}}\mathbf{V}_f \\ -\dfrac{Z_{42}}{Z_{22}}\mathbf{V}_f \end{bmatrix} \qquad (12–19)$$

If we assume that the power system was running at no-load conditions before the fault, it is easy to calculate the voltages at every bus during the fault. At no load, the voltage will be the same on every bus in the power system, so the voltage on every bus in the power system is $\mathbf{V}_f$. The *change* in voltage on every bus caused by the fault current $-\mathbf{I}_f''$ is given by Equation (12–19), so the total voltage during the fault is:

$$\begin{bmatrix} \mathbf{V}_1 \\ \mathbf{V}_2 \\ \mathbf{V}_3 \\ \mathbf{V}_4 \end{bmatrix} = \begin{bmatrix} \mathbf{V}_f \\ \mathbf{V}_f \\ \mathbf{V}_f \\ \mathbf{V}_f \end{bmatrix} + \begin{bmatrix} \Delta\mathbf{V}_1 \\ \Delta\mathbf{V}_2 \\ \Delta\mathbf{V}_3 \\ \Delta\mathbf{V}_4 \end{bmatrix} \qquad (12–20)$$

$$\begin{bmatrix} \mathbf{V}_1 \\ \mathbf{V}_2 \\ \mathbf{V}_3 \\ \mathbf{V}_4 \end{bmatrix} = \begin{bmatrix} \mathbf{V}_f \\ \mathbf{V}_f \\ \mathbf{V}_f \\ \mathbf{V}_f \end{bmatrix} + \begin{bmatrix} -\dfrac{Z_{12}}{Z_{22}}\mathbf{V}_f \\ -\mathbf{V}_f \\ -\dfrac{Z_{32}}{Z_{22}}\mathbf{V}_f \\ -\dfrac{Z_{42}}{Z_{22}}\mathbf{V}_f \end{bmatrix} = \begin{bmatrix} 1-\dfrac{Z_{12}}{Z_{22}} \\ 0 \\ 1-\dfrac{Z_{32}}{Z_{22}} \\ 1-\dfrac{Z_{42}}{Z_{22}} \end{bmatrix} \mathbf{V}_f \qquad (12–21)$$

Thus we can calculate the voltage at every bus in the power system during the fault from a knowledge of the pre-fault voltage at the faulted bus and the bus impedance matrix!

Once these bus voltages are known, we can also calculate the fault current flowing in any transmission line using the bus voltages and the bus admittance matrix $\mathbf{Y}_{bus}$.

The general procedure for finding the bus voltages and line currents during a symmetrical three-phase fault is as follows:

1. *Create a per-phase, per-unit equivalent circuit for the power system.* Include subtransient reactances X'' of each synchronous machine if you are looking for subtransient fault currents, and include the transient reactances X' of each synchronous machine if you are looking for transient fault currents. (You may also include the subtransient reactances of each induction motor if you are looking for subtransient fault currents, and ignore the induction motors completely if you are looking for transient fault currents.)

2. *Calculate the bus admittance matrix $\mathbf{Y}_{bus}$.* Include the admittances of all transmission lines, transformers, etc., between busses, including the admittances of the loads or generators themselves at each bus.

3. *Calculate the bus impedance matrix $\mathbf{Z}_{bus}$.* The bus impedance matrix $\mathbf{Z}_{bus}$ is the inverse of the bus admittance matrix $\mathbf{Y}_{bus}$. You can use MATLAB to perform this calculation.

4. *Assume that the power system is at no-load conditions and determine the voltage at every bus.* At no-load conditions, the voltage at every bus will be the same, since otherwise a current would flow between busses. This voltage will be the same as the internal voltage of the generators in the system. It is the prefault voltage in the power system, given the symbol $\mathbf{V}_f$.

5. *Calculate the current at the faulted bus.* The current at the faulted bus can be calculated directly from a knowledge of $\mathbf{V}_f$ and $\mathbf{Z}_{bus}$. If bus i is faulted, the fault current will be

$$\mathbf{I}''_{f,i} = \frac{\mathbf{V}_f}{Z_{ii}} \tag{12–22}$$

6. *Calculate the voltages at each bus during the fault.* The voltage at bus j during a symmetrical three-phase fault at bus i is given by the equation

$$\mathbf{V}_j = \left(1 - \frac{Z_{ji}}{Z_{ii}}\right)\mathbf{V}_f \tag{12–23}$$

7. *Calculate the currents in any desired transmission lines during the fault.* The current flowing through a transmission line between bus i and bus j can be calculated from the equation

$$\mathbf{I}_{ij} = (\mathbf{V}_i - \mathbf{V}_j)Y_{ij} \tag{12–24}$$

EXAMPLE 12–4

Assume that the power system shown in Figure 12–9 is operating a no-load conditions, and that it develops a symmetrical three-phase fault on bus 2. Calculate the per-unit subtransient fault current I_f'' at bus 2. Calculate the per-unit voltages at every bus in the power system during the subtransient period. Finally, calculate the per-unit current I_1 flowing in line 1 during the subtransient period of the fault.

■ Solution

We will follow the seven steps outlined to solve this problem, using MATLAB to help solve the problem by calculating the bus admittance matrix $\mathbf{Z}_{bus}$ from the bus admittance matrix $\mathbf{Y}_{bus}$.

1. *Create a per-phase, per-unit equivalent circuit for the power system.* This step was completed on page 573. The resulting per-phase, per-unit equivalent circuit is shown in Figure 12–10.

2. *Calculate the bus admittance matrix* $\mathbf{Y}_{bus}$. *The bus admittance matrix was calculated earlier. It is*

$$\mathbf{Y}_{bus} = \begin{bmatrix} -j16.212 & j5.0 & 0 & j6.667 \\ j5.0 & -j12.5 & j5.0 & j2.5 \\ 0 & j5.0 & -j13.333 & j5.0 \\ j6.667 & j2.5 & j5.0 & -j14.167 \end{bmatrix} \quad (12\text{–}14)$$

   The MATLAB statement required to initialize $\mathbf{Y}_{bus}$ to the values in Equation (12–14) is

```
% Create Y-bus
Ybus = ...
[ -j*16.212    j*5.000     0            j*6.667;  ...
   j*5.000    -j*12.500    j*5.000      j*2.500;  ...
   0           j*5.000    -j*13.333     j*5.000;  ...
   j*6.667     j*2.500     j*5.000     -j*14.167];
```

3. *Calculate the bus impedance matrix* $\mathbf{Z}_{bus}$. The bus impedance matrix $\mathbf{Z}_{bus}$ is the inverse of the bus admittance matrix $\mathbf{Y}_{bus}$. You can use MATLAB to perform this calculation.

```
» Zbus = inv(Ybus)
Zbus =
    0 + 0.1515i    0 + 0.1232i    0 + 0.0934i    0 + 0.1260i
    0 + 0.1232i    0 + 0.2104i    0 + 0.1321i    0 + 0.1417i
    0 + 0.0934i    0 + 0.1321i    0 + 0.1726i    0 + 0.1282i
    0 + 0.1260i    0 + 0.1417i    0 + 0.1282i    0 + 0.2001i
```

4. *Assume that the power system is at no-load conditions and determine the voltage at every bus.* For this power system, the no-load voltage at every bus, which will be equal to the pre-fault voltage at the bus, will be $\mathbf{V}_f = 1.00\angle 0°$ pu.

5. *Calculate the current at the faulted bus.* The current at the faulted bus is

$$I_{f,2}'' = \frac{\mathbf{V}_f}{Z_{22}} = \frac{1.00\angle 0°}{j0.2104} = 4.753\angle -90° \text{ pu} \quad (12\text{–}22)$$

6. *Calculate the voltages at each bus during the fault.* The voltage at bus j during a symmetrical three-phase fault at bus i is given by the equation

$$V_j = \left(1 - \frac{Z_{ji}}{Z_{ii}}\right)V_f \tag{12-23}$$

$$V_1 = \left(1 - \frac{Z_{12}}{Z_{22}}\right)V_f = \left(1 - \frac{j0.1232}{j0.2104}\right)(1.00\angle 0°) = 0.414\angle 0° \text{ pu}$$

$$V_2 = 0.0\angle 0° \text{ pu}$$

$$V_3 = \left(1 - \frac{Z_{32}}{Z_{22}}\right)V_f = \left(1 - \frac{j0.1321}{j0.2104}\right)(1.00\angle 0°) = 0.372\angle 0° \text{ pu}$$

$$V_4 = \left(1 - \frac{Z_{42}}{Z_{22}}\right)V_f = \left(1 - \frac{j0.1417}{j0.2104}\right)(1.00\angle 0°) = 0.327\angle 0° \text{ pu}$$

7. Calculate the currents in any desired transmission lines during the fault. The current flowing through transmission line 1 is given by

$$I_{12} = (V_1 - V_2)Y_{12} = (0.414\angle 0° - 0.0\angle 0°)(j5.0) = 2.07\angle 90° \text{ pu} \quad \blacksquare$$

As you can see, it is relatively easy to calculate the fault currents and the voltage on faulted power lines for symmetrical three-phase faults. The only really difficult part of the calculation is inverting the bus admittance matrix $\mathbf{Y}_{bus}$ to calculate the bus impedance matrix $\mathbf{Z}_{bus}$. This calculation is easy for small power systems when we have MATLAB to perform the inversion, but it becomes difficult and time-consuming when working with a large power system containing 1000 or more busses. For such large power systems, there are alternative techniques to build $\mathbf{Z}_{bus}$ directly instead of building $\mathbf{Y}_{bus}$ and inverting it. These alternative techniques are not discussed in this survey text—they are described in Reference 3.

12.4 | A SIMPLE FAULT CURRENT ANALYSIS PROGRAM

The software accompanying this book includes a simple MATLAB fault current analysis program, called `faults.m`. This program calculates the fault current that flows when a fault develops at a bus, as well as the voltages at every bus and the current flows in every transmission line during the fault. It works for subtransient, transient, or steady-state faults.

The source code for this program is available at this book's website. It can be used to find the voltages and fault currents in a power system, where the voltages, currents, impedances, and admittances are specified in per-unit.

The input data for program `faults` is placed in an input file, which can be created by using any available editor. The file may contain up to six types of lines:

- A **SYSTEM** line to define the system name and base MVA.
- Two or more **BUS** lines to define the bus names and pre-fault voltage.
- One or more **LINE** lines to define the transmission lines connecting the various busses together.
- One or more **GENERATOR** lines to define the impedances of generators in the system

Figure 12–14 | A sample input file for the MATLAB program `faults`. This input file solves the problem in Example 12–4.

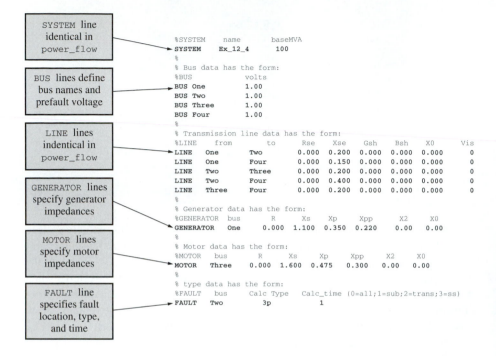

- Zero or more **MOTOR** lines to define the impedances of motors in the system
- A **FAULT** line to specify the location, type, and time period (subtransient, etc.) of the fault

A typical input file is shown in Figure 12–14. This file contains all six types of lines, plus comment lines that begin with a % symbol in column 1.

The **SYSTEM** card specifies the name of the power system case being studied, and the base apparent power of the power system (in MVA). The format of a **SYSTEM** line is shown below (note that the data are free format—the values do not have to appear in any particular columns.), and the values in the line are defined in Table 12–2.

```
%SYSTEM     name         baseMVA
SYSTEM   Test_System       100
```

Table 12–2 | Fields defined on the **SYSTEM** line

Field	Description
SYSTEM	Line identifier
Name	System study name (spaces are not allowed within the name)
base MVA	Base apparent power of system

The format of a **BUS** line is shown below, and the values in the line are defined in Table 12–3.

```
%BUS name      volts
BUS One        1.00
```

Table 12–3 I Fields defined on the **BUS** line

Field	Description
BUS	Line identifier
Name	Bus name (spaces are not allowed within the name)
Volts	Prefault bus voltage

The format of a **LINE** line is shown below, and the values in the line are defined in Table 12–4.

```
%LINE      from      to      Rse     Xse     Gsh     Bsh      X0 Vis
LINE       One       Two     0.100   0.400   0.000   0.000   0.000   0
```

Table 12–4 I Fields defined on the **LINE** line

Field	Description
LINE	Line identifier
From Bus Name	Bus where transmission line starts.
To Bus Name	Bus where transmission line ends.
Rse	Line series resistance, in per unit
Xse	Line series reactance, in per unit
Gsh	Line shunt conductance, in per unit
Bsh	Line shunt susceptance, in per unit
X0	Zero-sequence series impedance. (Unused for three-phase faults—set to 0 for now. This value will be defined in the next chapter.)
Vis	Zero sequence visibility flag (Unused for three-phase faults—set to 0 for now. This value will be defined in the next chapter.)

The formats of **GENERATOR** and **MOTOR** lines are shown below, and the values in the lines are defined in Table 12–5.

```
%GENERATOR     bus       R       Xs      Xp      Xpp     X2      X0
GENERATOR      One     0.000   1.100   0.350   0.220   0.000   0.000
%
%MOTOR    bus       R       Xs      Xp      Xpp     X2      X0
MOTOR     Three   0.000   1.600   0.475   0.300   0.000   0.000
```

Table 12–5 | Fields defined on the **GENERATOR** and **MOTOR** lines

Field	Description
GENERATOR	Generator line identifier
MOTOR	Motor line identifier
bus	Bus that machine is connected to
R	Machine resistance (pu)
Xs	Machine synchronous reactance X_S (pu)
Xp	Machine transient reactance X' (pu)
Xpp	Machine subtransient reactance X'' (pu)
X2	Negative sequence reactance X_2 (pu) (Unused for three-phase faults—set to 0 for now. This value will be defined in the next chapter.)
X0	Zero sequence reactance X_0 (pu) (Unused for three-phase faults—set to 0 for now. This value will be defined in the next chapter.)

The format of a **FAULT** line is shown below, and the values in the lines are defined in Table 12–6.

```
%FAULT    bus       Calc Type    Calc_time (0=all;1=sub;2=trans;3=ss)
FAULT     Two          3P             1
```

Table 12–6 | Fields defined on the **FAULT** line

Field	Description
FAULT	Line identifier
bus	Bus where fault occurs
Calc type	Fault type:
	3P – Three phase symmetrical fault
	LG – Line to ground fault
	LL – Line to line fault
	LLG – Double line to ground fault
Calc time	Time at which to calculate fault currents:
	0– All
	1– Subtransient period
	2– Transient period
	3– Steady-state period

To solve a fault current problem using this program, you must first calculate the per-unit impedances and admittances of all components to a common system base, and then create a file containing those input parameters. Note that each "line" in the power system should include *all* of the impedances in the line, whether they come from the line itself or from transformers at the ends of the line. In general, the prefault voltages at all busses should be assumed to be a constant value (typically 1.0) to correspond to prefault no-load conditions.

When program faults is executed with the input file shown in Figure 12–14, the results are as shown in Figure 12–15. Note that the bus voltage and the fault currents match the values we calculated by hand in Example 12–4.

Figure 12–15 | Result of executing **faults** with the input file shown in Figure 12–14. Note that the bus voltages and fault currents match the values that we calculated in Example 12–4.

```
Symmetrical Three-Phase Fault at Bus Two
Calculating Subtransient Currents
|==================Bus Information=============|=====Line Information=====|
Bus           Volts / angle   Amps / angle  |   To  | Amps / angle  |
no.     Name    (pu)   (deg)   (pu)   (deg)  |  Bus  |  (pu)   (deg)  |
|==============================================================|
   1     One   0.415/   0.00  0.000/   0.00     Two   2.073/  90.00
                                                Four  0.587/  90.00
   2     Two   0.000/   0.00  4.752/ -90.00     One   2.073/ -90.00
                                                Three 1.862/ -90.00
                                                Four  0.817/ -90.00
   3    Three  0.372/   0.00  0.000/   0.00     Two   1.862/  90.00
                                                Four  0.229/  90.00
   4     Four  0.327/   0.00  0.000/   0.00     One   0.587/ -90.00
                                                Two   0.817/  90.00
                                                Three 0.229/ -90.00

|==============================================================|
```

Figure 12–16 | The Ozzie Outback Electric Power Company system.

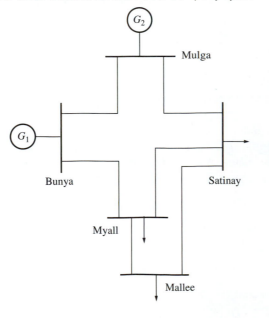

EXAMPLE 12–5

The Ozzie Outback Electric Power (OOEP) Company maintains the power system shown in Figure 12–16. This power system contains five busses, with generators attached to two of them and loads attached to the remaining ones. The power system has six transmission lines connecting the busses together, with the characteristics shown in Table 12–7. There are generators at busses Bunya and Mulga, and loads at all other busses. Note that this is the same power system that we studied in the last chapter.

Table 12–7 | Transmission lines in the OOEP system

From	To	R_{SE} (pu)	X_{SE} (pu)	Rating (MVA)
Bunya	Mulga	0.001	0.051	100
Mulga	Satinay	0.007	0.035	250
Bunya	Myall	0.007	0.035	200
Myall	Satinay	0.007	0.035	100
Myall	Mallee	0.011	0.051	60
Mallee	Satinay	0.011	0.051	60

You are an engineer in the company, and you have been asked to calculate the subtransient and transient fault currents and voltages to be expected when a symmetrical three-phase fault occurs at the Mallee bus. The per-unit impedances of the company's generators on a 100 MVA base are shown in Table 12–8. Determine the voltages and currents present during subtransient and transient periods of a symmetrical three-phase fault at Mallee.

Table 12–8 | Generators in the OOEP system

Name	Bus	R (pu)	X_S (pu)	X' (pu)	X'' (pu)
G1	Bunya	0.02	1.5	0.35	0.20
G2	Mulga	0.01	1.0	0.25	0.12

■ **Solution**

To solve this problem, we must create an input file, and execute program **faults**. The input file must contain the system, bus, line, generator, and fault data describing this system.

An appropriate input file for the subtransient current calculation in this power system is shown below. Note that the calculation time on the **FAULT** line is set to 0, meaning that the fault currents for all three time periods will be calculated.

```
% File describing a possible fault at bus Mallee on the
% Ozzie Outback Electric Power system.
%
% System data has the form:
%SYSTEM    name              baseMVA
SYSTEM  OOEP_Fault1            100
%
% Bus data has the form:
%BUS name       volts
BUS Bunya       1.00
BUS Mulga       1.00
BUS Mallee      1.00
BUS Myall       1.00
BUS Satinay     1.00
%
```

```
% Transmission line data has the form:
%LINE     from         to      Rse    Xse    Gsh    Bsh    X0   Vis
LINE     Bunya      Mulga     0.011  0.051  0.000  0.000  0.000   0
LINE     Mulga      Satinay   0.007  0.035  0.000  0.000  0.000   0
LINE     Bunya      Myall     0.007  0.035  0.000  0.000  0.000   0
LINE     Myall      Satinay   0.007  0.035  0.000  0.000  0.000   0
LINE     Myall      Mallee    0.011  0.051  0.000  0.000  0.000   0
LINE     Mallee     Satinay   0.011  0.051  0.000  0.000  0.000   0
%
% Generator data has the form:
%GENERATOR  bus       R      Xs     Xp     Xpp    X2     X0
GENERATOR   Bunya    0.02   1.5    0.35   0.20   0.00   0.00
GENERATOR   Mulga    0.01   1.0    0.25   0.12   0.00   0.00
%
% type data has the form:
%FAULT   bus       Calc Type    Calc_time (0=all;1=sub;2=trans;3=ss)
FAULT    Mallee       3P            0
```

When program **faults** is executed, the results are as shown in Figure 12–17. Note the dramatic difference between the subtransient, transient, and steady-state fault currents. ∎

Figure 12–17 | Result of executing faults for a symmetrical three-phase fault at bus Mallee in the OOEP power system.

```
                    Results for Case OOEP_Fault1

Symmetrical Three-Phase Fault at Bus Two
Calculating Subtransient Currents
|===================Bus Information===========|=====Line Information=====|
Bus            Volts / angle   Amps / angle  |   To  | Amps / angle |
no.     Name   (pu)   (deg)    (pu)   (deg)  | Bus   | (pu)   (deg)  |
|=============================================================================|
  1    Bunya   0.351/  -4.15   0.000/  0.00   Mulga   0.620/ -83.25
                                              Myall   3.583/  97.76
  2    Mulga   0.384/  -4.26   0.000/  0.00   Bunya   0.620/  96.75
                                              Satinay 4.513/  97.49
  3    Mallee  0.000/   0.00   8.366/ -82.38  Myall   4.099/ -82.36
                                              Satinay 4.267/ -82.40
  4    Myall   0.214/  -4.53   0.000/  0.00   Bunya   3.583/ -82.24
                                              Satinay 0.246/ -84.27
                                              Mallee  4.099/  97.64
  5    Satinay 0.223/  -4.58   0.000/  0.00   Mulga   4.513/ -82.51
                                              Myall   0.246/  95.73
                                              Mallee  4.267/  97.60
|=============================================================================|
```

(continued)

Figure 12–17 | *(concluded)*

```
                      Results for Case OOEP_Fault1

Symmetrical Three-Phase Fault at Bus Two
Calculating Transient Currents
|====================Bus Information============|=====Line Information=====|
Bus              Volts / angle    Amps / angle  |   To  |  Amps / angle  |
no.      Name     (pu)   (deg)    (pu)   (deg)   |  Bus  |  (pu)   (deg)  |
|==============================================================================|
  1      Bunya   0.224/  -6.92   0.000/   0.00    Mulga    0.272/ -87.27
                                                  Myall    2.490/  94.99
  2      Mulga   0.239/  -7.07   0.000/   0.00    Bunya    0.272/  92.73
                                                  Satinay  2.780/  94.67
  3      Mallee  0.000/   0.00   5.272/ -85.18    Myall    2.598/ -85.15
                                                  Satinay  2.672/ -85.21
  4      Myall   0.136/  -7.32   0.000/   0.00    Bunya    2.490/ -85.01
                                                  Satinay  0.108/ -88.29
                                                  Mallee   2.598/  94.85
  5      Satinay 0.139/  -7.38   0.000/   0.00    Mulga    2.780/ -85.33
                                                  Myall    0.108/  91.71
                                                  Mallee   2.672/  94.79

|==============================================================================|

                      Results for Case OOEP_Fault1

SSymmetrical Three-Phase Fault at Bus Mallee
Calculating Steady-State Currents
|====================Bus Information============|=====Line Information=====|
Bus              Volts / angle    Amps / angle  |   To  |  Amps / angle  |
no.      Name     (pu)   (deg)    (pu)   (deg)   |  Bus  |  (pu)   (deg)  |
|==============================================================================|
  1      Bunya   0.066/ -10.39   0.000/   0.00    Mulga    0.100/ -88.62
                                                  Myall    0.724/  91.47
  2      Mulga   0.071/ -10.42   0.000/   0.00    Bunya    0.100/  91.38
                                                  Satinay  0.830/  91.36
  3      Mallee  0.000/   0.00   1.554/ -88.59    Myall    0.763/ -88.58
                                                  Satinay  0.791/ -88.59
  4      Myall   0.040/ -10.75   0.000/   0.00    Bunya    0.724/ -88.53
                                                  Satinay  0.040/ -89.64
                                                  Mallee   0.763/  91.42
  5      Satinay 0.041/ -10.76   0.000/   0.00    Mulga    0.830/ -88.64
                                                  Myall    0.040/  90.36
                                                  Mallee   0.791/  91.41

|==============================================================================|
```

12.5 | COMMENTS ON FAULT CURRENT CALCULATIONS

The fault current calculation techniques that we have introduced in this chapter find the *AC symmetrical component* of the fault current that flows when a fault occurs. As we noted at the beginning of the chapter, there is also a transient DC component that is created at the moment of the fault in order for the current to be continuous at that time. The DC component is different in each phase, since it depends on the instantaneous value of current in each phase at the moment of the fault.

Immediately after the fault, both the subtransient AC currents and the transient DC currents are present in the power system. As a rule of thumb, the total rms current is taken to be about 1.5 times the size of the subtransient currents alone. All power equipment (transformers, circuit breakers, etc.) must be designed to withstand this extremely high current level for a brief period without damage. This is known as the *withstand* capability of a device.

Circuit breakers are designed to open when a fault is detected, clearing the fault and allowing the remaining part of the power system to continue operating. The current that a circuit breaker must interrupt will be less than the instantaneous withstand level, because circuit breakers do not act instantaneously. The longer the delay before a circuit breaker opens, the lower the fault current will become. This lower current level is the *interrupting* capability of a circuit breaker.

The techniques used to calculate the withstand and interrupting capabilities required for a circuit breaker are described in Chapter 10 of Reference 3.

12.6 | SHORT-CIRCUIT MVA

A power engineer working for an industrial plant or an industrial power system must be able to determine the fault currents at various points within the system in order to specify the proper sizes for circuit breakers within the plant. This can be a problem, since the fault current flow depends on the properties of the power system that supplies the plant, and the engineer will not want to model the entire power system just to determine the fault currents within his or her plant! What the engineer needs is the Thevenin equivalent of the power system at the point where the plant is attached.

Power systems typically supply this information, but they do it in a roundabout way by specifying the "short-circuit megavolt-amperes (MVA)" of the power system at the point of the connection. Short-circuit MVA is defined by the equation

$$\text{Short-circuit MVA} = \sqrt{3}V_{\text{LL, nominal}}I_{\text{SC}} \tag{12–25}$$

where $V_{\text{LL, nominal}}$ is the nominal line voltage of the power system and I_{SC} is the rms short-circuit current in a three-phase fault. Thus, specifying the short-circuit MVA at a connection is equivalent to specifying the short-circuit current at the connection as long as we know the bus voltage at the connection point.

If Equation (12–25) is expressed in per-unit, it becomes

$$\text{Short-circuit MVA}_{\text{pu}} = V_{\text{nominal, pu}}I_{\text{SC, pu}} \tag{12–26}$$

If the nominal voltage is the base voltage of the system at that point, then the voltage will be $1.0\angle 0°$ pu, so this equation reduces to

$$\text{Short-circuit MVA}_{\text{pu}} = I_{\text{SC, pu}} \tag{12–27}$$

The short-circuit MVA in per-unit is equal to the three-phase short-circuit current in per-unit! Therefore, the magnitude of the per-unit Thevenin impedance of the power system is

$$|Z_{TH}| = \frac{|\mathbf{V}|}{|\mathbf{I}_{SC}|} = \frac{1.0}{I_{SC}} \text{ pu} \qquad (12\text{--}28)$$

If the resistance and shunt capacitance of the power system are neglected, then $Z_{TH} \approx X_{TH}$. Thus, specifying the short-circuit MVA of the power system at a point is sufficient for the engineer to determine the Thevenin equivalent of the power system at that point, and therefore the fault currents at various points within the plant.

Note that the short-circuit current determined from this calculation can be either the subtransient or the transient current, depending on whether the short-circuit MVA applies to the subtransient or the transient period. Both quantities will usually be supplied by the power company, because both types of fault currents will need to be calculated. The subtransient currents are needed to specify the "withstand" capability of any circuit breakers in the system, while the transient currents are needed to specify the "interrupting" capability of the circuit breakers.

EXAMPLE 12–6

A power company supplies a nominal 2 MVA of apparent power at a power factor of 0.9 lagging to a 4160-V industrial power system. The company states that the subtransient short-circuit MVA of the power system at that point is 50 MVA. Answer the following questions about this power system.

a.   What is the nominal full-load current supplied to the industrial power plant?

b.   What is the symmetrical three-phase short-circuit current at the point where the power system connects to the industrial power plant.

■ **Solution**

If we select the base line-to-line voltage of the industrial power system to be 4160 V and the base apparent power to be 2 MVA, the base current will be

$$I_{L,\text{ base}} = \frac{S_{\text{base}}}{\sqrt{3}V_{L,\text{ base}}} \qquad (10\text{--}4)$$

$$= \frac{2 \text{ MVA}}{\sqrt{3}(4.16 \text{ kV})} = 278 \text{ A}$$

a.   At full load, the current supplied to the industrial power system is

$$I_L = \frac{S}{\sqrt{3}V_{LL}} = \frac{2 \text{ MVA}}{\sqrt{3}(4.16 \text{ kV})} = 278 \text{ A} \qquad (10\text{--}4)$$

b.   The per-unit short-circuit MVA is

$$\text{Short-circuit MVA}_{\text{pu}} = \frac{50 \text{ MVA}}{2 \text{ MVA}} = 25 \qquad (12\text{--}29)$$

Therefore the symmetrical three-phase short circuit current is $I_{SC} = 25$ pu, and the actual current is

$$I_{SC} = I_{SC, pu} I_{base} = (25)(278 \text{ A}) = 6950 \text{ A} \tag{12–30}$$

The per-unit Thevenin impedance of the power system can also be found easily. ∎

12.7 | SUMMARY

A *fault* in a circuit is any failure that interferes with the normal flow of current to the loads. In most faults, a current path forms between two or more phases, or between one or more phases and the neutral (ground). This current path has relatively low impedance, resulting in excessive current flows.

When a fault occurs, both a symmetrical AC fault current and an asymmetrical DC transient current are created. The level of the DC transient current depends on the instantaneous current in each phase at the instance of the fault. The AC symmetrical component of current starts out very high, and decays to a steady state level in two stages. The first period of very rapid decay is known as the subtransient period. It lasts only a few cycles. The subtransient currents are basically associated with the time constants of the amortisseur or damper windings of generators and motors. The second period of slower decay is known as the transient period. It lasts for 0.5 to 1.0 second. The transient currents are basically associated with time constants of the field windings of generators and motors.

The reactances of generators and motors in the subtransient period are known as subtransient reactances (X''). The reactances of generators and motors in the transient period are known as transient reactances (X'). The reactances of generators and motors in the steady-state period are the machines' synchronous reactances (X_S). Note that the resistances and reactances of transmission lines, transformers, and other static components do not change during the fault transients.

Fault currents can be easily calculated from the bus impedance matrix $\mathbf{Z}_{bus}$. A bus admittance matrix $\mathbf{Y}_{bus}$ is calculated separately for the subtransient, transient, and steady-state periods by using the appropriate reactances, and then inverted to get $\mathbf{Z}_{bus}$. Once $\mathbf{Z}_{bus}$ has been calculated, the fault current at a particular bus i can be found by dividing the prefault voltage at the bus by the driving-point impedance of the bus.

$$I_f = \frac{\mathbf{V}_f}{Z_{ii}}$$

If the subtransient bus impedance matrix is used, I_f will be the subtransient fault current. Similarly, if the transient bus impedance matrix is used, I_f will be the transient fault current.

If we assume that the power system was unloaded before the fault occurred, then we can also calculate the voltage at each bus by the equation

$$\mathbf{V}_j = \left(1 - \frac{Z_{ji}}{Z_{ii}}\right)\mathbf{V}_f$$

where bus i is the faulted bus.

We also introduced the MATLAB-based program **faults**, which can be used to solve for fault currents in simple power systems.

12.8 | QUESTIONS

12–1. What is a fault?

12–2. What is the difference between a transient fault and a permanent fault? What causes each type of fault?

12–3. What is the difference between a symmetrical and an unsymmetrical fault? What types of faults fall into each category?

12–4. Why is there a transient DC component of current right after a fault?

12–5. What causes the behavior observed during subtransient, transient, and steady-state periods after a fault occurs?

12–6. How is the bus impedance matrix $\mathbf{Z}_{bus}$ used to calculate the fault currents?

12–7. What is short-circuit MVA? How does it allow an engineer to calculate the Thevenin impedance of a power system at a bus?

12.9 | PROBLEMS

12–1. A 200-MVA, 20-kV, 60-Hz, three-phase synchronous generator is connected through a 200-MVA, 20/138-kV, Δ–Y transformer to a 138-kV transmission line. The generator reactances to the machine's own base are

$$X_S = 1.40 \qquad X' = 0.30 \qquad X'' = 0.15$$

The initial transient DC component in this machine averages 50 percent of the initial symmetrical AC component. The transformer's series reactance is 0.10 pu, and the resistance of both the generator and the transformer may be ignored. Assume that a symmetrical three-phase fault occurs on the 138-kV transmission line near the point where it is connected to the transformer.

a. What is the AC component of current in this generator the instant after the fault occurs?

b. What is the total current (AC plus DC) flowing in the generator right after the fault occurs?

c. What is the transient fault current I_f' in this fault?

d. What is the steady-state fault current I_f in this fault?

Figure P12–1 I The simple power system of Problem 12–2.

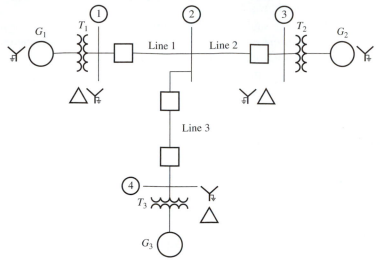

12–2. A simple three-phase power system is shown in Figure P12–1. Assume that the ratings of the various devices in this system are as follows:

Generator G_1:   250 MVA, 13.8 kV, $R = 0.1$ pu, $X'' = 0.18$ pu, $X' = 0.40$ pu
Generator G_2:   500 MVA, 20.0 kV, $R = 0.1$ pu, $X'' = 0.15$ pu, $X' = 0.35$ pu
Generator G_3:   250 MVA, 13.8 kV, $R = 0.15$ pu, $X'' = 0.20$ pu, $X' = 0.40$ pu
Transformer T_1:   250 MVA, 13.8-Δ/240-Y kV, $R = 0.01$ pu, $X = 0.10$ pu
Transformer T_2:   500 MVA, 20.0-Δ/240-Y kV, $R = 0.01$ pu, $X = 0.08$ pu
Transformer T_3:   250 MVA, 13.8-Δ/240-Y kV, $R = 0.01$ pu, $X = 0.10$ pu
Each line: $R = 20\ \Omega$, $X = 100\ \Omega$

Assume that the power system is initially unloaded, that the voltage at bus 4 is 250 kV, and that *all resistances may be neglected*.

a. Convert this power system to per unit on a base of 500 MVA at 20 kV at generator G_2.

b. Calculate $\mathbf{Y}_{bus}$ and $\mathbf{Z}_{bus}$ for this power system using the generator subtransient reactances.

c. Suppose that a three-phase symmetrical fault occurs at bus 4. What is the subtransient fault current? What is the voltage on each bus in the power system? What is the subtransient current flowing in each of the three transmission lines?

d. Which circuit breaker in the power system sees the highest instantaneous current when a fault occurs at bus 4?

e. What is the transient fault current when a fault occurs at bus 4? What is the voltage on each bus in the power system? What is the transient current flowing in each of the three transmission lines?

e. What is the steady-state fault current when a fault occurs at bus 4? What is the voltage on each bus in the power system? What is the steady-state current flowing in each of the three transmission lines?

   f.   Determine the subtransient short-circuit MVA of this power system at bus 4.

12–3. Calculate the subtransient fault current at bus 4 *if the power system resistances are not neglected.* How much difference does including the resistances make to the amount of fault current flowing?

12–4. Suppose that a three-bus power system has the subtransient bus impedance matrix given below, and that the power system is initially unloaded with a prefault bus voltage of 0.98 pu. Assume that a symmetrical three-phase fault occurs at bus 1, and find:

   a.   The current flowing in the transmission line from bus 2 to bus 1 during the subtransient period.

   b.   The current flowing in the transmission line from bus 3 to bus 1 during the subtransient period.

$$\mathbf{Z}_{bus} = \begin{bmatrix} j0.20 & j0.10 & j0.15 \\ j0.10 & j0.50 & j0.30 \\ j0.15 & j0.30 & j0.80 \end{bmatrix} \quad \text{per unit}$$

12–5. A 500-kVA, 480-V, 60-Hz, Y-connected synchronous generator has a subtransient reactance X'' of 0.10 pu, a transient reactance X' of 0.22 pu, and a synchronous reactance X_S of 1.0 pu. Suppose that this machine is supplying rated load at rated voltage and 0.9 PF lagging.

   a.   What is the magnitude of the internal generated voltage E_A of this machine?

   b.   A fault occurs at the terminals of the generator. What is the magnitude of the voltage behind subtransient reactance E''_A of this machine?

   c.   What is the subtransient current I_f'' flowing from this generator?

   d.   What is the magnitude of the voltage behind transient reactance E'_A of this machine?

   e.   What is the transient current I_f' flowing from this generator during the transient period?

   f.   Suppose that we ignore the loads on the generator and assume that it was unloaded before the fault occurred. What is the subtransient current I_f'' flowing from this generator? How much error is caused by ignoring the initial loading of the generator?

12–6. A synchronous generator is connected through a transformer and a transmission line to a synchronous motor. The rated voltage and apparent power of the generator serve as the base quantities for this power system. On the system base, the per-unit subtransient reactances of the generator and the motor are 0.15 and 0.40 respectively, the series reactance of the transformer is 0.10, and the series reactance of the transmission line is 0.20.

Figure P12–2 | One-line diagram of the power system in Problem 12–7.

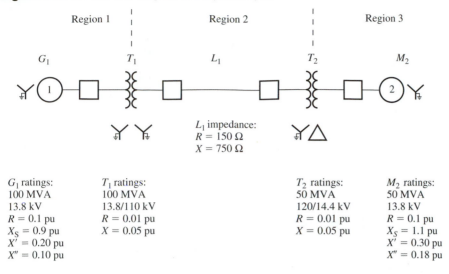

G_1 ratings:	T_1 ratings:	T_2 ratings:	M_2 ratings:
100 MVA	100 MVA	50 MVA	50 MVA
13.8 kV	13.8/110 kV	120/14.4 kV	13.8 kV
$R = 0.1$ pu	$R = 0.01$ pu	$R = 0.01$ pu	$R = 0.1$ pu
$X_S = 0.9$ pu	$X = 0.05$ pu	$X = 0.05$ pu	$X_S = 1.1$ pu
$X' = 0.20$ pu			$X' = 0.30$ pu
$X'' = 0.10$ pu			$X'' = 0.18$ pu

All resistances may be ignored. The generator is supplying rated apparent power and voltage at a power factor of 0.85 lagging, when a three-phase synchronous fault occurs at the terminals of the motor. Find the per-unit subtransient current at the fault, in the generator, and in the motor.

12–7. Assume that a symmetrical three-phase fault occurs on the high-voltage side of transformer T_2 in the power system shown in Figure P12–2. Make the assumption that the generator is operating at rated voltage, and that the power system is initially unloaded.

    a.   Calculate the subtransient, transient, and steady-state fault current, generator current, and motor current *ignoring resistances.*

    b.   Calculate the subtransient, transient, and steady-state fault current, generator current, and motor current *including resistances.*

    c.   How much effect does the inclusion of resistances have on the fault current calculations?

12–8. Assume that a symmetrical three-phase fault occurs at the terminals of motor M_1 on the low-voltage side of transformer T_1 in the power system shown in Figure P12–3. Make the assumption that the power system is operating at rated voltage, and that it is initially unloaded. Calculate the subtransient, transient, and steady-state fault current on the high-voltage side of the transformer, the low-voltage side of the transformer, and in the motor.

Figure P12–3 | One-line diagram of the power system in Problem 12–8.

Power System:
$V = 138\text{kV}$
Short-circuit MVA = 500 MVA

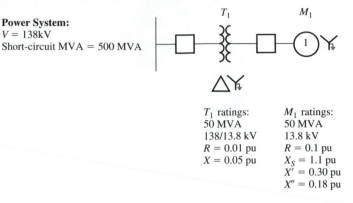

T_1 ratings:
50 MVA
138/13.8 kV
$R = 0.01$ pu
$X = 0.05$ pu

M_1 ratings:
50 MVA
13.8 kV
$R = 0.1$ pu
$X_S = 1.1$ pu
$X' = 0.30$ pu
$X'' = 0.18$ pu

12.10 | REFERENCES

1. Elgerd, Olle I.: *Electric Energy Systems Theory: An Introduction*, McGraw-Hill, New York, 1971.

2. Glover, J. Duncan, and Mulukutla Sarma: *Power Systems Analysis and Design*, PWS Publishers, Boston, 1987.

3. Grainger, John J., and William D. Stevenson, Jr.: *Power Systems Analysis*, McGraw-Hill, New York, 1994.

4. Gross, Charles A.: *Power Systems Analysis*, John Wiley and Sons, New York, 1979.

5. Stagg, Glen W., and Ahmed H. El-Abiad: *Computer Methods in Power Systems Analysis*, McGraw-Hill, New York, 1968.

CHAPTER 13

Unsymmetrical Faults

Most faults that occur on power systems are unsymmetrical faults, meaning that the magnitudes of the voltages and current flows in the three phases of the power system will differ. The most common example of an unsymmetrical fault is the single line-to-ground fault, in which there are high currents in one phase and the ground (or neutral) only. Other examples include line-to-line faults and double line-to-ground faults. In all three cases, the magnitudes of the voltages and the currents differ for different phases of the power system.

This is a serious problem, because the basic analysis tool that we have used to study power systems is the per-phase equivalent circuit. As we saw in Chapter 2, the per-phase equivalent circuit *assumes that the magnitudes of voltages and currents are identical in each phase*, and only works if that assumption is true. If it is not true, analysis techniques such as the simple Y–Δ transform either fail or become much more complicated.

In this chapter, we will learn a technique that allows us to extend per-phase equivalent circuits to analyze unbalanced power systems. This technique is known as the *method of symmetrical components*.

13.1 | SYMMETRICAL COMPONENTS

In 1918, C. L. Fortescue[1] invented a clever way to treat an unbalanced three-phase power system as though it were three balanced power systems. He proved that any unsymmetrical set of three-phase voltages or currents could be broken down into *three symmetrical sets* of balanced three-phase components. The three sets of components are:

1. A set of *positive-sequence* components consisting of three phasors equal in magnitude, displaced from each other by 120°, and having the *same* phase sequence as the original power system.

[1]C. L. Fortescue, "Method of Symmetrical Coordinates Applied to the Solution of Polyphase Networks," *Transactions of the AIEE*, vol. 37, pp. 1027–1140, 1918.

2. A set of *negative-sequence* components consisting of three phasors equal in magnitude, displaced from each other by 120°, and having the *opposite* phase sequence from the original power system.

3. A set of *zero-sequence* components consisting of three phasors equal in magnitude and phase.

When working with power systems, it is traditional to designate the three phases of the power system as *a*, *b*, and *c*, with phase *a* being the reference phase, phase *b* being 120° behind phase *a*, and phase *c* being 120° behind phase *b*. Such power systems are said to have phase sequence *abc*, because the three phases peak in that order (see Figure 13–1). In such a power system, the positive-sequence components will have phase sequence *abc*, while the negative sequence components will have phase sequence *acb*. The zero-sequence components do not have a phase sequence, since all three phases peak at the same time.

Figure 13–1 | The voltages in each phase of a power system with an *abc* phase sequence.

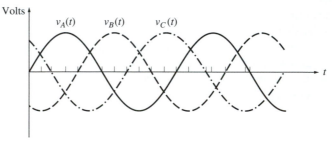

Traditionally, the positive-sequence components of the voltages in a power system are designated with the subscript 1, the negative-sequence components are designated with the subscript 2, and the zero-sequence components are designated with the subscript 0 (see Figure 13–2). If the voltages in the phases of the original power system are designated $\mathbf{V}_A$, $\mathbf{V}_B$, and $\mathbf{V}_C$, then the positive-sequence components of the voltages will be $\mathbf{V}_{A1}$, $\mathbf{V}_{B1}$, and $\mathbf{V}_{C1}$; the negative-sequence components of the voltages will be $\mathbf{V}_{A2}$, $\mathbf{V}_{B2}$, and $\mathbf{V}_{C2}$; and the zero-sequence components of the voltages will be $\mathbf{V}_{A0}$, $\mathbf{V}_{B0}$, and $\mathbf{V}_{C0}$. The total voltage in each phase will be the sum of the three components for that phase.

$$\mathbf{V}_A = \mathbf{V}_{A1} + \mathbf{V}_{A2} + \mathbf{V}_{A0}$$
$$\mathbf{V}_B = \mathbf{V}_{B1} + \mathbf{V}_{B2} + \mathbf{V}_{B0}$$
$$\mathbf{V}_C = \mathbf{V}_{C1} + \mathbf{V}_{C2} + \mathbf{V}_{C0} \tag{13–1}$$

The *a* Constant

In working with the symmetrical components of an unbalanced three-phase power system, the operation of shifting a phasor through an angle of 120° occurs over and over again. This operation is equivalent to multiplying the phasor by the quantity

Figure 13–2 | (a) Positive-sequence components have an *abc* phase sequence. (b) Negative-sequence components have an *acb* phase sequence. (c) Zero-sequence components are in phase with each other.

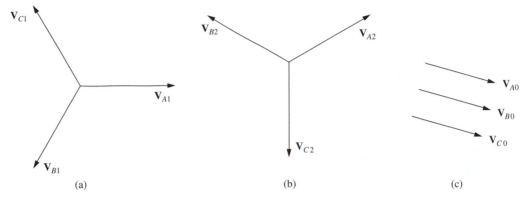

(a)   (b)   (c)

$1\angle 120°$. Because multiplication by this quantity occurs so often, we will introduce the constant *a* to represent it.

$$a = 1\angle 120°$$

Each multiplication by *a* rotates a phasor by 120° without changing its magnitude. Therefore,

$$a = 1\angle 120° \tag{13–2}$$
$$a^2 = 1\angle 240° \tag{13–3}$$
$$a^3 = 1\angle 360° = 1\angle 0° = 1 \tag{13–4}$$

Figure 13–3 shows the relationships among the positive and negative powers of *a*.

Figure 13–3 | Phasor diagram showing the relationships among the various powers of *a*.

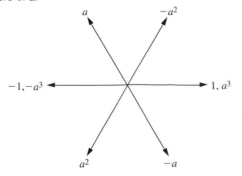

Representing Symmetrical Components with the *a* Constant

We can now represent the symmetrical components of a three phase circuit in terms of the *a* constant. Recall that the positive-sequence components have phase sequence

abc, meaning that the phases will peak in order a, b, c. Therefore the relationships among the positive-sequence components will be

$$\begin{aligned}\mathbf{V}_{B1} &= a^2\mathbf{V}_{A1}\\ \mathbf{V}_{C1} &= a\mathbf{V}_{A1}\end{aligned} \tag{13–5}$$

The negative-sequence components have phase sequence acb, meaning that the phases will peak in order a, c, b. Therefore the relationships among the negative-sequence components will be

$$\begin{aligned}\mathbf{V}_{B2} &= a\mathbf{V}_{A2}\\ \mathbf{V}_{C2} &= a^2\mathbf{V}_{A2}\end{aligned} \tag{13–6}$$

The zero-sequence components are the same in all phases. Therefore the relationships among the zero-sequence components will be

$$\begin{aligned}\mathbf{V}_{B0} &= \mathbf{V}_{A0}\\ \mathbf{V}_{C0} &= \mathbf{V}_{A0}\end{aligned} \tag{13–7}$$

Finding the Symmetrical Components of an Unbalanced Three-Phase Voltage

It is now possible to determine the symmetrical components of an unbalanced three-phase voltage. The unbalanced set of three-phase voltages is given in terms of symmetrical components by Equations (13–1), and the relationships between the phases of the positive-, negative-, and zero-sequence components is given by Equations (13–5) to (13–7). Substituting Equations (13–5) to (13–7) into Equations (13–1) yields

$$\begin{aligned}\mathbf{V}_A &= \mathbf{V}_{A1} &+ \mathbf{V}_{A2} &+ \mathbf{V}_{A0} \tag{13–8}\\ \mathbf{V}_B &= a^2\mathbf{V}_{A1} &+ a\mathbf{V}_{A2} &+ \mathbf{V}_{A0} \tag{13–9}\\ \mathbf{V}_C &= a\mathbf{V}_{A1} &+ a^2\mathbf{V}_{A2} &+ \mathbf{V}_{A0} \tag{13–10}\end{aligned}$$

In matrix form, these equations become

$$\begin{bmatrix}\mathbf{V}_A\\ \mathbf{V}_B\\ \mathbf{V}_C\end{bmatrix} = \begin{bmatrix}1 & 1 & 1\\ 1 & a^2 & a\\ 1 & a & a^2\end{bmatrix}\begin{bmatrix}\mathbf{V}_{A0}\\ \mathbf{V}_{A1}\\ \mathbf{V}_{A2}\end{bmatrix} \tag{13–11}$$

If we make the definition

$$A = \begin{bmatrix}1 & 1 & 1\\ 1 & a^2 & a\\ 1 & a & a^2\end{bmatrix} \tag{13–12}$$

we can write Equation (13–11) as

$$\begin{bmatrix}\mathbf{V}_A\\ \mathbf{V}_B\\ \mathbf{V}_C\end{bmatrix} = A\begin{bmatrix}\mathbf{V}_{A0}\\ \mathbf{V}_{A1}\\ \mathbf{V}_{A2}\end{bmatrix} \tag{13–13}$$

Therefore, the symmetrical components of the unbalanced three-phase voltage can be expressed as

$$\begin{bmatrix} \mathbf{V}_{A0} \\ \mathbf{V}_{A1} \\ \mathbf{V}_{A2} \end{bmatrix} = A^{-1} \begin{bmatrix} \mathbf{V}_A \\ \mathbf{V}_B \\ \mathbf{V}_C \end{bmatrix} \tag{13–14}$$

Since

$$A^{-1} = \begin{bmatrix} 1 & 1 & 1 \\ 1 & a^2 & a \\ 1 & a & a^2 \end{bmatrix}^{-1} = \frac{1}{3} \begin{bmatrix} 1 & 1 & 1 \\ 1 & a & a^2 \\ 1 & a^2 & a \end{bmatrix} \tag{13–15}$$

Equation (13–14) becomes

$$\begin{bmatrix} \mathbf{V}_{A0} \\ \mathbf{V}_{A1} \\ \mathbf{V}_{A2} \end{bmatrix} = \frac{1}{3} \begin{bmatrix} 1 & 1 & 1 \\ 1 & a & a^2 \\ 1 & a^2 & a \end{bmatrix} \begin{bmatrix} \mathbf{V}_A \\ \mathbf{V}_B \\ \mathbf{V}_C \end{bmatrix} \tag{13–16}$$

Equation (13–16) shows us how to calculate the symmetrical components of phase a of an unbalanced three-phase voltage. These components are

$$\mathbf{V}_{A0} = \tfrac{1}{3}(\mathbf{V}_A + \mathbf{V}_B + \mathbf{V}_C) \qquad \text{zero sequence} \tag{13–17}$$

$$\mathbf{V}_{A1} = \tfrac{1}{3}(\mathbf{V}_A + a\mathbf{V}_B + a^2\mathbf{V}_C) \qquad \text{positive sequence} \tag{13–18}$$

$$\mathbf{V}_{A2} = \tfrac{1}{3}(\mathbf{V}_A + a^2\mathbf{V}_B + a\mathbf{V}_C) \qquad \text{negative sequence} \tag{13–19}$$

The symmetrical components of the other two phases can be calculated from Equations (13–5) to (13–7).

Finding the Symmetrical Components of an Unbalanced Three-Phase Current

The relationships among unbalanced three-phase currents have the same forms as the relationships among unbalanced three-phase voltages. The currents in each phase can be represented in terms of symmetrical components as

$$\mathbf{I}_A = \mathbf{I}_{A1} + \mathbf{I}_{A2} + \mathbf{I}_{A0} \tag{13–20}$$

$$\mathbf{I}_B = \mathbf{I}_{B1} + \mathbf{I}_{B2} + \mathbf{I}_{B0} \tag{13–21}$$

$$\mathbf{I}_C = \mathbf{I}_{C1} + \mathbf{I}_{C2} + \mathbf{I}_{C0} \tag{13–22}$$

Substituting the phase relationships among the phases for the positive-, negative-, and zero-sequence components yields

$$\mathbf{I}_A = \mathbf{I}_{A1} + \mathbf{I}_{A2} + \mathbf{I}_{A0} \tag{13–23}$$

$$\mathbf{I}_B = a^2\mathbf{I}_{A1} + a\mathbf{I}_{A2} + \mathbf{I}_{A0} \tag{13–24}$$

$$\mathbf{I}_C = a\mathbf{I}_{A1} + a^2\mathbf{I}_{A2} + \mathbf{I}_{A0} \tag{13–25}$$

and the symmetrical components can be represented in terms of the currents in each phase as

$$\mathbf{I}_{A0} = \tfrac{1}{3}(\mathbf{I}_A + \mathbf{I}_B + \mathbf{I}_C) \qquad \text{zero sequence} \tag{13–26}$$

$$\mathbf{I}_{A1} = \tfrac{1}{3}(\mathbf{I}_A + a\mathbf{I}_B + a^2\mathbf{I}_C) \qquad \text{positive sequence} \tag{13–27}$$

$$\mathbf{I}_{A2} = \tfrac{1}{3}(\mathbf{I}_A + a^2\mathbf{I}_B + a\mathbf{I}_C) \qquad \text{negative sequence} \tag{13–28}$$

Equation (13–26) provides an important insight into the operation of unsymmetrical power systems. Recall that current in the neutral of a power system is given by

$$\mathbf{I}_N = \mathbf{I}_A + \mathbf{I}_B + \mathbf{I}_C \qquad (2\text{–}5)$$

Comparing Equation (13–26) to Equation (2–5) shows that

$$\mathbf{I}_N = 3\mathbf{I}_{A0} \qquad (13\text{–}29)$$

The current in the neutral (or ground) of the power system is equal to 3 times the zero-sequence current. Therefore, *if a component has no neutral current, then there can be no zero-sequence currents in the component.* Thus Δ-connected and ungrounded Y-connected components cannot have zero-sequence components of currents.

EXAMPLE 13–1

An ungrounded Y-connected three-phase load is shown in Figure 13–4. The currents in each line leading to the load are

$$\mathbf{I}_A = 10\angle 0° \text{ A} \qquad \mathbf{I}_B = 10\angle -180° \text{ A} \qquad \mathbf{I}_C = 0\angle 0° \text{ A}$$

Determine the symmetrical components of current in each phase in this load.

Figure 13–4 | An ungrounded Y-connected three-phase load with unbalanced
currents.

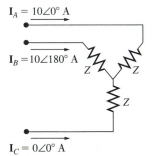

$$\mathbf{I}_A = 10\angle 0° \text{ A}$$

$$\mathbf{I}_B = 10\angle 180° \text{ A}$$

$$\mathbf{I}_C = 0\angle 0° \text{ A}$$

Solution
The zero-sequence current for phase *a* of this load is given by Equation (13–26)

$$\mathbf{I}_{A0} = \tfrac{1}{3}(\mathbf{I}_A + \mathbf{I}_B + \mathbf{I}_C) \qquad \text{zero sequence} \qquad (13\text{–}26)$$

$$= \tfrac{1}{3}(10\angle 0° + 10\angle 180° + 0)$$

$$= 0 \text{ A}$$

The positive-sequence current for phase *a* of this load is given by Equation (13–27)

$$I_{A1} = \tfrac{1}{3}(I_A + aI_B + a^2I_C) \qquad \text{positive sequence} \qquad (13\text{-}27)$$
$$= \tfrac{1}{3}[10\angle 0° + a(10\angle 180°) + a^2(0)]$$
$$= \tfrac{1}{3}[10\angle 0° \; 10\angle -60° + 0]$$
$$= 5.77\angle -30° \text{ A}$$

The negative-sequence current for phase a of this load is given by Equation (13-28)

$$I_{A2} = \tfrac{1}{3}(I_A + a^2I_B + aI_C) \qquad \text{negative sequence} \qquad (13\text{-}28)$$
$$= \tfrac{1}{3}[10\angle 0° + a^2(10\angle 180°) + a(0)]$$
$$= \tfrac{1}{3}[10\angle 0° + 10\angle 60° + 0]$$
$$= 5.77\angle 30° \text{ A}$$

The zero-, positive-, and negative-sequence currents for phase b are

$$I_{B0} = I_{A0} = 0 \text{ A}$$
$$I_{B1} = a^2I_{A1} = a^2(5.77\angle -30° \text{ A}) = 5.77\angle -150° \text{ A}$$
$$I_{B2} = aI_{A2} = a(5.77\angle 30° \text{ A}) = 5.77\angle 150° \text{ A}$$

The zero-, positive-, and negative-sequence currents for phase c are

$$I_{C0} = I_{A0} = 0 \text{ A}$$
$$I_{C1} = aI_{A1} = a(5.77\angle -30° \text{ A}) = 5.77\angle 90° \text{ A}$$
$$I_{C2} = a^2I_{A2} = a^2(5.77\angle 30° \text{ A}) = 5.77\angle -90° \text{ A}$$

Note that the zero-sequence components of current in the load are zero, which makes sense, since an ungrounded Y-connected load cannot have a current flowing in the neutral. ∎

13.2 | CALCULATING REAL AND REACTIVE POWER FROM SYMMETRICAL COMPONENTS

If the symmetrical components of voltage and current are known at a point in a power system, then it is possible to calculate the power flow at that point directly from the symmetrical components. Recall that the power in a single phase is given by

$$\mathbf{S} = \mathbf{VI}^* \qquad (2\text{-}28)$$

where $\mathbf{S}$ is the complex power, $\mathbf{V}$ is the voltage between the phase and neutral, and $\mathbf{I}$ is the current flowing in the phase. Then the power in the unsymmetrical three-phase circuit is given by

$$\boxed{\mathbf{S} = P + jQ = \mathbf{V}_A\mathbf{I}_A^* + \mathbf{V}_B\mathbf{I}_B^* + \mathbf{V}_C\mathbf{I}_C^*} \qquad (13\text{-}30)$$

where $\mathbf{V}_A$, $\mathbf{V}_B$, and $\mathbf{V}_C$ are line-to-neutral voltages in each phase, and $\mathbf{I}_A$, $\mathbf{I}_B$, and $\mathbf{I}_C$ are currents in each phase.

Equation (13–30) can be expressed in matrix notation as

$$\mathbf{S} = [\mathbf{V}_A \quad \mathbf{V}_B \quad \mathbf{V}_C] \begin{bmatrix} \mathbf{I}_A \\ \mathbf{I}_B \\ \mathbf{I}_C \end{bmatrix}^* = \begin{bmatrix} \mathbf{V}_A \\ \mathbf{V}_B \\ \mathbf{V}_C \end{bmatrix}^T \begin{bmatrix} \mathbf{I}_A \\ \mathbf{I}_B \\ \mathbf{I}_C \end{bmatrix}^* \tag{13–31}$$

Both the phase voltages and currents can be expressed in terms of symmetrical components as shown below:

$$\begin{bmatrix} \mathbf{V}_A \\ \mathbf{V}_B \\ \mathbf{V}_C \end{bmatrix} = \begin{bmatrix} 1 & 1 & 1 \\ 1 & a^2 & a \\ 1 & a & a^2 \end{bmatrix} \begin{bmatrix} \mathbf{V}_{A0} \\ \mathbf{V}_{A1} \\ \mathbf{V}_{A2} \end{bmatrix} \tag{13–11}$$

$$\begin{bmatrix} \mathbf{I}_A \\ \mathbf{I}_B \\ \mathbf{I}_C \end{bmatrix} = \begin{bmatrix} 1 & 1 & 1 \\ 1 & a^2 & a \\ 1 & a & a^2 \end{bmatrix} \begin{bmatrix} \mathbf{I}_{A0} \\ \mathbf{I}_{A1} \\ \mathbf{I}_{A2} \end{bmatrix}$$

Therefore, Equation (13–31) becomes

$$\mathbf{S} = \left(\begin{bmatrix} 1 & 1 & 1 \\ 1 & a^2 & a \\ 1 & a & a^2 \end{bmatrix} \begin{bmatrix} \mathbf{V}_{A0} \\ \mathbf{V}_{A1} \\ \mathbf{V}_{A2} \end{bmatrix} \right)^T \left(\begin{bmatrix} 1 & 1 & 1 \\ 1 & a^2 & a \\ 1 & a & a^2 \end{bmatrix} \begin{bmatrix} \mathbf{I}_{A0} \\ \mathbf{I}_{A1} \\ \mathbf{I}_{A2} \end{bmatrix} \right)^* \tag{13–32}$$

In matrix algebra, $(XY)^T = Y^T X^T$ and $(XY)^* = X^* Y^*$, so this equation becomes

$$\mathbf{S} = [\mathbf{V}_{A0} \quad \mathbf{V}_{A1} \quad \mathbf{V}_{A2}] \begin{bmatrix} 1 & 1 & 1 \\ 1 & a^2 & a \\ 1 & a & a^2 \end{bmatrix}^T \begin{bmatrix} 1 & 1 & 1 \\ 1 & a^2 & a \\ 1 & a & a^2 \end{bmatrix}^* \begin{bmatrix} \mathbf{I}_{A0} \\ \mathbf{I}_{A1} \\ \mathbf{I}_{A2} \end{bmatrix}^* \tag{13–33}$$

However, the transpose of A is just the same as A because of the symmetry of the matrix. In addition $a^* = a^2$, $(a^2)^* = a$, and $1^* = 1$ (we can see this directly on Figure 13–3), so Equation (13–33) reduces to

$$\mathbf{S} = [\mathbf{V}_{A0} \quad \mathbf{V}_{A1} \quad \mathbf{V}_{A2}] \begin{bmatrix} 1 & 1 & 1 \\ 1 & a^2 & a \\ 1 & a & a^2 \end{bmatrix} \begin{bmatrix} 1 & 1 & 1 \\ 1 & a & a^2 \\ 1 & a^2 & a \end{bmatrix} \begin{bmatrix} \mathbf{I}_{A0} \\ \mathbf{I}_{A1} \\ \mathbf{I}_{A2} \end{bmatrix}^* \tag{13–34}$$

$$\mathbf{S} = [\mathbf{V}_{A0} \quad \mathbf{V}_{A1} \quad \mathbf{V}_{A2}] \begin{bmatrix} 3 & 0 & 0 \\ 0 & 3 & 0 \\ 0 & 0 & 3 \end{bmatrix} \begin{bmatrix} \mathbf{I}_{A0} \\ \mathbf{I}_{A1} \\ \mathbf{I}_{A2} \end{bmatrix}^* \tag{13–35}$$

And finally

$$\boxed{\mathbf{S} = P + jQ = 3\mathbf{V}_{A0}\mathbf{I}_{A0}^* + 3\mathbf{V}_{A1}\mathbf{I}_{A1}^* + 3\mathbf{V}_{A1}\mathbf{I}_{A1}^*} \tag{13–36}$$

Equation (13–36) can be used to calculate the real and reactive power from the symmetrical components of the voltages and currents in an unbalanced three-phase circuit.

13.3 | SEQUENCE IMPEDANCES AND SEQUENCE NETWORKS

In any circuit, the voltage drop caused by the currents of a particular sequence depends on the impedance of the circuit to currents of that sequence. In general, the impedance of a circuit can differ for positive-sequence, negative-sequence, and zero-sequence currents.

The impedance of a circuit when only positive-sequence current is flowing is known as the *positive-sequence impedance* of the circuit. Similarly, the impedance of the circuit when only negative-sequence current is flowing is known as the *negative-sequence impedance* of the circuit, and the impedance of the circuit when only zero-sequence current is flowing is known as the *zero-sequence impedance* of the circuit.

To analyze an unsymmetrical fault, we must construct three different per-phase equivalent circuits, one for each type of symmetrical component. The *positive-sequence network* is a per-phase equivalent circuit containing only the positive-sequence impedances and sources, which means that it is identical to the per-phase equivalent circuit used for symmetrical fault analysis in Chapter 12. The *negative-sequence network* is a per-phase equivalent circuit containing only the negative-sequence impedances, and the *zero-sequence network* is a per-phase equivalent circuit containing only the zero-sequence impedances.

Once these per-phase equivalent circuits are constructed, they are interconnected in different ways, depending on the type of fault being analyzed, and the positive-, negative-, and zero-sequence components of current are calculated. These components can then be combined by using Equations (13–23) through (13–25) to calculate the actual currents flowing in the fault and elsewhere within the power system.

Therefore, to solve for any unsymmetrical fault, we must be able to create the per-phase equivalent circuits for each of the symmetrical component sequences. We will learn how to create the per-phase equivalent circuits in this section.

Positive-, Negative-, and Zero-Sequence Equivalent Circuits of Generators

Figure 13–5 shows the equivalent circuit of a Y-connected, three-phase synchronous generator grounded through an inductive reactance. Figure 13–6b shows the positive-sequence network for this synchronous generator. It is very familiar—it is just the same as the per-phase equivalent circuit that we used in Chapter 5 for steady-state analysis and in Chapter 12 for symmetrical three-phase fault analysis. The positive-sequence impedance Z_1 is the sum of the machine's armature resistance R_A plus the appropriate series reactance. For subtransient fault analysis, the reactance is the subtransient reactance X''. For transient fault analysis, the reactance is the transient reactance X'. For steady-state fault analysis, the reactance is the machine's synchronous reactance X_S.

Figure 13–6d shows the negative-sequence network for this synchronous generator. There are no voltage sources inside the generator that generate negative-sequence voltages, so the negative-sequence network does not contain voltage

Figure 13–5 | A Y-connected three-phase generator grounded through an inductive reactance.

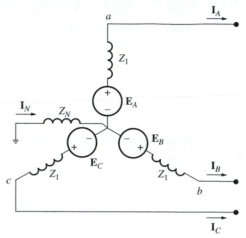

sources. The negative-sequence impedance Z_2 is the sum of the machine's armature resistance R_A plus the appropriate series reactance.

Remember that the negative-sequence currents have the opposite phase sequence to the positive-sequence currents. As we learned in Chapter 4, changing the phase sequence of the currents applied to a three-phase stator reverses the direction of rotation of the magnetic fields produced in the machine. As a result, the negative-sequence currents produce a rotating magnetic field *opposite to the direction of rotation of the machine*. This backward rotation induces a very high voltage in the amortisseur or damper windings of the generator, just as we would see at the very beginning of a fault. Furthermore, this voltage doesn't die out after a few cycles, because the magnetic fields continue to turn opposite to the rotation of the generator. Therefore, the negative-sequence currents see a reactance that is about the size of the *subtransient reactance* of the generator at all times.

Figure 13–6f shows the zero-sequence network for this synchronous generator. There are no voltage sources inside the generator that generate zero-sequence voltages, so the zero-sequence network does not contain voltage sources. The impedance Z_N between the neutral and the ground of the generator needs special attention here. As Figure 13–6e shows, the impedance Z_N contains the total current $3\mathbf{I}_{A0}$, since *the zero-sequence current from all three phases flows through it*, and the current is identical in each phase. Therefore, the voltage drop from point a to ground will be $-3\mathbf{I}_{A0}Z_N - \mathbf{I}_{A0}Z_{g0}$. To represent this voltage drop in the per-phase equivalent circuit, which contains only the current $\mathbf{I}_{A0}$, we must use an impedance 3 times larger than the physical impedance between neutral and ground. The total zero-sequence impedance of the generator is

$$Z_0 = Z_{g0} + 3Z_N \tag{13–37}$$

The zero-sequence impedance Z_{g0} of each winding in the generator consists of the series resistance R_A plus a relatively small reactance. Since the zero-sequence currents in all three phases increase and decrease together, their magnetic fields tend

Figure 13–6 | (a) A synchronous generator as seen by positive-sequence currents. (b) The positive-sequence network for the generator. (c) A synchronous generator as seen by negative-sequence currents. (d) The negative-sequence network for the generator. (e) A synchronous generator as seen by zero-sequence currents. (f) The zero-sequence network for the generator.

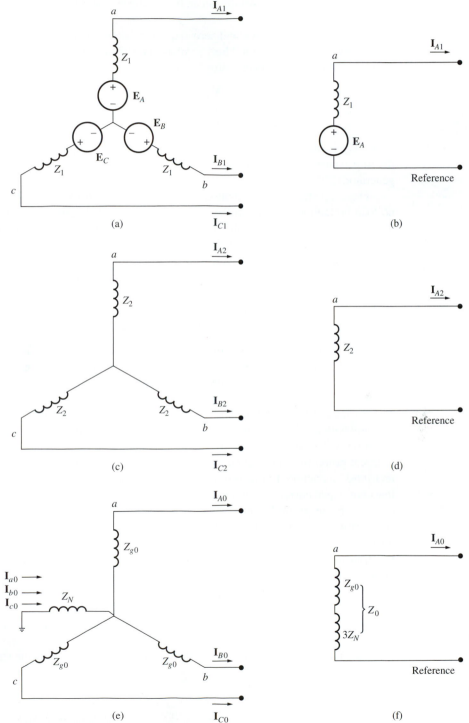

to cancel each other out, producing only a small reactance. The zero-sequence reactance X_{g0} of a generator's windings can be as little as one-quarter the machine's subtransient reactance.

The positive-, negative-, and zero-sequence voltages in phase a of this generator can be found by applying Kirchhoff's voltage law to each of the per-phase equivalent circuits. The resulting voltages are

$$\mathbf{V}_{A1} = \mathbf{E}_{A1} - \mathbf{I}_{A1}Z_1 \qquad (13\text{–}38)$$
$$\mathbf{V}_{A2} = -\mathbf{I}_{A2}Z_2 \qquad (13\text{–}39)$$
$$\mathbf{V}_{A0} = -\mathbf{I}_{A0}Z_0 \qquad (13\text{–}40)$$

where $\mathbf{E}_{A1}$ is the positive-sequence internal generated voltage of the generator, and the impedances are the positive-, negative-, and zero-sequence impedances of the generator.

These equations can be applied to unbalanced generators under either loaded or no-load prefault conditions. If the generators were loaded before the fault, then $\mathbf{E}_{A1}$ must be the voltage behind subtransient reactance, the voltage behind transient reactance, or the voltage behind synchronous reactance, as appropriate.

Also, note that this discussion applies equally to synchronous motors and synchronous generators, since they are physically the same machine.

Positive-, Negative-, and Zero-Sequence Equivalent Circuits of Transmission Lines

Since transmission lines consist of static components, there is no difference between how they behave to positive-sequence and negative-sequence components. Thus the equivalent circuit for a transmission line is identical for both positive- and negative-sequence networks.

However, overhead transmission lines behave very differently for zero-phase currents compared to positive- and negative-sequence currents. As we saw from Equation (9–22) in Chapter 9, the series inductance of a transmission line increases in direct proportion the natural logarithm of the distance between the sending and returning conductors. For zero-sequence currents, all three phases of the transmission line carry equal currents, and the return path is either through overhead ground wires or through the ground itself. Since both the ground wires and the ground itself are usually well separated from the phases, the zero-sequence reactance of a transmission line is usually higher than the positive- and negative-sequence reactance. For transmission lines without overhead ground wires, the return path must be through the ground itself, and the zero-phase reactance will be even higher because the phases are so far above the ground. The series reactance of a transmission line for zero-sequence currents will be 2 to 3.5 times higher than the series reactance for positive- or negative-sequence currents, with lines without ground wires at the upper end of this range.

Note that the series resistance of the transmission line remains the same for all three symmetrical components.

Zero-Sequence Equivalent Circuits of Y and Δ Connections

Zero-sequence currents have the same phase in all three phases of a three-phase power system, so they can *only* flow if there is a path to the ground for the return current flow. If there is no return path, then there can't be any zero-sequence current flow.

Figure 13–7a shows an ungrounded Y-connected load. Since there is no path to ground in this load, no zero-sequence currents can flow. The corresponding equivalent circuit has an impedance Z between the terminals of the load and the neutral, but an open circuit from there to ground.

Figure 13–7b shows a grounded Y-connected load. Since there is a path to ground in this load, zero-sequence currents can flow. The corresponding equivalent circuit has an impedance Z between the terminals of the load and the neutral, and a short circuit from there to ground.

Figure 13–7c shows a Y-connected load that is grounded through an impedance. Since there is a path to ground in this load, zero-sequence currents can flow. However, the current flowing in the path between the neutral and ground is 3 times the zero-sequence current, so the corresponding impedance in the equivalent circuit is $3Z_N$. The equivalent circuit has an impedance Z between the terminals of the load and the neutral, an impedance $3Z_N$ from there to ground.

Figure 13–7d shows a Δ-connected load. Since there is no path to ground in this load, no zero-sequence currents can flow into the load. However, if there happens to be any zero-sequence components *inside* the load, they can flow freely, because there is a closed path inside the load. The corresponding equivalent circuit has an open circuit at the terminals, and a closed loop containing the impedance Z.

Positive-, Negative-, and Zero-Sequence Equivalent Circuits of Transformers

Since transformers are static devices, there is no difference between how they behave for positive-sequence and negative-sequence components. Thus the equivalent circuit for a transformer is identical for both positive- and negative-sequence networks. This equivalent circuit consists of just the series impedance of the transformer. (Note that the shunt components representing the excitation branch of the transformer are usually ignored in creating per-phase equivalent circuits for fault current analysis.)

Zero-sequence equivalent circuits of transformers are a special and much more complicated case. Each side of the transformer is either Y connected or Δ connected, so the equivalent circuits of a side will look like the Y- or Δ-connected equivalent circuits shown in Figure 13–7. However, the current flow on the two sides of a transformer are related by the turns ration of the transformer

$$N_P \mathbf{I}_P = N_S \mathbf{I}_S \qquad (13\text{–}41)$$

so *if there is no path for zero-sequence currents on one side of the transformer, there must also be no zero-sequence currents on the other side of the transformer.*

Figure 13–7 | (a) An ungrounded Y-connected load has no path between the neutral and ground, so the zero-sequence equivalent circuit has an open circuit between those points. (b) A grounded Y-connected load has a short circuit between the neutral and ground, so the zero-sequence equivalent circuit has a short circuit between the neutral and the ground. (c) A Y-connected load grounded through an impedance. The zero-sequence equivalent circuit has the impedance $3Z_N$ between the neutral and the ground. (d) A Δ-connected load has no path between the neutral and ground, so no zero-sequence currents can flow into the load. However, any zero-sequence currents inside the load can flow freely within the Δ.

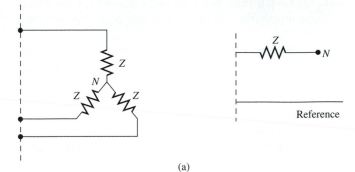

(a)

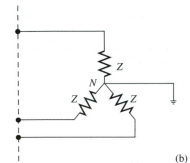

(b)

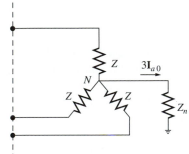

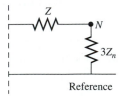

(c)

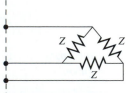

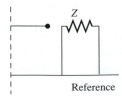

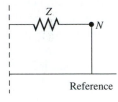

(d)

There are five possible cases of zero-sequence equivalent circuits for transformers, depending on their connections. They are described on page 606, and shown in Figure 13–8. Figure 13–8 shows the one-line diagram symbols for the various

Figure 13–8 | Symbols, connection diagrams, and zero-sequence equivalent circuits for various types of three-phase transformers.

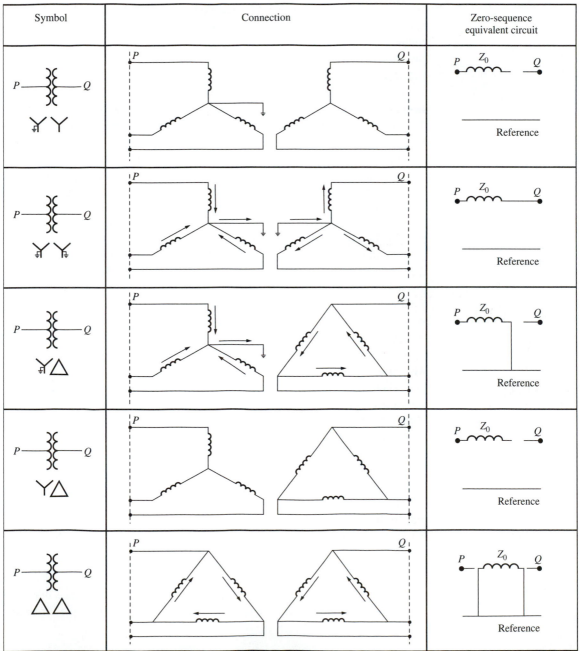

types of connections, together with the connection diagrams and zero-phase equivalent circuits.

1. *Y–Y three-phase transformer, one neutral grounded.* If one of the neutrals in a three-phase Y–Y-connected transformer is ungrounded, then no zero-sequence currents can flow in either winding. The absence of a ground path in one winding forces the zero-sequence currents to be zero on that side, and Equation (13–41) implies that no current can flow on the other side of the transformer either. Thus the zero-sequence per-phase equivalent circuit contains the impedance Z_0 in series with an open circuit, showing that no current can flow.

2. *Y–Y three-phase transformer, both neutrals grounded.* If both of the neutrals in a three-phase Y–Y-connected transformer are ungrounded, then zero-sequence currents can flow freely to and from ground in both windings. Thus the zero-sequence per-phase equivalent circuit contains the impedance Z_0.

3. *Y–Δ three-phase transformer, grounded Y.* If the neutral of the Y connection in a three-phase Y–Δ-connected transformer is grounded, then zero-sequence currents can flow to ground in the Y-connected winding. If such currents flow, then by Equation (13–41) zero-sequence currents must also flow in the Δ-connected winding. Since the phases of the Δ-connected winding are connected head to tail in a closed loop, this zero-sequence current can flow freely within the loop, and therefore zero-sequence currents can flow freely in the Y connection. However, no zero-sequence currents can enter a Δ connection from the lines tied to it. Therefore, the equivalent circuit of the transformer has an impedance Z_0 connected to the reference bus as seen from the Y side, and is an open circuit as seen from the Δ side.

4. *Y–Δ three-phase transformer, ungrounded Y.* If the neutral of the Y connection in a three-phase Y–Δ-connected transformer is not grounded, then no zero-sequence currents can flow to ground in the Y-connected winding. Also, no zero-sequence currents can enter a Δ connection from the lines tied to it. Therefore, the equivalent circuit of the transformer has an impedance Z_0 connected to the reference bus but not grounded as seen from the Y side, and is an open circuit as seen from the Δ side.

5. *Δ–Δ three-phase transformer.* In a three-phase Δ–Δ connected transformer, and zero-sequence currents flowing in the windings can circulate freely within the loops, since the phases of the Δ-connected windings are connected head to tail in a closed loop. However, no zero-sequence currents can enter a Δ connection from the lines tied to it. Therefore, the equivalent circuit of the transformer has an impedance Z_0 connected in a closed loop to represent the windings, and open circuits as seen from lines connected to either side.

Building Zero-Sequence Equivalent Circuits

It is simple to combine the components that we have presented to create the zero-phase equivalent circuit of a complete network. The following example illustrates this process.

EXAMPLE 13-2

Sketch the positive-, negative-, and zero-sequence equivalent circuits for the power system shown in Figure 13–9 during the subtransient period of a fault.

Figure 13–9 | A simple three-phase power system with transformer and generator connections shown.

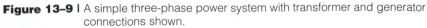
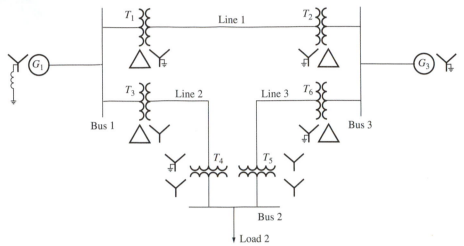

Solution

The positive–sequence equivalent circuit of this power system is just the standard per-phase equivalent circuit that we used in Chapter 12. Because this equivalent circuit is for the subtransient period, we will use the subtransient reactance X'' of the generators in the equivalent circuit. The resulting per-phase equivalent circuit is shown in Figure 13–10a.

The negative–sequence equivalent circuit of this power system looks like the positive-sequence equivalent circuit, except that there are no voltage sources and the negative-sequence reactance is used for the generators instead of the subtransient reactance. Note that the impedances of the transformers and lines are unchanged, since these components can't tell the difference between positive-sequence and negative-sequence currents. The resulting per-phase equivalent circuit is shown in Figure 13–10b.

The zero-sequence equivalent circuit is quite complex by comparison to the other two. Generator G_1 is grounded through an impedance, so its total zero-sequence impedance will be $Z_{g0} + 3Z_N$. Generator G_2 is solidly grounded, so its total zero-sequence impedance will be just Z_{g0}. Transformers T_1, T_2, and T_6 are Y–Δ transformers with the Y *grounded*, the Δ side toward the busses, and the Y side toward the transmission lines. These transformers will look like an open circuit to the busses, and will look like their zero-sequence impedance Z_0 to the transmission lines. Transformer T_3 is a Y–Δ transformer with the Y *ungrounded*, the Δ side toward the busses, and the Y side toward the transmission lines. This

Figure 13–10 | (a) Positive-sequence equivalent circuit for Example 13–2.
(b) Negative-sequence equivalent circuit for Example 13–2.
(c) Zero-sequence equivalent circuit for Example 13–2.

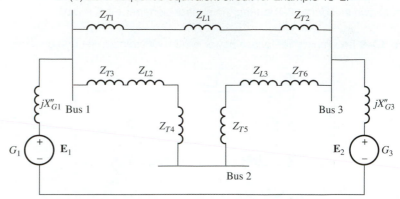

(a)

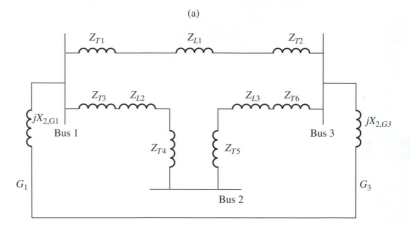

(b)

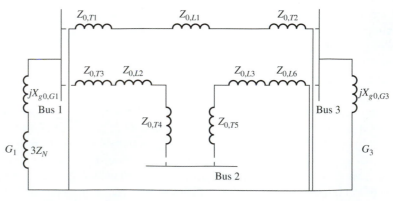

(c)

transformer will look like an open circuit to both the busses and the transmission lines. Transformer T_4 is a Y–Y transformer with one side grounded and one side ungrounded, and transformer T_5 is a Y–Y transformer with both sides ungrounded. These transformers will look like an open circuit to both the busses and the transmission lines. The resulting per-phase equivalent circuit is shown in Figure 13–10c. ■

13.4 | SINGLE LINE-TO-GROUND FAULT ON AN UNLOADED GENERATOR

Most faults that occur on power systems are unsymmetrical faults, which result in unbalanced current flows, and therefore require symmetrical components to solve for the voltages and currents during the fault. We will now analyze the three major types of unsymmetrical faults (single line-to-ground, line-to-line, and double line-to-ground) on a single unloaded generator. This section deals with the single line-to-ground fault on a single generator, while the following two sections deal with line-to-line and double line-to-ground faults on a single generator. Then, we will extend the results to apply to entire power systems.

An unloaded generator with a single line-to-ground fault on phase a is shown in Figure 13–11. The voltages and currents in phase a of this generator are given by Equations (13–38) to (13–40)

$$\mathbf{V}_{A1} = \mathbf{E}_{A1} - \mathbf{I}_{A1}Z_1 \qquad (13\text{--}38)$$
$$\mathbf{V}_{A2} = -\mathbf{I}_{A2}Z_2 \qquad (13\text{--}39)$$
$$\mathbf{V}_{A0} = -\mathbf{I}_{A0}Z_0 \qquad (13\text{--}40)$$

Figure 13–11 | A single line-to-ground fault on phase a of an unloaded generator. The generator's neutral is grounded through a reactance.

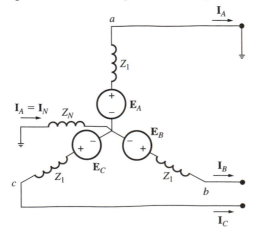

When a single line-to-ground fault occurs on phase a of a previously unloaded generator, the voltage at the fault goes to zero, and the currents in phases b and c remain zero.

$$\left. \begin{array}{l} \mathbf{I}_B = 0 \\ \mathbf{I}_C = 0 \\ \mathbf{V}_A = 0 \end{array} \right\} \text{ single-line-to-ground fault} \tag{13–42}$$

Since $\mathbf{I}_B = 0$ and $\mathbf{I}_C = 0$, the symmetrical components of current in the fault can be found from Equations (13–26) to (13–28):

$$\mathbf{I}_{A0} = \tfrac{1}{3}(\mathbf{I}_A + \mathbf{I}_B + \mathbf{I}_C) \qquad \text{zero sequence} \tag{13–26}$$

$$\mathbf{I}_{A1} = \tfrac{1}{3}[(\mathbf{I}_A + a\mathbf{I}_B + a^2\mathbf{I}_C)] \quad \text{positive sequence} \tag{13–27}$$

$$\mathbf{I}_{A2} = \tfrac{1}{3}[(\mathbf{I}_A + a^2\mathbf{I}_B + a\mathbf{I}_C)] \quad \text{negative sequence} \tag{13–28}$$

Substituting Equations (13–42) into Equations (13–26) to (13–28), we get

$$\mathbf{I}_{A0} = \tfrac{1}{3}(\mathbf{I}_A + 0 + 0) = \tfrac{1}{3}\mathbf{I}_A \tag{13–43}$$

$$\mathbf{I}_{A1} = \tfrac{1}{3}[\mathbf{I}_A + a(0) + a^2(0)] = \tfrac{1}{3}\mathbf{I}_A \tag{13–44}$$

$$\mathbf{I}_{A2} = \tfrac{1}{3}[\mathbf{I}_A + a^2(0) + a(0)] = \tfrac{1}{3}\mathbf{I}_A \tag{13–45}$$

Therefore,
$$\mathbf{I}_{A0} = \mathbf{I}_{A1} = \mathbf{I}_{A2} = \tfrac{1}{3}\mathbf{I}_A \tag{13–46}$$

This result shows that *the symmetrical components of current are all equal to each other in a single line-to-ground fault.*

If we substitute $\mathbf{I}_{A1}$ for $\mathbf{I}_{A2}$ and $\mathbf{I}_{A0}$ in Equations (13–38) to (13–40), the results are

$$\mathbf{V}_{A1} = \mathbf{E}_{A1} - \mathbf{I}_{A1}Z_1 \tag{13–47}$$

$$\mathbf{V}_{A2} = -\mathbf{I}_{A1}Z_2 \tag{13–48}$$

$$\mathbf{V}_{A0} = -\mathbf{I}_{A1}Z_0 \tag{13–49}$$

We know from Equation (13–8) that $\mathbf{V}_A = \mathbf{V}_{A1} + \mathbf{V}_{A2} + \mathbf{V}_{A0}$, so adding up these equations yields the result

$$\mathbf{V}_A = \mathbf{V}_{A0} + \mathbf{V}_{A1} + \mathbf{V}_{A2} = Z_0\mathbf{I}_{A1} + \mathbf{E}_A - Z_1\mathbf{I}_{A1} - Z_2\mathbf{I}_{A1} \tag{13–50}$$

Since the voltage in phase a is zero, we can solve for the positive-sequence current in the fault. It is

$$\boxed{\mathbf{I}_{A1} = \frac{\mathbf{E}_A}{Z_0 + Z_1 + Z_2}} \tag{13–51}$$

For a single line-to-ground fault, the positive-sequence current, negative-sequence current, and zero-sequence current are all equal, and can be found from Equation (13–51). Note that this result is equivalent to connecting the three sequence networks in *series* (see Figure 13–12) at the terminals representing the location of the fault. Such a connection would make the three currents equal, and would cause the total impedance of the circuit to be $Z_0 + Z_1 + Z_2$. Note that the voltages across each of the sequence networks will be the symmetrical components of $\mathbf{V}_A$ for that sequence.

Once the symmetrical components $\mathbf{I}_{A0}$, $\mathbf{I}_{A1}$, and $\mathbf{I}_{A2}$ are known, we can calculate the symmetrical components of voltage $\mathbf{V}_{A0}$, $\mathbf{V}_{A1}$, and $\mathbf{V}_{A2}$ from Equations (13–38) to (13–40). Finally, we can calculate the total phase voltages for all phases in the fault from Equations (13–8) to (13–10), and the total phase currents for all phases in the fault from Equations (13–23) to (13–25).

Figure 13–12 | For a single line-to-ground fault, the positive-, negative-, and zero-sequence networks are connected in series to determine the sequence voltages and currents in the fault.

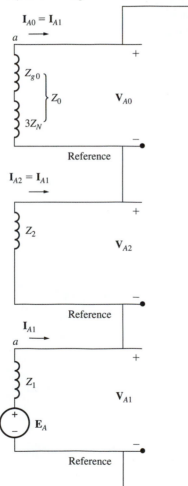

In summary, to calculate the voltages and currents in a single line-to-ground fault on a synchronous generator:

1.   Create the positive-, negative-, and zero-sequence networks for the generator.
2.   Connect the networks in series at the terminals representing the location of the fault.
3.   Calculate $\mathbf{I}_{A0} = \mathbf{I}_{A1} = \mathbf{I}_{A2}$ from this network.
4.   Calculate $\mathbf{V}_{A0}$, $\mathbf{V}_{A1}$, and $\mathbf{V}_{A2}$ from Equations (13–38) to (13–40).
5.   Calculate any required voltages and currents by combining the sequence components calculated above using the equations given in this chapter.

EXAMPLE 13–3

A 100-MVA, 13.8-kV, Y-connected synchronous generator has a subtransient reactance X'' of 0.20 pu, a negative-sequence reactance X_2 of 0.25 pu, a zero-sequence reactance of 0.10 pu, and negligible resistance. The neutral of the generator is solidly grounded. Assume that the generator is initially unloaded and operating at rated voltage, and that a single line-to-ground fault occurs on phase a at the terminals of the generator. Find the voltages and currents in each phase during the subtransient period immediately after the fault occurs. Also, find the line-to-line voltages at that time.

Solution

The base apparent power for this machine is 100 MVA and the base line-to-line voltage is 13.8 kV, so the base line-to-neutral voltage will be

$$V_{\phi,\,\text{base}} = \frac{V_{LL,\,\text{base}}}{\sqrt{3}} = \frac{13.8 \text{ kV}}{\sqrt{3}} = 7.967 \text{ kV}$$

and the base current will be

$$I_{\text{base}} = \frac{S_{3\phi,\,\text{base}}}{\sqrt{3}V_{LL,\,\text{base}}} = \frac{100 \text{ MVA}}{\sqrt{3}(13.8 \text{ kV})} = 4184 \text{ A} \tag{10–4}$$

Since this machine initially operating at no load and rated voltage, the voltage behind subtransient reactance will be $\mathbf{E}_A = 1\angle 0°$ pu. The sequence networks for this generator will be as shown in Figure 13–12, with $Z_1 = j0.20$, $Z_2 = j0.25$, and $Z_0 = j0.10$. The resulting sequence currents in phase a will be

$$\mathbf{I}_{A1} = \mathbf{I}_{A2} = \mathbf{I}_{A0} = \frac{\mathbf{E}_A}{Z_0 + Z_1 + Z_2} \tag{13–51}$$

$$\mathbf{I}_{A1} = \mathbf{I}_{A2} = \mathbf{I}_{A0} = \frac{1\angle 0°}{j0.10 + j0.20 + j0.25} = 1.818\angle -90° \text{ pu}$$

The sequence voltages in phase a will be:

$$\mathbf{V}_{A1} = \mathbf{E}_{A1} - \mathbf{I}_{A1}Z_1 \tag{13-38}$$
$$= 1.0\angle 0° - (1.818\angle -90°)(j0.20) = 0.636\angle 0° \text{ pu}$$
$$\mathbf{V}_{A2} = -\mathbf{I}_{A2}Z_2 \tag{13-39}$$
$$= -(1.818\angle -90°)(j0.25) = 0.455\angle 180° \text{ pu}$$
$$\mathbf{V}_{A0} = -\mathbf{I}_{A0}Z_0 \tag{13-40}$$
$$= -(1.818\angle -90°)(j0.10) = 0.182\angle 180° \text{ pu}$$

Therefore, the phase currents in this generator will be

$$\mathbf{I}_A = \mathbf{I}_{A1} + \mathbf{I}_{A2} + \mathbf{I}_{A0} \tag{13-23}$$
$$= 1.818\angle -90° \text{ pu} + 1.818\angle -90° \text{ pu} + 1.818\angle -90° \text{ pu}$$
$$= 5.455\angle -90° \text{ pu}$$
$$\mathbf{I}_B = a^2\mathbf{I}_{A1} + a\mathbf{I}_{A2} + \mathbf{I}_{A0} \tag{13-24}$$
$$= a^2(1.818\angle -90° \text{ pu}) + a(1.818\angle -90° \text{ pu}) + 1.818\angle -90° \text{ pu}$$
$$= 0.0 \text{ pu}$$
$$\mathbf{I}_C = a\mathbf{I}_{A1} + a^2\mathbf{I}_{A2} + \mathbf{I}_{A0} \tag{13-25}$$
$$= a(1.818\angle -90° \text{ pu}) + a^2(1.818\angle -90° \text{ pu}) + 1.818\angle -90° \text{ pu}$$
$$= 0.0 \text{ pu}$$

Note that the phase currents in phases *b* and *c* are zero, which makes sense, since they are still open circuited! The actual fault current during the subtransient period in amperes is

$$I_A = I_{A,pu}I_{base} = (5.455)(4184 \text{ A}) = 22,820 \text{ A}$$

The phase voltages in this generator are

$$\mathbf{V}_A = \mathbf{V}_{A1} + \mathbf{V}_{A2} + \mathbf{V}_{A0} \tag{13-8}$$
$$= 0.636\angle 0° + 0.455\angle 180° + 0.182\angle 180°$$
$$= 0.0\angle 0°$$
$$\mathbf{V}_B = a^2\mathbf{V}_{A1} + a\mathbf{V}_{A2} + \mathbf{V}_{A0} \tag{13-9}$$
$$= a^2(0.636\angle 0°) + a(0.455\angle 180°) + 0.182\angle 180°$$
$$= 0.983\angle -106°$$
$$\mathbf{V}_C = a\mathbf{V}_{A1} + a^2\mathbf{V}_{A2} + \mathbf{V}_{A0} \tag{13-10}$$
$$= a(0.636\angle 0°) + a^2(0.455\angle 180°) + 0.182\angle 180°$$
$$= 0.983\angle 106°$$

The line-to-line voltages are therefore

$$\mathbf{V}_{AB} = \mathbf{V}_A - \mathbf{V}_B = 0.0\angle 0° - 0.983\angle -106° = 0.986\angle 74° \text{ pu}$$
$$\mathbf{V}_{BC} = \mathbf{V}_B - \mathbf{V}_C = 0.983\angle -106° - 0.983\angle 106° = 1.890\angle -90° \text{ pu}$$
$$\mathbf{V}_{CA} = \mathbf{V}_C - \mathbf{V}_A = 0.983\angle 106° - 0.0\angle 0° = 0.986\angle 106° \text{ pu}$$

The line-to-line voltages expressed in volts are equal to the per-unit values multiplied by the base voltage of a phase

$$\mathbf{V}_{AB} = (0.986 \text{ pu})(7.967 \text{ kV}) = 7.85 \text{ kV}$$
$$\mathbf{V}_{BC} = (1.890 \text{ pu})(7.967 \text{ kV}) = 15.06 \text{ kV}$$
$$\mathbf{V}_{CA} = (0.986 \text{ pu})(7.967 \text{ kV}) = 7.85 \text{ kV}$$

The line and phase voltage relationships just before and just after the fault are shown in Figure 13–13. ∎

Figure 13–13 | (a) Relationship between phase and line voltages in the generator just before the single line-to-ground fault. (b) Relationship between phase and line voltages in the generator just after the single line-to-ground fault.

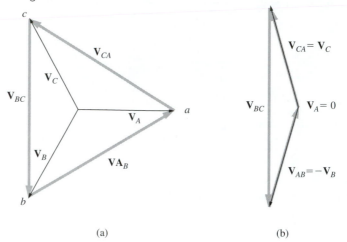

(a)   (b)

13.5 | LINE-TO-LINE FAULT ON AN UNLOADED GENERATOR

An unloaded generator with a line-to-line fault between phases b and c is shown in Figure 13–14. When a line-to-line fault occurs between phases b and c of a previously unloaded generator, the voltages at phases b and c must be equal (because they are tied together), the current in phase a must be zero (because that phase is not shorted), and the currents in phases b and c must be the negative of each other.

$$\left. \begin{array}{l} \mathbf{V}_B = \mathbf{V}_C \\ \mathbf{I}_A = 0 \\ \mathbf{I}_B = -\mathbf{I}_C \end{array} \right\} \text{ line-to-line fault} \tag{13–52}$$

The fact that $\mathbf{V}_C = \mathbf{V}_B$ tells us something about the relationships among the sequence voltages in this fault. Substituting $\mathbf{V}_C = \mathbf{V}_B$ into Equations (13–17) to (13–19), we get

$$\mathbf{V}_{A0} = \tfrac{1}{3}(\mathbf{V}_A + \mathbf{V}_B + \mathbf{V}_B) \tag{13–17}$$

$$\mathbf{V}_{A1} = \tfrac{1}{3}(\mathbf{V}_A + a\mathbf{V}_B + a^2\mathbf{V}_B) \tag{13–18}$$

$$\mathbf{V}_{A2} = \tfrac{1}{3}(\mathbf{V}_A + a^2\mathbf{V}_B + a\mathbf{V}_B) \tag{13–19}$$

Notice that *the last two equations are identical,* so for a line-to-line fault, $\mathbf{V}_{A1} = \mathbf{V}_{A2}$.

Figure 13–14 I A line-to-line fault between phases *b* and *c* of an unloaded generator.

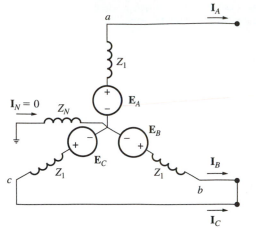

Also, since $\mathbf{I}_A = 0$ and $\mathbf{I}_B = -\mathbf{I}_C$, the symmetrical components of current in the fault can be found from Equations (13–26) to (13–28):

$$\mathbf{I}_{A0} = \tfrac{1}{3}(\mathbf{I}_A + \mathbf{I}_B + \mathbf{I}_C) = \tfrac{1}{3}(0 - \mathbf{I}_C + \mathbf{I}_C) \qquad (13\text{–}26)$$
$$\mathbf{I}_{A1} = \tfrac{1}{3}(\mathbf{I}_A + a\mathbf{I}_B + a^2\mathbf{I}_C) = \tfrac{1}{3}(0 - a\mathbf{I}_C + a^2\mathbf{I}_C) \qquad (13\text{–}27)$$
$$\mathbf{I}_{A2} = \tfrac{1}{3}(\mathbf{I}_A + a^2\mathbf{I}_B + a\mathbf{I}_C) = \tfrac{1}{3}(0 - a^2\mathbf{I}_C + a\mathbf{I}_C) \qquad (13\text{–}28)$$

Note from these equations that $\mathbf{I}_{A0} = 0$ and $\mathbf{I}_{A2} = -\mathbf{I}_{A1}$.

Also, $\mathbf{I}_{A0} = 0$ implies that

$$\mathbf{V}_{A0} = -\mathbf{I}_{A0}Z_0 = (0)Z_0 = 0 \qquad (13\text{–}40)$$

We now know that $\mathbf{V}_{A0} = 0$, $\mathbf{V}_{A1} = \mathbf{V}_{A2}$, $\mathbf{I}_{A0} = 0$, and $\mathbf{I}_{A2} = -\mathbf{I}_{A1}$. If we substitute this information in Equations (13–38) to (13–40), the results are

$$\mathbf{V}_{A1} = \mathbf{E}_{A1} - \mathbf{I}_{A1}Z_1 \qquad (13\text{–}53)$$
$$\mathbf{V}_{A1} = \mathbf{I}_{A1}Z_2 \qquad (13\text{–}54)$$
$$0 = -(0)Z_0 \qquad (13\text{–}55)$$

From Equations (13–53) and (13–54), we see that

$$\mathbf{I}_{A1}Z_2 = \mathbf{E}_{A1} - \mathbf{I}_{A1}Z_1 \qquad (13\text{–}56)$$

$$\boxed{\mathbf{I}_{A1} = \frac{\mathbf{E}_{A1}}{Z_1 + Z_2}} \qquad (13\text{–}57)$$

For a line-to-line fault, the positive-sequence current is the negative of the negative-sequence current, and the zero-sequence current is zero. In addition, the positive-sequence phase voltage is equal to the negative-sequence phase voltage. Note that this result is equivalent to connecting the positive- and negative-sequence networks in *parallel* at the terminals representing the location of the fault (see Figure 13–15).

Figure 13–15 I For a line-to-line fault, the positive- and negative-sequence networks
are connected in parallel to determine the sequence voltages and
currents in the fault.

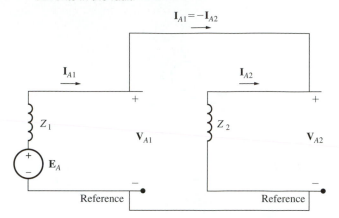

$$I_{A1} = -I_{A2}$$

Note that the zero-sequence current was zero in this kind of fault. This makes
sense physically, because the fault is not connected to the ground, so there is no complete circuit to ground for the zero-sequence currents to flow through.

In summary, to calculate the voltages and currents in a line-to-line fault on a synchronous
generator:

1. Create the positive- and negative-sequence networks for the generator. The zero-
   sequence network is not required.
2. Connect the two networks in parallel at the terminals representing the location of
   the fault.
3. Calculate $I_{A1} = -I_{A2}$ from this network. Note that $I_{A0} = 0$.
4. Calculate V_{A0}, V_{A1}, and V_{A2} from Equations (13–38) to (13–40).
5. Calculate any required voltages and currents by combining the sequence
   components calculated above using the equations given in this chapter.

EXAMPLE 13–4

A 100-MVA, 13.8-kV, Y-connected synchronous generator has a subtransient reactance X'' of 0.20 pu, a negative-sequence reactance X_2 of 0.25 pu, a zero-sequence
reactance of 0.10 pu, and negligible resistance. The neutral of the generator is
solidly grounded. Assume that the generator is initially unloaded and operating at
rated voltage, and that a line-to-line fault occurs between phases *b* and *c* at the terminals of the generator. Find the voltages and currents in each phase during the subtransient period immediately after the fault occurs. Also, find the line-to-line voltages
at that time.

Solution
The base apparent power for this machine is 100 MVA and the base line-to-line voltage is 13.8 kV, so the base line to neutral voltage will be

$$V_{\phi,\,base} = \frac{V_{LL,\,base}}{\sqrt{3}} = \frac{13.8 \text{ kV}}{\sqrt{3}} = 7.967 \text{ kV}$$

and the base current will be

$$I_{base} = \frac{S_{3\phi,\,base}}{\sqrt{3} V_{LL,\,base}} = \frac{100 \text{ MVA}}{\sqrt{3}(13.8 \text{ kV})} = 4184 \text{ A} \qquad (10\text{--}4)$$

Since this machine was initially operating at no load and rated voltage, the voltage behind subtransient reactance will be $\mathbf{E}_A = 1\angle 0°$ pu. The sequence networks for this generator will be as shown in Figure 13–15, with $Z_1 = j0.20$ and $Z_2 = j0.25$. (Note that Z_0 is unused!). The resulting sequence currents in phase a will be

$$\mathbf{I}_{A1} = \frac{\mathbf{E}_{A1}}{Z_1 + Z_2} \qquad (13\text{--}57)$$

$$= \frac{1\angle 0°}{j0.20 + j0.25} = 2.222\angle -90° \text{ pu}$$

$$\mathbf{I}_{A2} = -\mathbf{I}_{A1} = 2.222\angle 90° \text{ pu}$$

$$\mathbf{I}_{A0} = 0\angle 0° \text{ pu}$$

The sequence voltages in phase a will be:

$$\mathbf{V}_{A1} = \mathbf{E}_{A1} - \mathbf{I}_{A1}Z_1 \qquad (13\text{--}38)$$

$$= 1.0\angle 0° - (2.222\angle -90°)(j0.20) = 0.556\angle 0° \text{ pu}$$

$$\mathbf{V}_{A2} = -\mathbf{I}_{A2}Z_2 \qquad (13\text{--}39)$$

$$= -(2.222\angle 90°)(j0.25) = 0.556\angle 0° \text{ pu}$$

$$\mathbf{V}_{A0} = -\mathbf{I}_{A0}Z_0 \qquad (13\text{--}40)$$

$$= 0\angle 0° \text{ pu}$$

Note that $\mathbf{V}_{A1} = \mathbf{V}_{A2}$, as it should be. The phase currents in this generator will be

$$\mathbf{I}_A = \mathbf{I}_{A1} + \mathbf{I}_{A2} + \mathbf{I}_{A0} \qquad (13\text{--}23)$$

$$= 2.222\angle -90° + 2.222\angle 90° + 0.0\angle 0°$$

$$= 0.0\angle 0° \text{ pu}$$

$$\mathbf{I}_B = a^2\mathbf{I}_{A1} + a\mathbf{I}_{A2} + \mathbf{I}_{A0} \qquad (13\text{--}24)$$

$$= a^2(2.222\angle -90°) + a(2.222\angle 90°) + 0.0\angle 0°$$

$$= 3.849\angle 180° \text{ pu}$$

$$\mathbf{I}_C = a\mathbf{I}_{A1} + a^2\mathbf{I}_{A2} + \mathbf{I}_{A0} \qquad (13\text{--}25)$$

$$= a(2.222\angle -90°\text{pu}) + a^2(2.222\angle 90°\text{pu}) + 0.0\angle 0° \text{ pu}$$

$$= 3.849\angle 0° \text{ pu}$$

Note that the phase currents in phases b and c are the negative of each other, as they should be. The actual fault current during the subtransient period in amperes is

$$I_A = I_{A,\,\text{pu}} I_{\text{base}} = (3.849)(4184\text{ A}) = 16{,}100\text{ A}$$

The phase voltages in this generator are

$$
\begin{aligned}
\mathbf{V}_A &= \mathbf{V}_{A1} + \mathbf{V}_{A2} + \mathbf{V}_{A0} && (13\text{–}8)\\
&= 0.556\angle 0° + 0.556\angle 0° + 0.0\angle 0°\\
&= 1.112\angle 0°\\
\mathbf{V}_B &= a^2\mathbf{V}_{A1} + a\mathbf{V}_{A2} + \mathbf{V}_{A0} && (13\text{–}9)\\
&= a^2(0.556\angle 0°) + a(0.556\angle 0°) + 0.0\angle 0°\\
&= 0.556\angle 180°\\
\mathbf{V}_C &= a\mathbf{V}_{A1} + a^2\mathbf{V}_{A2} + \mathbf{V}_{A0} && (13\text{–}10)\\
&= a(0.556\angle 0°) + a^2(0.556\angle 0°) + 0.0\angle 0°\\
&= 0.556\angle 180°
\end{aligned}
$$

The line-to-line voltages are therefore

$$
\begin{aligned}
\mathbf{V}_{AB} &= \mathbf{V}_A - \mathbf{V}_B = 1.112\angle 0° - 0.556\angle 180° = 1.668\angle 0°\text{ pu}\\
\mathbf{V}_{BC} &= \mathbf{V}_B - \mathbf{V}_C = 0.556\angle 180° - 0.556\angle 180° = 0\angle 0°\text{ pu}\\
\mathbf{V}_{CA} &= \mathbf{V}_C - \mathbf{V}_A = 0.556\angle 180° - 1.112\angle 0° = 1.668\angle 180°\text{ pu}
\end{aligned}
$$

The line-to-line voltages expressed in volts are equal to the per-unit values multiplied by the base voltage of a phase:

$$
\begin{aligned}
\mathbf{V}_{AB} &= (1.668\text{ pu})(7.967\text{ kV}) = 13.3\text{ kV}\\
\mathbf{V}_{BC} &= (0.0\text{ pu})(7.967\text{ kV}) = 0\text{ kV}\\
\mathbf{V}_{CA} &= (1.668\text{ pu})(7.967\text{ kV}) = 13.3\text{ kV}
\end{aligned}
$$

13.6 | DOUBLE LINE-TO-GROUND FAULT ON AN UNLOADED GENERATOR

An unloaded generator with a double line-to-ground fault between phases b and c is shown in Figure 13–16. When a double line-to-ground fault occurs between phases b, c, and the ground of a previously unloaded generator, the voltages at phases b and c must both be equal to zero, and the current in phase a must be zero (because that phase is not shorted).

$$
\left.
\begin{array}{l}
\mathbf{V}_B = 0\\
\mathbf{V}_C = 0\\
\mathbf{I}_A = 0
\end{array}
\right\} \text{double line-to-ground fault} \qquad (13\text{–}58)
$$

The fact that $\mathbf{V}_B = \mathbf{V}_C = 0$ tells us something about the relationships among the sequence voltages in this fault. Substituting $\mathbf{V}_B = \mathbf{V}_C = 0$ into Equations (13–17) to (13–19), we get

Figure 13–16 | A double line-to-ground fault between phases *b, c,* and the ground of an unloaded generator.

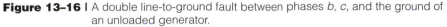
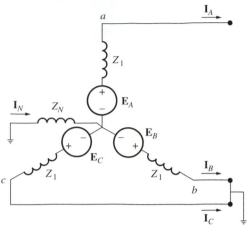

$$\mathbf{V}_{A0} = \tfrac{1}{3}(\mathbf{V}_A + 0 + 0) = \tfrac{1}{3}\,\mathbf{V}_A \qquad\qquad (13\text{--}17)$$

$$\mathbf{V}_{A1} = \tfrac{1}{3}[\mathbf{V}_A + a(0) + a^2(0)] = \tfrac{1}{3}\,\mathbf{V}_A \qquad (13\text{--}18)$$

$$\mathbf{V}_{A2} = \tfrac{1}{3}[\mathbf{V}_A + a^2(0) + a(0)] = \tfrac{1}{3}\,\mathbf{V}_A \qquad (13\text{--}19)$$

Therefore, $\mathbf{V}_{A1} = \mathbf{V}_{A2} = \mathbf{V}_{A0}$ for this type of fault.

Furthermore, the fault currents flowing out through phases *b* and *c* must return through the neutral, so

$$\mathbf{I}_N = \mathbf{I}_B + \mathbf{I}_C \qquad\qquad (13\text{--}59)$$

Substituting Equations (13–24), (13–25), and (13–29) in Equation (13–59) yields

$$\mathbf{I}_N = \mathbf{I}_B + \mathbf{I}_C \qquad\qquad (13\text{--}59)$$

$$3\mathbf{I}_{A0} = (a^2\mathbf{I}_{A1} + a\mathbf{I}_{A2} + \mathbf{I}_{A0}) + (a\mathbf{I}_{A1} + a^2\mathbf{I}_{A2} + \mathbf{I}_{A0})$$

$$\mathbf{I}_{A0} = (a^2 + a)\mathbf{I}_{A1} + (a^2 + a)\mathbf{I}_{A2}$$

$$\mathbf{I}_{A0} = -\mathbf{I}_{A1} - \mathbf{I}_{A2} \qquad\qquad (13\text{--}60)$$

We also know that

$$\mathbf{V}_{A1} = \mathbf{E}_{A1} - \mathbf{I}_{A1}Z_1 \qquad\qquad (13\text{--}38)$$

$$\mathbf{V}_{A2} = -\mathbf{I}_{A2}Z_2 \qquad\qquad (13\text{--}39)$$

$$\mathbf{V}_{A0} = -\mathbf{I}_{A0}Z_0 \qquad\qquad (13\text{--}40)$$

We can solve Equations (13–38) through (13–40) for $\mathbf{I}_{A0}$ and $\mathbf{I}_{A2}$ using the fact that the three voltages are equal, and then substitute that result into Equation (13–60) to calculate the current $\mathbf{I}_{A1}$. From Equations (13–38) and (13–39),

$$\mathbf{E}_{A1} - \mathbf{I}_{A1}Z_1 = -\mathbf{I}_{A2}Z_2$$

$$\mathbf{I}_{A2} = \frac{\mathbf{I}_{A1}Z_1 - \mathbf{E}_{A1}}{Z_2} \qquad\qquad (13\text{--}61)$$

From Equations (13–38) and (13–40),

$$\mathbf{E}_{A1} - \mathbf{I}_{A1}Z_1 = -\mathbf{I}_{A0}Z_0$$

$$\mathbf{I}_{A0} = \frac{\mathbf{I}_{A1}Z_1 - \mathbf{E}_{A1}}{Z_0} \qquad (13\text{–}62)$$

Substituting these values into Equation (13–60) yields

$$\mathbf{I}_{A0} = -\mathbf{I}_{A1} - \mathbf{I}_{A2} \qquad (13\text{–}60)$$

$$\frac{\mathbf{I}_{A1}Z_1 - \mathbf{E}_{A1}}{Z_0} = -\mathbf{I}_{A1} - \frac{\mathbf{I}_{A1}Z_1 - \mathbf{E}_{A1}}{Z_2}$$

$$\mathbf{I}_{A1}Z_1Z_2 - \mathbf{E}_{A1}Z_2 = -\mathbf{I}_{A1}Z_0Z_2 - (\mathbf{I}_{A1}Z_0Z_1 - \mathbf{E}_{A1}Z_0)$$

$$\mathbf{I}_{A1}Z_1Z_2 + \mathbf{I}_{A1}Z_0Z_2 + \mathbf{I}_{A1}Z_0Z_1 = \mathbf{E}_{A1}Z_2 + \mathbf{E}_{A1}Z_0$$

$$\mathbf{I}_{A1} = \frac{\mathbf{E}_{A1}\,(Z_2 + Z_0)}{Z_1Z_2 + Z_0Z_1 + Z_0Z_2}$$

$$\boxed{\mathbf{I}_{A1} = \frac{\mathbf{E}_{A1}}{Z_1 + \dfrac{Z_2Z_0}{Z_2 + Z_0}}} \qquad (13\text{–}63)$$

For a double line-to-ground fault, the voltages $\mathbf{V}_{A1} = \mathbf{V}_{A2} = \mathbf{V}_{A0}$, and the impedance is the sum of Z_1 plus the parallel combination of Z_2 and Z_0. Note that this result is equivalent to connecting all three sequence networks in *parallel* at the terminals representing the location of the fault (see Figure 13–17). Such a connection would make the three sequence voltages equal, and would make $\mathbf{I}_{A1} = -(\mathbf{I}_{A2} + \mathbf{I}_{A0})$. Once $\mathbf{I}_{A1}$ is known, we can calculate $\mathbf{I}_{A2}$ and $\mathbf{I}_{A0}$ either by using the current divider rule, or from Equations (13–38) through (13–40).

Finally, we can calculate the total phase voltages for all phases in the fault from Equations (13–8) to (13–10), and the total phase currents for all phases in the fault from Equations (13–23) to (13–25).

Figure 13–17 | For a double line-to-line fault, the positive-, negative-, and zero-sequence networks are connected in parallel to determine the sequence voltages and currents in the fault.

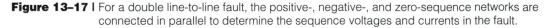

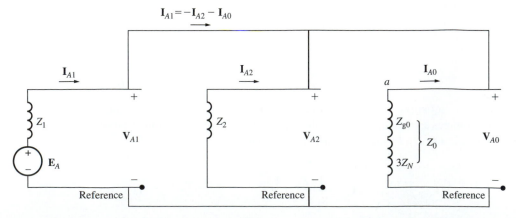

In summary, to calculate the voltages and currents in a single line-to-ground fault on a synchronous generator:

1. Create the positive-, negative-, and zero-sequence networks for the generator.
2. Connect the networks in parallel at the terminals representing the location of the fault.
3. Calculate $\mathbf{I}_{A1}$ from this network.
4. Calculate $\mathbf{V}_{A0}$, $\mathbf{V}_{A1}$, and $\mathbf{V}_{A2}$ from Equation (13–38), noting that the three voltages are equal.
5. Calculate $\mathbf{I}_{A0}$ and $\mathbf{I}_{A2}$ from Equations (13–39) and (13–40).
6. Calculate any required voltages and currents by combining the sequence components calculated above using the equations given in this chapter.

EXAMPLE 13–5

A 100-MVA, 13.8-kV, Y-connected synchronous generator has a subtransient reactance X'' of 0.20 pu, a negative-sequence reactance X_2 of 0.25 pu, and a zero-sequence reactance of 0.10 pu, and negligible resistance. The neutral of the generator is solidly grounded. Assume that the generator is initially unloaded and operating at rated voltage, and that a double line-to-ground fault occurs on phase *a* at the terminals of the generator. Find the voltages and currents in each phase during the subtransient period immediately after the fault occurs. Also, find the line-to-line voltages at that time.

Solution

The base apparent power for this machine is 100 MVA and the base line-to-line voltage is 13.8 kV, so the base line to neutral voltage will be

$$V_{\phi,\,base} = \frac{V_{LL,\,base}}{\sqrt{3}} = \frac{13.8\text{ kV}}{\sqrt{3}} = 7.967\text{ kV}$$

and the base current will be

$$I_{base} = \frac{S_{3\phi,\,base}}{\sqrt{3}V_{LL,\,base}} = \frac{100\text{ MVA}}{\sqrt{3}(13.8\text{ kV})} = 4184\text{ A} \qquad (10\text{–}4)$$

Since this machine initially operating at no load and rated voltage, the voltage behind subtransient reactance will be $\mathbf{E}_A = 1\angle0°$ pu. The sequence networks for this generator will be as shown in Figure 13–17, with $Z_1 = j0.20$, $Z_2 = j0.25$, and $Z_0 = j0.10$. The resulting positive-sequence currents $\mathbf{I}_{A1}$ will be

$$\mathbf{I}_{A1} = \frac{\mathbf{E}_{A1}}{Z_1 + \dfrac{Z_2 Z_0}{Z_2 + Z_0}} \tag{13-63}$$

$$= \frac{1\angle 0°}{j0.20 + \dfrac{(j0.25)(j0.10)}{j0.25 + j0.10}}$$

The sequence voltages in phase a will be:

$$\mathbf{V}_{A1} = \mathbf{E}_{A1} - \mathbf{I}_{A1} Z_1 \tag{13-38}$$
$$= 1.0\angle 0° - (3.684\angle -90°)(j0.20) = 0.263\angle 0° \text{ pu}$$
$$\mathbf{V}_{A2} = \mathbf{V}_{A1} = 0.263\angle 0° \text{ pu}$$
$$\mathbf{V}_{A0} = \mathbf{V}_{A1} = 0.263\angle 0° \text{ pu}$$

The negative- and zero-sequence currents will be

$$\mathbf{I}_{A2} = -\frac{\mathbf{V}_{A2}}{Z_2} = -\frac{0.263\angle 0°}{j0.25} = 1.052\angle 90° \tag{13-39}$$

$$\mathbf{I}_{A0} = -\frac{\mathbf{V}_{A0}}{Z_0} = -\frac{0.263\angle 0°}{j0.10} = 2.630\angle 90° \tag{13-40}$$

Therefore, the phase and neutral currents in this generator will be

$$\mathbf{I}_A = \mathbf{I}_{A1} + \mathbf{I}_{A2} + \mathbf{I}_{A0} \tag{13-23}$$
$$= 3.684\angle -90° + 1.052\angle 90° + 2.630\angle 90°$$
$$= 0.0\angle 0° \text{ pu}$$
$$\mathbf{I}_B = a^2 \mathbf{I}_{A1} + a\mathbf{I}_{A2} + \mathbf{I}_{A0} \tag{13-24}$$
$$= a^2(3.684\angle -90°) + a(1.052\angle 90°) + 2.630\angle 90°$$
$$= 5.692\angle 136.1° \text{ pu}$$
$$\mathbf{I}_C = a\mathbf{I}_{A1} + a^2 \mathbf{I}_{A2} + \mathbf{I}_{A0} \tag{13-25}$$
$$= a(3.684\angle -90°) + a^2(1.052\angle 90°) + 2.630\angle 90°$$
$$= 5.692\angle 43.9° \text{ pu}$$
$$\mathbf{I}_N = 3\mathbf{I}_{A0} = 3(2.630\angle 90°) = 7.890\angle 90°$$

Note that the phase currents in phases a is zero, which makes sense, it was not shorted, and there was no current flow before the fault occurred. The actual fault current during the subtransient period in amps is

$$I_B = I_C = I_{B,\text{pu}} I_{\text{base}} = (5.692)(4184 \text{ A}) = 23,820 \text{ A}$$

The phase voltages in this generator are

$$\mathbf{V}_A = \mathbf{V}_{A1} + \mathbf{V}_{A2} + \mathbf{V}_{A0} \tag{13-8}$$
$$= 0.263\angle 0° + 0.263\angle 0° + 0.263\angle 0°$$
$$= 0.789\angle 0°$$

$$\mathbf{V}_B = a^2 \mathbf{V}_{A1} + a\mathbf{V}_{A2} + \mathbf{V}_{A0} \tag{13-9}$$
$$= a^2(0.263\angle 0°) + a(0.263\angle 0°) + 0.263\angle 0°$$
$$= 0.0\angle 0°$$

$$\mathbf{V}_C = a\mathbf{V}_{A1} + a^2\mathbf{V}_{A2} + \mathbf{V}_{A0} \qquad (13\text{–}10)$$
$$= a^2(0.263\angle 0°) + a(0.263\angle 0°) + 0.263\angle 0°$$
$$= 0.0\angle 0°$$

The line-to-line voltages are therefore

$$\mathbf{V}_{AB} = \mathbf{V}_A - \mathbf{V}_B = 0.789\angle 0° - 0.0\angle 0° = 0.789\angle 0° \text{ pu}$$
$$\mathbf{V}_{BC} = \mathbf{V}_B - \mathbf{V}_C = 0.0\angle 0° - 0.0\angle 0° = 0.0\angle 0° \text{ pu}$$
$$\mathbf{V}_{CA} = \mathbf{V}_C - \mathbf{V}_A = 0.0\angle 0° - 0.789\angle 0° = 0.789\angle 180° \text{ pu}$$

The line-to-line voltages expressed in volts are equal to the per-unit values multiplied by the base voltage of a phase

$$V_{AB} = (0.789 \text{ pu})(7.967 \text{ kV}) = 6.29 \text{ kV}$$
$$V_{BC} = (0.0 \text{ pu})(7.967 \text{ kV}) = 0.0 \text{ kV}$$
$$V_{CA} = (0.789 \text{ pu})(7.967 \text{ kV}) = 6.29 \text{ kV}$$

■

13.7 | UNSYMMETRICAL FAULTS ON POWER SYSTEMS

In the last three sections, we learned how to calculate the voltages and currents in an unsymmetrical fault on a three-phase synchronous generator. We will now extend that discussion to consider calculating the voltages and currents in unsymmetrical faults on a more complex power system.

The key to extending our treatment of generators to more complex power systems is Thevenin's theorem. Thevenin's theorem states that any linear circuit that can be separated by two terminals from the rest of the circuit can be replaced by a single voltage source in series with an impedance.

To understand the significance of this theorem for fault analysis, consider the power system shown in Figure 13–18. Assume that there is a small stub of wire

Figure 13–18 | A simple power system with a fault at the point indicated by the ×.

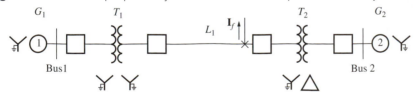

attached to each phase of the power system at point ×, and that a fault occurs on that stub.[2] If a fault occurs on the stub, then a current I_f will flow out of the fault. The positive-, negative-, and zero-sequence networks for this power system are shown in Figure 13–19, with the location of the fault labeled on each network. Note that each

---

[2]This stub of wire does not really have to exist, but it is a convenient concept, since all of the fault current from everywhere in the network will be concentrated and flowing through the stub. The stub represents a place to isolate the power system from the outside world.

Figure 13–19 | The positive-, negative-, and zero-sequence networks for the power system shown in Figure 13–18, together with their Thevenin equivalent circuits. Note that $\mathbf{V}_f$ is the prefault voltage at the location of the fault.

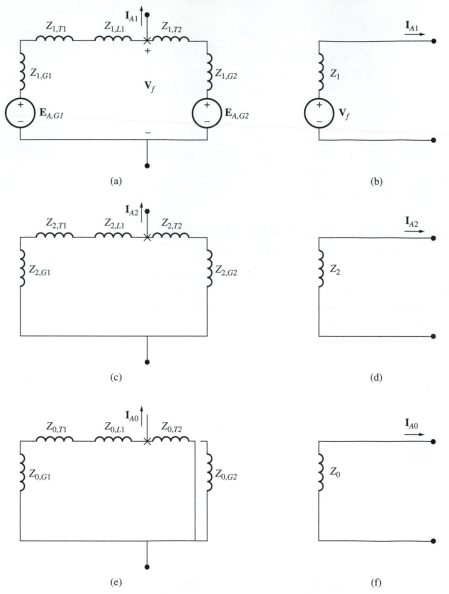

sequence network can be isolated from the rest of the circuit by two terminals, one at the stub and one at the reference. Therefore, we can create a Thevenin equivalent circuit for each network, as seen from the location of the fault. These Thevenin equivalent circuits are shown to the right of the corresponding sequence networks. Notice that *the Thevenin equivalents of the positive-, negative-, and zero-sequence circuits look exactly like the positive-, negative-, and zero-sequence equivalent circuits of a*

Figure 13–20 | Connection diagram for a single line-to-ground fault on a power system.

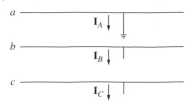

generator. Therefore, everything that we have learned in the last three sections about faults on synchronous generators will also apply to power systems, if we represent the power systems by their Thevenin equivalents.

Single Line-to-Ground Faults on a Power System

For a single line-to-ground fault, the stubs of wire are connected as shown in Figure 13–20. The following relationships exist at the fault

$$
\left.\begin{array}{l}
\mathbf{I}_B = 0 \\
\mathbf{I}_C = 0 \\
\mathbf{V}_A = 0
\end{array}\right\} \text{ single line-to-ground fault} \qquad (13\text{–}64)
$$

These relationships are the same as those for a single line-to-ground fault on a synchronous generator, so the symmetrical components of the fault voltages and currents will have the same solutions as for the synchronous generator, except that the prefault voltage $\mathbf{V}_f$ replaces $\mathbf{E}_A$. Thus, the current $\mathbf{I}_{A1}$ will be given by the equation

$$
\mathbf{I}_{A1} = \frac{\mathbf{V}_f}{Z_0 + Z_1 + Z_2} \qquad (13\text{–}65)
$$

where $\mathbf{V}_f$ is the *prefault* voltage at the location of the fault and Z_0, Z_1, and Z_2 are the Thevenin equivalent impedances of the power system sequence diagrams as seen from the location of the fault. The three sequence networks should be connected in series *at the fault point P* to calculate the effects of a single line-to-ground fault, as shown in Figure 13–21.

Line-to-Line Faults on a Power System

For a line-to-line fault, the stubs of wire are connected as shown in Figure 13–22. The following relationships exist at the fault

$$
\left.\begin{array}{l}
\mathbf{V}_B = \mathbf{V}_C \\
\mathbf{I}_A = 0 \\
\mathbf{I}_B = \mathbf{I}_C
\end{array}\right\} \text{ line-to-line fault} \qquad (13\text{–}66)
$$

Figure 13–21 I For a single line-to-ground fault on a power system, the positive-, negative-, and zero-sequence networks are connected in series *at the point of the fault P* to determine the sequence voltages and currents in the fault.

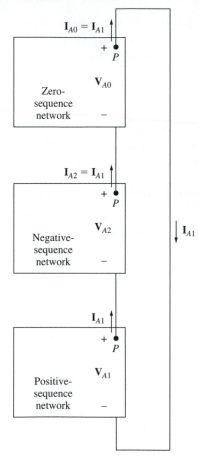

Figure 13–22 I Connection diagram for a line-to-line fault on a power system.

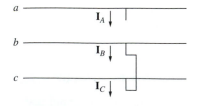

These relationships are the same as those for a single line-to-line fault on a synchronous generator, so the symmetrical components of the fault voltages and currents will have the same solutions as for the synchronous generator, except that the pre-fault voltage $\mathbf{V}_f$ replaces $\mathbf{E}_A$. Thus, the current $\mathbf{I}_{A1}$ will be given by the equation

$$\boxed{\mathbf{I}_{A1} = \frac{\mathbf{V}_f}{Z_1 + Z_2}} \qquad (13\text{--}67)$$

where $\mathbf{V}_f$ is the *prefault* voltage at the location of the fault and Z_1 and Z_2 are the Thevenin equivalent impedances of the power system sequence diagrams as seen from the location of the fault. The positive- and negative-sequence networks should be connected in parallel *at the fault point P* to calculate the effects of a line-to-line fault, as shown in Figure 13–23.

Figure 13–23 | For a line-to-line fault on a power system, the positive- and negative-sequence networks are connected in parallel *at the point of the fault P* to determine the sequence voltages and currents in the fault.

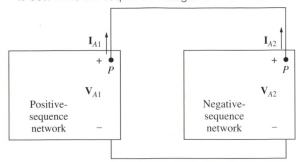

Double Line-to-Ground Faults on a Power System

For a double line-to-ground fault, the stubs of wire are connected as shown in Figure 13–24. The following relationships exist at the fault

$$\left. \begin{array}{l} \mathbf{V}_B = 0 \\ \mathbf{V}_C = 0 \\ \mathbf{I}_A = 0 \end{array} \right\} \text{ double line-to-ground fault} \qquad (13\text{--}68)$$

Figure 13–24 | Connection diagram for a double line-to-ground fault on a power system.

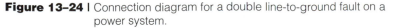
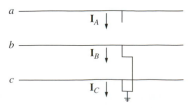

These relationships are the same as those for a double line-to-ground fault on a synchronous generator, so the symmetrical components of the fault voltages and currents will have the same solutions as for the synchronous generator, except that the prefault voltage $\mathbf{V}_f$ replaces $\mathbf{E}_A$. Thus, the current $\mathbf{I}_{A1}$ will be given by the equation

$$\mathbf{I}_{A1} = \frac{\mathbf{V}_f}{Z_1 + \dfrac{Z_2 Z_0}{Z_2 + Z_0}} \tag{13-69}$$

where $\mathbf{V}_f$ is the *prefault* voltage at the location of the fault and Z_0, Z_1, and Z_2 are the Thevenin equivalent impedances of the power system sequence diagrams as seen from the location of the fault. The three sequence networks should be connected in parallel *at the fault point P* to calculate the effects of a single line-to-ground fault, as shown in Figure 13–25.

Figure 13–25 I For a double line-to-ground fault on a power system, the positive-, negative-, and zero-sequence networks are connected in parallel *at the point of the fault P* to determine the sequence voltages and currents in the fault.

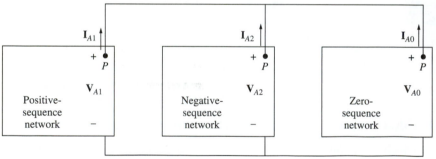

EXAMPLE 13–6

The devices in power system shown in Figure 13–26 have the following ratings:

G_1: 100 MVA, 13.8 kV, grounded Y, $X'' = 0.15$ pu, $X_2 = 0.15$ pu, $X_0 = 0.05$ pu
G_2: 50 MVA, 20.0 kV, grounded Y, $X'' = 0.15$ pu, $X_2 = 0.15$ pu, $X_0 = 0.05$ pu
T_1: 100 MVA, 13.8/138 kV, Y–Y (both grounded), $X_1 = 0.08$ pu, $X_2 = 0.08$ pu, $X_0 = 0.08$ pu
T_2: 50 MVA, 20.0/138 kV, Δ–Y (grounded), $X_1 = 0.08$ pu, $X_2 = 0.08$ pu, $X_0 = 0.08$ pu
L_1: 100 MVA, 138 kV, $X_1 = 19\ \Omega$, $X_2 = 19\ \Omega$, $X_0 = 50\ \Omega$

Assume that all resistances in the power system can be neglected. The system base apparent power is 100 MVA and base line voltage is 13.8 kV at bus 1. Assume that the voltage at bus 1 is 13.8 kV and that the power system is initially unloaded. Answer the following questions about this power system.

a. Find the subtransient per-phase equivalent circuits of this power system for positive-, negative-, and zero-sequence components.

Figure 13–26 | The power system of Example 13–6.

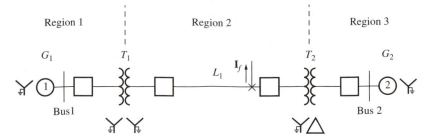

b. Find the subtransient fault current at the fault for a symmetrical three-phase fault at the location indicated by an ×.

c. Find the subtransient fault current at the fault for a single line-to-ground fault at the location indicated by an ×.

d. Find the subtransient fault current at the fault for a line-to-line fault at the location indicated by an ×.

e. Find the subtransient fault current at the fault for a double line-to-ground fault at the location indicated by an ×.

Solution

The base values for this system can be found from Equations (10–1) to (10–6). They are:

	Region 1	**Region 2**	**Region 3**
S_{base}	100 MVA	100 MVA	100 MVA
$V_{LL, base}$	13.8 kV	138 kV	20 kV
$V_{\phi, base}$	7.967 kV	79.67 kV	11.55 kV
I_{base}	4184 A	418.4 A	2887 A
Z_{base}	1.904 Ω	190.4 Ω	4.000 Ω

a. The per-unit values for generator G_1 and transformer T_1 are already to the proper base. The impedances for generator G_2 are:

$$\text{Per-unit } Z_{new} = \text{per-unit } Z_{given} \left(\frac{V_{given}}{V_{new}}\right)^2 \left(\frac{S_{new}}{S_{given}}\right) \qquad (10\text{–}8)$$

$$X''_{G2} = 0.15 \left(\frac{20 \text{ kV}}{20 \text{ kV}}\right)^2 \left(\frac{100 \text{ MVA}}{50 \text{ MVA}}\right) = 0.30$$

$$X_{2, G2} = 0.15 \left(\frac{20 \text{ kV}}{20 \text{ kV}}\right)^2 \left(\frac{100 \text{ MVA}}{50 \text{ MVA}}\right) = 0.30$$

$$X_{0, G2} = 0.05 \left(\frac{20 \text{ kV}}{20 \text{ kV}}\right)^2 \left(\frac{100 \text{ MVA}}{50 \text{ MVA}}\right) = 0.10$$

The impedances for transformer T_2 are:

$$\text{Per-unit } Z_{new} = \text{per-unit } Z_{given} \left(\frac{V_{given}}{V_{new}}\right)^2 \left(\frac{S_{new}}{S_{given}}\right) \tag{10–8}$$

$$X_{1,\,T2} = X_{2,\,T2} = X_{0,\,T2} = 0.08 \left(\frac{20\ kV}{20\ kV}\right)^2 \left(\frac{100\ MVA}{50\ MVA}\right) = 0.16$$

The impedances for the transmission line L_1 are:

$$\text{Quantity in per unit} = \frac{\text{actual value}}{\text{base value of quantity}} \tag{10–7}$$

$$X_{1,\,L1} = X_{2,\,L2} = \frac{19\ \Omega}{190.4\ \Omega} = 0.10\ \text{pu}$$

$$X_{0,\,L1} = \frac{50\ \Omega}{190.4\ \Omega} = 0.263\ \text{pu}$$

Also, since we assume that the power system was initially unloaded before the fault occurred, the voltage sources in the positive-sequence network will be $1.0\angle 0°$. Therefore, the sequence networks are as shown in Figure 13–27.

b.   To calculate the symmetrical three-phase fault current at the fault, only the positive-sequence network is needed, and a short will be connected from the point of the fault back to the reference (see Figure 13–28a). To find the current $\mathbf{I}_{A1}$, we will calculate the Thevenin equivalent circuit of the positive sequence network between the location of the fault and the reference bus. To determine the Thevenin equivalent voltage, open-circuit the terminals of the network and measure the voltage at the terminals, which is $\mathbf{V}_f$. Since the network was initially unloaded, $\mathbf{V}_f = 1.0\angle 0°$. Then, short-circuit the voltages sources and look for the impedance between the terminals. This impedance is given by the parallel combination of $j0.15 + j0.08 + j0.10$ and $j0.16 + j0.30$:

$$Z_{1,\,TH} = \frac{(j0.15 + j0.08 + j0.10)(j0.016 + j0.30)}{j0.15 + j0.08 + j0.10 + j0.016 + j0.30} = j0.192\ \text{pu}$$

The resulting equivalent circuit is shown in Figure 13–28b. The positive-sequence current is thus

$$\mathbf{I}_{A1} = \frac{1.0\angle 0°}{j0.192} = 5.208\angle -90°\ \text{pu}$$

Since the positive-sequence current is the only current present in a symmetrical three-phase fault, $\mathbf{I}_A = \mathbf{I}_{A1} = 5.208$ pu, and the actual fault current in amperes is

Figure 13–27 | The positive-, negative-, and zero-sequence networks for the power
system in Example 13–6.

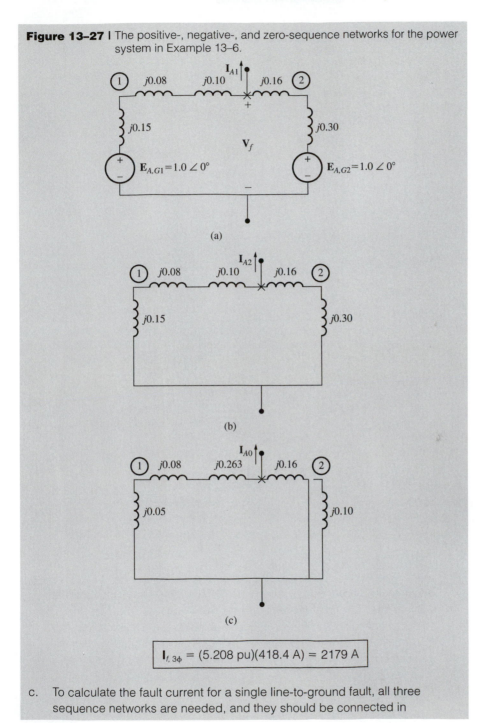

$$\mathbf{I}_{f,\,3\phi} = (5.208 \text{ pu})(418.4 \text{ A}) = 2179 \text{ A}$$

c. To calculate the fault current for a single line-to-ground fault, all three
sequence networks are needed, and they should be connected in

Figure 13–28 | (a) The connections for a symmetrical three-phase fault. (b) The Thevenin equivalent circuit for the positive-sequence network.

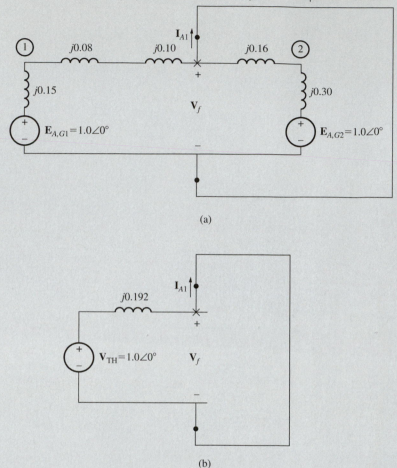

(a)

(b)

series (see Figure 13–29a). To find the current $I_{A1} = I_{A2} = I_{A0}$, we will calculate the Thevenin equivalent circuit of the three sequence networks between the location of the fault and the reference bus. The Thevenin equivalent circuit of the positive-sequence network was calculated above. To calculate the Thevenin equivalent of the negative-sequence network, we need only worry about the impedance, since there are no voltage sources. This impedance is given by the parallel combination of $j0.15 + j0.08 + j0.10$ and $j0.16 + j0.30$:

$$Z_{2,\,TH} = \frac{(j0.15 + j0.08 + j0.10)(j0.16 + j0.30)}{j0.15 + j0.08 + j0.10 + j0.16 + j0.30} = 0.192 \text{ pu}$$

Figure 13–29 | (a) The connections for a single line-to-ground fault.
(b) The Thevenin equivalent circuit for the sequence networks.

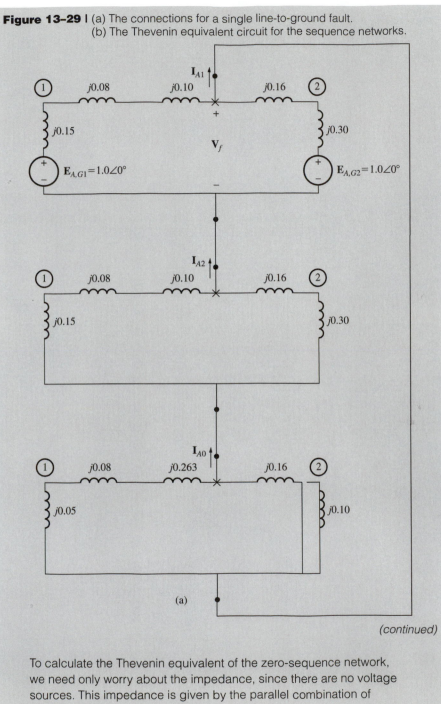

(a)

(continued)

To calculate the Thevenin equivalent of the zero-sequence network, we need only worry about the impedance, since there are no voltage sources. This impedance is given by the parallel combination of $j0.05 + j0.08 + j0.263$ and $j0.16$:

Figure 13–29 I *(concluded)*

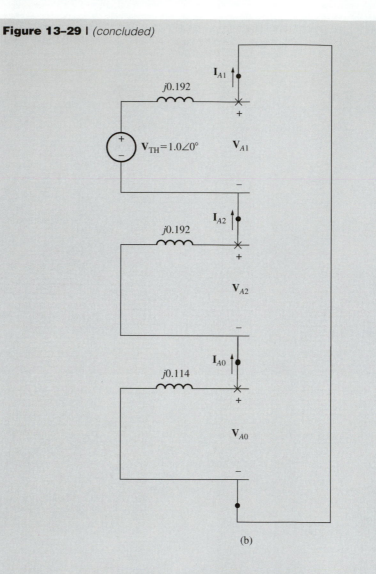

(b)

$$Z_{0,\,TH} = \frac{(j0.05 + j0.08 + j0.263)(j0.16)}{j0.05 + j0.08 + j0.263 + j0.16} = j0.114 \text{ pu}$$

The resulting equivalent circuit is shown in Figure 13–29b. The positive-sequence current is thus

$$\mathbf{I}_{A1} = \mathbf{I}_{A2} = \mathbf{I}_{A0} = \frac{1.0\angle 0°}{j0.192 + j0.192 + j0.114} = 2.008\angle -90° \text{ pu}$$

Figure 13–30 | (a) The connections for a line-to-line fault. (b) The Thevenin equivalent circuit for the sequence networks.

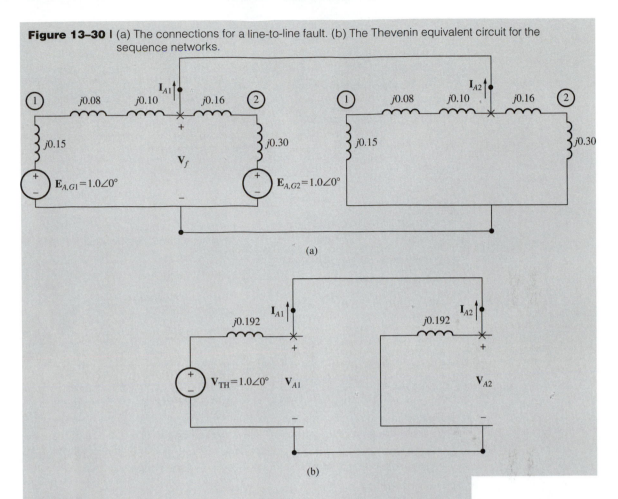

(a)

(b)

The fault current in phase a is

$$\boxed{I_A = I_{A1} + I_{A2} + I_{A0} = 6.024\angle{-90°}\ \text{pu}}$$

The actual fault current in amperes is

$$\boxed{I_{f,\,SLG} = (6.024\ \text{pu})(418.4\ \text{A}) = 2520\ \text{A}}$$

d. To calculate the fault current for a line-to-line fault, the positive- and negative-sequence networks are needed, and they should be connected in parallel (see Figure 13–30a). To find the current $I_{A1} = -I_{A2}$, we will calculate the Thevenin equivalent circuit of the two sequence networks between the location of the fault and the reference bus. This calculation was already done above. The resulting equivalent circuit is shown in Figure 13–30b. The positive-sequence current is thus

$$\mathbf{I}_{A1} = \frac{\mathbf{V}_f}{Z_1 + Z_2} \tag{13-67}$$

$$= \frac{1.0\angle 0°}{j0.192 + j0.192} = 2.604\angle -90° \text{ pu}$$

$$\mathbf{I}_{A2} = -\mathbf{I}_{A2} = 2.604\angle -90° \text{ pu}$$

The fault current in phase b is

$$\mathbf{I}_B = a^2\mathbf{I}_{A1} + a\mathbf{I}_{A2} + \mathbf{I}_{A0} \tag{13-24}$$

$$= a^2(2.604\angle -90°) + a(2.604\angle 90°) + 0.0\angle 0°$$

$$\boxed{\mathbf{I}_B = 4.510\angle 180° \text{ pu}}$$

and the fault current in phase c is $\mathbf{I}_C = 4.510\angle 0°$ pu. The actual fault current in amperes is

$$\boxed{\mathbf{I}_{f,\,LL} = (4.510 \text{ pu})(418.4 \text{ A}) = 1887 \text{ A}}$$

e. To calculate the fault current for a double line-to-ground fault, all three sequence networks are needed, and they should be connected in parallel. To find the current $\mathbf{I}_{A1}$, we must calculate the Thevenin equivalent circuit of the three sequence networks between the location of the fault and the reference bus. This calculation was already done above. The resulting equivalent circuit is shown in Figure 13–31. The positive-sequence current is thus

Figure 13–31 | The Thevenin equivalent circuit for the sequence networks in a double line-to-ground fault.

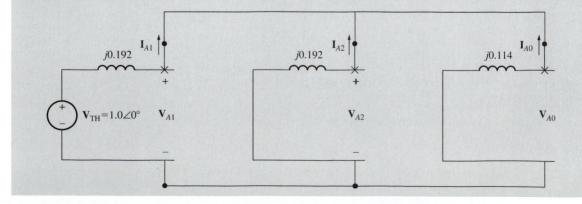

$$I_{A1} = \frac{V_f}{Z_1 + \dfrac{Z_2 Z_0}{Z_2 + Z_0}} \qquad (13\text{–}69)$$

$$= \frac{1\angle 0°}{j0.192 + \dfrac{(j0.192)(j0.114)}{j0.192 + j0.114}} = 3.80\angle -90° \text{ pu}$$

The sequence voltages in phase a will be:

$$V_{A1} = E_{A1} - I_{A1}Z_1 \qquad (13\text{–}38)$$
$$= 1.0\angle 0° - (3.80\angle -90°)(j0.192) = 0.271\angle 0° \text{ pu}$$
$$V_{A2} = V_{A1} = 0.271\angle 0° \text{ pu}$$
$$V_{A0} = V_{A1} = 0.271\angle 0° \text{ pu}$$

The negative- and zero-sequence currents will be

$$I_{A2} = -\frac{V_{A2}}{Z_2} = -\frac{0.271\angle 0°}{j0.192} = 1.41\angle 90° \qquad (13\text{–}39)$$

$$I_{A0} = -\frac{V_{A0}}{Z_0} = -\frac{0.271\angle 0°}{j0.114} = 2.38\angle 90° \qquad (13\text{–}40)$$

The fault current in phase b is

$$I_B = a^2 I_{A1} + a I_{A2} + I_{A0} \qquad (13\text{–}24)$$
$$= a^2(3.80\angle -90°) + a(1.41\angle 90°) + 2.38\angle 90°$$

$$\boxed{I_B = 5.76\angle 141.6° \text{ pu}}$$

The fault current in phase c is

$$I_C = a I_{A1} + a^2 I_{A2} + I_{A0} \qquad (13\text{–}24)$$
$$= a(3.80\angle -90°) + a^2(1.41\angle 90°) + 2.38\angle 90°$$

$$\boxed{I_C = 5.76\angle 38.4° \text{ pu}}$$

The actual fault current in amperes is

$$\boxed{I_{f,\,DLG} = (5.76 \text{ pu})(418.4 \text{ A}) = 2410 \text{ A}} \qquad \blacksquare$$

13.8 | ANALYSIS OF UNSYMMETRICAL FAULTS USING THE BUS IMPEDANCE MATRIX

In Chapter 12, we used the positive-sequence bus impedance matrix Z_{bus} to solve for the fault current in a symmetrical three-phase fault, and the voltages and currents at various other busses during the fault. It is also possible to create bus impedance matrices for negative- and zero-sequence networks, and they can be strung together in different ways to calculate voltages and currents for unsymmetrical faults in a power system.

For example, consider a single line-to-ground fault. If a single line-to-ground fault occurs at Bus i in a power system, then we know that the positive-sequence fault current at Bus i can be found by connecting the three sequence diagrams in series at the point of the fault. The positive-sequence current I_{A1} will be equal to the pre-fault voltage $\mathbf{V}_f$ divided by the sum of the Thevenin equivalent impedances of each sequence network.

$$\mathbf{I}_{A1} = \frac{\mathbf{V}_f}{Z_0 + Z_1 + Z_2} \tag{13–65}$$

If a sequence network is represented by a bus impedance matrix $\mathbf{Z}_{\text{bus}}$, then *the on-diagonal terms of the bus impedance matrix Z_{ii} are just the Thevenin equivalent impedances of the network at the corresponding busses.* Therefore, if we have calculated the bus impedance matrices of the three sequence networks, the positive-sequence current $\mathbf{I}_{A1}$ from a single line-to-ground fault at Bus i can be found using the Z_{ii} terms from the bus impedance matrices:

$$\mathbf{I}_{A1} = \frac{\mathbf{V}_f}{Z_{ii,\,1} + Z_{ii,\,2} + Z_{ii,\,0}} \tag{13–70}$$

where $Z_{ii,\,1}$ is the impedance of element ii of the positive-sequence bus impedance matrix, $Z_{ii,\,2}$ is the impedance of element ii of the negative-sequence bus impedance matrix, and $Z_{ii,\,0}$ is the impedance of element ii of the zero-sequence bus impedance matrix.

Similarly, if a line-to-line fault occurs at Bus i, the positive-sequence fault current at Bus i can be found by the equation:

$$\mathbf{I}_{A1} = \frac{\mathbf{V}_f}{Z_{ii,\,1} + Z_{ii,\,2}} \tag{13–71}$$

Finally, if a double line-to-ground fault occurs at Bus i, then the positive-sequence fault current at Bus i can be found by the equation:

$$\mathbf{I}_{A1} = \frac{\mathbf{V}_f}{Z_{ii,\,1} + \dfrac{Z_{ii,\,2}Z_{ii,\,0}}{Z_{ii,\,2} + Z_{ii,\,0}}} \tag{13–72}$$

Once the positive-sequence fault current at Bus i ($\mathbf{I}_{A1}$) has been found, then the negative- and zero-sequence currents can be found from the equations appropriate to the type of fault being analyzed, and the actual fault currents in each phase can be found from Equations (13–20) to (13–22).

EXAMPLE 13–7

Perform the following calculations for the power system in Example 13–6.

(a) Calculate the subtransient positive-, negative-, and zero-sequence bus admittance matrices ($\mathbf{Y}_{\text{bus}}$) and impedance matrices ($\mathbf{Z}_{\text{bus}}$) of the power system.

(b) Find the subtransient current flowing in a single line-to-ground fault at the location indicated by the "×."

Solution

To solve this problem, we must calculate the bus admittance matrices $\mathbf{Y}_{bus}$ for each sequence network, and then invert them to get the corresponding bus impedance matrices $\mathbf{Z}_{bus}$. Figure 13–32 shows the sequence networks for this power system with three busses defined. Note that we have defined a new "Bus 3" at the location of the fault—the system must include a bus at the location of the fault, or else the impedances that we need won't get calculated.

(a) The bus admittance matrices for this three-bus system can be found as described in Chapter 10. The positive-sequence bus admittance matrix for this system is

$$\mathbf{Y}_{bus,\,1} = \begin{bmatrix} -j12.223 & 0 & j5.556 \\ 0 & -j9.583 & j6.25 \\ j5.556 & j6.25 & -j11.806 \end{bmatrix}$$

The negative-sequence bus admittance matrix is just the same as the positive-sequence matrix.

$$\mathbf{Y}_{bus,\,2} = \begin{bmatrix} -j12.223 & 0 & j5.556 \\ 0 & -j9.583 & j6.25 \\ j5.556 & j6.25 & -j11.806 \end{bmatrix}$$

The zero-sequence bus admittance matrix is a little trickier. Note that for zero-sequence currents, Bus 2 is not connected to any other bus in the power system. The bus admittance matrix is

$$\mathbf{Y}_{bus,\,0} = \begin{bmatrix} -j22.914 & 0 & j2.914 \\ 0 & -j10.0 & 0 \\ j2.914 & 0 & -j9.164 \end{bmatrix}$$

The bus impedance matrices are the reciprocals of the bus admittance matrices.

$$\mathbf{Z}_{bus,\,1} = \begin{bmatrix} j0.1215 & j0.0570 & j0.0873 \\ j0.0570 & j0.1861 & j0.1253 \\ j0.0873 & j0.1253 & j0.1921 \end{bmatrix}$$

$$\mathbf{Z}_{bus,\,2} = \begin{bmatrix} j0.1215 & j0.0570 & j0.0873 \\ j0.0570 & j0.1861 & j0.1253 \\ j0.0873 & j0.1253 & j0.1921 \end{bmatrix}$$

$$\mathbf{Z}_{bus,\,0} = \begin{bmatrix} j0.0455 & 0 & j0.0145 \\ 0 & j0.1000 & 0 \\ j0.0145 & 0 & j0.1137 \end{bmatrix}$$

(b) The positive-sequence current at the fault (Bus 3) is

Figure 13–32 I The subtransient positive-, negative-, and zero-sequence networks with the impedances expressed as admittances. (Note that the equivalent admittance of the series combinations of admittances are also shown to simplify calculations.)

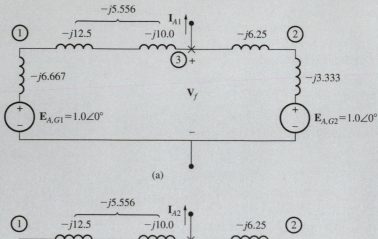

(a)

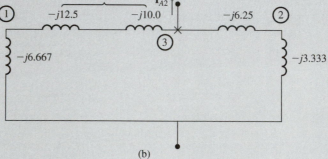

(b)

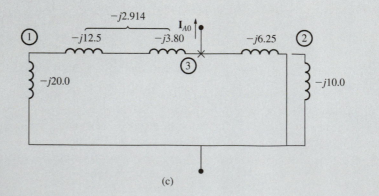

(c)

$$I_{A1} = \frac{V_f}{Z_{33,1} + Z_{33,2} + Z_{33,0}}$$ (13–70)

$$I_{A1} = \frac{1\angle 0°}{j0.1921 + j0.1921 + j0.1137} = 2.008\angle -90° \text{ pu}$$

Because the sequence networks are connected in series for a single line-to-ground fault, the negative- and zero-sequence currents are the same as the positive-sequence current.

$$I_{A1} = I_{A2} = I_{A0} = 2.008\angle -90° \text{ pu}$$

Therefore, the total subtransient fault current in the fault is

$$I_A = I_{A1} + I_{A2} + I_{A0} = 6.024\angle -90° \text{ pu}$$
$$I_B = a^2 I_{A1} + a I_{A2} + I_{A0} = 0\angle 0° \text{ pu}$$
$$I_C = a I_{A1} + a^2 I_{A2} + I_{A0} = 0\angle 0° \text{ pu}$$

Current flows in phase a, and not in phases b and c. This is what we would expect from a single line-to-ground fault. The fault occurs in Region 2, where the base current is $I_{base} = 418.4$ A. Therefore, the total subtransient fault current is

$$\boxed{I_{f,\,SLG} = (6.024 \text{ pu})(418.4 \text{ A}) = 2520 \text{ A}}$$

This is the same answer as in the previous exercise. ∎

In this example, we had to do a lot of work calculating $\mathbf{Y}_{bus}$ and $\mathbf{Z}_{bus}$. In fact, we worked harder using the bus impedance method than we did solving for the single line-to-ground fault current in Exercise 13–6. However, once we have created $\mathbf{Z}_{bus}$, it would be possible to solve for the fault currents at every bus in the power system without recalculating impedances. In addition, the regular nature of this solution makes it easy to code in a computer program. The fault current analysis program introduced in the next section finds fault currents using the bus impedance method.

13.9 | FAULT CURRENT ANALYSIS PROGRAM FOR UNSYMMETRICAL FAULTS

The fault current analysis program from Chapter 12 can also be used to calculate unsymmetrical fault currents. The program calculates the fault current that flows when a fault develops at any bus, as well as the voltages at every bus and the current flows in every transmission line during the fault. It works for three-phase, single line-to-ground, line-to-line, or double line-to-ground faults in the subtransient, transient, or steady-state periods. The MATLAB source code for this program is available at the book's website.

The input data for program `faults` is placed in an input file, which can be created by using any available editor. The input values for each line were defined in Chapter 12, except for the negative- and zero-sequence reactances X_2 and X_0 on **GENERATOR** and **MOTOR** cards, and the zero-sequence reactance X_0 and the visibility flag `Vis` on **LINE** cards.

A typical input file is shown in Figure 13–33. The visibility flag added to the **LINE** line is used to determine whether or not a zero-phase line impedance is visible from a particular bus. As we learned, a transformer with a Δ connection or an ungrounded Y connection cannot pass zero-sequence currents, so lines containing them will have open circuits in the zero-sequence diagram. If a line (including any transformers in it) appears connected to a bus so that zero-sequence currents can flow, it is said to be "visible" to that bus. If the zero-sequence representation of a transmission line is visible to the "from" bus only, the visibility code is 1. If the zero-sequence representation of a transmission line is visible to the "to" bus, the visibility code is 2. If it is visible to both, the visibility code is 3, and if it is visible to neither, the visibility code is 0.

Figure 13–33 | A sample input file for the MATLAB program `faults`. This input file solves the for single line-to-ground fault in Example 13–6.

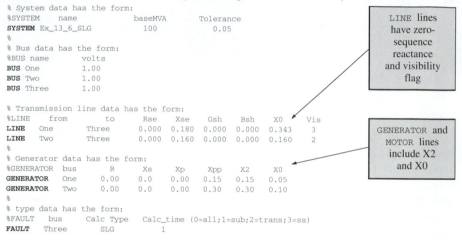

```
% System data has the form:
%SYSTEM    name              baseMVA         Tolerance
SYSTEM Ex_13_6_SLG            100             0.05
%
% Bus data has the form:
%BUS name        volts
BUS One          1.00
BUS Two          1.00
BUS Three        1.00

% Transmission line data has the form:
%LINE     from        to      Rse    Xse    Gsh    Bsh    X0      Vis
LINE    One        Three    0.000  0.180  0.000  0.000  0.343   3
LINE    Two        Three    0.000  0.160  0.000  0.000  0.160   2
%
% Generator data has the form:
%GENERATOR  bus      R      Xs     Xp     Xpp    X2     X0
GENERATOR   One     0.00   0.0    0.00   0.15   0.15   0.05
GENERATOR   Two     0.00   0.0    0.00   0.30   0.30   0.10
%
% type data has the form:
%FAULT    bus      Calc Type   Calc_time (0=all;1=sub;2=trans;3=ss)
FAULT    Three      SLG          1
```

LINE lines have zero-sequence reactance and visibility flag

GENERATOR and MOTOR lines include X2 and X0

To understand the visibility code better, let's reexamine the power system in Example 13–6. To solve for the current at the point of the fault using program `faults`, that point must be defined as a "bus." Figure 13–34 shows the power system with the

Figure 13–34 | The power system of Example 13–6, with the point of the fault defined as "bus 3".

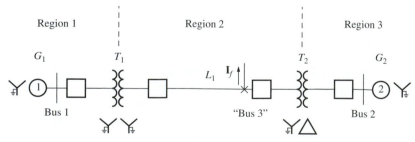

point of the fault defined as "Bus 3." The zero-sequence diagram of the power system is shown in Figure 13–35. Note that transmission line L_1 connects bus 1 and bus 3 and "line" L_2 (really just a transformer in this case) connects bus 2 to bus 3. Transmission line L_1 connects bus 1 and bus 3, and the zero-phase impedance is visible from both, so the visibility code for this transmission line is 3 in Figure 13–33. By contrast, "line" L_2 connects bus 2 and bus 3, but the zero-phase impedance appears to be an open circuit as seen from bus 2. Since the line is visible from the "to" bus but not from the "from" bus, it has a visibility code of 2 in Figure 13–33.

Figure 13–35 | The zero-sequence diagram for the power system of Example 13–6, with the point of the fault defined as "bus 3".

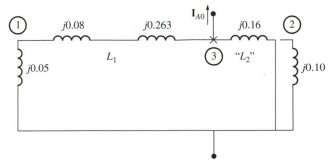

To solve a fault current problem using this program, you must first calculate the per-unit impedances and admittances of all components to a common system base, and then create a file containing those input parameters. Note that you must define a "bus" at every point in the power system where you wish to calculate voltages, including the point of the fault. These busses are interconnected by "lines," which contain all of the series impedances in the path between the busses, whether they come from the line itself or from transformers at the ends of the line. In general, the prefault voltages at all busses should be assumed to be a constant value (typically 1.0) to correspond to prefault no-load conditions.

The output from this program calculates the voltages and faults currents in *each phase* at each bus and in each transmission line, since unsymmetrical faults will have different effects on different phases.

When program `faults` is executed with the input file shown in Figure 13–33, the results are as shown in Figure 13–36. Note that the bus voltage and the fault currents match the values we calculated by hand in Example 13–6.

13.10 | SUMMARY

Unbalanced three-phase voltages and currents can be resolved into their symmetrical components. There are three symmetrical components: a positive-sequence (*abc*) set, a negative-sequence (*acb*) set, and a zero-sequence (in phase) set. Positive-sequence components peak in phase order *abc*, while negative-sequence components peak in phase order *acb*, and zero-sequence components all peak at the same time. Problems

Figure 13–36 | Result of executing `faults` with the input file shown in Figure 13–33. Note that the fault currents match the values that we calculated in Example 13–6 for a single line-to-ground fault.

```
» faults E_13_6_SLG
Input summary statistics:
  23 lines in system file
   1 SYSTEM lines
   3 BUS lines
   2 LINE lines
   2 GENERATOR lines
   0 MOTOR lines
   1 TYPE lines


                   Results for Case Ex_13_6_SLG


Single Line-to-Ground Fault at Bus Three
Calculating Subtransient Currents
|=====================Bus Information=============|======Line Information======|
Bus           P   Volts / angle    Amps / angle  |   To  | P  Amps / angle  |
no.    Name   h   (pu)    (deg)     (pu)    (deg) |  Bus  | h  (pu)    (deg)  |
|=============================================================================|
  1     One a    0.620/   0.00    0.000/   0.00
            b    0.935/-112.21    0.000/   0.00
            c    0.935/ 112.21    0.000/   0.00
                                                   Three    a 2.919/   90.00
                                                            b 0.588/  -90.00
                                                            c 0.588/  -90.00
  2     Two a    0.497/   0.00    0.000/   0.00
            b    0.901/-106.00    0.000/   0.00
            c    0.901/ 106.00    0.000/   0.00
                                                   Three    a 1.678/   90.00
                                                            b 0.839/  -90.00
                                                            c 0.839/  -90.00
  3   Three a    0.000/   0.00    6.024/ -90.00
            b    0.931/-111.58    0.000/   0.00
            c    0.931/ 111.58    0.000/   0.00
                                                   One      a 2.919/  -90.00
                                                            b 0.588/   90.00
                                                            c 0.588/   90.00
                                                   Two      a 1.678/  -90.00
                                                            b 0.839/   90.00
                                                            c 0.839/   90.00
|=============================================================================|
```

are solved with symmetrical components by treating each set of components separately and superimposing the results.

Currents of a given sequence produce voltages of that sequence only, so it is possible to create separate positive-, negative-, and zero-sequence per-unit equivalent circuits (called sequence diagrams), and to solve for the voltage and current relationships in each one separately.

The positive-sequence network alone is sufficient to solve problems involving power flows or three-phase faults, because those power systems are balanced. If the power system becomes unbalanced for some reason (such as a single line-to-ground fault), then all the phase sequences are needed.

The fault currents in a single line-to-ground fault can be calculated by connecting the three sequence networks in series so that $\mathbf{I}_{A1} = \mathbf{I}_{A2} = \mathbf{I}_{A0}$. Then, the positive-sequence current in the fault becomes

$$\mathbf{I}_{A1} = \frac{\mathbf{V}_f}{Z_0 + Z_1 + Z_2} \tag{13–65}$$

where $\mathbf{V}_f$ is the prefault voltage on the bus where the fault occurs, and Z_1, Z_2, and Z_0 are the Thevenin impedances of the positive-, negative-, and zero-sequence networks at the point of the fault.

The fault currents in a line-to-line fault can be calculated by connecting the positive- and negative-sequence networks in parallel so that $\mathbf{V}_{A1} = \mathbf{V}_{A2}$ and $\mathbf{I}_{A1} = -\mathbf{I}_{A2}$. Then, the positive-sequence current in the fault becomes

$$\mathbf{I}_{A1} = \frac{\mathbf{V}_f}{Z_1 + Z_2} \tag{13–67}$$

where $\mathbf{V}_f$ is the prefault voltage on the bus where the fault occurs, and Z_1 and Z_2 are the Thevenin impedances of the positive- and negative-sequence networks at the point of the fault.

The fault currents in a double line-to-ground fault can be calculated by connecting the positive-, negative-, and zero-sequence networks in parallel so that $\mathbf{V}_{A1} = \mathbf{V}_{A2} = \mathbf{V}_{A0}$ and $\mathbf{I}_{A1} = -\mathbf{I}_{A2} - \mathbf{I}_{A0}$. Then, the positive-sequence current in the fault becomes

$$\mathbf{I}_{A1} = \frac{\mathbf{V}_f}{Z_1 + \dfrac{Z_2 Z_0}{Z_2 + Z_0}} \tag{13–69}$$

where $\mathbf{V}_f$ is the prefault voltage on the bus where the fault occurs and Z_1, Z_2, and Z_0 are the Thevenin impedances of the positive-, negative-, and zero-sequence networks at the point of the fault.

In any type of fault, once $\mathbf{I}_{A1}$ has been calculated, it is easy to calculate $\mathbf{I}_{A2}$ and $\mathbf{I}_{A0}$ from the special relationships that apply to the particular type of fault. Then the actual fault currents in each phase can be calculated from the equations

$$\mathbf{I}_A = \mathbf{I}_{A1} + \mathbf{I}_{A2} + \mathbf{I}_{A0} \tag{13–23}$$
$$\mathbf{I}_B = a^2\mathbf{I}_{A1} + a\mathbf{I}_{A2} + \mathbf{I}_{A0} \tag{13–24}$$
$$\mathbf{I}_C = a\mathbf{I}_{A1} + a^2\mathbf{I}_{A2} + \mathbf{I}_{A0} \tag{13–26}$$

It is possible to create separate positive-, negative-, and zero-sequence bus admittance matrices $\mathbf{Y}_{bus}$ and bus impedance matrices $\mathbf{Z}_{bus}$ for a power system. Since the on-diagonal terms of the bus impedance matrix contains the Thevenin impedance of the power system at that point, these terms can be used in the equations for $\mathbf{I}_{A1}$ to calculate the fault currents that would flow at each bus in the power system for each type of fault.

13.11 | QUESTIONS

13–1. What are symmetrical components? Why are they used?

13–2. What are sequence diagrams?

13–3. Why is the magnitude of the neutral-to-ground impedance Z_N in a Y-connected generator trebled in a zero-sequence diagram of the generator?

13–4. What does the zero-sequence diagram of a three-phase transformer look like if it is a Y–Δ transformer with a grounded Y connection? What if the Y connection is ungrounded?

13–5. How are the sequence networks connected to find the currents flowing in a three-phase fault?

13–6. How are the sequence networks connected to find the currents flowing in a single line-to-ground fault?

13–7. How are the sequence networks connected to find the currents flowing in a line-to-line fault?

13–8. How are the sequence networks connected to find the currents flowing in a double line-to-ground fault?

13.12 | PROBLEMS

Problems 13–1 to 13–5 refer to a 200-MVA, 20-kV, 60-Hz, Y-connected, solidly grounded, three-phase synchronous generator connected through a 200-MVA, 20/138-kV Y–Y transformer to a 138-kV transmission line. The generator reactances to the machine's own base are

$$X_S = 1.40 \qquad X' = 0.30 \qquad X'' = 0.15 \qquad X_2 = 0.15 \qquad X_{g0} = 0.10$$

Both of the transformer's Y connections are solidly grounded, and its positive-, negative-, and zero-sequence series reactances are all 0.10 pu. The resistance of both the generator and the transformer may be ignored.

13–1. Assume that a symmetrical three-phase fault occurs on the 138-kV transmission line near the point where it is connected to the transformer.
   a. How much current flows at the point of the fault during the subtransient period?
   b. What is the voltage at the terminals of the generator during the subtransient period?
   c. How much current flows at the point of the fault during the transient period?
   d. What is the voltage at the terminals of the generator during the transient period?

13–2. Assume that a single-line-to-ground fault occurs on the 138-kV transmission line near the point where it is connected to the transformer.

    a.   How much current flows at the point of the fault during the subtransient period?

    b.   What is the voltage at the terminals of the generator during the subtransient period?

    c.   How much current flows at the point of the fault during the transient period?

    d.   What is the voltage at the terminals of the generator during the transient period?

13–3. Assume that a line-to-line fault occurs on the 138-kV transmission line near the point where it is connected to the transformer.

    a.   How much current flows at the point of the fault during the subtransient period?

    b.   What is the voltage at the terminals of the generator during the subtransient period?

    c.   How much current flows at the point of the fault during the transient period?

    d.   What is the voltage at the terminals of the generator during the transient period?

13–4. Assume that a double line-to-ground fault occurs on the 138-kV transmission line near the point where it is connected to the transformer.

    a.   How much current flows at the point of the fault during the subtransient period?

    b.   What is the voltage at the terminals of the generator during the subtransient period?

    c.   How much current flows at the point of the fault during the transient period?

    d.   What is the voltage at the terminals of the generator during the transient period?

13–5. Suppose the transformer in this power system is changed from a Y-Y to a Y-Δ, with the Y-connection solidly grounded. Nothing else changes in the power system. Now suppose that a single-line-to-ground fault occurs on the 138-kV transmission line near the point where it is connected to the transformer.

    a.   How much current flows at the point of the fault during the subtransient period?

    b.   What is the voltage at the terminals of the generator during the subtransient period?

    c.   How much current flows at the point of the fault during the transient period?

    d.   What is the voltage at the terminals of the generator during the transient period?

13–6. A simple three-phase power system is shown in Figure P13–1. Assume that the ratings of the various devices in this system are as follows:

Generator G_1: 250 MVA, 13.8 kV, $R = 0.1$ pu, $X'' = 0.18$ pu, $X' = 0.40$ pu, $X_2 = 0.15$ pu, $X_{g0} = 0.10$ pu. This generator is grounded through an impedance $Z_N = j0.20$ pu.

Generator G_2: 500 MVA, 20.0 kV, $R = 0.1$ pu, $X'' = 0.15$ pu, $X' = 0.35$ pu $X_2 = 0.15$ pu, $X_{g0} = 0.10$ pu. This generator is grounded through an impedance $Z_N = j0.20$ pu.

Generator G_3: 250 MVA, 13.8 kV, $R = 0.15$ pu, $X'' = 0.20$ pu, $X' = 0.40$ pu, $X_2 = 0.20$ pu, $X_{g0} = 0.15$ pu. This generator is grounded through an impedance $Z_N = j0.25$ pu.

Transformer T_1: 250 MVA, 13.8-Δ/240-Y kV, $R = 0.01$ pu, $X_1 = X_2 = X_0 = 0.10$ pu
Transformer T_2: 500 MVA, 20.0-Δ/240-Y kV, $R = 0.01$ pu, $X_1 = X_2 = X_0 = 0.08$ pu
Transformer T_3: 250 MVA, 13.8-Δ/240-Y kV, $R = 0.01$ pu, $X_1 = X_2 = X_0 = 0.10$ pu
Each line: $R = 20\ \Omega$, $X_1 = X_2 = 100\ \Omega$, $X_0 = 200\ \Omega$.

Assume that the power system is initially unloaded, that the voltage at bus 4 is 250 kV, and that *all resistances may be neglected*.

a. Convert this power system to per unit on a base of 500 MVA at 20 kV at generator G_2. Create the positive-, negative-, and zero-phase sequence diagrams.

b. Calculate $\mathbf{Y}_{bus}$ and $\mathbf{Z}_{bus}$ for this power system using the generator subtransient reactances.

c. Suppose that a three-phase symmetrical fault occurs at bus 4. What is the subtransient fault current? What is the voltage on each bus in the power system? What is the subtransient current flowing in each of the three transmission lines?

c. Suppose that a single line-to-ground fault occurs at bus 4. What is the subtransient fault current? What is the voltage on each bus in the power system? What is the subtransient current flowing in each of the three transmission lines?

Figure P13–1 I The simple power system of Problem 13–6.

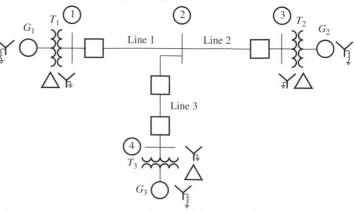

13–7. For the power system of Problem 13–6, calculate the subtransient fault current for a single line-to-ground fault at bus 4 *if the neutrals of the three generators are all solidly grounded*. How much difference did the inclusion of the impedances to ground in the generator neutrals make to the total fault current? Why would a power company wish to ground the neutrals of their generators through an impedance?

13–8. Assume that a fault occurs on the high-voltage side of transformer T_2 in the power system shown in Figure P13–2. Make the assumption that the generator is operating at rated voltage, and that the power system is initially unloaded.

    a.   Calculate the subtransient fault current, generator current, and motor current for a symmetrical three-phase fault.

    b.   Calculate the subtransient fault current, generator current, and motor current for a single-line-to-ground fault.

    c.   How will the answers in part (b) change if the Y connection of transformer T_2 is solidly grounded?

Figure P13–2 | One-line diagram of the power system in Problem 13–8.

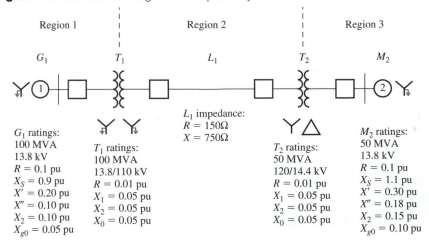

| Region 1 | Region 2 | Region 3 |

G_1 T_1 L_1 T_2 M_2

G_1 ratings:
100 MVA
13.8 kV
$R = 0.1$ pu
$X_S = 0.9$ pu
$X' = 0.20$ pu
$X'' = 0.10$ pu
$X_2 = 0.10$ pu
$X_{g0} = 0.05$ pu

T_1 ratings:
100 MVA
13.8/110 kV
$R = 0.01$ pu
$X_1 = 0.05$ pu
$X_2 = 0.05$ pu
$X_0 = 0.05$ pu

L_1 impedance:
$R = 150\Omega$
$X = 750\Omega$

T_2 ratings:
50 MVA
120/14.4 kV
$R = 0.01$ pu
$X_1 = 0.05$ pu
$X_2 = 0.05$ pu
$X_0 = 0.05$ pu

M_2 ratings:
50 MVA
13.8 kV
$R = 0.1$ pu
$X_S = 1.1$ pu
$X' = 0.30$ pu
$X'' = 0.18$ pu
$X_2 = 0.15$ pu
$X_{g0} = 0.10$ pu

13–9 The Ozzie Outback Electric PoZwer (OOEP) Company maintains the power system shown in Figure P13–3. This power system contains five busses, with generators attached to two of them and loads attached to the remaining ones. The power system has six transmission lines connecting the busses together, with the characteristics shown in Table 13–1. There are generators at busses Bunya and Mulga, and loads at all other busses. The characteristics of the two generators are shown in Table 13–2. Note that the neutrals of the two generators are both grounded through an inductive reactance of 0.60 pu. Note that all values are in per-unit on a 100-MVA base.

Figure P13–3 | The Ozzie Outback Electric Power Company system.

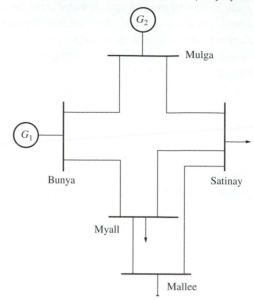

Table 13–1 | Transmission lines in the OOEP system

From	To	R_{SE} (pu)	X_1 (pu)	X_2 (pu)	X_0 (pu)
Bunya	Mulga	0.001	0.051	0.051	0.090
Mulga	Satinay	0.007	0.035	0.035	0.090
Bunya	Myall	0.007	0.035	0.035	0.060
Myall	Satinay	0.007	0.035	0.035	0.060
Myall	Mallee	0.011	0.051	0.051	0.110
Mallee	Satinay	0.011	0.051	0.051	0.090

Table 13–2 | Generators in the OOEP system

Name	Bus	R (pu)	X_S (pu)	X' (pu)	X'' (pu)	X_2 (pu)	X_{g0} (pu)	X_N (pu)
G1	Bunya	0.02	1.5	0.35	0.20	0.20	0.10	0.25
G2	Mulga	0.01	1.0	0.25	0.12	0.12	0.05	0.25

a. Assume that a single line-to-ground fault occurs at the Mallee bus. What is the per-unit fault current during the subtransient period? During the transient period?

b. Assume that the neutrals of the generators are now solidly grounded, and that a single line-to-ground fault occurs at the Mallee bus. What is the per-unit fault current during the subtransient period now? During the transient period?

c. Assume that a line-to-line fault occurs at the Mallee bus. What is the per-unit fault current during the subtransient period? During the transient period?

d. How much does the per-unit fault current change for a line-to-line fault if the generators are either solidly grounded or grounded through an impedance?

13.13 | REFERENCES

1. Elgerd, Olle I.: *Electric Energy Systems Theory: An Introduction,* McGraw-Hill, New York, 1971.

2. Glover, J. Duncan, and Mulukutla Sarma: *Power Systems Analysis and Design,* PWS Publishers, Boston, 1987.

3. Grainger, John J., and William D. Stevenson, Jr.: *Power Systems Analysis*, McGraw-Hill, New York, 1994.

4. Gross, Charles A.: *Power Systems Analysis*, John Wiley and Sons, New York, 1979.

5. Stagg, Glen W., and Ahmed H. El-Abiad: *Computer Methods in Power Systems Analysis,* McGraw-Hill, New York, 1968.

Tables of Constants and Conversion Factors

Constants

Acceleration due to gravity	$g = 9.807 \text{ m/s}^2$
Charge of the electron	$e = -1.6 \times 10^{-19} \text{ C}$
Permeability of free space	$\mu_0 = 4\pi \times 10^{-7} \text{ H/m}$
Permitivity of free space	$\epsilon_0 = 8.854 \times 10^{-12} \text{ F/m}$
Resistivity of annealed copper	$\rho = 1.72 \times 10^{-8} \text{ }\Omega\text{-m}$

Conversion factors

Length	1 meter (m)	= 3.281 ft
		= 39.37 in
	1 mile (mi)	= 1.609 kilometers (km)
Mass	1 kilogram (kg)	= 0.0685 slug
		= 2.205 lb mass (lbm)
Force	1 newton (N)	= 0.2248 lb force (lbf)
		= 7.233 poundals
		= 0.102 kg (force)
Torque	1 newton-meter (N-m)	= 0.738 pound-feet (lb-ft)
Energy	1 joule (J)	= 0.738 foot-pounds (ft-lb)
		= 3.725 × 10⁻⁷ horsepower-hour (hp-h)
		= 2.778 × 10⁻⁷ kilowatt-hour (kWh)
Power	1 watt (W)	= 1.341 × 10⁻³ hp
		= 0.7376 ft-lbf/s
	1 horsepower	= 746 W
Magnetic flux	1 weber (Wb)	= 10⁸ maxwells (lines)
Magnetic flux density	1 tesla (T)	= 1 Wb/m²
		= 10,000 gauss (G)
		= 64.5 kilolines/in²
Magnetizing intensity	1 ampere·turn/meter (A-turn/m)	= 0.0254 A-turns/in
		= 0.0126 oersted (Oe)

B

Thevenin's and Norton's Theorems

Thevenin's theorem and Norton's theorem are two of the most important tools in elementary circuit analysis. Thevenin's theorem states that any linear circuit that can be separated by two terminals from the rest of the system can be replaced a single voltage source in *series* with an equivalent impedance. Norton's theorem states that any linear circuit that can be separated by two terminals from the rest of the system can be replaced a single current source in *parallel* with an equivalent impedance. The procedures for creating the equivalent circuits are illustrated in Figure B–1.

To create the Thevenin equivalent of a circuit, first open-circuit the terminals of the circuit and measure the open-circuit voltage $\mathbf{V}_O$. Then, short-circuit the terminals of the circuit and measure the short-circuit current $\mathbf{I}_S$. The Thevenin equivalent of the circuit will consist of a voltage source equal to the open-circuit voltage $\mathbf{V}_O$ in series with an impedance given by the equation

$$Z_{\text{TH}} = \frac{\mathbf{V}_O}{\mathbf{I}_S}$$

To create the Norton equivalent of a circuit, measure $\mathbf{V}_O$ and $\mathbf{I}_S$ as before. The Norton equivalent of the circuit will consist of a current source equal to the short-circuit current $\mathbf{I}_S$ in parallel with the Thevenin equivalent impedance Z_{TH}.

Figure B–1 | Creating the Thevenin and Norton equivalents of a linear circuit.

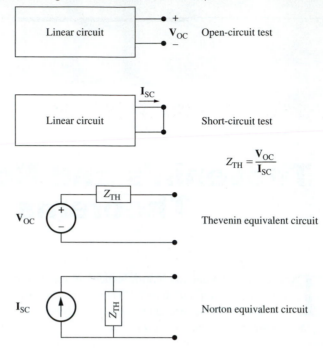

Open-circuit test

Short-circuit test

$$Z_{TH} = \frac{V_{OC}}{I_{SC}}$$

Thevenin equivalent circuit

Norton equivalent circuit

C

Faults Through an Impedance

Chapter 13 describes how to calculate the fault currents in a power system if the faulted phases are solidly connected to either each other or to ground, as appropriate for a particular fault type. Such direct short circuits produce the highest possible current flows for a given type of fault, so calculations like this are used to determine the maximum possible fault currents that a power system must be able to withstand without being destroyed.

In the real world, however, a majority of faults have a significant impedance. Most faults are caused by insulator flashover after a lightning strike, and the path between the phase and ground includes the impedance of the arc, the impedance of the tower structure itself, and the impedance of the ground connection if there are no overhead ground wires. These impedances significantly reduce the total current flow compared to the direct short case.

It is important to be able to estimate the fault currents through an impedance, since we must know the expected amount of fault current to properly set the trigger current levels on circuit breakers and other protective gear. How can we estimate the current flow if there is an impedance in the fault? We will now explore that question for each type of fault.

C.1 | THREE-PHASE FAULTS THROUGH AN IMPEDANCE

Figure C–1a shows a symmetrical three-phase fault through an impedance Z_f. Note that this impedance appears in each phase, so the three-phase power system remains balanced during the fault. Since the system remains balanced, only positive-sequence components of voltage and current are present in the fault.

If the current flowing in phase a of the fault is $\mathbf{I}_A$ then the voltage on phase a during the fault will be

$$\mathbf{V}_A = \mathbf{I}_A Z_f \tag{C–1}$$

Figure C–1 | Connection diagrams for various faults through an impedance on a power system: (a) three-phase fault; (b) single line-to-ground fault; (c) line-to-line fault; (d) double line-to-ground fault.

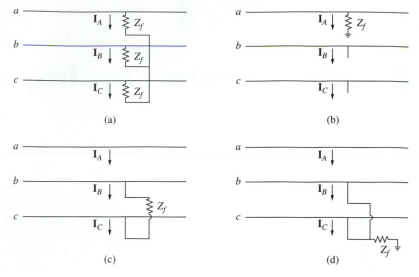

But since this is a symmetrical three-phase fault, $\mathbf{V}_A = \mathbf{V}_{A1}$ and $\mathbf{I}_A = \mathbf{I}_{A1}$. Therefore,

$$\mathbf{V}_{A1} = \mathbf{I}_{A1}Z_f \qquad \text{(C–2)}$$

However, we already know that the positive sequence voltage in the fault is given by the equation

$$\mathbf{V}_{A1} = \mathbf{V}_f - \mathbf{I}_{A1}Z_1 \qquad \text{(13–47)}$$

Combining Equations (C–2) and (13–47) and solving for $\mathbf{I}_{A1}$ yields the result

$$\boxed{\mathbf{I}_{A1} = \frac{\mathbf{V}_f}{Z_1 + Z_f}} \qquad \text{(C–3)}$$

Thus, the fault impedance Z_f should be connected in series with the positive-sequence network for a three-phase fault. This connection is shown in Figure C–2a.

C.2 | SINGLE LINE-TO-GROUND FAULTS THROUGH AN IMPEDANCE

Figure C–1b shows a single line-to-ground fault through an impedance Z_f. Here, the impedance Z_f appears in the path to ground. Since it is in the path to ground, only *zero-sequence components* of current can flow through it, and the zero-sequence components from all three phases will flow through the same impedance Z_f. Therefore, the impedance must be included in series with the zero-sequence network, so

Figure C–2 | Connections of sequence networks to simulate faults through an impedance on a power system: (a) three-phase fault; (b) single line-to-ground fault; (c) line-to-line fault; (d) double line-to-ground fault.

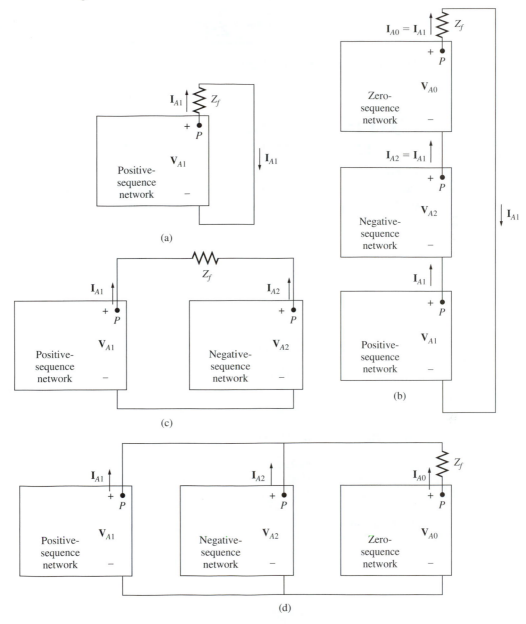

(a)

(b)

(c)

(d)

that zero-sequence current will flow through it. The size of the impedance in the diagram must be $3Z_f$ to account for the fact that the zero-sequence currents from all three phases are flowing through it. Since the positive-, negative-, and zero-sequence networks are all in series for this type of fault, the impedances of each network are in series with $3Z_f$.

$$\boxed{\mathbf{I}_{A1} = \frac{\mathbf{V}_f}{Z_1 + Z_2 + Z_0 + Z_f}} \tag{C–4}$$

This connection is shown in Figure C–2b.

C.3 | LINE-TO-LINE FAULTS THROUGH AN IMPEDANCE

Figure C–1c shows a line-to-line fault through an impedance Z_f. This impedance causes a voltage drop between phases a and b, so the following relationships exist at the fault

$$\begin{aligned}
\mathbf{V}_C &= \mathbf{V}_B - \mathbf{I}_B Z_f \\
\mathbf{I}_A &= 0 \\
\mathbf{I}_B &= -\mathbf{I}_C
\end{aligned} \tag{C–5}$$

Currents $\mathbf{I}_A$, $\mathbf{I}_B$, and $\mathbf{I}_C$ have the same relationships as they have in a line-to-line fault without impedance, so once more

$$\mathbf{I}_{A1} = -\mathbf{I}_{A2}$$

The sequence components of voltage are

$$\begin{aligned}
\mathbf{V}_{A0} &= \tfrac{1}{3}(\mathbf{V}_A + \mathbf{V}_B + \mathbf{V}_C) &&\text{zero-sequence} &&(13\text{–}17) \\
\mathbf{V}_{A1} &= \tfrac{1}{3}(\mathbf{V}_A + a\mathbf{V}_B + a^2\mathbf{V}_C) &&\text{positive-sequence} &&(13\text{–}18) \\
\mathbf{V}_{A2} &= \tfrac{1}{3}(\mathbf{V}_A + a^2\mathbf{V}_B + a\mathbf{V}_C) &&\text{negative-sequence} &&(13\text{–}19)
\end{aligned}$$

Substituting Equations (C–5) into them yields

$$\begin{aligned}
\mathbf{V}_{A0} &= \tfrac{1}{3}[\mathbf{V}_A + \mathbf{V}_B + (\mathbf{V}_B - \mathbf{I}_B Z_f)] \\
\mathbf{V}_{A1} &= \tfrac{1}{3}[\mathbf{V}_A + a\mathbf{V}_B + a^2(\mathbf{V}_B - \mathbf{I}_B Z_f)] \\
\mathbf{V}_{A2} &= \tfrac{1}{3}[\mathbf{V}_A + a^2\mathbf{V}_B + a(\mathbf{V}_B - \mathbf{I}_B Z_f)]
\end{aligned} \tag{C–6}$$

Working with the positive- and negative-sequence equations, we get

$$\begin{aligned}
3\mathbf{V}_{A1} &= \mathbf{V}_A + (a + a^2)\mathbf{V}_B - a^2\mathbf{I}_B Z_f &&(C\text{–}7) \\
3\mathbf{V}_{A2} &= \mathbf{V}_A + (a + a^2)\mathbf{V}_B - a\mathbf{I}_B Z_f &&(C\text{–}8)
\end{aligned}$$

Subtracting these equations to eliminate $\mathbf{V}_B$ yields

$$3(\mathbf{V}_{A1} - \mathbf{V}_{A2}) = (a - a^2)\mathbf{I}_B Z_f = j\sqrt{3}\mathbf{I}_B Z_f \tag{C–9}$$

But because $\mathbf{I}_{A1} = -\mathbf{I}_{A2}$

$$\mathbf{I}_B = a^2\mathbf{I}_{A1} + a\mathbf{I}_{A2} = a^2\mathbf{I}_{A1} - a\mathbf{I}_{A1} = (a^2 - a)\mathbf{I}_{A1} = -j\sqrt{3}\mathbf{I}_{A1} \tag{C–10}$$

so substituting Equation (C–10) into Equation (C–9) yields

$$3(\mathbf{V}_{A1} - \mathbf{V}_{A2}) = j\sqrt{3}(-j\sqrt{3}\mathbf{I}_{A1})\,Z_f$$
$$3(\mathbf{V}_{A1} - \mathbf{V}_{A2}) = 3\mathbf{I}_{A1}\,Z_f$$
$$\mathbf{V}_{A1} - \mathbf{V}_{A2} = \mathbf{I}_{A1}\,Z_f \qquad\qquad \text{(C–11)}$$

Equation (C–11) indicates that we need to insert the impedance Z_f between the positive- and negative-sequence networks, so that the voltages will differ by just the voltage drop across the impedance. This connection is shown in Figure C–2b. The resulting positive sequence current will be

$$\boxed{\mathbf{I}_{A1} = \frac{\mathbf{V}_f}{Z_1 + Z_2 + Z_f}} \qquad\qquad \text{(C–12)}$$

C.4 | DOUBLE LINE-TO-GROUND FAULTS THROUGH AN IMPEDANCE

Figure C–1d shows a double line-to-ground fault through an impedance Z_f. Here, the impedance Z_f appears in the path to ground. Since it is in the path to ground, only *zero-sequence components* of current can flow through it, and the zero-sequence components from all three phases will flow through the same impedance Z_f. Therefore, the impedance must be included *in series* with the zero-sequence network, so that zero-sequence current will flow through it. The size of the impedance in the diagram must be $3Z_f$ to account for the fact that the zero-sequence currents from all three phases are flowing through it. Since the positive-, negative-, and zero-sequence networks are all in parallel for this type of fault, the positive-sequence total current will be

$$\boxed{\mathbf{I}_{A1} = \frac{\mathbf{V}_f}{Z_1 + \dfrac{Z_2(Z_0 + 3Z_f)}{Z_2 + Z_0 + 3Z_f}}} \qquad\qquad \text{(C–13)}$$

This connection is shown in Figure C–2d.

Trigonometric Identities

1. $\sin^2 \theta + \cos^2 \theta = 1$
2. $\cos 2\theta = \cos^2 \theta - \sin^2 \theta = 2\cos^2 \theta - 1 = 1 - 2\sin^2 \theta$
3. $\cos^2 \theta = \frac{1}{2}(1 + \cos 2\theta)$
4. $\sin^2 \theta = \frac{1}{2}(1 - \cos 2\theta)$
5. $\cos(\alpha + \beta) = \cos \alpha \cos \beta - \sin \alpha \sin \beta$
6. $\cos(\alpha - \beta) = \cos \alpha \cos \beta + \sin \alpha \sin \beta$
7. $\sin(\alpha + \beta) = \sin \alpha \cos \beta + \cos \alpha \sin \beta$
8. $\sin(\alpha - \beta) = \sin \alpha \cos \beta - \cos \alpha \sin \beta$
9. $\cos \alpha \cos \beta = \frac{1}{2}[\cos(\alpha - \beta) + \cos(\alpha + \beta)]$
10. $\sin \alpha \sin \beta = \frac{1}{2}[\cos(\alpha - \beta) - \cos(\alpha + \beta)]$
11. $\sin \alpha \cos \beta = \frac{1}{2}[\sin(\alpha + \beta) + \sin(\alpha - \beta)]$

INDEX